AF806026

Ecological Studies, Vol. 227

Analysis and Synthesis

Edited by

Martyn M. Caldwell, Logan, UT, USA
Sandra Díaz, Cordoba, Argentina
Gerhard Heldmaier, Marburg, Germany
Robert B. Jackson, Durham, NC, USA
Otto L. Lange, Würzburg, Germany
Delphis F. Levia, Newark, DE, USA
Harold A. Mooney, Stanford, CA, USA
Ernst-Detlef Schulze, Jena, Germany
Ulrich Sommer, Kiel, Germany

Ecological Studies is Springer's premier book series treating all aspects of ecology. These volumes, either authored or edited collections, appear several times each year. They are intended to analyze and synthesize our understanding of natural and managed ecosystems and their constituent organisms and resources at different scales from the biosphere to communities, populations, individual organisms and molecular interactions. Many volumes constitute case studies illustrating and synthesizing ecological principles for an intended audience of scientists, students, environmental managers and policy experts. Recent volumes address biodiversity, global change, landscape ecology, air pollution, ecosystem analysis, microbial ecology, ecophysiology and molecular ecology.

More information about this series at http://www.springer.com/series/86

Laszlo Nagy • Bruce R. Forsberg • Paulo Artaxo
Editors

Interactions Between Biosphere, Atmosphere and Human Land Use in the Amazon Basin

Springer

Editors
Laszlo Nagy
Departamento de Biologia Vegetal
Universidade Estadual de Campinas
Campinas
Brazil

Bruce R. Forsberg
Instituto Nacional de Pesquisas da Amazonia
Manaus
Brazil

Paulo Artaxo
Instituto de Física
Universidade de São Paulo
São Paulo
Brazil

ISSN 0070-8356 ISSN 2196-971X (electronic)
Ecological Studies
ISBN 978-3-662-49900-9 ISBN 978-3-662-49902-3 (eBook)
DOI 10.1007/978-3-662-49902-3

Library of Congress Control Number: 2016944616

© Springer-Verlag Berlin Heidelberg 2016
This work is subject to copyright. All rights are reserved by the Publisher, whether the whole or part of the material is concerned, specifically the rights of translation, reprinting, reuse of illustrations, recitation, broadcasting, reproduction on microfilms or in any other physical way, and transmission or information storage and retrieval, electronic adaptation, computer software, or by similar or dissimilar methodology now known or hereafter developed.
The use of general descriptive names, registered names, trademarks, service marks, etc. in this publication does not imply, even in the absence of a specific statement, that such names are exempt from the relevant protective laws and regulations and therefore free for general use.
The publisher, the authors and the editors are safe to assume that the advice and information in this book are believed to be true and accurate at the date of publication. Neither the publisher nor the authors or the editors give a warranty, express or implied, with respect to the material contained herein or for any errors or omissions that may have been made.

Printed on acid-free paper

This Springer imprint is published by Springer Nature
The registered company is Springer-Verlag GmbH Berlin Heidelberg

Preface

The Amazon basin is considered a treasure trove of biodiversity, which has yet to be fully explored. The basin is also an important provider of a number of ecosystem services, both provisioning and supporting. These two facets often clash as provisioning services that require land cover change can radically change the capacity of natural ecosystems to maintain their supporting services. A well-known example is the much-debated impact of forest conversion to grazing and cropland that may impact the regional transport of moisture from the Amazon basin to south-western Brazil and the Plata basin in Argentina, currently contributing an estimated 20–25 % of precipitation in these areas. Along with water, carbon sequestration by the native forest is thought to be offsetting anthropogenic CO_2 emissions at both regional and global scales. However, it is important that none of these or any other monetisable 'services' are taken for granted as their 'delivery' is function of short-term climate variation, itself likely to be influenced by anthropogenic greenhouse gas emissions. For example, the more or less neutral carbon balance at the basin scale can be tipped to make the forest a carbon source in extreme drought years, which are often combined with increased fire occurrence. The frequency of such extreme years is projected to increase. The Amazon forest is unique in the strong coupling between forest functioning and climate. Many of the biological processes in the forest influence precipitation and radiation balance, making this coupling very unique in the ecosystems of our planet.

Amazonia, notwithstanding national sovereignties, is a world heritage asset, be it for its biodiversity or for its many services to humankind. This includes cultural services such as the scientific research presented in this book, which contributes to an ever-increasing body of scientific knowledge on the Amazon (still disproportionally small when compared with the vastness of the region). Today, it is imperative to use this knowledge to inform decisions on land use and development in Amazonia (and the policy incentives behind them). Such decisions, irrespective of the scale in question, require careful consideration and the participation of interdisciplinary teams of experts to help comply with national and international commitments to maintain the integrity of the remaining natural ecosystems so that future generations may also experience the joy that Amazonia offers and the responsibilities that its conservation demands.

The editors would like to express their sincere gratitude to all authors of the chapters in this book for their contributions. We are indebted to Andrea Schlitzberger, commissioning editor, for her encouragement in organising this book and for her help during the process. We thank Detlef Schultze for his comments and the large number of colleagues who acted as chapter reviewers.

Campinas, Brazil — Laszlo Nagy
Manaus, Brazil — Bruce R. Forsberg
São Paulo, Brazil — Paulo Artaxo

Contents

Part I
Introduction

Chapter 1
Interactions Between Biosphere, Atmosphere, and Human Land Use in the Amazon Basin: An Introduction

Laszlo Nagy, Paulo Artaxo, and Bruce R. Forsberg

1.1 The Amazon Basin

The Amazon basin, with its nearly 6 million km^2 extent (Fig. 1.1) is about 60 % of the size of the continental United States or Europe taken together with eastern Russia to the Ural Mountains. Of the total area of the basin, 75 % is covered by lowland evergreen rainforest types and the rest is composed of a mixture of seasonal forest, savanna, and various montane forests, alpine formations, and an increasing proportion of grazed/cultivated land, estimated at over 7 % (Eva et al. 2002).

The history of the Amazon rainforest is thought to span ca. 55 M years, during which time large changes in the geomorphology, soils, climate, and biota of the basin have taken place (Hoorn et al. 2010). The last major change in climate occurred after the end of the last glacial maximum, when the earth's climate was much drier and cooler than it is today. After this point, the global climate became gradually warmer, continuing through the last century where average air temperatures have increased sharply due to human activities (Malhi et al. 2014). The threat of accelerated global warming and the incomplete knowledge about the integrated effect of the complex mosaic of terrestrial and aquatic ecosystems in the Amazon on the atmosphere were the driving motive for the establishment of an

L. Nagy (✉)
Departamento de Biologia Vegetal, Universidade Estadual de Campinas, Campinas, Brazil
e-mail: lnagy@unicamp.br

P. Artaxo
Instituto de Física, Universidade de São Paulo, São Paulo, Brazil

B.R. Forsberg
Instituto Nacional de Pesquisas da Amazonia, Manaus, Brazil

© Springer-Verlag Berlin Heidelberg 2016

L. Nagy et al. (eds.), *Interactions Between Biosphere, Atmosphere and Human Land Use in the Amazon Basin*, Ecological Studies 227,
DOI 10.1007/978-3-662-49902-3_1

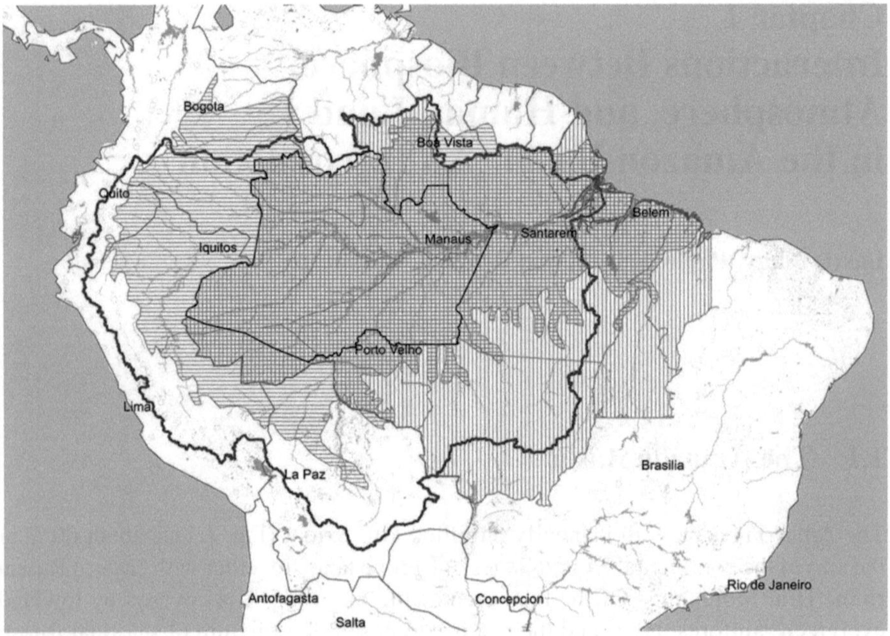

Fig. 1.1 The term Amazonia has been used in a number of ways by researchers and this multi-authored book is no exception. The *thick line* indicates the outline of the Amazon river basin, i.e. the catchment area of the Amazon river system. *Horizontal hatching* indicates the Amazon biogeographic provinces after Navarro and Maldonado (2002): Loreto, Acre and Madre dos Dios, Rio Negro and high Orinoco, Roraima and the lower Amazon, the Amazon delta region, and Madeira and Tapajos. *Vertical hatching* indicates Brazilian Amazonia, the so-called '*Amazonia Legal*', with cross-hatching indicating the intersection of areas of Brazilian Amazonia with the Amazonian biogeographic provinces of Navarro and Maldonado (2002) which are of lowland evergreen rainforest; the vertically hatched area is biogeographically different, with original vegetation cover being largely eastern cerrado (savanna) and seasonal forest occurring between the evergreen forest and savanna. Within Brazilian Amazonia the heavier outline indicates the boundary of the state of Amazonas. The unhatched areas in the south and west include the biogeographic provinces of the western cerrado, Beni (wet savannas), the northern tip of the Chaco and of the Bolivian-Tucuman province, and of the xeric puna that extends south into north-west Argentina; along the slopes of the Andes the evergreen montane forests ('yungas') and above the treeline the alpine vegetation ('puna' in Bolivia and Peru and 'páramo' in Ecuador and Colombia)

ambitious research programme, known as the Large-scale Biosphere–Atmosphere (LBA) experiment, in 1998.

Given the vast extension of Amazonia (ca. 5 % of the total land surface of the Earth) and its large potential influence on tropical weather systems, it could affect energy, water, and carbon balances, regionally in South America and perhaps globally too. The fundamental question that arises in relation to global change, land use being the primary driver in tropical forests, and also in relation to forecasting climate change is how Amazonia, until recently little perturbed (but see e.g. Balée 1998; Denevan 2001), functions as an entity and what environmental services it contributes both regionally and globally. More precisely, how will land use change in the Amazon affect the biogeochemistry of terrestrial and aquatic ecosystems and their interactions with the atmosphere. In addition, how forecast climate change (increase in temperature and drying, increased frequency of extreme weather/climate events) will modify the functioning of both remaining natural ecosystems and increasing anthropogenic systems (Fig. 1.2). The final and most important question concerns future plans for land use and conservation in the Amazon: can a governance framework be found to optimise the benefit at the planetary level derived from functions and services that the various ecosystems of Amazonian biosphere and their interaction with atmosphere provide? Before this last point can be tackled, one needs scientific evidence that quantifies these

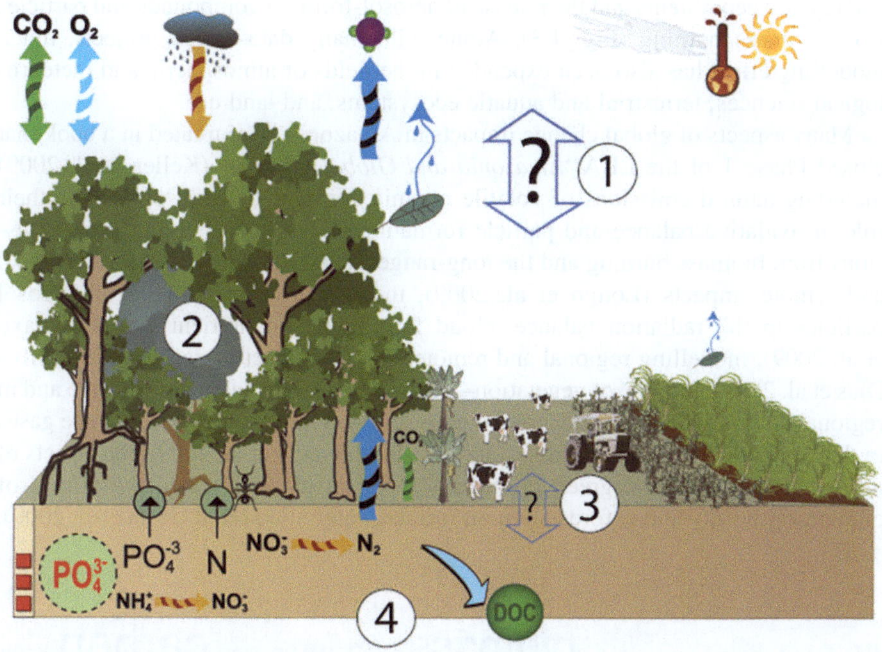

Fig. 1.2 Human land use in the Amazon basin produces ecosystems with reduced structural and functional properties (Malhi et al. 2014). Such (agro-)ecosystems have very different rates of exchanges of energy and matter along the soil–plant–atmosphere continuum. The main focus of research within LBA has been on biosphere–atmosphere interactions: exchanges with the atmosphere (*1*); aerosol and cloud formation; resource uptake, use, and storage by vegetation; (*2*) internal recycling (litterfall, root turnover, root exudates, canopy leaching); decomposition of organic matter; mineralisation and nutrient mobility/retention in the soil; (*3*) export via microbial gas emissions; BVOC formation (see e.g. Jardine and Jardine 2016); and (4) leaching (mineral nutrients, DOC). Erosion, fire, and harvest are important components for establishing the balance of elemental cycles, but these factors are often poorly quantified. (copyright L. Nagy—Courtesy of the integration and application network, University of Maryland Center for Environmental Science (ian.umces.edu/symbols/))

environmental functions and services. This is what LBA set out to do in 1998. In its current second phase, LBA is continuing the science in that vein with an increasing focus on the impact of human interventions on ecosystem functions, the use of environmental services, and feedbacks between the socio-economic and biophysical domains in this complex socio-ecological system.

1.2 The LBA Projects 1998–2007

The main topics of the large number of publications produced by LBA projects have concerned carbon stored in forest biomass and the soil, and climatic and atmospheric properties, including the emission of greenhouse gases from terrestrial and aquatic ecosystems and the release of aerosol-forming compounds and particles from biomass burning (Fig. 1.3). Along with many data-driven projects, much modelling effort has also been expended in the fields of atmospheric and meteorological sciences, terrestrial and aquatic ecosystems, and land use.

Many aspects of global change impacts in Amazonia were treated in a book that closed Phase 1 of the LBA '*Amazonia and Global Change*' (Keller et al. 2009) including natural emissions of volatile organic compounds from plants and their role in oxidative balance and particle formation (Kesselmeier et al. 2009), emissions from biomass burning and the long-range transport of smoke and its regional and remote impacts (Longo et al. 2009), the composition and role of aerosol particles in the radiation balance, cloud formation and nutrient cycles (Artaxo et al. 2009), modelling regional and remote climate impact of deforestation (Silva Dias et al. 2009), the role of vegetation—climate feedbacks in climate change and in regional and global impacts (Marengo et al. 2009), sources and sinks of trace gases in the Amazon forest and Cerrado 'biomes' (Bustamante et al. 2009), the effects of drought in lowland evergreen rainforests (Meir et al. 2009), and the effects of climate variability and deforestation on surface water yield (Heil Costa et al. 2009).

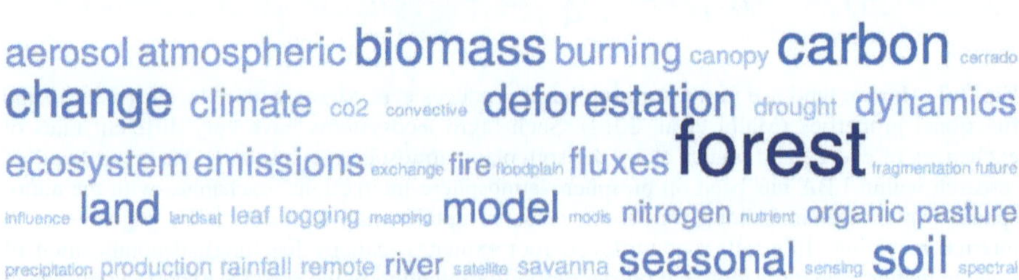

Fig. 1.3 Word cloud generated in TagCrowd (http://tagcrowd.com/) by using the titles and keywords of ISI journal articles from the LBA database of publications between 1998 and 2012. The size of the lettering is proportional to the frequency at which the words appeared

1.3 This Book

This book presents an overview of recent work, focusing on climate change and impacts of land use in the socio-ecological system of the Amazon, with a focus on the biophysical components (Table 1.1). Targeting an interdisciplinary audience, this new book also complements the research undertaken by LBA, reporting work carried out in the Amazon basin by various research programmes funded by the Ministry of Science, Technology and Innovation of the Brazilian Government (climate, *Rede Clima*; various thematic topics, National Institutes of Science and Technology, *INCT*s; and the Forest Dynamics in Forest Fragments project, *PDBFF*).

Biosphere–Atmosphere Exchanges The biological bases of emissions of volatile organic compounds (BVOC) from plants and their potential functional role in the Amazon basin are exposed by Jardine and Jardine (2016). Their findings point to the necessity of a better understanding of two-way exchanges between vegetation (biosphere) and atmosphere and the inclusion of this two-way relationship in earth system models. In addition, it is of considerable interest how BVOC production is species/vegetation type specific and how the production of various compounds will change with climate change and land use change. For example, it has been demonstrated how vastly palm oil plantations are different in their magnitude of BVOC production in south-east Asia (MacKenzie et al. 2011).

Climate Extremes and Changes in Hydrology Field observations and numerical models have indicated that large-scale deforestation decreases evapotranspiration, thereby increasing sensible heat flux and water yield and discharge in many places, particularly at the agricultural frontier of south-eastern Brazilian Amazonia. Accelerating greenhouse gas emissions are expected to increase the intensity and frequency of droughts, causing further disruption to the energy and water cycles in the basin (Marengo et al. 2009; Coe et al. 2016). The last 10 years have been characterised by intense climatic and hydrological variations in parts of the Amazon region, with extreme impacts, such as droughts and large-scale floods. These events have been linked to natural climate variability and have had impacts on the socio-ecological system. The number of fires has increased during the years of drought, and the resulting smoke has had a negative impact on the health of local populations. Model projections suggest that such extremes may happen more frequently and with more intensity in the future and that the impacts of human activities, mainly in the form of changes in land use, can aggravate these extremes and enhance their impact on populations (Marengo et al. 2016). It is worth noting that a rain exclusion experiment ($-50\,\%$) designed to simulate extreme long-term (12 years) soil drought (but without accompanying atmospheric drought) in Caxiuanã, Pará, Brazil, has shown an interesting trajectory of forest response. The initial effects of rain exclusion that caused tree mortality (Meir et al. 2009) and a reduction in primary production appear to have diminished after 12 years,

Table 1.1 An overview of the content of the chapters that make up the book and the geographical coverage of each of them

Chapter	Topic	Geography	Methods/Approaches/ Keywords
Nagy et al.	Introduction	Amazon basin	The Amazon basin; Large-scale Biosphere–Atmosphere programme; the structure of the book
Coe et al.	Hydrology and energy balance	Amazon basin	Review of cycling of energy and water
Marengo et al.	Climate and its extremes	Amazon basin; Brazilian Amazonia; northern central South America	Review of extreme events and climate modelling in relation to extreme events
Aragão et al.	Fire	Brazilian Amazonia	Review of remote sensing, modelling, experimental burning, forecasting
Jardine and Jardine	Biogenic volatile compounds	Local case studies, primary forest near Manaus	Experimental/instrumental investigation
Grace	Greenhouse gas (GHG) balance	Amazon basin	Synthesis of GHG accounting
Gloor	Overall carbon balance	Northern South America between the Atlantic Ocean and the Andes, including the Amazon basin/Brazilian Amazonia	Synthesis of aircraft-based measurement of atmospheric CO_2 concentrations
Melack	Aquatic systems and their carbon balance	Amazon basin	Review of developments in studying the carbon cycle in aquatic and seasonally flooded environments
Araújo et al.	Primary net ecosystem exchange productivity and water availability	Three locations in Brazilian Amazonia	Micrometeorology/flux tower data (eddy covariance, carbon, seasonality, inter-annual NEE)
Higuchi et al.	Forest biomass and carbon stored	Amazonas State	New data from sample plots; destructive sampling and allometric equations
Buscardo et al.	Biogeochemistry and vegetation diversity	Amazon basin	Analysis of published data per forest type; statistics of components of C, N, P cycles
Quesada and Lloyd	Forest vegetation—soil	Amazon basin	Review of primary forest vegetation—soil interactions; permanent sample plots (RAINFOR)
Kruijt et al.	Dynamic vegetation modelling	Brazilian Amazonia	Review of ongoing modelling activities
Ometto et al.	Recent land use history	Brazilian Amazonia	Review of land use and land cover change (1960–2010)

(continued)

Table 1.1 (continued)

Chapter	Topic	Geography	Methods/Approaches/ Keywords
Fearnside	Land use impact on carbon sequestration	Brazilian Amazonia	Review of land use and land use change impacts on greenhouse gas emissions
Laurence et al.	Fragmentation	Local, primary forest fragments near Manaus	Updated review on habitat fragmentation and climate impacts on biodiversity and ecosystem services; permanent surveillance plots (PDBFF)
Vieira et al.	Land use dynamics and sustainability	Brazilian Amazonia, with focus on the State of Pará	Policy assessment for ecological and economic sustainability
Artaxo et al.	Concluding chapter		Perspectives for Amazonia in transformation

As indicated in the third and fourth columns, different authors have used different concepts of Amazonia and applied a variety of methods for estimating the variables reported. This has inevitably resulted in some variation in values reported across the chapters in this book, especially with regard to the carbon cycle

when comparing gross and net primary production between control and rain excluded plots (da Costa et al. 2014).

The Carbon Cycle and Carbon Balance—Multi-Scale Measurements and Estimates Grace (2016) reviews the approaches and methods used to characterise the carbon balance of the Amazon basin: (i) the 'bottom-up' approach, using repeat census data from field plots (e.g. Phillips et al. 2016; Galbraith et al. 2014) and remote sensing, and (ii) the 'top-down' approach, using aircraft-based measurements at the planetary boundary layer. Plot data provide 'insights' into processes that aircraft data cannot provide, but upscaling data from the plot scale to the entire basin involves many assumptions and uncertainties. To do this, it is necessary to estimate the carbon fluxes associated with deforestation, forest degradation, logging, and other activities and processes separately and then sum them up. Reviewing data derived from plots, Grace (2016) estimated a carbon balance close to zero. In other words, the loss of carbon to the atmosphere through deforestation is balanced by the growth of primary and secondary forests in a 'normal' year. Extreme events such as the droughts experienced in 2005 or 2010 can make the basin a strong source of CO_2. Aircraft flights have provided a more direct measure of the carbon balance. Flights need to be made several times a year, and the atmosphere profiles obtained over the Amazon basin need to be compared to the measured concentration of CO_2 in the air at the Atlantic edge of the basin. Such measurements showed that terrestrial vegetation was a sink in a normal year, but the region as a whole was about neutral because of carbon loss through burning (e.g. Gatti et al. 2014; Gloor 2016). In general, the results from aircraft profiles do not show much difference in the estimates from those derived from plots.

Carbon dioxide and methane emissions to the atmosphere from aquatic ecosystems, including wetlands, are an important component of the carbon cycle in the Amazon basin. Most of the carbon dioxide loss from wetlands is probably derived from atmospheric CO_2 fixed photosynthetically by angiosperms in aquatic environments. More research is needed on the carbon balance in seasonally flooded forests and in aquatic ecosystems dominated by floating herbaceous plants. In addition, quantifying the area of all water courses and their carbon dioxide emissions needs to be undertaken (Melack et al. 2009a, b; Melack 2016) so that the contribution of aquatic ecosystems to the greenhouse gas (GHG) balance at the basin scale can be made more reliable.

A study comparing net ecosystem exchange (NEE) in three locations across the Brazilian Amazonia challenges the general view that Amazon forest ecosystems are water limited in the dry season (1–3 months with <100 mm rainfall). Araújo et al. (2016) showed that forests appear to have high/increasing levels of photosynthetic capacity in the dry season. This pattern is also reflected in the inter-annual fluctuations in NEE and evapotranspiration (ET), with annual NEE being negatively correlated with annual rainfall and ET being largely related to the annual course of solar radiation.

The portion of NEE that turns into biomass (above- and below-ground) and stores carbon over the vast area of Amazonas State (1.5 M km^{-2}) is addressed by Higuchi et al. (2016). They sampled over 1800 plots in non-flooded (*terra firme*) primary forests at 18 sites and applied an allometric equation, developed after destructive sampling and adapted to each site based on dominant tree height. They estimated mean above-ground (149.9 ± 8.8 Mg C ha^{-1}) and total carbon stocks (159.8 ± 9.2 Mg C ha^{-1}). This adds yet another important estimate of Amazonian forest carbon stocks—a somewhat lower figure than most published estimates. The extensive data on coarse root biomass carbon are a welcome new contribution and the values presented for many sites challenge the generally used but hardly ever measured value (21 % of the value of that above-ground). If the presented inter-site variation in below-ground carbon ranges of 2–20 % of above-ground biomass is accepted, efforts are required to establish the underlying causes of this variation so that uncertainties in future estimates of carbon sequestration by the Amazon biome can be adequately addressed.

Environmental Variation and Global Change Tropical forests, including those in the Amazon basin, play a role in the global carbon balance and hydrological cycle that affects the pace and nature of climate change. Understanding ecosystem processes, such as the internal carbon cycle, alternative trajectories of succession in secondary forests, and the impacts of disturbance on forest structure, is essential to better understanding of the functioning of the tropical forests of the Amazon basin and to predict how they will respond to global environmental change (Galbraith et al. 2014). Climate change and biodiversity are linked by feedback mechanisms, for example, how different plant species respond to climate change and the resulting changes in species composition can determine whether a forest sequesters or releases carbon. The characteristics of winner species (with advantageous adaptation characteristics in terms of morphology, physiology, and genetics) in response to environmental changes can either exacerbate or mitigate the impacts of climate change (Phillips et al. 2016).

An area that until now has received relatively little consideration is how biodiversity losses—whether because of climate change or through forest conversion to agriculture—could affect the functioning of Amazonian ecosystems and the provisioning of ecosystem services. The progress made by the Brazilian Research Programme for Biodiversity (*PPBio*) in understanding the spatial distribution of biodiversity and the relationship between forest ecosystems and environmental variability in the Amazon basin can contribute to this in the future. As relationships between Amazon forest ecosystems and environmental variation (topography, soil, hydrology) show strong spatial and temporal variability (e.g. Buscardo et al. 2016; Quesada and Lloyd 2016), maintenance of the *PPBio* long-term ecological research (LTER) sites should be a priority for understanding the functioning of biodiversity and the impacts of climate change on the ecosystems of the Amazon basin.

Fire has become a widely used tool for clearing natural vegetation and for maintaining pastures in the Amazon region. The incidence and intensity of these fires increases during extreme droughts, which have become more frequent in the last decade (major droughts occurred in 2005 and 2010). Recently, eastern Amazonia has been heavily affected by forest fires, while before 2013–2014, most of the forest fires occurred in the region called arc of deforestation. Understanding the temporal and spatial patterns of fire and its consequences on forest structure, species composition, and the carbon cycle is critical for minimising the impacts of global change on Amazon ecosystems and human populations. Aragão et al. (2016) review the current state of knowledge about the spatial and temporal patterns of fire incidence in the Amazon basin, their relationship with land use and land cover, and their responses to climate, including regular seasonal droughts. They identify research priorities to better understand the long-term consequences of fire in the Amazon biome.

The quantification of the impact of fire on the carbon balance in the Amazon basin is fundamental (Gatti et al. 2014; Aragão et al. 2016; Gloor 2016; Grace 2016; Fearnside 2016). The impact of fire on the transport and deposition of nutrients, aerosol formation (SAMBBA, Artaxo et al. 2009; Longo et al. 2009), and human health (Reddington et al. 2015) is also of great importance, as is its impact on changes in the distribution of precipitation and the frequency of extreme drought events (Aragão et al. 2016; Marengo et al. 2016).

Water relations, temperature dependence, CO_2 and nutrient relations, growth and mortality patterns, and fire make up the main issues in contemporary modelling of the carbon balance and dynamics of Amazonian terrestrial vegetation (Kruijt et al. 2016). Despite remarkable advances, a number of issues appear limiting progress. As for water relations, a better understanding and empirical underpinning of different 'strategies' that plants use to prevent or compensate water stress are required. Future models must include the temperature sensitivity of primary production and respiration in current and predicted future tropical conditions.

To simulate realistic responses to rising CO_2 levels, it is essential that models incorporate the dynamic dependence of CO_2 exchange on soil nutrient availability, including feedbacks between vegetation change and nutrient availability. Finally, all realistic models must incorporate the complex influence of fire.

One of the major scientific challenges is to incorporate such dynamic vegetation models into complex socio-ecological models, including the generation and use of environmental services, and the behaviour of the socio-economic subsystem (policy, use, and feedback loops), in response to changes in the availability of environmental services caused by climate change.

Land Use and Human Dimensions—Integrating Considerations Between Biophysical and Social Aspects In the recent past, parts of the Amazon basin and the majority of Brazilian Amazonia that lies outside of the basin (see Fig. 1.1) have experienced a conversion to agriculture that totalled 18 % and which has not only affected the environment but had large-scale impacts on society. Deforestation has impacted biodiversity, soil structure, and the hydrological regime and, through a positive feedback, has probably caused changes in local and regional climate (Ometto et al. 2016). Deforestation and associated GHG emissions have declined substantially since their peak in 2004, but an increase is expected again due to the implementation of infrastructure projects already planned and through the recent weakening of the Brazilian Forest Law (Fearnside 2016). Along with deforestation, increased attention should be paid to GHG emissions from forest degradation caused by logging and the use of fire, which are growing.

The ecological impacts of forest fragmentation have been investigated at a long-term research site (Forest Dynamics in Forest Fragments project, *PDBFF*), run by the Smithsonian Institute and the National Institute of Amazon Research (INPA) over the last 35 years in 11 forest fragments, 1–100 ha in size, which have experienced a wide variety of environmental changes. Edge effects have been a dominant 'driver' of the dynamics of the fragments, strongly affecting microclimate, tree mortality, and carbon storage. The vegetation matrix surrounding the fragments changed considerably over time (from grassland to a mosaic of abandoned pastures and secondary forest), and this, in turn, strongly influenced the dynamics of the fragments and their fauna. Rare extreme weather events and key global change factors significantly influenced the structure and dynamics of the forest across the study area, both in forest fragments and intact forest nearby. The main factors of large-scale changes seem likely to interact synergistically with habitat fragmentation (Laurance et al. 2016).

The search for sustainable development in order to conserve and maintain ecosystem functions and services in the Brazilian Amazon has led to a series of public policy decisions and legislative actions regarding land use. 'Socio-environmentalism' emerged in the 1980s in Brazil to address the negative impacts of forest conversion and poorly planned infrastructure projects on biological and cultural diversity. However, the rapid development of agribusiness in large parts of the region, along with a large number of major infrastructural development projects, including mega-dams, major roads, and large mining projects, continues to generate conflicts with traditional populations and smallholders, in addition to causing major ecological impacts on Amazonian ecosystems and landscapes (Vieira et al. 2016). The construction of appropriate frameworks to consider the socio-economic system as part of an integrated socio-ecological system is a priority to help predict the outcomes of the impacts of land use and climate change along with society's responses to these changes in today's and tomorrow's Amazonia.

We hope that after reading this book, the reader will engage in furthering research in Amazonia, whatever their interest, to contribute to maintaining the region's planetary functions, as well as its natural and cultural diversity, charm, and enchantment.

References

Aragão LEOC, Anderson LO, Lima A, Arai E (2016) Fires in Amazonia. In: Nagy L, Forsberg B, Artaxo P (eds) Interactions between biosphere, atmosphere and human land use in the Amazon basin. Springer, Heidelberg, pp 301–329

Araújo AC, von Randow C, Restrepo-Coupe N (2016) Ecosystem–atmosphere exchanges of CO2 in dense and open 'Terra Firme' rainforests in Amazonia. In: Nagy L, Forsberg B, Artaxo P (eds) Interactions between biosphere, atmosphere and human land use in the Amazon basin. Springer, Heidelberg, pp 147–167

Artaxo P et al (2009) Aerosol particles in Amazonia: their composition, role in the radiation balance, cloud formation, and nutrient cycles. In: Keller M, Bustamante M, Gash J, Silva Dias P (eds) Amazonia and global change. American Geophysical Union, Washington, DC, pp 233–250

Artaxo P, Forsberg B, Nagy L (2016) Amazonia in perspective as a changing environment. In: Nagy L, Forsberg B, Artaxo P (eds) Interactions between biosphere, atmosphere and human land use in the Amazon basin. Springer, Heidelberg, pp 465–469

Balée W (1998) Advances in historical ecology. Columbia University Press, New York, NY

Buscardo E, Nardoto G, Luizao F, Piedade MTF, Schöngart J, Wittmann F, Doughty C, Quesada CA, Nagy L (2016) The biogeochemistry of the main forest vegetation types in Amazonia. In: Nagy L, Forsberg B, Artaxo P (eds) Interactions between biosphere, atmosphere and human land use in the Amazon basin. Springer, Heidelberg, pp 223–265

Bustamante MMC, Keller M, Silva DA (2009) Sources and sinks of trace gases in Amazonia and the Cerrado. In: Keller M, Bustamante M, Gash J, Silva Dias P (eds) Amazonia and global change. American Geophysical Union, Washington, DC, pp 337–354

Coe MT, Macedo MN, Brando PM, Lefebvre P, Panday P, Divino S (2016) The hydrology and energy balance of the Amazon basin. In: Nagy L, Forsberg B, Artaxo P (eds) Interactions between biosphere, atmosphere and human land use in the Amazon basin. Springer, Heidelberg, pp 33–51

da Costa ACL et al (2014) Ecosystem respiration and net primary productivity after 8–10 years of experimental through-fall reduction in an eastern Amazon forest. Plant Ecol Divers 7:7–24. doi:10.1080/17550874.2013.798366

Denevan WM (2001) Cultivated landscapes of native Amazonia and the Andes. Oxford University Press, Oxford

Eva HD, de Miranda EE, Di Bella CM, Gond V et al (2002) A vegetation map of South America. European Commission, Luxembourg

Fearnside P (2016) The impact of land use on carbon stocks and fluxes in Brazilian Amazonia: implications for policy. In: Nagy L, Forsberg B, Artaxo P (eds) Interactions between biosphere, atmosphere and human land use in the Amazon basin. Springer, Heidelberg, pp 385–405

Galbraith D, Malhi Y, Aragão L, Baker T (2014) The ecosystem dynamics of Amazonian and Andean forests. Plant Ecol Divers 7:1–6. doi:10.1080/17550874.2013.826744

Gatti LV et al (2014) Drought sensitivity of Amazonian carbon balance revealed by atmospheric measurements. Nature 506:76–80. doi:10.1038/nature12957

Gloor E (2016) Climate and the Amazonian carbon balance. In: Nagy L, Forsberg B, Artaxo P (eds) Interactions between biosphere, atmosphere and human land use in the Amazon basin. Springer, Heidelberg, pp 99–115

Grace J (2016) The Amazon carbon balance: an evaluation of methods and results. In: Nagy L, Forsberg B, Artaxo P (eds) Interactions between biosphere, atmosphere and human land use in the Amazon basin. Springer, Heidelberg, pp 77–97

Heil Costa M, Coe MT, Loup Guyot J (2009) Effects of climatic variability and deforestation on surface water regimes. In: Keller M, Bustamante M, Gash J, Silva Dias P (eds) Amazonia and global change. American Geophysical Union, Washington, DC, pp 543–553

Higuchi N, Suwa R, Higuchi FG, Lima AJN, Santos J, Noguchi H, Kajimoto T, Ishizuka M (2016) Overview of forest carbon stocks study in Amazonas State, Brazil. In: Nagy L, Forsberg B, Artaxo P (eds) Interactions between biosphere, atmosphere and human land use in the Amazon basin. Springer, Heidelberg, pp 169–187

Hoorn C et al (2010) Amazonia through time: Andean uplift, climate change, landscape evolution, and biodiversity. Science 330:927–931. doi:10.1126/science.1194585

Jardine K, Jardine A (2016) Biogenic volatile organic compounds in Amazonian forest ecosystems. In: Nagy L, Forsberg B, Artaxo P (eds) Interactions between biosphere, atmosphere and human land use in the Amazon basin. Springer, Heidelberg, pp 17–31

Keller M, Bustamante MMC, Gash J, Dias PS (2009) Amazonia and global change. American Geophysical Union, Washington, DC

Kesselmeier J, Guenther A, Hoffmann T, Piedade MT, Warnke J (2009) Natural volatile organic compound emissions from plants and their roles in oxidant balance and particle formation. In: Keller M, Bustamante M, Gash J, Silva Dias P (eds) Amazonia and global change. American Geophysical Union, Washington, DC, pp 183–206

Kruijt B, Meir P, Johnson M, Rammig A, Fauset S, Baker T, Galbraith D, von Randow C, Verbeeck H (2016) Modelling Amazonian carbon budgets and vegetation dynamics in a changing climate. In: Nagy L, Forsberg B, Artaxo P (eds) Interactions between biosphere, atmosphere and human land use in the Amazon basin. Springer, Heidelberg, pp 331–366

Laurance WF, Camargo JLC, Fearnside PM, Lovejoy TE, Williamson GB, Mesquita RCG, Meyer CFJ, Bobrowiec PED, Laurance SGW (2016) An Amazonian forest and its fragments as a laboratory of global change. In: Nagy L, Forsberg B, Artaxo P (eds) Interactions between biosphere, atmosphere and human land use in the Amazon basin. Springer, Heidelberg, pp 407–439

Longo KM, Freitas SR, Andreae MO, Yokelson R, Artaxo P (2009) Biomass burning in Amazonia: emissions, long-range transport of smoke and its regional and remote impacts. In: Keller M, Bustamante M, Gash J, Silva Dias P (eds) Amazonia and global change. American Geophysical Union, Washington, DC, pp 207–232

MacKenzie AR, Langford B, Pugh TAM, Robinson N, Misztal PK, Heard DE, Lee JD, Lewis AC, Jones CE, Hopkins JR, Phillips G, Monks PS, Karunaharan A, Hornsby KE, Nicolas-Perea V, Coe H, Gabey AM, Gallagher MW, Whalley LK, Edwards PM, Evans MJ, Stone D, Ingham T, Commane R, Furneaux KL, McQuaid JB, Nemitz E, Seng YK, Fowler D, Pyle JA, Hewitt CN (2011) The atmospheric chemistry of trace gases and particulate matter emitted by different land uses in Borneo. Philos Trans R Soc Lond B: Biol Sci 366:3177–3195

Malhi Y, Gardner TA, Goldsmith GR, Silman MR, Zelazowski P (2014) Tropical forests in the Anthropocene. Annu Rev Environ Resour 39:125–159

Marengo J, Nobre CA, Betts RA, Cox PM, Sampaio G, Salazar L (2009) Global warming and climate change in Amazonia: climate-vegetation feedback and impacts on water resources. In: Keller M, Bustamante M, Gash J, Silva Dias P (eds) Amazonia and global change. American Geophysical Union, Washington, DC, pp 273–292

Meir P et al (2009) The effects of drought on Amazonian rain forests. In: Keller M, Bustamante M, Gash J, Silva Dias P (eds) Amazonia and global change. American Geophysical Union, Washington, DC, pp 429–449

Melack JM, Novo EMLM, Forsberg BR, Piedade MTF, Maurice L (2009a) Floodplain ecosystem processes. In: Keller M, Bustamante M, Gash J, Silva Dias P (eds) Amazonia and global change. American Geophysical Union, Washington, DC, pp 525–541

Melack JM, Victoria RL, Tomasella J (2009b) Surface waters in Amazonia: key findings and perspectives. In: Keller M, Bustamante M, Gash J, Silva Dias P (eds) Amazonia and global change. American Geophysical Union, Washington, DC, pp 485–488

Melack J (2016) Aquatic ecosystems. In: Nagy L, Forsberg B, Artaxo P (eds) Interactions between biosphere, atmosphere and human land use in the Amazon basin. Springer, Heidelberg, pp 117–145

Nagy L, Artaxo P, Forsberg B (2016) Interactions between biosphere, atmosphere, and human land use in the Amazon basin: an introduction. In: Nagy L, Forsberg B, Artaxo P (eds) Interactions between biosphere, atmosphere and human land use in the Amazon basin. Springer, Heidelberg, pp 3–14

Navarro G, Maldonado M (2002) Geografía Ecológica de Bolivia. Vegetación y Ambientes Acuaticos. Fundación Simón I, Patiño, Cochabamba

Ometto JP, Sousa-Neto ER, Tejada G (2016) Land use, land cover and land use change in the Brazilian Amazon (1960–2013). In: Nagy L, Forsberg B, Artaxo P (eds) Interactions between biosphere, atmosphere and human land use in the Amazon basin. Springer, Heidelberg, pp 369–383

Quesada CA, Lloyd J (2016) Soil–vegetation interactions in Amazonia. In: Nagy L, Forsberg B, Artaxo P (eds) Interactions between biosphere, atmosphere and human land use in the Amazon basin. Springer, Heidelberg

Reddington CL et al (2015) Air quality and human health improvements from reductions in deforestation-related fire in Brazil. Nat Geosci 8:768–771. doi:10.1038/ngeo2535, http://www.nature.com/ngeo/journal/v8/n10/abs/ngeo2535.html#supplementary-information

Silva Dias MA, Avissar R, Silva Dias P (2009) Modeling the regional and remote climatic impact of deforestation. In: Keller M, Bustamante M, Gash J, Dias PS (eds) Amazonia and global change. American Geophysical Union, Washington, DC, pp 251–260

Part II
Biosphere–Atmosphere Interactions

Chapter 2
Biogenic Volatile Organic Compounds in Amazonian Forest Ecosystems

Kolby Jardine and Angela Jardine

2.1 Introduction

Biogenic volatile organic compounds (BVOCs) are produced directly within plants via biochemical pathways associated with primary and secondary metabolic processes. Although non-volatile metabolites are typically bound within specific cellular organelles in lipids or aqueous phases, BVOC volatile metabolites can readily partition between these phases and the intracellular airspace (Fall 2003). Thus, many BVOCs may freely exchange among cellular organelles, cells, and tissues, contributing to an integration of whole organism carbon and energy metabolism. Moreover, exchange of the intracellular airspace with the atmosphere may help coordinate the metabolisms of different plants within an ecosystem in response to environmental and biological factors (Yan and Wang 2006). In addition, land–atmosphere exchange of VOCs integrates local and regional atmospheric chemistry with plant metabolism (Jardine et al. 2011a).

The emerging field of volatile ecosystem metabolomics integrates the volatile component into the chemical, physical, and biological processes involved in the processing of metabolites within the land–atmosphere interface including potential perturbations of the system by anthropogenic activities (e.g. VOC emissions from

K. Jardine (✉)
Earth and Environmental Sciences, Climate Sciences Department, Lawrence Berkeley National Laboratory, Berkeley, CA, USA

Programa Clima e Ambiente, Instituto Nacional de Pesquisas da Amazônia e Universidade do Estado do Amazonas, Manaus, Amazonas, Brazil
e-mail: kjjardine@lbl.gov

A. Jardine
Programa Clima e Ambiente, Instituto Nacional de Pesquisas da Amazônia e Universidade do Estado do Amazonas, Manaus, Amazonas, Brazil
e-mail: angela.jardine@inpa.gov.br

© Springer-Verlag Berlin Heidelberg 2016

L. Nagy et al. (eds.), *Interactions Between Biosphere, Atmosphere and Human Land Use in the Amazon Basin*, Ecological Studies 227,
DOI 10.1007/978-3-662-49902-3_2

biomass and fossil fuel burning). The power of volatile metabolomics comes from the fact that many cellular processes leave unique chemical fingerprints (biomarkers) behind in the atmosphere (e.g. BVOCs, such as volatile isoprenoids, organic acids, alcohols, esters, aldehydes, ketones, aromatics, sulphides, nitriles). Therefore, volatile metabolomics provides non-invasive techniques to study plant metabolism from a variety of spatial and temporal scales. The application of these methods in the tropics may improve our mechanistic understanding of how environmental and biological variables associated with climate and land use change affect the carbon and energy metabolism of natural and managed forests.

Tropical ecosystems cycle more carbon than any other biome (~1.4 versus 0.5 Pg C year^{-1} for temperate ecosystems (Sarmiento et al. 2010)) and are estimated to account for nearly 80 % of global BVOC emissions (Levis et al. 2003). The Amazon basin is a major tropical source of BVOCs to the global atmosphere (Greenberg et al. 2004), yet less is known about BVOCs from its vegetation than from that in other regions of the world (Guenther 2013). This is due to a number of factors, among which are the extremely high tree species diversity (ter Steege et al. 2013; Macarthur 1965) and difficulties in conducting and sustaining remote field studies with highly sensitive analytical chemistry instrumentation for even short periods of time. What we do know about Amazonian BVOCs primarily results from a few important short-term plant enclosure studies, above-canopy measurements, and aircraft observations (Kesselmeier et al. 2002; Karl et al. 2007; Jardine et al. 2011b).

Historically, BVOCs have been studied with respect to their significant impacts on the chemistry and physics of the atmosphere, which remains an active area of research (e.g. Artaxo et al. 2009). The focus of this chapter is towards developing a mechanistic understanding of BVOC dynamics within plants and ecosystems. First is a brief overview of BVOC production within plants followed by an introduction to the field of volatile metabolomics. Next, examples of Amazonian plant and ecosystem functional roles of several BVOCs are discussed. The chapter ends with suggested directions for Amazonian BVOC research including detailed plant VOC emission surveys, the identification of new BVOCs, and the characterisation of potential BVOC ecosystem–atmosphere bidirectional exchange (i.e. emission and consumption by vegetation).

2.2 BVOC Biosynthesis in Plants

During photosynthesis, plants assimilate atmospheric carbon dioxide (CO_2) into primary metabolites which are essential components required for growth and development. Primary metabolites can be used as substrate for the biosynthesis of new biomass including proteins, nucleic acids, carbohydrates, and lipids or can be used as a carbon and energy source during respiration (Bourgaud et al. 2001; Tegeder and Weber 2008). Secondary metabolites are a diverse set of compounds which are intricately involved in many physiological and ecological processes

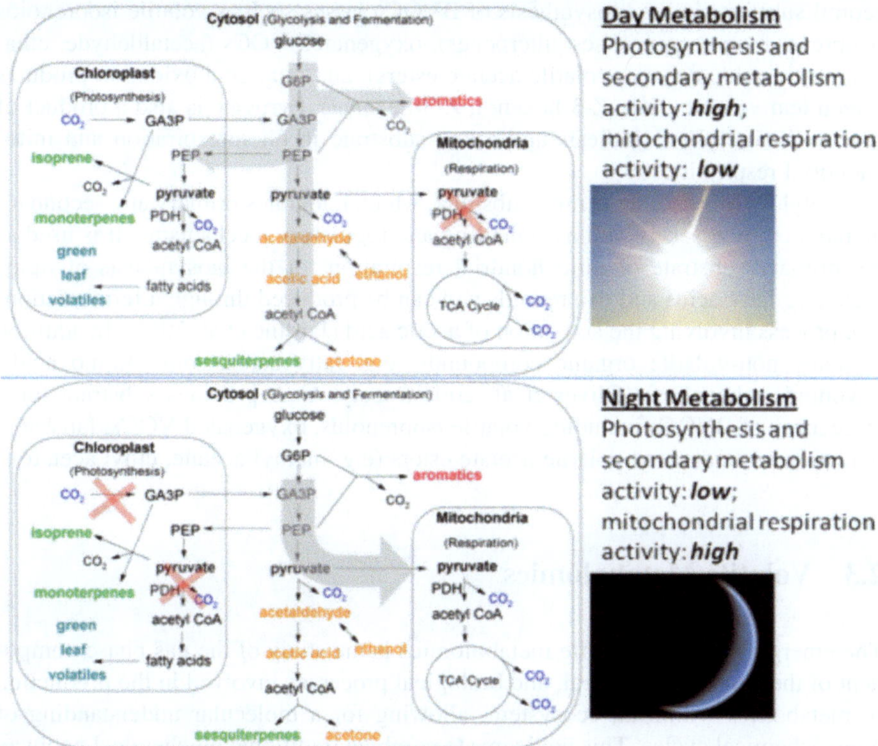

Fig. 2.1 A simplified representation of the flow of carbon in the cytosol, chloroplast, and mitochondria, which produce classes of biogenic volatile organic compounds (BVOCs) in photosynthetic cells during primary and secondary metabolism

within plants (Weng et al. 2012). For example, secondary metabolites have been identified as plant defences against abiotic and biotic stresses and chemical communication within and between species (Weng et al. 2012; Jardine et al. 2008; Karl et al. 2008; Pophof et al. 2005). Both primary and secondary metabolic pathways within plants create intermediates with sufficient vapour pressures to be directly emitted into the atmosphere in the gas phase as BVOCs.

Plant metabolic pathways involved in BVOC biosynthesis occur in and across a number of cellular organelles including the cytosol, chloroplast, mitochondria, and peroxisome (Fig. 2.1). Some BVOCs, such as isoprene, are strictly light dependent with a strong connection to recently assimilated carbon (Jones and Rasmussen 1975). In contrast, instantaneous emissions of other BVOCs, such as methanol, are much less light dependent and can be produced at night (Fall and Benson 1996; Harley et al. 2007).

Key substrates (i.e. glyceraldehyde-3-phosphate, phosphoenol pyruvate, pyruvate, and acetyl-CoA) can be produced independently within plant compartments or can be transported across compartmental boundaries. For example, pyruvate is a central substrate in the biosynthesis of BVOC classes such as volatile isoprenoids (isoprene, monoterpenes, sesquiterpenes), oxygenated VOCs (acetaldehyde, ethanol, acetic acid, acetone, volatile acetate esters), and fatty acid oxidation products (green leaf volatiles, e.g. Z-3-hexenol, Z-3-hexenal). Pyruvate is also a product of glycolysis and photosynthesis and a key substrate in photorespiration and mitochondrial respiration.

Acetyl CoA is another central substrate, which integrates primary and secondary metabolic processes as well as signalling and regulatory mechanisms. It is used as the primary substrate of mitochondrial respiration for the biosynthesis of lipids including fatty acids and isoprenoids and can be produced through a fermentation-like process involving the activation of acetic acid (Jardine et al. 2013). In addition to many non-volatile organic compounds (e.g. fatty acids, some amino acids, flavonoids, phenolics) (Oliver et al. 2009), acetyl CoA provides substrate for a large array of BVOCs including volatile isoprenoids, oxygenated VOCs, fatty acid oxidation products, and volatile acetate esters (e.g. methyl acetate, ethyl acetate).

2.3 Volatile Metabolomics

The emerging field of volatile metabolomics is the study of the gas phase component of the chemical, physical, and biological processes involved in the production of metabolites within an ecosystem, allowing for a molecular understanding of biogeochemical cycles. This field aims to combine traditional biochemical pathway studies involving destructive tissue sampling and metabolite extraction and analysis with non-invasive atmospheric analytical chemistry techniques used for the identification and quantification of BVOCs within and above forest canopies to gain new insights into within-plant carbon and energy allocation to primary and secondary metabolic processes. For example, plant BVOC exchange fluxes using enclosures and ambient BVOC concentrations and fluxes within and above forest canopies can be made in situ in real time and used as biomarkers of carbon allocation processes such as photosynthesis, photorespiration, respiration, and fermentation (Jardine et al. 2010b; Loreto et al. 2007; Kesselmeier et al. 1997; Bracho-Nunez et al. 2012), cell wall expansion and growth (Harley et al. 2007), acetyl-CoA and fatty acid biosynthesis and degradation (Fall et al. 1999; Jardine et al. 2012b), and signalling and defence against abiotic and biotic stresses (Niinemets 2010; Karl et al. 2008; Jardine et al. 2014).

For example, Jardine et al. (2010b) used positional ^{13}C-labelled pyruvate to trace the metabolic pathways responsible for the biosynthesis of volatile isoprenoids and oxygenated VOCs in real time for individual mango leaves (*Mangifera indica*). While the metabolic pathways leading to isoprenoid biosynthesis are well documented, those leading to the production of oxygenated VOCs are still uncertain. In their study, leaves fed with the pyruvate-2-^{13}C solution resulted in large enrichments ($^{13}C/^{12}C$) of both ^{13}C-labelled isoprenoids and oxygenated VOCs (up to 266 % for sesquiterpenes and 154 % for acetaldehyde for example).

However, when mango leaves were fed with pyruvate-1-^{13}C, ^{13}C labelling of BVOCs was greatly reduced. Positional pyruvate labelling was then used to demonstrate that the $C_{2,3}$ atoms of pyruvate are directly utilised for the biosynthesis of volatile and non-volatile metabolites and biopolymers. However, the C_1 of pyruvate has a completely different fate and is decarboxylated to CO_2, representing a new source of CO_2 not previously considered in studies of plant CO_2 sources and sinks (Jardine et al. 2013). Thus, BVOCs can be used to track cellular processes that imprint a unique chemical fingerprint on the atmosphere surrounding individual plants. When applied at the ecosystem level, volatile metabolomics has the potential to advance mechanistic understanding of BVOC biosynthesis as a function of changing environmental conditions.

2.4 BVOCs as Biomarkers of Lipid Peroxidation Under Oxidative Stress

Lipids serve numerous critical functions in plant biology including providing membrane structure and participating in the light reactions of photosynthesis, antioxidant, and signalling processes. Saturated lipids are extremely resistant to oxidation within plants and the environment, where plant alkanes with ages greater than one billion years have been detected in lake sediments (Oro et al. 1965). In contrast, unsaturated lipids including isoprenoids and fatty acids are highly susceptible to oxidisation with their pools rapidly turned over in oxidising conditions. Moreover, the oxidative power of the lower atmosphere is strongly influenced by the emission of unsaturated volatile lipids from vegetation, especially isoprenoids and reactive volatile lipids which can be emitted at high rates from many plants fuelling atmospheric chemistry through photo-oxidation reactions (Monson 2002).

Reactive oxygen species (ROS) including singlet oxygen ($^1O_2$), superoxide anion (O_2^-), hydrogen peroxide (H_2O_2), and the hydroxyl radical (OH) are continuously generated in plants by the incomplete reduction of oxygen (O_2). While ROS concentrations within plants are generally kept low by ROS quenching and scavenging systems, excessive ROS accumulation can result in extensive oxidation of plant lipids (Apel and Hirt 2004; Jardine et al. 2010a). While traditionally described as the 'Oxygen Paradox' where ROS are a toxic by-product of aerobic metabolism, ROS-lipid signalling is now recognised as an integral component of plant response to abiotic and biotic stress as well as regulation of growth, development, and programmed cell death (Suzuki et al. 2011; Mittler et al. 2011).

In a changing global climate with increasing air pollution and rapid land use changes, plants are exposed to a wide variety of biotic (e.g. microbes, insects) and abiotic (e.g. thermal, radiative, drought) stressors. In plant tissues, these stressors can cause the accumulation of ROS, which if left unchecked can overwhelm cellular antioxidant defences including enzyme-mediated ROS quenching reactions, internal systems for ROS scavenging, and defence gene activation (Møller

2001). This can lead to extensive ROS-mediated oxidation of important components, such as nucleic acids, proteins, and lipids leading to cell death (Apel and Hirt 2004). Therefore, plants with a diverse suite of antioxidant defences may better tolerate stressful environmental conditions occurring in response to local and global changes in climate.

The oxidation of plant fatty acids via non-enzymatic (Durand et al. 2009; Mene-Saffrane et al. 2009) and enzymatic (Andreou and Feussner 2009; Gigot et al. 2010; Heiden et al. 2003) mechanisms produces a broad range of oxidation product biomarkers termed oxylipins. The accumulation of ROS in plant tissues initiates fatty acid (e.g. α-linolenic acid) peroxidation, yielding a large array of 'oxidative stress' biomarkers. Lipid peroxidation generates a number of products, which have been extensively used as quantitative indicators of oxidative damage in plants (Gutteridge 1995; Shulaev and Oliver 2006). For example, 4-hydroxy-2-nonenal (HNE), 4-hydroxy-2-hexenal (HHE), and malondialdehyde are widely used as biomarkers of non-enzymatic lipid peroxidation (Hartley et al. 1999; Halliwell and Gutteridge 1999; Long and Picklo 2010). However, the extraction from plant tissues, derivatisation, and compound-specific analysis (GC-MS or HPLC) of these reactive carbonyl compounds remains a challenge due to their trace abundances, high reactivity, water solubility, and volatility (Shibamoto 2006). Nonetheless, a number of classes of lipid peroxidation products have been identified including hydrocarbons, ketones, furans, alkanals, 2-alkenals, 2,4-alkadienals, 2-hydroxyalkanals, 4-hydroxy-2-alkenals, and dicarbonyls (Kawai et al. 2007; Steeghs et al. 2006; Frankel et al. 1989; Mark et al. 1997; Moseley et al. 2003; Nielsen et al. 1997).

Given the volatile nature of many of these biomarkers, it is interesting to speculate on the potential of detecting them in atmospheric samples as non-invasive indicators of oxidative stress at a variety of temporal and spatial scales. Numerous volatile oxylipins have been recently observed as direct emissions from plants under oxidative stress generated by freeze–thaw treatment of tropical leaves (Table 2.1). These include alkanals (e.g. propanal, butanal, pentanal,

Table 2.1 Example of isoprene (bold) and fatty acid peroxidation biomarkers from plants under oxidative stress

Parent lipid	Class	Oxidation Biomarkers
isoprene	**isoprene**	**methacrolein, methyl vinyl ketone**
isoprene	**isoprene**	**3-methyl furan, 2-methyl-3-buten-2-ol**
fatty acids	GLVs	3-hexenal, 3-hexen-1-ol, 3-hexen-1-yl acetate
fatty acids	furans and furanones	tetrahydrofuran, 2-ethyl furan, 5-ethyl 2(5H)-furanone
fatty acids	alkanes	propane, butane, pentane…undecane
fatty acids	2-alkenes	2-propene, 2-butene, 2-pentene…2-undecene
fatty acids	alkanals	propanal, butanal, pentanal…undecanal
fatty acids	2-alkenals	2-propenal, 2-butenal, 2-pentenal…2-undecenal
fatty acids	2,4-alkadienals	2,4-hexadienal, 2,4-heptadienal, 2,4-octadienal
fatty acids	2-ketones	2-butanone, 2-pentanone... 2-undecanone
fatty acids	alkenones	1-hexen-3-one, 1-penten-3-one, 1-octen-3-one, 6-methyl-5-hepten-2-one
fatty acids	4-hydroxy 2-alkenals	4-hydroxy-2-hexenal, 4-hydroxy-2-nonenal
fatty acids	dicarbonyls	malondialdehyde, gyloxal, methyl glyoxal, diacetyl

hexanal), 2-alkenals (e.g. 2-propenal, 2-butenal, 2-pentenal, 2-hexenal), 2-alkenes (e.g. 2-propene, 2-butene, 2-pentene, 2-hexene), 2,4-alkadienals (e.g. 2,4-hexadienal), furans and furanones (e.g., tetrahydrofuran, 2-ethyl furan), and dicarbonyls (e.g. malondialdehyde, gyloxal, methyl glyoxal, and diacetyl). In addition, the enzymatic peroxidation of plant fatty acids by lipoxygenase enzymes can lead to the formation and emission of characteristic oxidation products known as green leaf volatiles (GLVs) via the lipoxygenase pathway (Loreto and Schnitzler 2010; Hatanaka 1993; Fall et al. 1999). In this pathway, the formation of the classic 6-carbon (C6) GLVs in plants is initiated by the ubiquitous type 2 lipoxygenase enzymes (13-LOX) in chloroplasts which catalyse the oxygenation of α-linolenic acid (the dominant fatty acid in the aerial tissues of most plants) to form 13-hydroperoxy linolenic acid (HPLA) (Andreou and Feussner 2009). HPLA can be degraded (catalysed by hydroperoxide lyase) to form the primary GLV (Z)-3-hexenal which is then reduced and acetylated to form the corresponding alcohol (Z)-3-hexen-1-ol and acetate ester (Z)-3-hexen-1-yl acetate, respectively (D'Auria et al. 2007).

The emissions of GLVs from plants have been documented during processes known to be associated with ROS accumulation including programmed cell death during senescence (Holopainen et al. 2010) and a wide variety of biotic and abiotic stresses including pathogen attack (Jansen et al. 2009), high ambient ozone concentrations (Heiden et al. 2003; Beauchamp et al. 2005), herbivory (Arimura et al. 2009), desiccation (De Gouw et al. 2000), high light and temperature (Loreto et al. 2006), mechanical wounding (Fall et al. 1999), and freeze–thaw events (Fall et al. 2001). Thus, both enzymatic and non-enzymatic lipid peroxidation mechanisms lead to the formation of characteristic fatty acid peroxidation biomarkers that may be detectable as gas-phase emissions from plant tissue under stress at the ecosystem scale. Although not yet reported from Amazonian ecosystems, these and other lipid peroxidation biomarkers may be emitted under environmental extremes associated with changes in land use and climate and are therefore prime candidates for ecosystem scale volatile metabolomics studies (Kawai et al. 2007; Steeghs et al. 2006; Frankel et al. 1989; Mark et al. 1997; Moseley et al. 2003; Nielsen et al. 1997).

The five-carbon molecule, isoprene, is estimated to be the most abundant and well-studied BVOC emitted from terrestrial ecosystems (Rasmussen and Khalil 1988). Isoprene is a biomarker of photosynthesis and concentrations are known to positively correlate with light and temperature (Fig. 2.2) (Monson and Fall 1989). As reviewed by Vickers et al. (2009), a rich literature exists demonstrating the role of isoprene as a photosynthesis protector from oxidative stress caused by high temperature and light. Isoprene has also been shown to quench ozone and hydrogen peroxide (Loreto and Velikova 2001), singlet oxygen (Velikova et al. 2004), and nitric oxide (Velikova et al. 2005), suggesting an antioxidant role for isoprene. Recently, investigations using mango leaves and branches (*Mangifera indica*) as well as ambient air samples from an enclosed tropical mesocosm and a central Amazon forest support the role of antioxidant properties of isoprene (Jardine et al. 2012a, 2013). Observations of temperature-stressed leaves from isoprene-

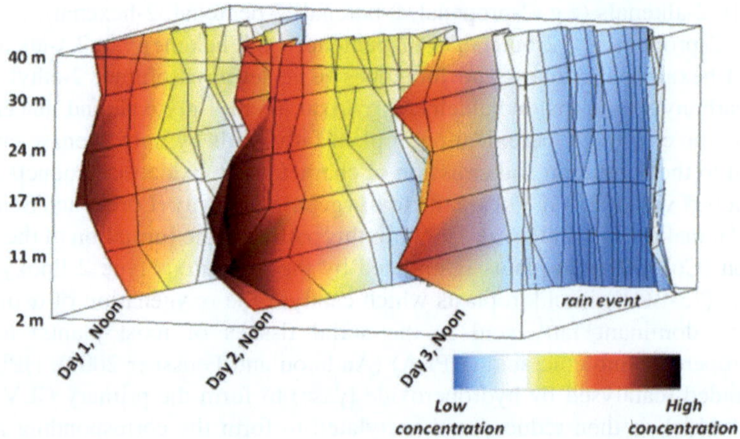

Fig. 2.2 A snapshot of the diurnal patterns of isoprene concentration variations over a 3-day period measured from a tower in a central Amazon forest. The *x*-axis represents time, while the *y*-axis shows measurement heights from 2 m above the ground to 10 m above the canopy. The *colour scale* ranges from minimum (*blue*) to maximum (*black*) concentrations. The vertical gradients show clear sources within the under-canopy (10–17 m) and canopy (30 m) layers corresponding with maximum temperature and light at midday and minimum concentrations at night. A sudden drop in concentration at all heights occurring during a rain event which decreased temperatures and light and likely diminished ecosystem photosynthesis is also observable

emitting plants showed production of first-generation oxidation products of isoprene: methyl vinyl ketone (MVK), methacrolein (MAC), and 3-methyl furan (3-MF). The authors suggested that the emissions of isoprene oxidation products increased with temperature stress due to direct chemical reactions with ROS or, in other words, in-plant oxidation of isoprene (Fig. 2.3).

To a large extent, the oxidative power of the lower atmosphere is controlled by ecosystem emissions of biogenic VOCs, especially those that contain carbon–carbon double bonds, such as volatile isoprenoids, which are readily available for oxidation through reaction with hydroxyl radicals, ozone, and nitrate radicals (Monson 2002). However, current models cannot adequately describe atmospheric oxidant levels in biogenically dominated areas like the Amazon basin (Lelieveld et al. 2008). The discovery of significant primary emissions of isoprene oxidation products in the central Brazilian Amazon (Jardine et al. 2012a) may reduce previous estimates of the effect of VOCs on the oxidising power of the troposphere and increase the role of leaves as a source for at least the first-generation products of VOC oxidation to the atmosphere. These findings may have important implications for characterising the oxidising capacity of the atmosphere and its impacts on atmospheric chemistry and climate. Understanding lipid production and oxidation dynamics within plants may be critical for predicting ecosystem response to the increasing temperature and light as a result of a changing climate. Investigating these mechanistic processes is complex, because of the very broad scale over which lipid production is influenced and over which they react. For example, isoprene

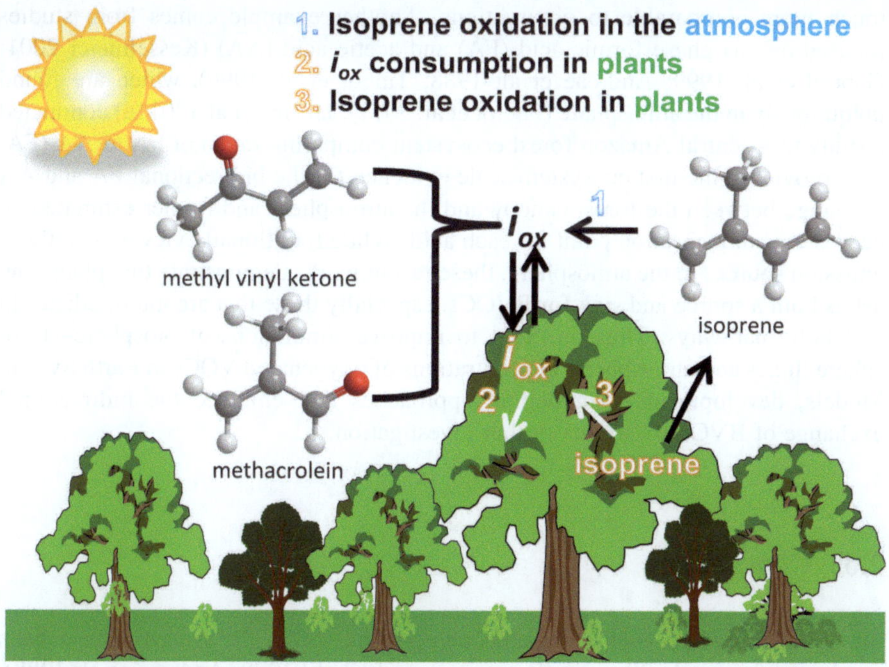

Fig. 2.3 A simplified schematic showing the production of isoprene and its oxidation products methyl vinyl ketone and methacrolein in plants as well as their emission to the atmosphere

starts at the subcellular level in chloroplasts and ultimately impacts key atmospheric processes, which in turn modify global climate, which in turn influences isoprene production in the biosphere. Investigation therefore requires a mixture of expertise and collaborating disciplines.

2.5 BVOC Bidirectional Biosphere–Atmosphere Exchange

There is a growing consensus that a large number of oxygenated compounds are both emitted and consumed by plants and that bidirectional exchange occurs between tropical forests and the atmosphere (Jardine et al. 2008, 2011a; Karl et al. 2005; Ganzeveld et al. 2008; Andreae et al. 1988). The compensation point is the point at which the ambient air mixing ratio of a BVOC results in a net zero flux, where consumption and emission are balanced.

A recent study focused on methanol and acetone Ganzeveld et al. (2008) showed that a commonly applied algorithm to simulate global acetone and methanol biogenic exchanges substantially overestimates ambient concentrations and emission strengths (compared with observations). In contrast, the use of a compensation point approach simulated ambient concentrations and exchange dynamics that were much more comparable to observations. Another example comes from studies focused on gas-phase formic acid (FA) and acetic acid (AA) (Kesselmeier 2001; Gabriel et al. 1999; Andreae et al. 1988; Talbot et al. 1990), which are found ubiquitously in the atmosphere (Paulot et al. 2011). Jardine et al. (2011a) conducted a study in a central Amazon forest ecosystem comparing ratios of FA to AA (FA: AA) providing the first ecosystem scale evidence for the bidirectional FA and AA exchange between the forest canopy and the atmosphere and further estimated an ecosystem compensation point for each acid. While traditionally viewed strictly as emission sources to the atmosphere, these recent works demonstrate that plants can act as both a source and sink for BVOCs, especially those that are intermediates in metabolic pathways. Thus, in order to improve simulations of biosphere–atmosphere fluxes and atmospheric concentrations of oxygenated VOCs in Earth System Models, development of modelling approaches that embrace the bidirectional exchange of BVOCs warrants further investigation.

2.6 Conclusions

Studies of biogenic volatile organic compounds within the Amazon have been conducted for nearly 30 years (Rasmussen and Khalil 1988; Andreae et al. 1988); however, we are still at the forefront of our understanding of what BVOCs are emitted from vegetation, the metabolic pathways that produce them within plants, their functional roles in terrestrial ecosystems, and how these roles may change under a changing climate. Some of the basic yet extremely important questions to address with respect to BVOCs in the Amazon are: (i) What are the identities, concentrations, and fluxes of BVOC emissions from individual tree species and whole ecosystems? (ii) What are the functional biological roles of BVOCs and what roles will they play under future land use and climate change? (iii) What controls the amount of assimilated carbon allocated to the production and emission of BVOCs in relation to non-volatile metabolites and respiratory processes?

Addressing these questions requires intensive observations at both the plant species and ecosystem scales within the Amazon basin. As highly sensitive analytical chemistry tools become available, the identification and quantification of novel BVOCs is also emerging. The promising area of research, volatile metabolomics, is beginning to provide non-invasive methodologies to develop a mechanistic understanding of BVOC metabolism and therefore may lead to new understanding of the functional roles of BVOCs at the plant and ecosystem scales. In turn, improvements in Earth System Models can further our ability to predict changes in BVOC impacts on atmospheric chemistry and climate.

Acknowledgements This material is based upon work supported as part of the GoAmazon 2014/5 and the Next Generation Ecosystem Experiments-Tropics (NGEE-Tropics) funded by the U.S. Department of Energy, Office of Science, Office of Biological and Environmental Research through contract no. DE-AC02-05CH11231 to LBNL, as part of DOE's Terrestrial Ecosystem Science Program.

References

Andreae M, Talbot R, Andreae T, Harriss R (1988) Formic and acetic acid over the central Amazon region, Brazil: 1. Dry season. J Geophys Res: Atmos (1984–2012) 93(D2):1616–1624

Andreou A, Feussner I (2009) Lipoxygenases—structure and reaction mechanism. Phytochemistry 70(13–14):1504–1510. doi:10.1016/j.phytochem.2009.05.008

Apel K, Hirt H (2004) Reactive oxygen species: metabolism, oxidative stress, and signal transduction. Annu Rev Plant Biol 55:373–399

Arimura G, Matsui K, Takabayashi J (2009) Chemical and molecular ecology of herbivore-induced plant volatiles: proximate factors and their ultimate functions. Plant Cell Physiol 50 (5):911–923. doi:10.1093/Pcp/Pcp030

Artaxo P, Rizzo LV, Paixão M, De Lucca S, Oliveira PH, Lara LL, Wiedemann KT, Andreae MO, Holben B, Schafer J, Correia AL, Pauliquevis TM (2009) Aerosol particles in Amazonia: their composition, role in the radiation balance, cloud formation, and nutrient cycles. In: Keller M, Bustamante M, Gash J, Silva Dias P (eds) Amazonia and global change. American Geophysical Union, Washington, DC, pp 233–250. doi:10.1029/2008gm000847

Beauchamp J, Wisthaler A, Hansel A, Kleist E, Miebach M, Niinemets U, Schurr U, Wildt J (2005) Ozone induced emissions of biogenic VOC from tobacco: relationships between ozone uptake and emission of LOX products. Plant Cell Environ 28(10):1334–1343

Bourgaud F, Gravot A, Milesi S, Gontier E (2001) Production of plant secondary metabolites: a historical perspective. Plant Sci 161(5):839–851. doi:10.1016/S0168-9452(01)00490-3

Bracho-Nunez A, Knothe NM, Costa WR, Astrid LRM, Kleiss B, Rottenberger S, Piedade MTF, Kesselmeier J (2012) Root anoxia effects on physiology and emissions of volatile organic compounds (VOC) under short-and long-term inundation of trees from Amazonian floodplains. SpringerPlus 1(1):1–16

D'Auria JC, Pichersky E, Schaub A, Hansel A, Gershenzon J (2007) Characterization of a BAHD acyltransferase responsible for producing the green leaf volatile (Z)-3-hexen-1-yl acetate in *Arabidopsis thaliana*. Plant J 49(2):194–207. doi:10.1111/j.1365-313X.2006.02946.x

De Gouw JA, Howard CJ, Custer TG, Baker BM, Fall R (2000) Proton-transfer chemical-ionization mass spectrometry allows real-time analysis of volatile organic compounds released from cutting and drying of crops. Environ Sci Technol 34(12):2640–2648

Durand T, Bultel-Ponce V, Guy A, Berger S, Mueller MJ, Galano JM (2009) New bioactive oxylipins formed by non-enzymatic free-radical-catalyzed pathways: the phytoprostanes. Lipids 44(10):875–888. doi:10.1007/s11745-009-3351-1

Fall R (2003) Abundant oxygenates in the atmosphere: a biochemical perspective. Chem Rev 103 (12):4941–4951

Fall R, Benson AA (1996) Leaf methanol—the simplest natural product from plants. Trends Plant Sci 1(9):296–301

Fall R, Karl T, Hansel A, Jordan A, Lindinger W (1999) Volatile organic compounds emitted after leaf wounding: on-line analysis by proton-transfer-reaction mass spectrometry. J Geophys Res: Atmos 104(D13):15963–15974

Fall R, Karl T, Jordon A, Lindinger W (2001) Biogenic C5VOCs: release from leaves after freeze-thaw wounding and occurrence in air at a high mountain observatory. Atmos Environ 35 (22):3905–3916

Frankel EN, Hu ML, Tappel AL (1989) Rapid headspace gas-chromatography of hexanal as a measure of lipid-peroxidation in biological samples. Lipids 24(11):976–981

Gabriel R, Schäfer L, Gerlach C, Rausch T, Kesselmeier J (1999) Factors controlling the emissions of volatile organic acids from leaves of *Quercus ilex* L. (Holm oak). Atmos Environ 33 (9):1347–1355

Ganzeveld L, Eerdekens G, Feig G, Fischer H, Harder H, Königstedt R, Kubistin D, Martinez M, Meixner F, Scheeren H (2008) Surface and boundary layer exchanges of volatile organic compounds, nitrogen oxides and ozone during the GABRIEL campaign. Atmos Chem Phys 8 (20):6223–6243

Gigot C, Ongena M, Fauconnier ML, Wathelet JP, Du Jardin P, Thonart P (2010) The lipoxygenase metabolic pathway in plants: potential for industrial production of natural green leaf volatiles. Biotechnol Agron Soc 14(3):451–460

Greenberg JP, Guenther AB, Pétron G, Wiedinmyer C, Vega O, Gatti LV, Tota J, Fisch G (2004) Biogenic VOC emissions from forested Amazonian landscapes. Glob Chang Biol 10 (5):651–662

Guenther A (2013) Biological and chemical diversity of biogenic volatile organic emissions into the atmosphere. ISRN Atmos Sci

Gutteridge JMC (1995) Lipid-peroxidation and antioxidants as biomarkers of tissue-damage. Clin Chem 41(12B):1819–1828

Halliwell B, Gutteridge JMC (1999) Free radicals in biology and medicine, 3rd edn. Oxford Science Publications, Clarendon Press, Oxford, New York, NY

Harley P, Greenberg J, Niinemets U, Guenther A (2007) Environmental controls over methanol emissions from leaves. Biogeosci Discuss 4:2593–2640

Hartley DP, Kolaja KL, Reichard J, Petersen DR (1999) 4-hydroxynonenal and malondialdehyde hepatic protein adducts in rats treated with carbon tetrachloride: Immunochemical detection and lobular localization. Toxicol Appl Pharmacol 161(1):23–33

Hatanaka A (1993) Studies on biogeneration and physiological-role of green odor by plant. Nippon Nogeik Kaishi 67(10):1391–1398

Heiden AC, Kobel K, Langebartels C, Schuh-Thomas G, Wildt J (2003) Emissions of oxygenated volatile organic compounds from plants—part I: emissions from lipoxygenase activity. J Atmos Chem 45(2):143–172

Holopainen JK, Heijari J, Oksanen E, Alessio GA (2010) Leaf volatile emissions of *Betula pendula* during autumn coloration and leaf fall. J Chem Ecol 36(10):1068–1075. doi:10.1007/s10886-010-9857-4

Jansen RMC, Miebach M, Kleist E, van Henten EJ, Wildt J (2009) Release of lipoxygenase products and monoterpenes by tomato plants as an indicator of *Botrytis cinerea*-induced stress. Plant Biol 11(6):859–868. doi:10.1111/j.1438-8677.2008.00183.x

Jardine K, Harley P, Karl T, Guenther A, Lerdau M, Mak J (2008) Plant physiological and environmental controls over the exchange of acetaldehyde between forest canopies and the atmosphere. Biogeosciences 5(6):1559–1572

Jardine K, Abrell L, Kurc SA, Huxman T, Ortega J, Guenther A (2010a) Volatile organic compound emissions from *Larrea tridentata* (creosotebush). Atmos Chem Phys 10 (24):12191–12206. doi:10.5194/acp-10-12191-2010

Jardine K, Sommer E, Saleska S, Huxman T, Harley P, Abrell L (2010b) Gas phase measurements of pyruvic acid and its volatile metabolites. Environ Sci Technol 44:2454–2460. doi:10.1021/es903544p

Jardine K, Abrell L, Yanez Serrano AM, Arneth A, Alves E, Kesselmeier J, Huxman T, Saleska S, Jardine A, Taylor T, Artaxo P (2011a) Ecosystem-scale compensation points of formic and acetic acid in the central Amazon. Biogeosciences 8:3709–3720

Jardine K, Abrell L, Yanez Serrano AM, Arneth A, Yoko Ishida F, Huxman T, Saleska S, Jardine A, Karl T, Artaxo P (2011b) Within-canopy sesquiterpene ozonolysis in Amazonia. J Geophys Res [Atmos] 116:D19301. doi:10.1029/2011JD016243

Jardine K, Abrell L, Jardine A, Huxman T, Saleska S, Arneth A, Monson R, Karl T, Fares S, Loreto F, Goldstein A (2012a) Within-plant isoprene oxidation confirmed by direct emissions of oxidation products methyl vinyl ketone and methacrolein. Glob Chang Biol 18:973–984

Jardine K, Barron-Gafford GA, Norman JP, Abrell L, Monson RK, Meyers KT, Pavao-Zuckerman M, Dontsova K, Kleist E, Werner C, Huxman TE (2012b) Green leaf volatiles and oxygenated metabolite emission bursts from mesquite branches following light–dark transitions. Photosynth Res 113(1–3):321–333. doi:10.1007/s11120-012-9746-5

Jardine KJ, Meyers K, Abrell L, Alves EG, Serrano AM, Kesselmeier J, Karl T, Guenther A, Chambers JQ, Vickers C (2013) Emissions of putative isoprene oxidation products from mango branches under abiotic stress. J Exp Bot 64(12):3697–3709. doi:10.1093/Jxb/Ert202

Jardine K, Wegener F, Abrell L, van Haren J, Werner C (2014) Phytogenic biosynthesis and emission of methyl acetate. Plant Cell Environ 37(2):414–424

Jones CA, Rasmussen RA (1975) Production of isoprene by leaf tissue. Plant Physiol 55 (6):982–987. doi:10.1104/Pp.55.6.982

Karl T, Harley P, Guenther A, Rasmussen R, Baker B, Jardine K, Nemitz E (2005) The bi-directional exchange of oxygenated VOCs between a loblolly pine (*Pinus taeda*) plantation and the atmosphere. Atmos Chem Phys 5(11):3015–3031

Karl T, Guenther A, Yokelson RJ, Greenberg J, Potosnak M, Blake DR, Artaxo P (2007) The tropical forest and fire emissions experiment: emission, chemistry, and transport of biogenic volatile organic compounds in the lower atmosphere over Amazonia. J Geophys Res: Atmos 112(D18). doi: 10.1029/2007jd008539. Artn D18302

Karl T, Guenther A, Turnipseed A, Patton E, Jardine K (2008) Chemical sensing of plant stress at the ecosystem scale. Biogeosciences 5(5):1287–1294

Kawai Y, Takeda S, Terao J (2007) Lipidomic analysis for lipid peroxidation-derived aldehydes using gas chromatography–mass spectrometry. Chem Res Toxicol 20(1):99–107. doi:10.1021/Tx060199e

Kesselmeier J (2001) Exchange of short-chain oxygenated volatile organic compounds (VOCs) between plants and the atmosphere: a compilation of field and laboratory studies. J Atmos Chem 39(3):219–233

Kesselmeier J, Bode K, Hofmann U, Muller H, Schafer L, Wolf A, Ciccioli P, Brancaleoni E, Cecinato A, Frattoni M, Foster P, Ferrari C, Jacob V, Fugit JL, Dutaur L, Simon V, Torres L (1997) Emission of short chained organic acids, aldehydes and monoterpenes from *Quercus ilex* L. and *Pinus pinea* L. in relation to physiological activities, carbon budget and emission algorithms. Atmos Environ 31:119–133. doi:10.10.16/S1352-2310(97)00079-4

Kesselmeier J, Kuhn U, Rottenberger S, Biesenthal T, Wolf A, Schebeske G, Andreae M, Ciccioli P, Brancaleoni E, Frattoni M (2002) Concentrations and species composition of atmospheric volatile organic compounds (VOCs) as observed during the wet and dry season in Rondonia (Amazonia). J Geophys Res: Atmos (1984–2012) 107(D20):LBA 20-21–LBA 20-13

Lelieveld J, Butler TM, Crowley JN, Dillon TJ, Fischer H, Ganzeveld L, Harder H, Lawrence MG, Martinez M, Taraborrelli D, Williams J (2008) Atmospheric oxidation capacity sustained by a tropical forest. Nature 452(7188):737–740. doi:10.1038/Nature06870

Levis S, Wiedinmyer C, Bonan GB, Guenther A (2003) Simulating biogenic volatile organic compound emissions in the community climate system model. J Geophys Res 108(D21):4659

Long EK, Picklo MJ (2010) Trans-4-hydroxy-2-hexenal, a product of n-3 fatty acid peroxidation: make some room HNE .... Free Radic Biol Med 49(1):1–8. doi:10.1016/j.freeradbiomed.2010.03.015

Loreto F, Schnitzler JP (2010) Abiotic stresses and induced BVOCs. Trends Plant Sci 15 (3):154–166. doi:10.1016/j.tplants.2009.12.006

Loreto F, Velikova V (2001) Isoprene produced by leaves protects the photosynthetic apparatus against ozone damage, quenches ozone products, and reduces lipid peroxidation of cellular membranes. Plant Physiol 127(4):1781–1787

Loreto F, Barta C, Brilli F, Nogues I (2006) On the induction of volatile organic compound emissions by plants as consequence of wounding or fluctuations of light and temperature. Plant Cell Environ 29(9):1820–1828. doi:10.1111/j.1365-3040.2006.01561.x

Loreto F, Centritto M, Barta C, Calfapietra C, Fares S, Monson RK (2007) The relationship between isoprene emission rate and dark respiration rate in white poplar (*Populus alba* L.) leaves. Plant Cell Environ 30(5):662–669. doi:10.1111/j.1365-3040.2007.01648.x

Macarthur RH (1965) Patterns of species diversity. Biol Rev 40(4):510–533. doi:10.1111/j.1469-185X.1965.tb00815.x

Mark RJ, Lovell MA, Markesbery WR, Uchida K, Mattson MP (1997) A role for 4-hydroxynonenal, an aldehydic product of lipid peroxidation, in disruption of ion homeostasis and neuronal death induced by amyloid beta-peptide. J Neurochem 68(1):255–264

Mene-Saffrane L, Dubugnon L, Chetelat A, Stolz S, Gouhier-Darimont C, Farmer EE (2009) Nonenzymatic oxidation of trienoic fatty acids contributes to reactive oxygen species management in Arabidopsis. J Biol Chem 284(3):1702–1708. doi:10.1074/jbc.M807114200

Mittler R, Vanderauwera S, Suzuki N, Miller G, Tognetti VB, Vandepoele K, Gollery M, Shulaev V, Van Breusegem F (2011) ROS signaling: the new wave? Trends Plant Sci 16 (6):300–309. doi:10.1016/j.tplants.2011.03.007

Møller IM (2001) Plant mitochondria and oxidative stress: electron transport, NADPH turnover, and metabolism of reactive oxygen species. Annu Rev Plant Biol 52(1):561–591

Monson RK (2002) Volatile organic compound emissions from terrestrial ecosystems: a primary biological control over atmospheric chemistry. Isr J Chem 42(1):29–42

Monson RK, Fall R (1989) Isoprene emission from aspen leaves—influence of environment and relation to photosynthesis and photorespiration. Plant Physiol 90(1):267–274

Moseley R, Hilton JR, Stephens P, Waddington RJ, Thomas DW (2003) Evaluation of oxidative stress biomarkers in chronic inflammatory disease prognoses. J Dent Res 82:545–545

Nielsen F, Mikkelsen BB, Nielsen JB, Andersen HR, Grandjean P (1997) Plasma malondialdehyde as biomarker for oxidative stress: reference interval and effects of life-style factors. Clin Chem 43(7):1209–1214

Niinemets U (2010) Mild versus severe stress and BVOCs: thresholds, priming and consequences. Trends Plant Sci 15(3):145–153. doi:10.1016/j.tplants.2009.11.008

Oliver DJ, Nikolau BJ, Wurtele ES (2009) Acetyl-CoA—life at the metabolic nexus. Plant Sci 176 (5):597–601

Oro J, Nooner DW, Zlatkis A, Wikstrom SA, Barghoor ES (1965) Hydrocarbons of biological origin in sediments about 2 billion years old. Science 148(3666):77–79

Paulot F, Wunch D, Crounse J, Toon G, Millet D, DeCarlo P, Vigouroux C, Deutscher NM, González Abad G, Notholt J (2011) Importance of secondary sources in the atmospheric budgets of formic and acetic acids. Atmos Chem Phys 11(5):1989–2013

Pophof B, Stange G, Abrell L (2005) Volatile organic compounds as signals in a plant–herbivore system: electrophysiological responses in olfactory sensilla of the moth *Cactoblastis cactorum*. Chem Senses 30(1):51–68

Rasmussen R, Khalil M (1988) Isoprene over the Amazon basin. J Geophys Res: Atmos (1984–2012) 93(D2):1417–1421

Sarmiento J, Gloor M, Gruber N, Beaulieu C, Jacobson A, Mikaloff Fletcher S, Pacala S, Rodgers K (2010) Trends and regional distributions of land and ocean carbon sinks. Biogeosciences 7 (8):2351–2367

Shibamoto T (2006) Analytical methods for trace levels of reactive carbonyl compounds formed in lipid peroxidation systems. J Pharmaceut Biomed 41(1):12–25

Shulaev V, Oliver DJ (2006) Metabolic and proteomic markers for oxidative stress. New tools for reactive oxygen species research. Plant Physiol 141(2):367–372. doi:10.1104/pp.106.077925

Steeghs MML, Moeskops BAWM, van Swam K, Cristescu SM, Scheepers PTJ, Harren FJM (2006) On-line monitoring of UV-induced lipid peroxidation products from human skin in vivo using proton-transfer reaction mass spectrometry. Int J Mass Spectrom 253(1–2):58–64. doi:10.1016/j.ijms.2006.02.015

Suzuki N, Miller G, Morales J, Shulaev V, Torres MA, Mittler R (2011) Respiratory burst oxidases: the engines of ROS signaling. Curr Opin Plant Biol 14(6):691–699. doi:10.1016/j.pbi.2011.07.014

Talbot R, Andreae M, Berresheim H, Jacob D, Beecher K (1990) Sources and sinks of formic, acetic, and pyruvic acids over central Amazonia 2. Wet season. J Geophys Res 95 (D10):16799–16716, 16811

Tegeder M, Weber AP (2008) Metabolite transporters in the control of plant primary metabolism. In: Plaxton WC, McManus MT (eds) Annual plant reviews, volume 22: Control of primary metabolism in plants. Blackwell Publishing Ltd, Oxford, pp 85–120 ter Steege H, Pitman NC, Sabatier D, Baraloto C, Salomão RP, Guevara JE, Phillips OL, Castilho CV, Magnusson WE, Molino J-F (2013) Hyperdominance in the Amazonian tree flora. Science 342(6156):1243092

Velikova V, Edreva A, Loreto F (2004) Endogenous isoprene protects *Phragmites australis* leaves against singlet oxygen. Physiol Plant 122(2):219–225. doi:10.1111/j.0031-9317.2004.00392.x

Velikova V, Pinelli P, Pasqualini S, Reale L, Ferranti F, Loreto F (2005) Isoprene decreases the concentration of nitric oxide in leaves exposed to elevated ozone. New Phytol 166(2):419–426. doi:10.1111/j.1469-8137.2005.01409.x

Vickers CE, Gershenzon J, Lerdau MT, Loreto F (2009) A unified mechanism of action for volatile isoprenoids in plant abiotic stress. Nat Chem Biol 5(5):283–291

Weng J-K, Philippe RN, Noel JP (2012) The rise of chemodiversity in plants. Science 336 (6089):1667–1670

Yan ZG, Wang CZ (2006) Wound-induced green leaf volatiles cause the release of acetylated derivatives and a terpenoid in maize. Phytochemistry 67(1):34–42. doi:10.1016/j.phytochem. 2005.10.005

Chapter 3
The Hydrology and Energy Balance of the Amazon Basin

Michael T. Coe, Marcia N. Macedo, Paulo M. Brando, Paul Lefebvre, Prajjwal Panday, and Divino Silvério

3.1 Introduction

The Amazon basin (Fig. 3.1) is the world's largest river basin, with an area of about 6.65×10^6 km^2 and average annual rainfall of 2200 mm year^{-1} [Tropical Rainfall Monitoring Mission (TRMM); Huffman et al. 2007]. Between 50 and 75 % of this precipitation is returned to the atmosphere via evapotranspiration (ET) (Shuttleworth 1988; Malhi et al. 2002; D'Almeida et al. 2006; Lathuillière et al. 2012), and the remainder is exported from the basin as river discharge. The discharge of the Amazon River represents almost 20 % of total global river run-off (Coe et al. 2007)—a volume three times that of the world's second largest river, the Congo, and greater than the sum of the discharge of the next nine largest rivers by volume. The Amazon also supports a vast area of wetlands and seasonally flooded forests (Fig. 3.1), totalling c. 800,000 km^2 (Melack and Hess 2010; Melack et al. 2009).

Deforestation and degradation are already causing significant changes to the energy and water cycles in the southeastern Amazon (Davidson et al. 2012; Coe et al. 2013). Projections of future deforestation, even under conditions of strong

M.T. Coe (✉) • M.N. Macedo • P. Lefebvre • P. Panday
The Woods Hole Research Center, 149 Woods Hole Rd, Falmouth, MA 02540, USA
e-mail: mtcoe@whrc.org

P.M. Brando
The Woods Hole Research Center, 149 Woods Hole Rd, Falmouth, MA 02540, USA

Instituto de Pesquisa Ambiental da Amazônia (IPAM), Belém, Brazil

D. Silvério
Instituto de Pesquisa Ambiental da Amazônia (IPAM), Belém, Brazil

Departamento de Ecologia, Universidade de Brasília, Distrito Federal, Brazil

© Springer-Verlag Berlin Heidelberg 2016

L. Nagy et al. (eds.), *Interactions Between Biosphere, Atmosphere and Human Land Use in the Amazon Basin*, Ecological Studies 227,
DOI 10.1007/978-3-662-49902-3_3

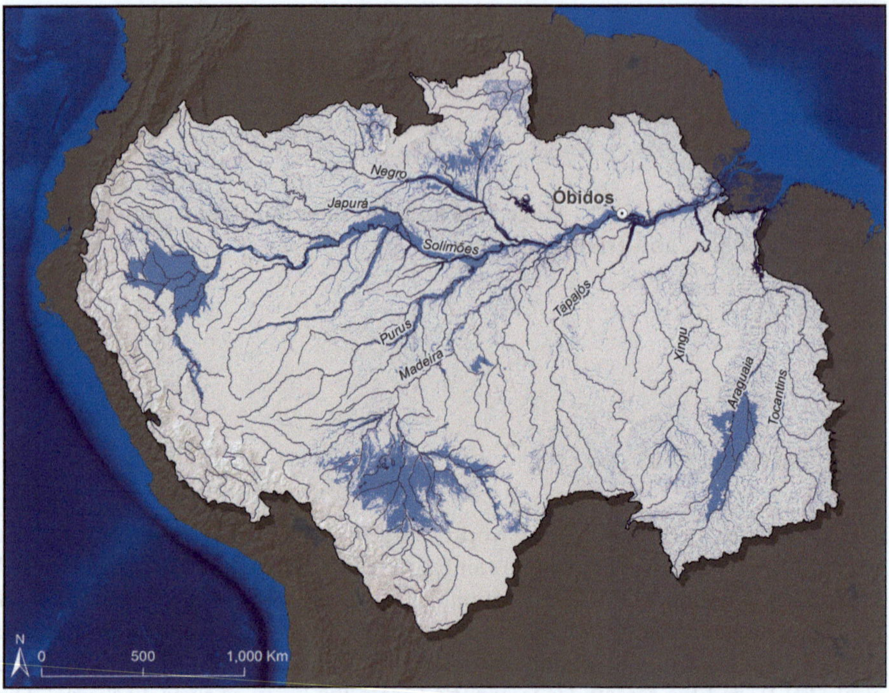

Fig. 3.1 The Amazon River and wetlands system. The river system (*dark blue*) is provided by the Brazilian National Water Agency. Wetlands (*pale blue*) are from Melack and Hess (2010)

governance, suggest that 20–30 % or more of the Amazon basin could be deforested within the next 40 years (Soares-Filho et al. 2006). Together these deforestation pressures have the potential to significantly alter the energy and water cycles and ultimately affect regional and global climate (Wang and Eltahir 2000; Saad et al. 2010; Snyder 2010; Butt et al. 2011; Knox et al. 2011; Spracklen et al. 2012, Bagley et al. 2014). Global climate change is likely to exacerbate the effects of deforestation, driving further changes to the water and energy cycles in the near future. Although uncertainty about its specific impacts on the Amazon remain high, increasing atmospheric greenhouse gases are expected to increase temperatures, drought frequency, and drought intensity and may cause significant disruptions to the cycling of energy and water in the Amazon (e.g. Malhi et al. 2009).

This chapter reviews the current state of our knowledge of the hydrology and energy balance of the Amazon basin. It describes the mean conditions, as well as the spatial and temporal variability of the energy and water cycles, examines how anthropogenic influences have affected the hydrology of the Basin, and evaluates the scale of potential future changes to Amazon basin hydrology in the twenty-first century.

3.2 Current Conditions

The Amazon basin (Fig. 3.1) is the planet's largest and most intense land-based convective centre, exerting strong influences on atmospheric circulation both within and outside the tropics. This convective system (Fig. 3.2) is driven by high net surface radiation (R_{net}), which is the sum of the absorbed solar shortwave and net longwave radiation fluxes to the land surface. Incoming solar radiation is absorbed, reflected, or transmitted by the surface, which includes the land, vegetation, and water. The net downward longwave flux is the sum of the longwave radiation absorbed by the land surface from the atmosphere and that emitted from the land surface back to the atmosphere. Over the long term, R_{net} is balanced (Fig. 3.2) by the fluxes of latent heat (ET, the energy released by evaporation from the soil and plant surfaces and by plant transpiration) and sensible heat (H, the flux of heat between the land surface and the atmosphere). Similarly, incoming precipitation over the basin is balanced by the evaporative flux of water to the atmosphere (ET) and river discharge, which returns excess water to the oceans (Fig. 3.2).

3.2.1 Mean Conditions and Spatial Variability

The Amazon basin receives a relatively large amount of incoming solar energy, due to its proximity to the equator, averaging 20 Mj m^{-2} throughout the year (Pereira et al. 2006). The mean reflectance (albedo) of the extensive forest in the basin is about 13 % (von Randow et al. 2004; Culf et al. 1995), which results in a surface absorption of 10–14 Mj m^{-2} day^{-1} of incoming solar energy (da Rocha et al. 2009; Souza et al. 2011; Andrade et al. 2009). The persistent presence of clouds (which increase albedo) during the wet season and their relative absence during the dry season generally result in less incoming solar radiation during the wet season. Between December and February, due to the southerly position of the sun and pervasive cloud cover in the northern basin, the southeastern Amazon receives proportionally more energy than the rest of the Basin (Pereira et al. 2006; da Rocha et al. 2009).

Precipitation in the Amazon basin varies from <1600 mm year^{-1} in the southeastern region to more than 2800 mm year^{-1} in the northwestern region (Fig. 3.3a). Whereas the northwestern basin receives relatively constant rainfall throughout the year, precipitation in the southeastern basin is strongly seasonal (Fig. 3.3d), with peak rainfall occurring in January and a long dry season centred on July. The lower mean annual precipitation and longer dry season (4–5 months with <100 mm rainfall, Fig. 3.3a, d), typical of the south and southeastern portions of the Basin, are associated with seasonal shifts in the position of the Intertropical Convergence Zone (ITCZ).

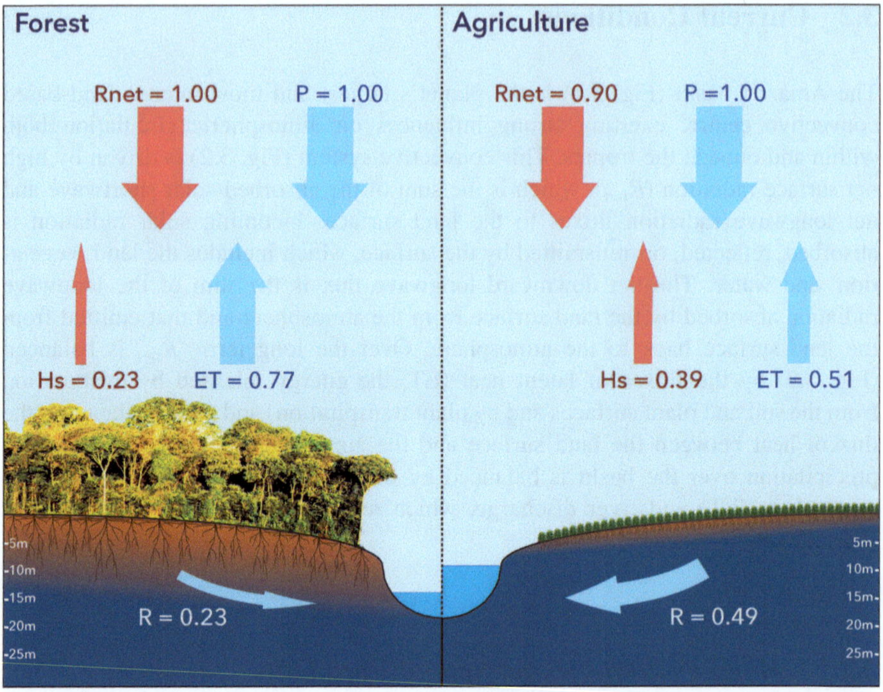

Fig. 3.2 Schematic of the linked water and energy cycles. In the hydrologic cycle, incoming precipitation (*P*) is balanced by outgoing evapotranspiration (ET) and run-off (*R*). In the energy cycle, incoming net radiation (R_{net}, the sum of the short- and longwave fluxes) is balanced by outgoing sensible (H_s) and latent (ET) heat fluxes. *Left*: Undisturbed forest example. ET from vegetation and land surfaces accounts for about 77 % of the incoming *P* and R_{net}, with the remainder of the *P* and R_{net} going to run-off and H_s, respectively. *Right*: Deforested example. Deforestation decreases R_{net} relative to the forested value due to increased upward longwave radiation and albedo. ET decreases because the plants that replace forest have lower annual mean water demands. *P* may not change significantly because the large-scale atmospheric circulation may not be adversely affected. As a result of these responses, the Bowen ratio (H_s/ET), run-off (*R*) and discharge (*D*) increase, compared to an undisturbed landscape (*left*)

Evapotranspiration (Fig. 3.3b) is the predominant pathway by which net incoming radiation (R_{net}) and precipitation are transferred from the land surface back to the atmosphere (Fig. 3.2; Zeng et al. 1999; Nepstad et al. 2004; Aragão et al. 2007; da Rocha et al. 2009; Lathuillière et al. 2012). The relatively low albedo of vegetated surfaces results in high R_{net} (Bonan 2002). The deep-rooted evergreen vegetation, relatively high leaf area index, generally abundant soil moisture, and high R_{net} associated with tropical forests result in high rates of ET (Fig. 3.3b) to meet high photosynthetic demand (e.g. Nepstad et al. 1994; Bonan 2002; Oliveira et al. 2005; Spracklen et al. 2012). In forested landscapes, ET averages about 75–110 mm month^{-1} (5.6–8.3 Mj m^{-2} day^{-1}) and is relatively constant throughout the year (e.g. Malhi et al. 2002; D'Almeida et al. 2006; Lathuillière et al. 2012), even in the south and southeastern Amazon, where there is a prolonged (i.e. 4–6 month; Juárez et al. 2007) dry season (Fig. 3.3b, d). Flux tower measurements at

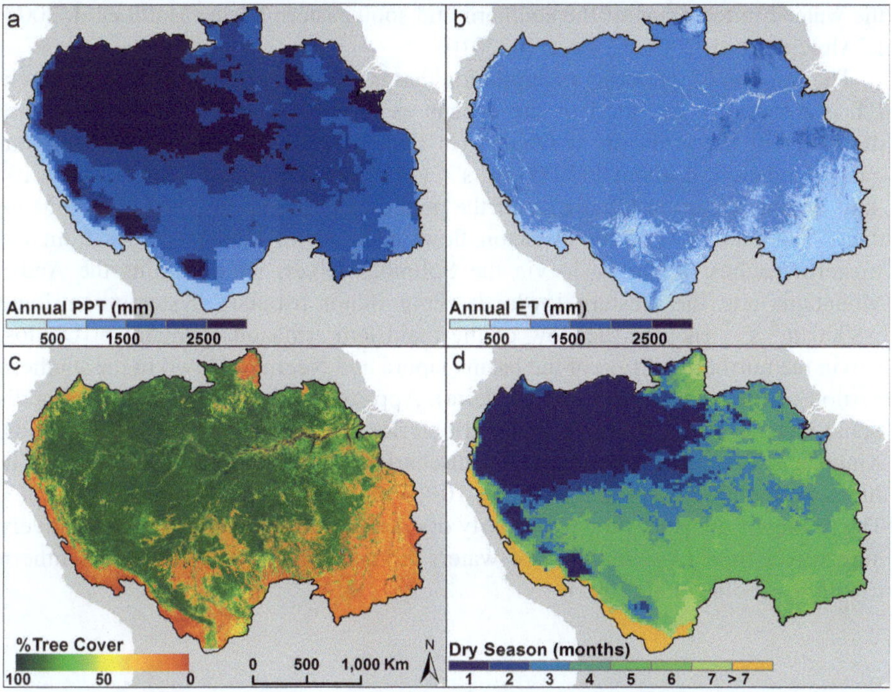

Fig. 3.3 Summary of Amazon basin hydrological variables. (**a**) Annual mean rainfall (mm year^{-1}) from 2001 to 2010, calculated from the Tropical Rainfall Monitoring Mission monthly precipitation data (TRMM 3B43 v7, Huffman et al. 2007). (**b**) Annual mean evapotranspiration (ET, mm year^{-1}) from 2001 to 2010, calculated from the Moderate Resolution Imaging Spectroradiometer (MODIS) ET data (MOD16, Mu et al. 2011). (**c**) Percent observed tree cover in 2010 from MODIS vegetation continuous fields data (MOD44B, DiMiceli et al. 2011). (**d**) Mean dry season length, calculated from TRMM monthly mean rainfall as the number of months with <100 mm month^{-1} of rainfall

various locations in the Amazon indicate that approximately 77 % of the R_{net} of undisturbed forests is converted to latent heat via evapotranspiration (Fig. 3.2; Priante-Filho et al. 2004; da Rocha et al. 2009; von Randow et al. 2004).

North of about 12° S, there is a tendency towards an ET maximum during the early dry season consistent with increased incoming radiation (because of decreased cloudiness) and abundant soil moisture. A weak minimum in ET during the late dry season generally occurs in these regions as soil moisture decreases (e.g. Costa et al. 2010). In the southeastern basin—where the dry season is particularly long and the vegetation shifts to drought deciduous species—the ET decreases early in the dry season and may be less than half of that of the wet season ET (Costa et al. 2010; Lathuillière et al. 2012). The sensible heat flux (H_s) increases proportionally to the ET decrease during the dry season. In total, the ET flux in undisturbed forests and savanna regions is equivalent to between 40 and 75 % of the incoming precipitation (Fig. 3.3a, b), with higher values relative to precipitation in the water-limited areas of the southern and southeastern basin (Malhi et al. 2002; D'Almeida et al. 2006; Costa et al. 2010).

The fraction of incoming precipitation that is not returned to the atmosphere via ET is eventually exported to the oceans as river discharge. The mean annual discharge of the Amazon at Óbidos (in Pará, Brazil), the furthest downstream gauged location, is about 175,000 $m^3 s^{-1}$ (1192 mm m^{-2} $year^{-1}$ for the 4.6×10^6 km^2 area upstream of Óbidos) over the period of observation (1968–2010). More than 75,000 $m^3 s^{-1}$ of the Amazon flow upstream of Óbidos comes from the western reaches of the basin via the Solimões River, which drains the Andes Mountains and the western lowlands. Four major tributary systems contribute >8000 $m^3 s^{-1}$ each to the flow of the mainstem Amazon upstream of Óbidos, two in the northern portion of the basin (Japurá and Negro) and two in the southern portion of the basin (Purus and Madeira). Approximately 20,000 $m^3 s^{-1}$ of additional flow is added by three major tributaries downstream of Óbidos (Tapajós, Xingu, and Tocantins/Araguaia). Peak discharge in the mainstem Amazon occurs in June, with a flood wave that averages 10–18 m in amplitude (Richey et al. 1986). The timing of the peak flow in the early dry season is consistent with the relatively long travel time (3–4 months) of waters from the major western and southern tributaries (Foley et al. 2002).

3.2.2 Temporal Variability

The two major modes of temporal variations of the Amazon energy and water balance (Botta et al. 2002) are strongly linked to sea surface temperature (SST) in the adjacent oceans (Marengo et al. 2008a, b, 2011a, b). The first mode operates on a sub-decadal timescale and is associated with short-term variations in SST in the tropical Pacific, the so-called El Niño/Southern Oscillation (ENSO). The second mode is a 28-year cycle that is likely driven by variations in SST in the tropical North Atlantic (Marengo et al. 2008a, b, 2011a, b; Botta et al. 2002; Foley et al. 2002). Together, these two modes explain the bulk of the observed climatological variability in the Amazon (Botta et al. 2002; Marengo et al. 2008a, b).

Warm SST anomalies in the eastern Pacific during the El Niño phase of ENSO tend to cause increased net solar radiation (due to reduced cloudiness), increased land surface temperature, reduced rainfall, and droughts in Amazonia. Cold SST anomalies in the eastern Pacific (La Niña) are generally associated with a cloudier, cooler, and wetter Amazon, particularly towards the end of the year (Foley et al. 2002; Zeng et al. 2008; Espinoza et al. 2009; da Rocha et al. 2009; Marengo et al. 2011a). Anomalously high rainfall and cooler temperatures during La Niña events are concentrated particularly in the north and northeast of the Amazon region and are associated with many large-scale floods (Marengo 1992; Poveda and Mesa 1993; Ronchail et al. 2002; Espinoza et al. 2009). El Niño events are associated with drier and warmer conditions and decreased discharge, particularly from the northern and western tributaries (Marengo and Tomasella 1998), and decreased flood height and inundated area on the mainstem Amazon River (Foley et al. 2002). During severe El Nino events, such as 1997/1998, much of the Amazon experiences reduced precipitation, increased forest fires, reduced river discharge, and disrupted river transportation.

Despite its demonstrated importance as a driver of short-term climatological variability in the Amazon, ENSO events explain <40 % of the variation in rainfall in northern Amazonia (Marengo and Tomasella 1998) and are not always the best predictor of extreme events in the region (Marengo et al. 2008a, b, 2011b). Anomalously warm SST in the tropical North Atlantic is associated with a more northern mean position of the ITCZ and, therefore, reductions in mean rainfall, particularly in the southern portions of the Basin. These tropical North Atlantic SST anomalies explain the greatest amount of variation in observed precipitation and are associated with some of the strongest droughts and lowest discharge rates and river levels in the historical record (e.g. 1926, 1963, 2005, 2010; Marengo et al. 2008a, b, 2011a, b; Gloor et al. 2013).

3.3 Anthropogenic Influences

Land cover/land use change and climate change due to increasing atmospheric greenhouse gases are among the most important anthropogenic drivers of change in the Amazon today (Malhi et al. 2009). Research indicates that both of these factors will likely increase land surface temperatures, decrease rainfall, and increase drought frequency and intensity in the Amazon (e.g. Malhi et al. 2009; Davidson et al. 2012; Coe et al. 2013).

3.3.1 Land Cover and Land Use Change

Deforestation in the Brazilian Amazon averaged 19,625 km^2 $year^{-1}$ from 1996 to 2005, dropped to 10,660 km^2 $year^{-1}$ from 2006 to 2010, and dropped even further from 2011 to 2014, reaching 5400 km^2 $year^{-1}$ (INPE 2015). Most of the newly deforested area was converted to pastures and croplands, particularly in the states of Mato Grosso, Pará, and Rondônia. The Brazilian cattle herd has grown substantially over the past two decades and today—at over 200 million heads of cattle—it is the world's largest commercial cattle herd (McAlpine et al. 2009). Likewise, soybean cultivation in Brazil has increased dramatically over the last decade. Today Brazil is the world's largest producer and exporter of soybeans, with over 94,000 km^2 of soybeans planted in the Legal Amazon in 2013 (IBGE 2015). Despite recently reduced rates of forest clearing (Ometto et al. 2011; Meyfroidt and Lambin 2011; Macedo et al. 2012), deforestation is expected to continue to expand as global demand for soy, beef, and other agricultural commodities continues to rise (Soares-Filho et al. 2006; Gibbs et al. 2010).

3.3.1.1 Vegetation Controls on Water/Energy Balance

Most of the Amazon basin is dominated by moist tropical rainforest, including many regions with a strong dry season (Fig. 3.3c, d). For example, in the southeast of the Amazon, evergreen forests predominate in regions where the dry season averages <100 mm $month^{-1}$ for three or more consecutive months (Fig. 3.3d). In the extreme southeast, where the dry season extends for 4–6 months, there is a shift to the savanna vegetation more typical of the Cerrado biome, but it represents a relatively small portion of the Amazon watershed (Fig. 3.3c, d).

One single large tree can transfer as much as 500 l of soil water per day to the atmosphere via evapotranspiration. Together, the millions of trees that grow in the Amazon transfer ~7400 km^3 of water to the atmosphere each year (Salati and Nobre 1991), more water than New York City would consume over 4500 years. This striking characteristic of tropical forests is largely associated with their capacity to absorb large amounts of solar radiation and to return most of this energy back to the atmosphere as latent heat (primarily) and sensible heat (secondarily) (da Rocha et al. 2004). What allows forests to absorb more solar energy than most ecosystems around the world is the low albedo of forest canopies, coupled with the high amount of solar radiation inherent to tropical regions (Hasler and Avissar 2007). Forests take advantage of the water stored in the deep soils of the Amazon, partitioning more energy into latent than sensible heat. Because of the domination of the energy balance by water fluxes (ET), the Amazon is often referred to as a green ocean.

Although the Amazon landscape may appear uniformly green at first glance, there is still a high spatial variability in canopy properties that affect ET across the basin (Fig. 3.3b). Two major environmental processes appear to control most of this spatial variability (Hasler and Avissar 2007). The first relates to solar radiation (Fisher et al. 2009). While tropical regions receive substantial amounts of energy throughout the year, wetter portions of the basin can be so cloudy that during rainy months the available net radiation is reduced (by as much as 13 %, da Rocha et al. 2009; Souza et al. 2011), thus driving less ET.

The second environmental process controlling spatial variability relates to soil water availability (da Rocha et al. 2004; Hasler and Avissar 2007). In regions with a strong dry season, trees may cope with soil water stress by investing a greater proportion of net primary productivity to growing fine roots, performing hydraulic lift (Oliveira et al. 2005), and even assimilating water by their leaves (Nepstad et al. 2002). In so doing, trees are able to evapotranspire large amounts of water even during dry climatic conditions (Fisher et al. 2009). But when soil and plant water stress becomes too large, many trees and lianas of the Amazon close their stomata and shed leaves, which reduces ET and increases sensible heat flux (Nepstad et al. 2004). Thus, synergies between climate and forest structure and functioning control much of the observed spatial variability in the water and energy balance in the Amazon.

3.3.1.2 Land Use Impacts on the Water/Energy Balance

Deforestation and land management practices influence the surface energy and water balance in two predominant ways. First, they change how incoming precipitation and radiation are partitioned among sensible and latent heat fluxes, run-off, and river discharge (Fig. 3.2; Costa and Foley 1997; Bonan et al. 2004; D'Almeida et al. 2007; Coe et al. 2009, 2011; Panday et al. 2015). Second, they can alter precipitation patterns at regional and continental scales (Butt et al. 2011; Costa and Foley 2000; Costa and Pires 2009; Delire et al. 2001; Dickinson and Henderson-Sellers 1988; Knox et al. 2011; Malhi et al. 2008; Nobre et al. 1991). Compared to native forests, croplands and pasturelands have a shallower rooting depth, lower leaf area index (for much of the year outside of the peak growing season), and up to 50 % higher surface albedo (von Randow et al. 2004)—all of which contribute to a large reduction in net radiation and evapotranspiration in deforested areas (e.g. Bruijnzeel 2004; Costa and Foley 1997). These changes reduce the system's capacity to cycle water, increasing the proportion of net radiation dissipated as sensible heat by as much as 40 % (Priante-Filho et al. 2004; von Randow et al. 2004; Souza et al. 2011).

A recent analysis in the southeastern Amazon (Silvério et al. 2015) illustrates how land use transitions from forests to pastures and soybean croplands have dramatic effects on the energy and water balance. Combining satellite-derived (MODIS) and weather station data (INMET 2012) with annual land use maps (Macedo et al. 2012), Silvério reconstructed the land use transitions from 2000 to 2010 and evaluated the effect of different land uses (e.g. pasture, forest, and soybeans) on the energy balance. Results for the Xingu basin indicate that in 2008 R_{net} in pastures and soybean areas were 12 and 19 % lower than forests, respectively (Table 3.1; Fig. 3.4). 70–85 % of the R_{net} decrease was from increased outgoing longwave radiation and 15–30 % from increased surface albedo (Table 3.1). In forested regions 69 % of R_{net} was converted to latent heat via ET, compared to 61 and 57 % in pasture and soybeans, respectively (Table 3.1, Fig. 3.4). As a result, despite lower R_{net} the amount of energy converted to sensible heat in

Table 3.1 Summary of the energy balance for the upper Xingu River Basin in 2008 in forests, croplands, and pasture

Variable	Forest	Pasture	Cropland
Albedo	0.13±0.01	0.15±0.01	0.18±0.02
R_s (net)	15.36±1.16	15.22±1.23	14.52±1.22
R_{net}	11.68±2.17	9.99±2.71	8.96±3.32
ET	8.08±2.19	5.42±3.37	4.98±3.55
H_s	3.60±2.76	4.49±2.34	3.98±1.93

All values are yearly mean±standard deviation and are expressed in MJ m^{-2} day^{-1} (excepted albedo). Land use map from Macedo et al. (2012)

Rs(net) net shortwave radiation; R_{net} net radiation; *ET* latent heat flux; H_s sensible heat flux

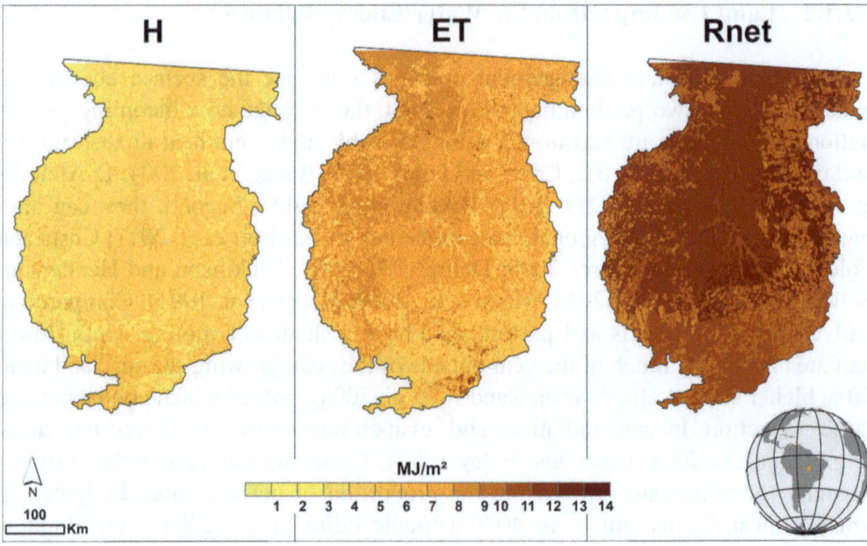

Fig. 3.4 Components of the energy balance for the upper Xingu River Basin in 2008. *Right*: R_{net}—net radiation (yearly mean calculated from multiple MODIS products and local weather station; method adapted from Bisht and Bras 2010). *Centre*: ET—latent heat flux [yearly mean from MODIS ET data (MOD16, Mu et al. 2011)]. *Left*: H_s—sensible heat flux ($H_s = R_{net} - \text{ET}$). All values are expressed in MJ m^{-2} day^{-1}

pasture and soybeans in 2008 was 8 and 10 % higher (respectively) compared to forests, resulting in substantial local climate changes (Table 3.1, Figs. 3.4 and 3.5).

Observations in small and large watersheds show that these land use-driven changes in R_{net}, ET, and sensible heat flux increase river discharge. Studies in 1-ha paired catchments in Rondonia and eastern Amazonia (Moraes et al. 2006; Chaves et al. 2008) found that the ratio of run-off to precipitation increased from 1 to 3 % in forested catchments to about 18 % in pasture catchments. In two paired 1-km^2 catchments in the central Amazonia, Trancoso (2006) found that the ratio of run-off to precipitation increased from 21 to 43 % in forest and pasture catchments, respectively. Likewise, a study comparing discharge from three forest watersheds and four soybean watersheds (ranging from 2 to 14 km^2) in southeastern Amazonia (Querencia, Mato Grosso) found a more than fourfold increase in run-off, which jumped from 7 % of precipitation in forest streams to 31 % in soybean streams (Fig. 3.6; Hayhoe et al. 2011). At a much larger scale, analysis of discharge data in the 175,000 km^2 Tocantins and 82,000 km^2 Araguaia River basins in eastern Amazonia suggests that extensive land cover changes since 1950 have been associated with a c. 20 % increase in the annual mean discharge (Costa et al. 2003; Coe et al. 2009, 2011).

Finally, global and meso-scale climate model studies suggest that if deforestation in the Amazon basin were to occur on a very large scale (>several 100,000s km^2), atmospheric feedbacks would lead to reduced precipitation (e.g. Oyama and Nobre 2003; Bounoua et al. 2002; Berbet and Costa 2003;

Fig. 3.5 Dust devil in a fallow soybean field in the southeastern Amazon (Mato Grosso Brazil). This location (about 51°43′6″ W, 11°28′15″ S) was formerly broadleaf evergreen forest. Forest removal significantly reduces evapotranspiration in the dry season (Fig. 3.3). As a result, a larger fraction of the incoming R_{net} is converted to a sensible heat flux compared to forested areas. These conditions favour dust-devil formation, something that does not occur in a forested environment where R_{net} is more strongly converted to ET

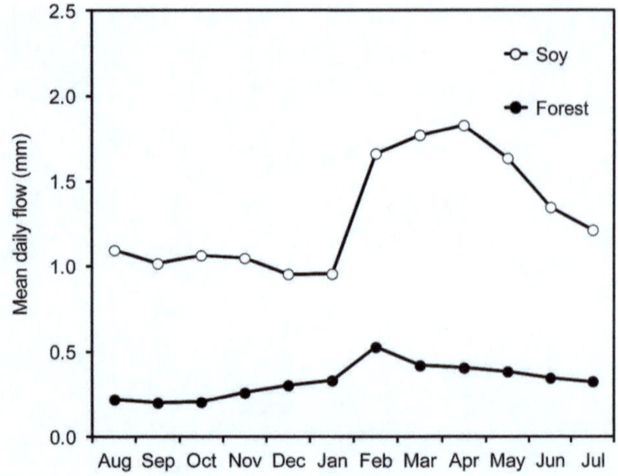

Fig. 3.6 Median daily water yield (mm day^{-1}) for three forested watersheds (*black dots*) and four soybean watersheds (*white dots*) in the southeastern Amazon (Mato Grosso, Brazil, located between 52°23′30″–52°18′50″ W and 13°9′12″–12°41′40″ S). Watershed areas range from 2.5 to 13.5 km^2 and flow data span the period from August 2007 to August 2008. Figure adapted from Hayhoe et al. (2011)

Nobre et al. 2009). Replacing large tracts of forest with higher albedo, less water-demanding crops and pastures would lead to reduced net surface radiation, decreased moisture convergence over the Basin, decreased water recycling, and ultimately reduced precipitation (Costa and Foley 2000; Dickinson and Henderson-Sellers 1988; Malhi et al. 2008; Nobre et al. 1991). Numerical model evaluations of the effects of partial deforestation on regional climate suggest that a large-scale change in the precipitation of the Amazon could occur after about 40–60 % of the basin has been deforested (Oyama and Nobre 2003; Costa et al. 2007; Sampaio et al. 2007; Walker et al. 2009; Oliveira et al. 2013; Stickler et al. 2013). Recent observational evidence suggests that a significant lengthening of the dry season may already be occurring in portions of the south and southeastern Amazon as a consequence of regional deforestation (Costa and Pires 2009; Butt et al. 2011; Knox et al. 2011; Marengo et al. 2011b).

3.3.2 Global Climate Change

Climate changes associated with increasing atmospheric greenhouse gas concentrations are expected to lead to significant changes in the energy and water cycle of the Amazon. There is generally high confidence that increasing greenhouse gas concentrations will create a significantly warmer and drier climate in the Amazon by the end of the century (Huntingford et al. 2008; Malhi et al. 2008, 2009). However, the magnitude and distribution of the expected climate changes are uncertain. For example, mean air temperature is predicted to increase between 2 and 10 °C in the twenty-first century. This large range in temperature predictions stems from many sources of uncertainty, including the climate model used, emissions scenario assumed in climate simulations, strength of carbon cycle feedbacks incorporated in the model, and degree to which anthropogenic deforestation is considered (Costa and Foley 2000; Betts 2004; Malhi et al. 2009). Despite these uncertainties, the majority of models considered in the fourth and fifth Intergovernmental Panel on Climate Change assessments suggest a potentially severe decline in dry season rainfall and the potential for more frequent and severe drought in the southern and eastern Amazon (Malhi et al. 2008; IPCC 2013).

The net effect of increased atmospheric CO_2 and a warmer and dryer climate on the Amazon water balance (ET, run-off, and discharge) is more uncertain. Although most models show an increase in ET and a decrease in run-off and discharge with increasing temperature, the results depend strongly on the biophysical response of the simulated forest to increasing CO_2. Theory suggests that plant water use efficiency will increase with increasing atmospheric CO_2 content. This improved efficiency could mean that under higher CO_2 concentrations plants could accomplish the same amount of photosynthesis with lower ET. Under scenarios of a warmer climate and increased plant water use efficiency, ET and discharge may actually decrease or be largely unchanged compared to the present (Lapola et al. 2009; Holtum and Winter 2010; Rammig et al. 2010) because of the competing effects of increasing surface temperature (increased ET) and atmospheric CO_2 (decreased ET per unit of photosynthetic activity). However, the actual biophysical response of tropical forest water use efficiency to increasing CO_2 is not known. Few field experiments exist and it would be difficult to capture the full range of species traits and complex interactions that might occur under a scenario with greatly increased CO_2 concentrations. As a result, there is high uncertainty in the response of the water and energy cycles of the Amazon to future climate and CO_2 change (Coe et al. 2013).

As discussed in Sect. 2.2, the temporal and spatial variability of precipitation in the Amazon depends, in part, on atmospheric responses to the El Niño-Southern Oscillation and sea surface temperature oscillations in the tropical North Atlantic. These SST phenomena are likely to be affected by increasing greenhouse gas concentrations (Malhi et al. 2009). However, future predictions of SST under increasing atmospheric CO_2 and the response of Amazon climate to SST changes remain highly uncertain, making it challenging to predict future energy and water balance responses and feedbacks in the Amazon.

3.4 Conclusions

The energy and water cycle of the Amazon basin supports the world's largest tropical evergreen forest ecosystem. This forest ecosystem supports the energy and water cycle through extensive water recycling throughout the year. The resulting powerful convective system is driven by high net surface radiation, which is dissipated by the fluxes of latent and sensible heat. Incoming precipitation over the basin is balanced over the long term by the evaporative flux of water to the atmosphere and discharge, which returns excess water to the oceans. Temporal variability of this cycle is largely controlled by oscillations of tropical Pacific and

North Atlantic SST with sub-decadal and multi-decadal periodicities. Synergies between climate and forest structure and functioning control much of the observed spatial variability in the water and energy balance.

Extensive field studies provide evidence of human influences on the hydrology and energy cycles of the Amazon basin. Deforestation has decreased net surface radiation and evapotranspiration, increased sensible heat flux, and increased water yield and stream discharge in many locations throughout the agricultural frontier. There is growing evidence that these changes to the surface energy and water balance are suppressing precipitation at the beginning and end of the dry season over some deforested regions. Numerical modelling studies suggest that deforestation and climate change from increasing atmospheric greenhouse gases are both likely to lead to a significantly drier climate and increased drought frequency and severity, particularly in the south and southeastern portion of the Amazon in the coming decades.

In the past decade, we have developed a much clearer understanding of the energy and water cycle of the Amazon, some of the large-scale drivers of change taking place, and the potential responses associated with human-induced changes. Research focused on clarifying some of the uncertainties in our knowledge of the climatological and biophysical processes governing land–atmosphere and land–ocean feedbacks are an important next step for understanding future trajectories for the Amazon basin.

Acknowledgements Funding from the US National Aeronautics and Space Administration, the US National Science Foundation, and the Gordon and Betty Moore Foundation has supported much of our research in the Amazon.

References

Andrade NLR, Aguiar RG, Sanches L, Freitas-Alves ECR, Nogueira JS (2009) Net radiation partition in Amazonian forest and transitional Amazonia-Cerrado forest areas. Rev Bras Meteorol 24(3):346–355. doi:10.1590/S0102-77862009000300008

Aragão LE, Malhi Y, Roman-Cuesta RM, Saatchi S, Anderson LO, Shimabukuro YE (2007) Spatial patterns and fire response of recent Amazonian droughts. Geophys Res Lett 34(7): L07701. doi:10.1029/2006GL028946

Bagley JE, Desai AR, Harding KJ, Snyder PK, Foley JA (2014) Drought and deforestation: has land cover change influenced recent precipitation extremes in the Amazon? J Climate 27:345–361

Berbet MLC, Costa MH (2003) Climate change after tropical deforestation: seasonal variability of surface albedo and its effects on precipitation change. J Climate 16:2099–2104

Betts RA (2004) Global vegetation and climate: self-beneficial effects, climate forcings and climate feedbacks. J Phys IV 121:37–60

Bisht G, Bras RL (2010) Estimation of net radiation from the moderate resolution imaging spectroradiometer over the Continental United States. IEEE Trans Geosci Remote Sens 49 (6):2448–2462. doi:10.1109/TGRS.2010.2096227

Bonan GB (2002) Ecological climatology: concepts and applications. Cambridge University Press, Cambridge

Bonan GB, DeFries RS, Coe MT, Ojima DS (2004) Land use and climate. In: Gutman G (ed) Land change science. Kluwer Academic Publishers, Dordrecht, pp 301–314

Botta A, Ramankutty N, Foley JA (2002) Long-term variations of climate and carbon fluxes over the Amazon basin. Geophys Res Lett 29(9). doi:10.1029/2001GL013607

Bounoua L, DeFries R, Collatz GJ, Sellers P, Khan H (2002) Effects of land cover conversion on surface climate. Clim Change 52:29–64

Bruijnzeel L (2004) Hydrological functions of tropical forests: not seeing the soil for the trees? Agric Ecosyst Environ 104:185–228

Butt N, Oliveira PA, Costa MH (2011) Evidence that deforestation affects the onset of the rainy season in Rondonia, Brazil. J Geophys Res 116(D11120):8. doi:10.1029/2010JD015174

Chaves J, Neill C, Germer S, Neto SG, Krusche A, Elsenbeer H (2008) Land management impacts on runoff sources in small Amazon watersheds. Hydrol Process 22(22):1766–1775

Coe MT, Costa MH, Howard EA (2007) Simulating the surface waters of the Amazon River Basin: impacts of new river geomorphic and dynamic flow parameterizations. Hydrol Process 21. doi:10.1002/hyp.6850

Coe MT, Costa MH, Soares-Filho BS (2009) The Influence of historical and potential future deforestation on the stream flow of the Amazon River—land surface processes and atmospheric feedbacks. J Hydrol 369:165–174. doi:10.1016/j.jhydrol.2009.1002.1043

Coe MT, Latrubesse EM, Ferreira ME, Amsler ML (2011) The Effects of deforestation and climate variability on the streamflow of the Araguaia River, Brazil. Biogeochemistry 105:119–131. doi:10.1007/s10533-011-9582-2

Coe M, Marthews TR, Costa MH, Galbraith D, Greenglass N, Acioli H, Levine N, Malhi Y, Moorcroft PR, Nobre Muza M, Powell T, Saleska SR, Solorzano LA, Wang J (2013) Deforestation and climate feedbacks threaten the ecological integrity of south-southeastern Amazonia. Philos Trans R Soc B 368:20120155. doi:10.1098/rstb.2012.0155

Costa MH, Foley JA (1997) Water balance of the Amazon Basin. J Geophys Res 102 (D20):23973–23989

Costa MH, Foley JA (2000) Combined effects of deforestation and doubled atmospheric CO2 concentrations on the climate of Amazonia. J Climate 13(1):18–34

Costa MH, Pires GF (2009) Effecs of Amazon and Central Brazil deforestation scenarios on the duration of the dry season in the arc of deforestation. Int J Climatol 30:1970–1979. doi:10.1002/joc.2048

Costa MH, Botta A, Cardille JA (2003) Effects of large-scale changes in land cover on the discharge of the Tocantins River, Southeastern Amazonia. J Hydrol 283:206–217

Costa MH, Yanagi SNM, Souza P, Ribeiro A, Rocha EJP (2007) Climate change in Amazonia caused by soybean cropland expansion, as compared to caused by pastureland expansion. Geophys Res Lett 34:L07706. doi:10.1029/2007GL029271

Costa MH, Biajoli MC, Sanches L, Malhado A, Hutyra LR, da Rocha HR, Aguiar RG, de Araújo AC (2010) Atmospheric versus vegetation controls of Amazonian tropical rain forest evapotranspiration: are the wet and seasonally dry rain forests any different? J Geophys Res Biogeosci 115:2005–2012

Culf AD, Fisch G, Hodnett MG (1995) The albedo of Amazonian forest and ranchland. J Climate 8:1544–1554. doi:10.1175/1520-0442(1995)008<1544:TAOAFA>2.0.CO;2

D'Almeida C, Vorosmarty CJ, Marengo JA, Hurtt GC, Dingman SL, Keim BD (2006) A water balance model to study the hydrological response to different scenarios of deforestation in Amazonia. J Hydrol 331(1–2):125–136

D'Almeida C, Vörösmarty CJ, Hurtt GC, Marengo JA, Dingman SL, Keim BD (2007) The effects of deforestation on the hydrological cycle in Amazonia: a review on scale and resolution. Int J Climatol 27(5):633–647

da Rocha HR, Goulden ML, Miller SD (2004) Seasonality of water and heat fluxes over a tropical forest in eastern Amazonia. Ecol Appl 14:22–32 da Rocha HR, Manzi AO, Cabral OM, Miller SD, Goulden ML, Saleska SR, Coupe NR (2009) Patterns of water and heat flux across a biome gradient from tropical forest to savanna in Brazil. J Geophys Res 114(G00B12):1–8

Davidson EA, de Araújo AC, Artaxo P, Balch JK, Brown IF, Bustamante MMC, Coe MT, DeFries RS, Keller M, Longo M, Munger JW, Schroeder W, Soares-Filho BS, Souza CM Jr, Wofsy SC (2012) The Amazon basin in transition. Nature 481:321–328. doi:10.1038/nature10717

Delire C, Behling P, Coe MT, Foley JA, Jacob R, Kutzbach J, Liu Z, Vavrus S (2001) Simulated response of the atmosphere–ocean system to deforestation in the Indonesian Archipelago. Geophys Res Lett 28(10):2081–2084

Dickinson RE, Henderson-Sellers A (1988) Modelling tropical deforestation: a study of GCM land-surface parametrizations. Q J Roy Meteorol Soc 114:439–462

DiMiceli CM, Carroll ML, Sohlberg RA, Huang C, Hansen MC, Townshend JGR (2011) Annual global automated MODIS vegetation continuous fields (MOD44B) at 250 m spatial resolution for data years beginning day 65, 2000–2010, collection 5 percent tree cover. University of Maryland, College Park, MD

Espinoza JC, Guyot JL, Ronchail J, Cocheneau G, Filizola N, Fraizy P, Labat D, de Oliveira E, Ordoñez JJ, Vauchel P (2009) Contrasting regional discharge evolutions in the Amazon basin (1974–2004). J Hydrol 375:297–311

Fisher JB, Malhi Y, Bonal D, Da Rocha HR, De AraÃšJo AC, Gamo M, Goulden ML, Hirano T, Huete AR, Kondo H, Kumagai TO, Loescher HW, Miller S, Nobre AD, Nouvellon Y, Oberbauer SF, Panuthai S, Roupsard O, Saleska S, Tanaka K, Tanaka N, Tu KP, Von Randow C (2009) The land atmosphere water flux in the tropics. Glob Chang Biol 15(11):2694–2714. doi:10.1111/j.1365-2486.2008.01813.x

Foley JA, Botta A, Coe MT, Costa MH (2002) El Nino–southern oscillation and the climate, ecosystems and rivers of Amazonia. Global Biogeochem Cycles 16(4):1132

Gibbs HK, Ruesch AS, Achard F, Clayton MK, Holmgren P, Ramankutty N, Foley JA (2010) Tropical forests were the primary sources of new agricultural land in the 1980s and 1990s. Proc Natl Acad Sci USA 107:16732–16737

Gloor M, Brienen R, Galbraith D, Feldpausch T, Schöngart J, Guyot JL, Espinoza J, Lloyd J, Phillips O (2013) Intensification of the Amazon hydrological cycle over the last two decades. Geophys Res Lett 40:1729–1733

Hasler N, Avissar R (2007) What controls evapotranspiration in the Amazon basin? J Hydrometeorol 8(3):380–395

Hayhoe S, Neill C, McHorney R, Porder S, Lefebvre P, Coe MT, Elsenbeer H, Krusche A (2011) Conversion to soy on the Amazonian agricultural frontier increases streamflow without affecting stormflow dynamics. Glob Chang Biol 17:1821–1833. doi:10.1111/j.1365-2486.2011.02392.x

Holtum JAM, Winter K (2010) Elevated [CO2] and forest vegetation: more a water issue than a carbon issue? Funct Plant Biol 37:694–702

Huffman GJ, Adler RF, Bolvin DT, Gu G, Nelkin EJ, Bowman KP, Hong Y, Stocker EF, Wolff DB (2007) The TRMM multi-satellite precipitation analysis: quasi-global, multi-year, combined-sensor precipitation estimates at fine scale. J Hydrometeorol 8(1):38–55

Huntingford C, Fisher RA, Mercado L, Booth BB, Sitch S, Harris PP, Cox PM, Jones CD, Betts RA, Malhi Y, Harris GR, Collins M, Moorcroft P (2008) Towards quantifying uncertainty in predictions of Amazon "dieback.". Philos Trans R Soc Lond B 363:1857–1864. doi:10.1098/rstb.2007.0028

IBGE, Brazilian Institute of Geography and Statistics (2015) Municipal agricultural production. Available at http://seriesestatisticas.ibge.gov.br/. Accessed 4 May 2015

INMET, Brazilian Nation Meteorological Institute (2012) Automatic weather station data. Available at http://www.inmet.gov.br/portal/

INPE, Brazilian National Space Agency (2015) PRODES project: satellite monitoring of deforestation in the Brazilian Amazon. Available at http://www.obt.inpe.br/prodes/prodes_1988_2014.htm

IPCC, Intergovernmental Panel on Climate Change (2013) The physical science basis. Contribution of working group I to the fifth assessment report of the Intergovernmental Panel on Climate Change. In: Stocker TF, Qin D, Plattner G-K et al (eds) IPCC climate change 2013. Cambridge University Press, Cambridge and New York, NY, p 1535

Juárez RIN, Hodnett MG, Fu R, Goulden ML, von Randow C (2007) Control of dry season evapotranspiration over the Amazonian forest as inferred from observations at a southern Amazon forest site. J Climate 20:2827–2839

Knox R, Bisht G, Wang J, Bras RL (2011) Precipitation variability over the forest to non-forest transition in southwestern Amazonia. J Climate 24(9):2368–2377. doi:10.1175/2010JCLI3815.1

Lapola DM, Oyama MD, Nobre CA (2009) Exploring the range of climate biome projections for tropical South America: the role of CO2fertilization and seasonality. Global Biogeochem Cycles 23(3). doi:10.1029/2008gb003357

Lathuillière MJ, Johnson MS, Donner SD (2012) Water use by terrestrial ecosystems: temporal variability in rainforest and agricultural contributions to evapotranspiration in Mato Grosso, Brazil. Environ Res Lett 7:24024. doi:10.1088/1748-9326/7/2/024024

Macedo M, Defries R, Morton D, Stickler C, Galford G, Shimabukuro Y (2012) Decoupling of deforestation and soy production in the southern Amazon during the late 2000s. Proc Natl Acad Sci USA 109(4):1341–1346. doi:10.1073/pnas.1111374109

Malhi Y, Pegoraro E, Nobre AD, Pereira MGP, Grace J, Culf AD, Clement R (2002) The water and energy dynamics of a central Amazonian rain forest. J Geophys Res Atmos 107:D20

Malhi Y, Roberts JT, Betts RA, Killeen TJ, Li W, Nobre CA (2008) Climate change, deforestation, and the fate of the Amazon. Science 319(5860):169–172. doi:10.1126/science.1146961

Malhi Y, Aragao LEOC, Galbraith D, Huntingford C, Fisher R, Zelazowski P, Sitch S, McSweeney C, Meir P (2009) Exploring the likelihood and mechanism of a climate-change-induced dieback of the Amazon rainforest. Proc Natl Acad Sci USA. doi:10.1073/pnas.0804619106

Marengo JA (1992) Interannual variability of surface climate in the Amazon basin. Int J Climatol 12:853–863

Marengo JA, Tomasella J (1998) Trends in streamflow and rainfall in tropical South America: Amazonia, eastern Brazil and northwestern Peru. J Geophys Res 103(D2):1775–1783

Marengo JA, Nobre CA, Tomasella J, Oyama MD, Oliveira GS, Oliveira R, Camargo H, Alves LM, Brown IF (2008a) The drought of Amazonia in 2005. J Climate 21:495–516

Marengo JA, Nobre CA, Tomasella J, Cardosa MF, Oyama MD (2008b) Hydro-climate and ecological behaviour of the drought of Amazonia in 2005. Philos Trans R Soc 363 (1498):1773–1778

Marengo JA, Tomasella J, Soares WR, Alves LM, Nobre CA (2011a) Extreme climatic events in the Amazon basin. Climatological and hydrological context of recent floods. Theor Appl Climatol. doi:10.1007/s00704-00011-00465-00701

Marengo JA, Tomasella J, Alves LM, Soares WR, Rodriguez DA (2011b) The drought of 2010 in the context of historical droughts in the Amazon region. Geophys Res Lett 38:L12703. doi:10.1029/2011GL047436

McAlpine CA, Etter A, Fearnside PM, Seabrook L, Laurance WF (2009) Increasing world consumption of beef as a driver of regional and global change: a call for policy action based evidence from Queensland (Australia), Colombia and Brazil. Glob Environ Chang 19 (1):21–33

Melack JM, Hess LL (2010) Remote sensing of the distribution and extent of wetlands in the Amazon basin. In: Junk WJ, Piedade MTF, Wittmann F, Schöngart J, Parolin P (eds) Amazonian floodplain forests: ecophysiology, biodiversity and sustainable management. Ecological studies, vol 210. Springer, New York, NY, pp 43–60. doi:10.1007/978-90-4818725-6_3

Melack JM, Novo EMLM, Forsberg BR, Piedade MTF, Maurice L (2009) Floodplain ecosystem processes. In: Keller MEA (ed) Amazonia and global change. American Geophysical Union Books, Washington, DC, pp 525–542

Meyfroidt P, Lambin EF (2011) Global forest transition: prospects for an end to deforestation. Annu Rev Environ Resour 36:343–371

Moraes JM, Schuler AE, Dunne T, Figueiredo RO, Victoria RL (2006) Water storage and runoff processes in plinthic soils under forest and pasture in eastern Amazonia. Hydrol Process 20:2509–2526

Mu QZ, Zhao MS, Running SW (2011) Improvements to a MODIS global terrestrial evapotranspiration algorithm. Remote Sens Environ 115:1781–1800

Nepstad DC, Carvalho CJR, Davidson EA, Jipp P, Lefebvre PA, Negreiros GH, da Silva ED, Stone TA, Trumbore SE, Vieira S (1994) The role of deep roots in the hydrological and carbon cycles of Amazonian forests and pastures. Nature 372:666–669

Nepstad DC, Moutinho PRS, Dias-Filho MB, Davidson EA, Cardinot G, Markewitz D, Figueiredo R, Viana N, Lefebvre PA, Ray DG, Chambers JQ, Barros L, Ishida FY, Belk E, Schwalbe K (2002) The effects of rainfall exclusion on canopy processes and biogeochemistry of an Amazon forest. J Geophys Res 107(D20):53-1–53-18

Nepstad DC, Lefebvre PA, Silva UL Jr, Tomasella J, Schlesinger P, Solorzano L, Moutinho PRS, Ray DG (2004) Amazon drought and its implications for forest flammability and tree growth: a basin-wide analysis. Glob Chang Biol 10:704–717

Nobre CA, Sellers PJ, Shukla J (1991) Amazonian deforestation and regional climate change. J Climate 4:957–988

Nobre P, Malagutti M, Urbano DF, Almeida RAF, Giarolla E (2009) Amazon deforestation and climate change in a coupled model simulation. J Climate 22:5686–5697

Oliveira RS, Bezerra L, Davidson EA, Pinto F, Klink CA, Nepstad DC, Moreira A (2005) Deep root function in soil water dynamics in cerrado savannas of central Brazil. Funct Ecol 19:574–581

Oliveira LJC, Costa MH, Soares-Filho BS, Coe MT (2013) Large-scale expansion of agriculture in Amazonia may be a no-win scenario. Environ Res Lett 8:024021. doi:10.1088/1748-9326/8/2/024021

Ometto JP, Aguiar APD, Martinelli LA (2011) Amazon deforestation in Brazil: effects, drivers and challenges. Future Sci 2:575–585

Oyama MD, Nobre CA (2003) A new climate-vegetation equilibrium state for tropical South America. Geophys Res Lett 30(23):2199

Panday PK, Coe MT, Macedo MN, Lefebvre P, Castanho ADA (2015) Deforestation offsets water balance changes due to climate variability in the Xingu River in eastern Amazonia. J Hydrol 523:822–829

Pereira EB, Martins FR, Abreu SL, Ruther R (2006) Brazilian atlas of solar energy. UNEP, São Jose dos Campos

Poveda G, Mesa OJ (1993) Metodologias de Prediccion de la Hidrologia Colombiana Considerando el Evento El Niño-Oscilacion del Sur (ENOS), vol 17, Atmosfera. Sociedad Colombiana de Meteorologia, Bogota

Priante-Filho N, Vourlitis GL, Hayashi MMS, Noguiera JS, Campelo JH Jr, Nunes PC, Sousa LSE, Couto EG, Hoeger W (2004) Comparison of the mass and energy exchange of a pasture and a mature transitional tropical forest of the southern Amazon Basin during a seasonal transition. Glob Chang Biol 10:863–876

Rammig A, Jupp T, Thonicke K, Tietjen B, Heinke J, Ostberg S, Lucht W, Cramer W, Cox P (2010) Estimating the risk of Amazonian forest dieback. New Phytol 187:694–706. doi:10.1111/j.1469-8137.2010.03318.x

Richey JE, Meade RH, Salati E, Devol AH, Nordin CF Jr, dos Santos U (1986) Water discharge and suspended sediment concentrations in the Amazon River: 1982–1984. Water Resour Res 22(5):756–764

Ronchail JG, Cochonneau M, Molinier JL, Guyot AG, de Miranda C, Guimarães V, de Oliveira E (2002) Interannual rainfall variability in the Amazon basin and sea-surface temperatures in the equatorial Pacific and tropical Atlantic Oceans. Int J Climatol 22:1663–1686

Saad SI, da Rocha HR, Silva Dias MAF, Rosolem R (2010) Can the deforestation breeze change the rainfall in Amazonia? A case study for the BR-163 highway region. Earth Interact 14 (18):1–25. doi:10.1175/2010EI351.1

Salati E, Nobre CA (1991) Possible climatic impacts of tropical deforestation. Clim Change 19:177–198

Sampaio G, Nobre C, Costa MH, Satyamurty P, Soares-Filho BS, Cardoso M (2007) Regional climate change over eastern Amazonia caused by pasture and soybean cropland expansion. Geophys Res Lett 34:L17709. doi:10.1029/2007GL030612

Shuttleworth WJ (1988) Evaporation from Amazonian forests. Philos Trans R Soc B 233:321–346

Silvério DV, Brando PM, Macedo MN, Beck PSA, Bustamante M, Coe MT (2015) Agricultural expansion dominates climate changes in southeastern Amazonia: the overlooked non-GHG forcing. Environ Res Lett 10:104105. doi:10.1088/1748-9326/10/10/104015

Snyder PK (2010) The influence of tropical deforestation on the northern hemisphere climate by atmospheric teleconnections. Earth Interact 14:1–34. doi:10.1175/2010EI280.1

Soares-Filho B, Nepstad D, Curran L, Cerqueira G, Garcia R, Ramos C, Voll E, McDonald A, Lefebvre P, Schlesinger P, McGrath D (2006) Modelling conservation in the Amazon basin. Nature 440:520–523

Souza PJ, Ribeiro A, da Rocha EJP, Bothelo MDM, Marlisom A, De Sousa L, De Souza EB, Farias JRB (2011) Impacts of soyabean expansion on the Amazon energy balance: a case study. Exp Agric 47(3):553–567

Spracklen DV, Arnold SR, Taylor CM (2012) Observations of increased tropical rainfall preceded by air passage over forests. Nature 489(7415):282–U127. doi:10.1038/nature11390

Stickler CM, Coe MT, Costa MH, Nepstad DC, McGrath DG, Dias LCP, Rodrigues HO, Soares-Filho BS (2013) Dependence of hydropower energy generation on forests in the Amazon Basin at local and regional scales. Proc Natl Acad Sci USA 110(23):9601–9606. doi:10.1073/pnas.1215331110

Trancoso R (2006) Changes in land cover and alterations in the hydrological response of catchments in Amazonia. INPA/UFAM, Brazil, p 139

von Randow C, Manzi AO, Kruijt B, De Oliveira PJ, Zanchi FB, Silva RL, Hodnett MG, Gash JHC, Elbers JA, Waterloo MJ, Cardoso FL, Kabat P (2004) Comparative measurements and seasonal variations in energy and carbon exchange over forest and pasture in South West Amazonia. Theor Appl Climatol 78:5–26. doi:10.1007/s00704-004-0041-z

Walker R, Moore NJ, Arima E, Perz S, Simmons C, Caldas M, Vergara D, Bohrere C (2009) Protecting the Amazon with protected areas. Proc Natl Acad Sci USA 106(26):10582–10586. doi:10.1073/pnas.0806059106

Wang G, Eltahir EAB (2000) Role of vegetation dynamics in enhancing the low frequency variability of the Sahel rainfall. Water Resour Res 36:1013–1021

Zeng N, Neelin JD, Lau K-M, Tucker CJ (1999) Enhancement of interdecadal climate variability in the Sahel by vegetation interaction. Science 286:1537–1540

Zeng N, Yoon J-H, Marengo JA, Subramaniam A, Nobre CA, Mariotti A, Neelin D (2008) Causes and impacts of the 2005 Amazon drought. Environ Res Lett 3. doi:10.1088/1748-9326/1083/1081/014002

Chapter 4
Extreme Seasonal Climate Variations in the Amazon Basin: Droughts and Floods

José A. Marengo, Earle R. Williams, Lincloln M. Alves, Wagner R. Soares, and Daniel A. Rodriguez

4.1 Introduction

The Amazon region is characterised by several rainfall regimes. In southern Amazonia, rainfall peaks during austral summer, in central Amazonia and near the Amazon delta it peaks in the autumn, and north of the Equator it peaks in austral winter. This is due to the alternating warming of the two hemispheres and to the annual cycle of the South American Monsoon System (e.g. Vera et al. 2006; Marengo et al. 2010a, b, 2012a, b), associated with the seasonal meridional migration of the Intertropical Convergence Zone (ITCZ).

As a consequence of the seasonal cycle in rainfall, the Amazon River main stem and tributaries show high/low river levels a few months after the peak rainfall season, and the river level/discharge depend on the precipitation in the rainy season during the previous wet season. Rivers that extend over southern Amazon basin (e.g. Solimões, Madeira) peak in April–May while rivers with basins in the central Amazon basin (e.g. Rio Negro) peak in May–June. For more details about rainfall and river regimes in the Amazon basin, see Salati et al. (1978), Figueroa and Nobre (1990), Meade et al. (1991), and Espinoza et al. (2009a, b).

J.A. Marengo (✉) • W.R. Soares
National Center for Monitoring and Early Warning of Natural Disasters (CEMADEN), Rodovia Dutra km 40, 12630-000 Cachoeira Paulista, São Paulo, Brazil
e-mail: jose.marengo@cemaden.gov.br

E.R. Williams
Parsons Laboratory, Department of Civil and Environmental Engineering, MIT, Cambridge, MA 02139, USA

L.M. Alves • D.A. Rodriguez
Earth System Science Center, National Institute for Space Research (CCST INPE), Rodovia Dutra km 40, 12630-000 Cachoeira Paulista, São Paulo, Brazil

© Springer-Verlag Berlin Heidelberg 2016

L. Nagy et al. (eds.), *Interactions Between Biosphere, Atmosphere and Human Land Use in the Amazon Basin*, Ecological Studies 227,
DOI 10.1007/978-3-662-49902-3_4

As part of its natural climate variability, the Amazon region is periodically subject to rainfall deficiency and excess across the basin, with consequences for the hydrology of the basin in the form of droughts or floods. The perception of drought and flood in this region may be different from that in other regions, since perhaps the best indictors of drought or floods are the river levels/discharges. During the last 10 years, the droughts in 2005 and 2010 and floods in 2009, 2012, and more recently in 2014–2015 were the most severe on record, and the corresponding record low and high river levels impacted human and natural systems (Pinho et al. 2014; Tomasella et al. 2013; Marengo et al. 2013a, b).

Previous observational studies (Sternberg 1987; Marengo 2004; Williams et al. 2005; Ronchail et al. 2002; Marengo et al. 2008a, b, 2010a, b, 2011, 2013a, b; Fernandes et al. 2011; Zeng et al. 2008; Tomasella et al. 2011, 2013; Jenkins 2009; Espinoza et al. 2009a, b, 2011, 2012, 2013; Yoon and Zeng 2010; Marengo and Espinoza 2015, and references therein) have identified deficiencies or excesses in rainfall that have produced droughts and floods in Amazonia. These hydrometeorological anomalies were a consequence of low and upper level atmospheric circulation anomalies, a consequence of warming or cooling of the tropical Pacific Ocean from the El Niño or La Nina, were attributable to strong warming or cooling in the tropical Atlantic, or were a result of the combination of the two.

The droughts of 2005 and 2010 were similar in terms of river impacts, but the distribution of rainfall deficits across the basin was different. While in 2005 the drying was more intense in southwestern Amazonia, in 2010 the drying out was stronger in central and eastern Amazonia mainly during the austral summer and autumn and later on during the beginning of austral spring (Marengo and Espinoza 2015). During the drought of 2005, the levels and streamflow in rivers such as the Madeira and Solimões, with basins extending over southern and western Amazonia, were very low, while the levels of the Rio Negro at Manaus were lower but not as low as in the drought of 2010 (Fig. 4.1). The 2010 drought started during an El Niño event in the early austral summer of 2010 and then became more intense during a La Niña, in spring and winter during the dry season (Marengo et al. 2011; Lewis et al. 2011; Espinoza et al. 2011).

In contrast, during the time of the so-called flood of the century in 2009 and later on in 2012 (Marengo et al. 2011; Sena et al. 2012; Marengo et al. 2013a, b), the Amazon basin was hit by heavy flooding, which resulted in water levels and river discharges with a magnitude and duration rarely observed in several decades in the basin. In July 2009, the levels of the Rio Negro in the port of Manaus reached a new record high level for the last 107 years (Fig. 4.1). The main characteristic of the 2008–2009 and 2012–2023 hydrological season was a premature onset of the rainy season in northern and northwestern Amazonia and a longer rainy season compared to the other wet years, where the onset was delayed by at least 1 month (Sena et al. 2012; Vale et al. 2011; Marengo et al. 2011, 2013a, b; Satyamurty et al. 2013). The hydrological consequences of this pattern were earlier-than-normal floods in the Amazon northern tributaries, in which peak discharges at their confluences with the main stem almost coincided with the peaks in the southern tributaries. More

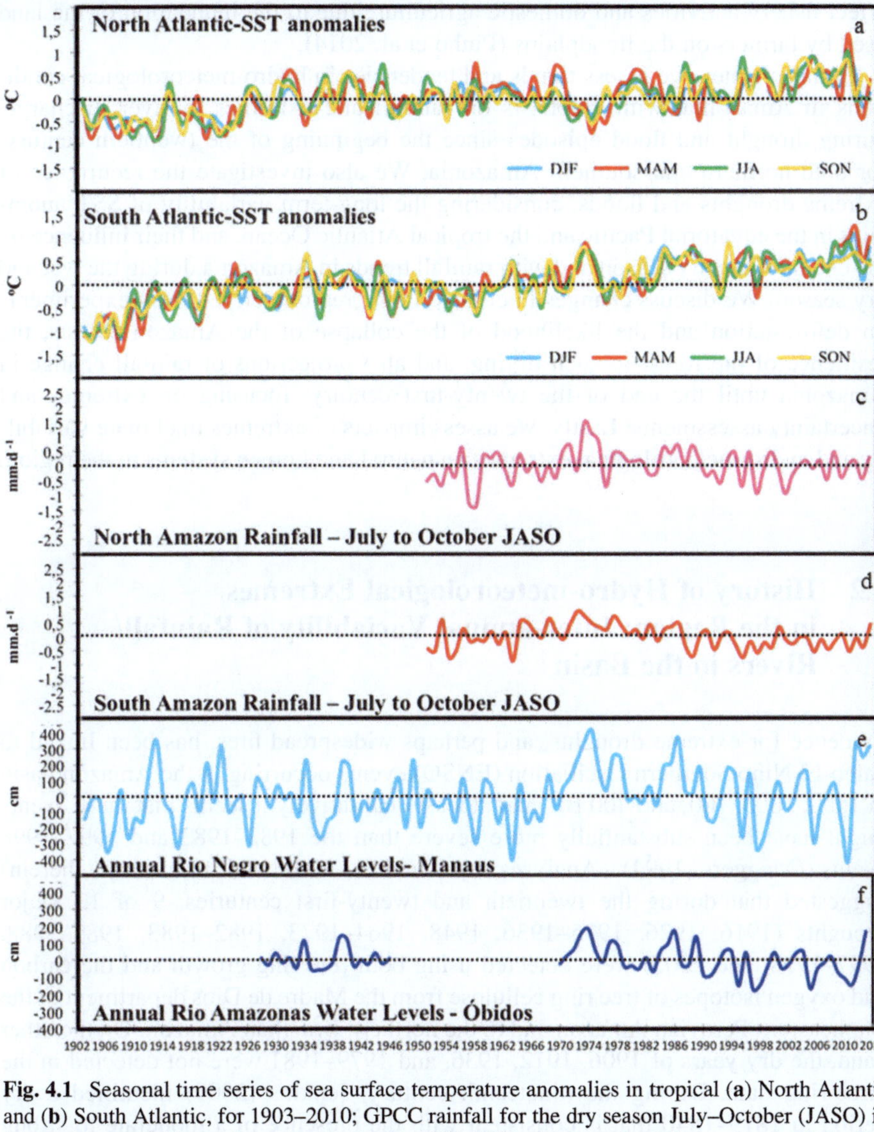

Fig. 4.1 Seasonal time series of sea surface temperature anomalies in tropical (**a**) North Atlantic and (**b**) South Atlantic, for 1903–2010; GPCC rainfall for the dry season July–October (JASO) in (**c**) northern and (**d**) southern Amazonia for 1951–2010; (**e**) annual minimum levels of the Rio Negro in Manaus for 1903–2010; and (**f**) discharges of the Amazon in Óbidos for 1928–1947 and 1970–2010. Units are in °C in figures (**a**) and (**b**), mm day^{-1} in figures (**c**) and (**d**), and cm in figures (**e**) and (**f**). Base period for calculations of the anomalies is from the mean of all datasets (Marengo et al. 2012a, b)

recently, in 2014–2015, floods in western Amazonia affected the levels of the Rio Madeira and Rio Branco and isolated cities in Acre and Rondonia states, Brazil, and by June 2014 the water levels of the Rio Negro at Manaus started to rise (Espinoza et al. 2014). Besides the risk for the people living near the banks of the rivers, floods affect fishery activities and domestic agriculture due to the inundation of the land used by farmers on the floodplains (Pinho et al. 2014).

In this chapter, we assess trends and tendencies in hydro-meteorological conditions in Amazonia, with emphasis on rainfall and extremes in river discharge, during drought and flood episodes since the beginning of the twentieth century, for both northern and southern Amazonia. We also investigate the recurrence of extreme droughts and floods, considering the long-term variability of SST anomalies in the equatorial Pacific and the tropical Atlantic Ocean, and their influence on the extent of the dry season and with rainfall trends in Amazonia during the wet and dry season. We discuss changes in climate in the region derived from experiments on deforestation and the likelihood of the collapse of the Amazon forests, the resilience of the forest to such drying, and also projections of rainfall change in Amazonia until the end of the twenty-first century, focusing on extremes and uncertainty assessments. Lastly, we assess impacts of extremes in climate variability and hydro-meteorological extremes on natural and human systems in the region.

4.2 History of Hydro-meteorological Extremes in the Region: Inter-annual Variability of Rainfall/Rivers in the Basin

Evidence for extreme droughts, and perhaps widespread fires, has been linked to paleo-El Niño Southern Oscillation (ENSO) events occurring in the Amazon basin in 1500, 1000, 700, and 400 BP, and at least qualitatively it seems that these events might have been substantially more severe than the 1982–1983 and 1997–1998 events (Meggers 1994). Analyses by Jenkins (2009 and references therein) suggested that during the twentieth and twenty-first centuries, 9 of 13 major droughts (1916, 1926, 1935–1936, 1948, 1964–1973, 1982–1983, 1987–1988, 1997–1998, and 2005) were detected using both tree ring growth and the carbon and oxygen isotopes of tree ring cellulose from the Madre de Dios department of the southeastern Peruvian Amazon and in the northern Brazilian Amazon. On the other hand, the dry years of 1906, 1912, 1936, and 1979–1981 were not detected in the three-ring data. During the nineteenth century, Jenkins (2009) identified a dry period in 1819–1840 that is consistent with the presence of a moderate to strong El Niño event.

Based on soil moisture indices, Sheffield and Wood (2011) showed long-term drought events in the 1960s, the 1980s, and 1990s. They identified dry conditions during the mid-1960s, and other low-index values coincided with strong El Niño events, occurring in 1957–1958, 1965, 1972–1973, 1992, and 1997. In fact, various observational studies have shown that after a relatively long wet period during the 1940s and 1950s, major droughts occurred in the Amazon region in 1963–1964, 1970, 1983, 1987, 1997–1998, 2005, and more recently in 2010 (Aceituno 1988;

Marengo et al. 2013a, b). Very intense El Niño events have been associated with the extreme droughts in 1925–1926, 1982–1983, and 1997–1998 and the last two also experienced an anomalously warm tropical North Atlantic along with warming in the equatorial Pacific.

Figure 4.1 shows the levels of the Rio Negro in Manaus during some extreme years with record low and high levels. The lowest levels were detected during El Niño years in 1912 and 1926 all year long, while in the other dry years the lower levels were noticed during the second half of the year, as in 2005. However, only a fraction of Amazon rainfall variability can be explained by ENSO (Yoon and Zeng 2010; Joetzjer et al. 2013, and references cited therein) and other factors linked to rainfall deficiency are related to an anomalously warm tropical North Atlantic Ocean. In fact, intense droughts in 1964, 1980, and 2005 were not linked to ENSO, and while the drought in 2010 started with an ENSO event it continued during La Niña. The spatial extent of each drought event was different (Ronchail et al. 2002; Williams et al. 2005; Saleska et al. 2007; Marengo et al. 2008a, b, 2011), and the impacts of rainfall anomalies were also detected in the water level records at Manaus and other sites across Amazonia (Tomasella et al. 2011, 2013).

The drought of 1925–1926 was studied in detail by Sternberg (1987), Meggers (1994), Williams et al. (2005), Marengo et al. (2008a, b), and Sheffield and Wood (2011). Different from those in 2005 and 2010, this drought occurred during the presence of one of the most intense El Niño events in modern history. Rainfall anomalies in central-northern Brazilian Amazonia and southern Venezuela in 1926 resulted in precipitation that was about 50 % lower than normal. During this particular drought, extensive fires prevailed in Venezuela and the upper Rio Negro basin. Unusually high air temperature anomalies were recorded in Venezuelan and northern Brazilian Amazonian towns for both 1925 and 1926, and it is plausible that the dryness in the northern portion of the Rio Negro basin in 1925 also contributed to the major drought in 1926 by a depletion of soil moisture.

While several studies have analysed the droughts of 1982–1983 (Aceituno 1988, Marengo et al. 1998) and 1997–1998 (Nepstad et al. 1999) and their impacts on climate, hydrology, and fires in Amazonia, there are only casual references to the drought event of 1963–1964 in some local newspapers. In fact, the drought of 1963–1964 has received little attention (Marengo et al. 2008a), despite being identified as the most severe with up to 12-month duration. Water levels of major Amazon tributaries fell drastically to unprecedented low values, and isolated the floodplain population, whose transportation depends upon local streams, which completely dried up.

Floods in 1953–1954 and 2008–2009 were not linked entirely to La Niña, and in fact were related to a warming in the tropical South Atlantic, similar to those in 1989 and 1999. The highest water levels were detected during 2009. We can say every dry or wet year, regardless of whether they are associated with El Nino or with SST anomalies in the tropical Atlantic, is different in terms of the geographical distribution of rainfall anomalies across the basin, and so is the response of the hydrology of the Amazon region, as shown in the levels of the Rio Negro at Manaus.

The levels of the Negro river at the port of Manaus during July 2009 reached a record high level at the time of 29.77 m, according to the Brazilian Geological Service (CPRM), with a return period of about 55 years. The six previous river record highs observed in Manaus were 1953 (29.69 m), 1976 (29.61 m), 1989 (29.42 m), 1922 (29.35 m), and 1999 (29.30 m). Other high-water years in Manaus were 1909 (29.17 m), 1971 (29.12 m), 1975 (29.11 m), and 1994 (29.05 m). The level of the Negro River reached 29.97 m in May 2012, the highest mark in 110 years on record, since the beginning in 1903. The high levels during the flood in 2009 exhibit a return period of c. 55 years. It is important to note that water level in Manaus at this station also is controlled by the backwater effect of the Solimões River (Meade et al. 1991).

In 2014 and 2015, the southwestern Amazon experienced severe floods due to summer rainfall that exceeded by more than 100 % that normally experienced over the Brazilian states of Acre and Rondonia and in Bolivian and Peruvian Amazonia, and the levels of the Madeira River and the Rio Branco reached record high levels, flooding cities, farms, and roads, damaging fisheries, and isolating residents (Espinoza et al. 2014; CPRM—http://www.cprm.gov.br).

As explained above, previous observational studies have identified rainfall deficiencies or excesses that have produced droughts and floods in Amazonia associated with atmospheric circulation anomalies, as a consequence of El Niño, or attributed to strong warming in the tropical Atlantic, or to a combination of both (Table 4.1).

Figure 4.2 shows a simplified scheme of the large-scale circulation in the tropical region during El Niño and La Niña years, and also during situations with warmer and colder than normal tropical North and South Atlantic, with subsequent rainfall anomalies in the Amazon basin. Figure 4.2a shows a situation during El Niño and warmer tropical North Atlantic: warmer sea surface temperature (SST) in the equatorial Pacific favours convection over that region, generating compensatory subsidence on the other side of the Andes over Amazonia; in the meantime warm SST anomalies in the tropical North Atlantic are related to weak near-surface northeast trade winds and moisture flux into Amazonia and together with subsidence make for less rainfall in the basin due to an anomalously northward position of the ITCZ. Figure 4.2b shows the situation during La Niña and colder tropical North Atlantic: cold SST anomalies over the equatorial Pacific are linked with subsidence and at the other side of the Andes, over Amazonia, convection is observed; the intense northeast trades over the colder tropical North Atlantic favoured intense moisture flux inside Amazonia, and with an anomalously southward position of the ITCZ, feeding convection and consequently producing more rainfall in the basin. Other situations depart from this representation, such as the drought of 2005, which was related to warmer tropical North Atlantic but an absence of El Niño.

While there is a general consensus on the occurrence of intense droughts and floods since the early twentieth century as shown by different datasets, there is some uncertainty in the estimates of precipitation, especially in the earlier part of the twentieth century. This is particularly true in remote regions of the Amazon region,

Table 4.1 History of droughts and floods in the Amazon basin, with indication if they are related to El Niño (EN), La Niña (LN), or to sea surface temperature conditions in the northern (TNA) or southern (TSA) tropical Atlantic (Marengo and Espinoza 2015)

Year	Extreme seasonal event	Related to	References
1906	Drought	EN	Sombroek (2001)
1912	Drought	EN	Williams et al. (2005), Marengo et al. (2008a, b)
1916	Drought	EN	Sombroek (2001), Jenkins (2009)
1925–1926[a]	Drought	EN	Sternberg (1968, 1987), Meggers (1994), Williams et al. (2005), Marengo et al. (2008a, b), Sheffield and Wood (2011)
1948	Drought	EN	Sombroek (2001)
1963–1964	Drought	Warm TNA	Sombroek (2001), Marengo et al. (2008a, b), Sheffield and Wood (2011)
1979–1981	Drought	Warm TNA	Sheffield and Wood (2011)
1982–1983	Drought	EN + warm TNA	Sombroek (2001), Ronchail et al. (2005), Marengo et al. (2008a, b)
1995	Drought	EN + warm TNA	Espinoza et al. (2011)
1997–1998	Drought	EN + warm TNA	Sombroek (2001), Moran et al. (2006), Marengo et al. (2008a, b, 2011), Zeng et al. (2008), Espinoza et al. (2011), Tomasella et al. (2011, 2013), Coelho et al. (2013)
2005[a]	Drought	Warm TNA	Marengo et al. (2008a, b, 2011), Zeng et al. (2008), Espinoza et al. (2011), Cox et al. (2008), Tomasella et al. (2013), Yoon and Zeng (2010), Aragão et al. (2007), Coelho et al. (2013)
2010[a]	Drought	EN + warm TNA	Lewis et al. (2011), Marengo et al. (2011), Espinoza et al. (2011), Coelho et al. (2013)
1953	Flood	?	Salati et al. (1978), Ronchail et al. (2005), Marengo et al. (2010a)
1976	Flood	LN	Marengo et al. (2010a), Satyamurty et al. (2013)
1989	Flood	LN	Ronchail et al. (2006), Marengo et al. (2011, 2013a, b), Espinoza et al. (2013)
1999	Flood	LN	Ronchail et al. (2006), Marengo et al. (2011, 2013a, b), Espinoza et al. (2013), Satyamurty et al. (2013)
2009[a]	Flood	Warm TSA	Marengo et al. (2010a), Filizola et al. (2014), Sena et al. (2012), Vale et al. (2011)
2012[a]	Flood	LN + warm TSA	Marengo et al. (2013a, b), Espinoza et al. (2013), Satyamurty et al. (2013)
2014	Flood	Warm IP + warm SSA	Espinoza et al. (2014)

SSA Subtropical South Atlantic; *IP* Indo-Pacific Ocean
[a]Events characterised at the time as 'once in a century'

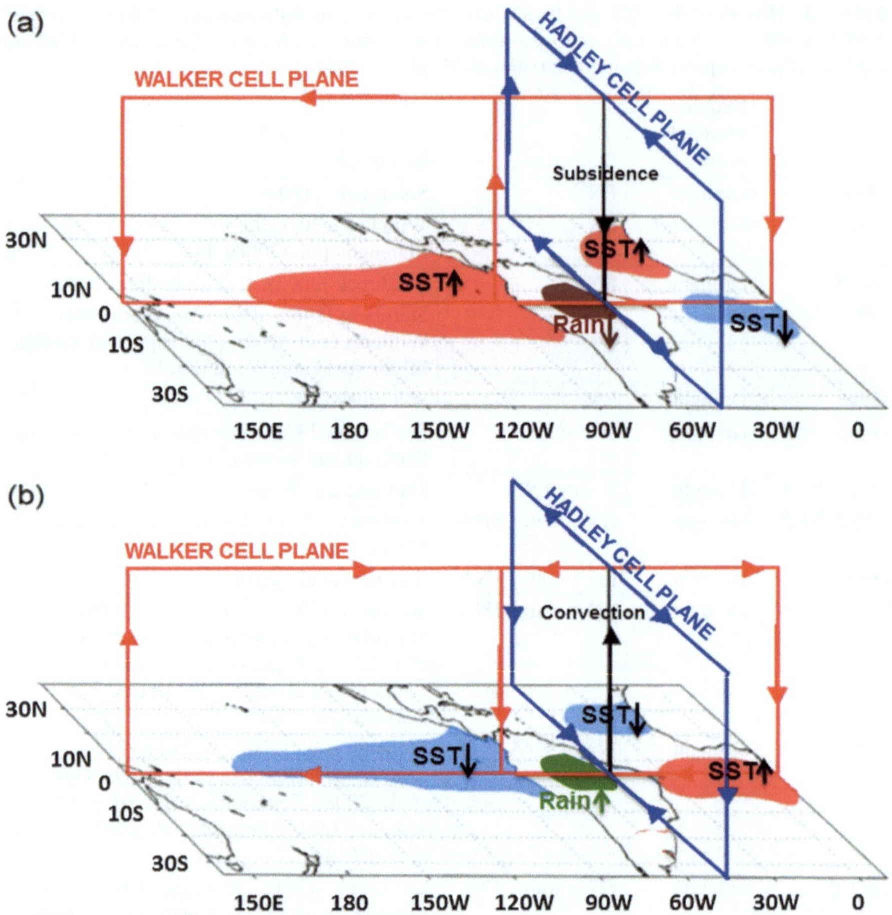

Fig. 4.2 Schematic diagrams showing the anomalous regional Walker and Hadley circulation for the (**a**) El Niño and warmer tropical North Atlantic, and (D) La Niña and cold tropical North Atlantic. The *blue* (*red*) *arrows* indicate the regional climatological circulation of the Walker (Hadley) cell and the *black thick arrows* show the anomalous vertical circulation. *Green/brown shades* indicate positive/rainfall anomalies in the basin (modified from Ambrizzi et al. 2004)

where the number of rain gauges and their spatial coverage were low (Marengo 2004). Thus, there are differences among various rainfall datasets that are based on station or gridded data. However, the availability of long-term river datasets has allowed the identification of droughts and floods in the past.

One of the best indicators of large-scale rainfall in Amazonia is the Manaus gauge itself, since it represents rainfall over nearly 50 % of the complete Amazon basin (Williams et al. 2005). Several studies have used the Rio Negro levels at the Manaus site (see Marengo and Espinoza 2015 and references therein) since the series shows clearly the impacts of El Niño or tropical Atlantic. Figure 4.3 shows the annual cycle of the mean water levels at Manaus during dry and wet years.

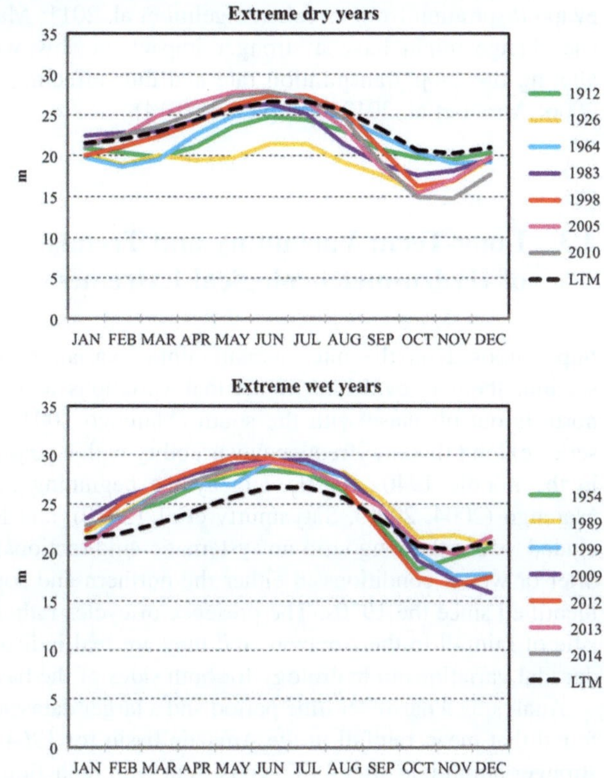

Fig. 4.3 Annual mean values of the levels of the Rio Negro in Manaus, Brazil (in metres), for some extreme dry years (1912, 1926, 1964, 1983, 1998, 2005 and 2010) and wet years (1954, 1989, 1999, 2009, 2012, 2013, 2014) as compared to the long term average LTM 1903–1986 (*black thick lines*). *Source* of data: CPRM-Manaus, Brazil

Together with Table 4.1, this figure shows very low levels during 1926 (the lowest during the entire year), 1912, and 1983 (all El Niño years) and in 1964 from January to September, while in 2005 and 2010 levels below normal were detected mainly during September–December. During wet years, most of the episodes with levels above normal were related to La Niña years. The difference among dry years is consistent with different spatial distribution of rainfall anomalies across the basin during each dry year, where droughts related to El Niño may have different rainfall anomaly patterns as compared to those related to a warm tropical North Atlantic (Marengo et al. 2011).

Significant scientific uncertainties remain regarding the attribution of causes of the droughts and floods, particularly regarding the extreme events of the last 10 years—at the time of occurrence thought to be of once-a-century severity. Might they yet be explained by natural climate variation in terms of SST anomalies in the tropical Pacific and/or the tropical North Atlantic Oceans or were they due to anthropogenic impacts?

The question arises because changes in precipitation over Amazonia are due to a combination of different regional processes and interactions that are partly influenced by large-scale circulation (including El Niño-La Niña and Tropical Atlantic related SST anomalies), as well as by local water sources via evapotraspiration from forests (Angelini et al. 2011; Makaireva et al. 2013). Land use change might have a stronger impact on downwind rainfall in Amazonia, altering the evapotranspiration rate and thus affecting the water cycle (Marengo 2006; Arraut et al. 2012; Zemp et al. 2014).

4.3 Long-Term Variability and Trends of Hydro-meteorological Extremes

Superimposed on the inter-annual rainfall variations discussed in the previous section, there is evidence of decadal variations across the basin, although the north is out of phase with the south (Marengo 2004). In fact, rainfall and river series exhibit this low-frequency variability with a peak near 30 years, with breaks in the middle 1940s, 1970s, and by the beginning of the twenty-first century. Marengo (2004, 2009), Satyamurty et al. (2010), and Buarque et al. (2010) concluded that in the long term no systematic unidirectional long-term trends towards drier or wetter conditions in either the northern and southern Amazon have been identified since the 1920s. The presence of cycles rather than a trend is characteristic of rainfall in the Amazon, and they are real indicators of decadal and multidecadal variations in hydrology for both sides of the basin.

Analysing a narrower time period and a larger dataset, Espinoza et al. (2009a, b) found that mean rainfall in the Amazon basin for 1964–2003 has decreased, with stronger amplitude after 1982, consistent with reductions in convection and cloudiness in the same region (Arias et al. 2011). Studies by Gloor et al. (2013) suggest that the Amazon precipitation is increasing since 1990 due to increasing atmospheric water vapour import from the warming tropical Atlantic, and this is also reflected on the Rio Negro levels at Manaus and the Amazon discharge at Óbidos.

An important aspect of rainfall variability in Amazonia is a possible delay of the onset of the rainy season or late end of the dry season, resulting in longer dry seasons (Butt et al. 2011)—or extension of the dry season (Marengo et al. 2011; Fu et al. 2013) that have been observed since the 1970s. Previous studies have suggested local and remote factors that affect the wet season onset in the Amazon, which could be initiated by increased evapotranspiration (Fu et al. 1999; Li and Fu 2004; Li et al. 2006) as a result of the response of rainforest to a seasonal increase in solar radiation (Myneni et al. 2007) on the local forcing, or due to changes of cross-equatorial atmospheric moisture transport influencing convection and thus the wet season onset linked to warming in the tropical Atlantic or Pacific Oceans (Marengo et al. 2010a).

Figure 4.1 shows time series of SST anomalies in the tropical South and North Atlantic, together with rainfall records over the dry season in northern and southern Amazonia, and annual minimum levels of the Negro and Amazon rivers. In general, a weak negative tendency during the dry season precipitation is apparent in both southern and northern Amazonia starting in the mid-1970s, while from 1951 to the mid-1970s a slight increase is apparent. On the longer term, since the 1920s rainfall has shown decadal variability, with periods with more or less rainfall in Amazonia, and the trends detected in recent decades shown are part of this natural decadal climate variability, with no signal of impacts of deforestation on the rainfall regime (Marengo 2009).

Changes in large-scale circulation associated with SST patterns are responsible for rainfall anomalies, particularly during the austral summer and autumn. The hydrological impacts of these anomalies are experienced in the form of lower-than-normal river discharge not only in the wet season but also during the following winter and spring dry seasons. Since the late 1970s (Fig. 4.1), the SST anomalies in the tropical North Atlantic have gradually increased, reaching high values during 1980, 1998, and 2005 and then in 2010, all of which coincide with drought years in the Amazon. During March–May of 2010, the seasonal temperature anomaly was the largest (1.5 °C) since 1923, exceeding the previous MAM record of about 0.9 °C in 2005. In the dry season (June–August), the temperature anomalies reached 1 and 0.9 °C in 2010 and 2005, respectively, again two records since 1902. The warming of the tropical North Atlantic exhibits inter-decadal SST trends.

It is noted that the negative rainfall trends were detected in the dry season precipitation in both northern and southern Amazonia. Rainfall anomalies were calculated for northern and southern Amazonia, using the Global Precipitation Climatology Centre (GPCC) gridded rainfall dataset (Rudolf et al. 2005), as explained in Marengo et al. (2011). Changes in rainfall are consistent with surface warming in the tropical North Atlantic, as well as a weakening in the moisture transport from the tropical north Atlantic. The tropical Atlantic was at least 1.5–2 °C above average all year long in 2005, which is much higher than in other drought years. Thus, minimum water levels in Manaus and Óbidos also show a negative trend, consistent with the reduction of dry season precipitation in both sections of the basin.

Regarding the impacts of land use changes on changes in the hydrology of South America, one of the distinctive features to consider is the relationship between the hydrological behaviour and vegetation—atmospheric feedbacks. Although feedback mechanisms are present at all scales, the atmospheric influence is more significant at large scales. Studies of land use change in the Brazilian southern Amazonia (Rodriguez et al. 2010) for recent decades have shown that the impact on the hydrological response is time lagged at larger scales. The extension of the dry season also exhibits inter-annual and decadal scale variations, linked either to natural climate variability or as suggested by Wang et al. (2011) as a result of land use change in deforestation and vegetation dynamics on decadal variability of rainfall in the region. This suggests that inter-annual and longer-term variations of rainfall in Amazonia are as much a consequence of natural climate variability in the form of SST-forced variability and also some influence of human activities in the form of land use change.

4.4 Projected Climate Change in Amazonia

A review by Marengo (2006) has shown an evolution in deforestation experiments with global models from the late 1970s to the present, and in all of them the total deforestation of Amazonia may lead to drier and warmer climates in the region, affecting the climate in the rest of South America and the world (Cox et al. 2000, 2004). Recent model experiments including dynamic vegetation models and transient deforestation have shown profound changes in rainfall and temperature in Amazonia, which may result in the collapse of the Amazon forest 'biome', and leading to the so-called *Amazon Dieback,* or 'savannisation'. In this scenario projected by the UK Met Office Hadley Centre Global Coupled climate model HadCM3 for the A2 high emission scenario, the Amazon region would change from a net sink to a net source of CO_2 by the end of the twenty-first century, thereby reinforcing global warming and a regional decrease in precipitation (see reviews in Cox et al. 2000; Sampaio et al. 2007; Betts et al. 2004; Malhi et al. 2009; Nobre and Borma 2009). As temperatures increase, the prospects of climate-change-induced increases in drought frequency and severity over the twenty-first century in Amazonia could increase, leaving the region and the entire South American continent under considerable environmental stress.

Even if the uncertainties related to such a dieback are still high (Malhi et al. 2009), and that such changes are not being detected at present, should it happen its possible impacts on regional and global climate could be extensive. Even though the Amazon forest may not necessarily become savanna type vegetation, the impacts on the functioning of the forest at regional and global scales may be important. In fact, Huntingford et al. (2013) analysed 22 CMIP3 models and found that the possibility of climate-induced (that is, not direct deforestation) damage to tropical rainforests in the period to the year 2100, even under the A2 emission scenario, might be lower than some earlier studies suggest, and in only one of the simulations are tropical forests projected to lose biomass by the end of the twenty-first century—and then only for the Americas, suggesting forest resiliency in the region.

Numerical experiments by Zhang et al. (2009) have suggested that biomass-burning aerosols can work against the seasonal monsoon circulation transition and thus reinforce the dry season rainfall pattern for southern Amazonia. For future climate change scenarios in seasonal extremes, the hypothesis is that droughts similar to that in 2005 may become more frequent and intense in a future climate change due to changes in aerosols in the northern hemisphere, as suggested by Cox et al. (2008).

Analyses from global IPCC models used in the Fourth Assessment Report AR4 (Meehl et al. 2007) and from the downscaling of some of these models using various regional models for the Amazon region suggest for the last decades of the twenty-first century an increase of precipitation in western Amazonia, while decreases are projected for eastern Amazonia. However, the individual models CMIP3 show mixed results (Marengo et al. 2010b, 2011; Mendes and Marengo

2010; Menendez et al. 2010; Nuñez et al. 2009; Seth et al. 2010, Meehl et al. 2007; Minvielle and Garreaud 2011; Urrutia and Vuille 2009; Vera et al. 2006). In the IPCC AR4 models, the future of precipitation over the Amazon depends on the model considered; some models show rainfall reductions while others show little change in rainfall. In addition, IPCC AR4 models show deficiencies in simulating present-day precipitation and its inter-annual variability over the Amazon basin (Vera et al. 2006; Dai 2006; Rojas et al. 2006; Malhi et al. 2009). Even more uncertainties arise when it is recognised that most of these models did not include some of the biogeochemical feedbacks that are at play in the Amazon basin (Rammig et al. 2010).

A closer comparison between models used in the IPCC Fifth Assessment Report (namely, AR5 and the AR4 models, IPCC 2013, 2014) highlights a weaker consensus on increased precipitation during the wet season, but a stronger consensus on a drying and lengthening of the dry season in eastern Amazonia. Significant improvements have been made from IPCC AR4 to AR5 models to capture present-day precipitation over the Amazon basin (IPCC 2013, 2014). The latter response is related to a northward shift of the boreal summer intertropical convergence zone in AR5 models, in line with a more asymmetric warming between the northern and southern hemispheres. Joetzjer et al. (2013) and Fu et al. (2013) have made a closer comparison between IPCC AR4 and AR5 models and highlight a weaker consensus on increased precipitation during the wet season, especially in western Amazonia, but a stronger consensus on drying and lengthening of the dry season, particularly in the northeastern portion of the basin. These authors have proposed that the large uncertainties that persist in the rainfall response arise from contrasted anomalies in both moisture convergence and evapotranspiration.

4.5 Regional Extremes and Impacts on Future Climate Change Scenarios

Downscaling experiments on climate change scenarios in South America have also shown a reduction of rainfall in Amazonia for 2071–2100 from the A2 high emission and B2 low emission scenarios for CO_2 (Marengo et al. 2010b, 2012b). This finding is in agreement with global projections of extremes from IPCC AR4 and AR5 projections shown in Tebaldi et al. (2006), Vera et al. (2006), Li et al. (2006) and Meehl et al. (2007), Joetzjer et al. (2013), and Fu et al. (2013). A recent study (Marengo et al. 2012b) has assessed the uncertainty and estimated the most likely future climate using the Eta-CPTEC regional model driven by four members of an ensemble of the HadCM3 model. The global model ensemble was run over the twenty-first century according to the A1B emissions scenario, but with each member having different climate sensitivity. The four members selected to drive the Eta regional model used at the Brazilian Center for Weather Forecasts and Climate Studies (CPTEC) spanned the range in the global model ensemble.

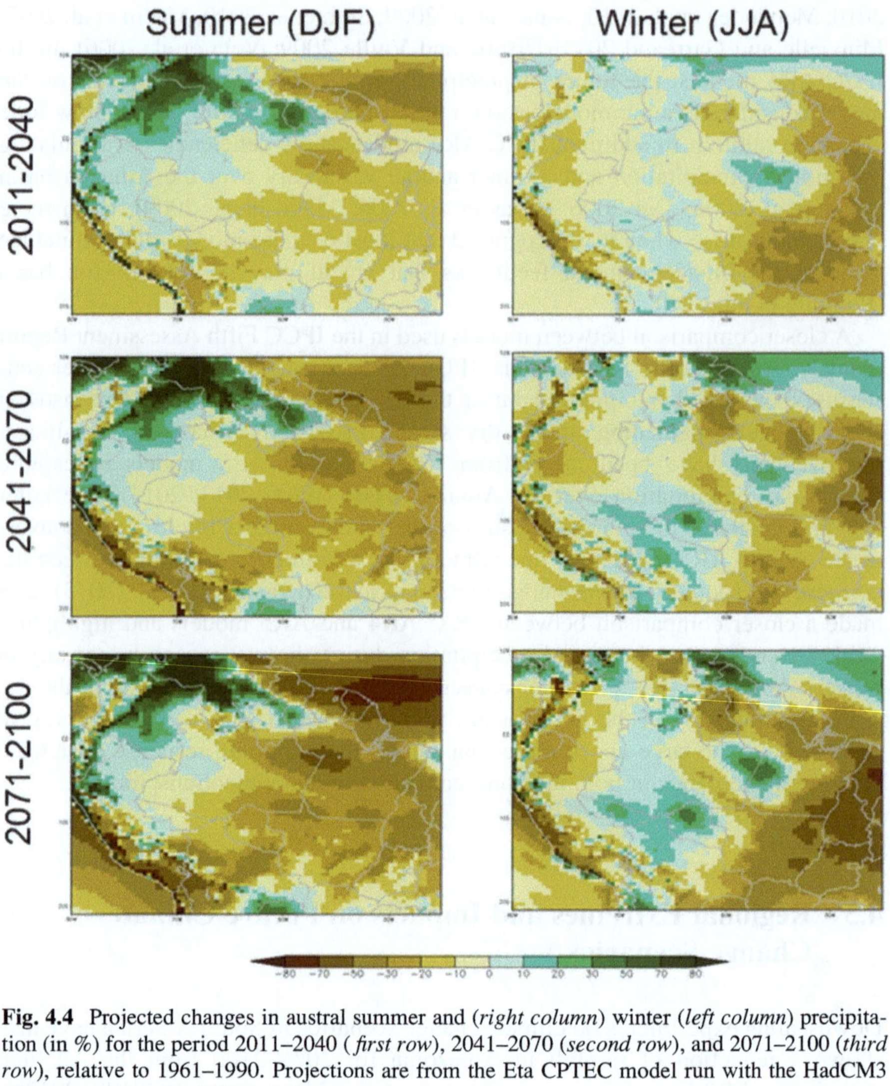

Fig. 4.4 Projected changes in austral summer and (*right column*) winter (*left column*) precipitation (in %) for the period 2011–2040 (*first row*), 2041–2070 (*second row*), and 2071–2100 (*third row*), relative to 1961–1990. Projections are from the Eta CPTEC model run with the HadCM3 global model for the A1B emission scenario

The results of the Precipitation–Evaporation (P–E) difference in the A1B downscaled scenario suggest water deficits and possible river run-off reductions in the eastern Amazon, making this region susceptible to drier conditions and droughts in the future. In the following, we show projections of mean seasonal rainfall changes for austral summer (December–February) and winter (June–August) for 2011–2040, 2041–2070, and 2071–2100 time slices relative to 1961–1990 (Fig. 4.4). In the Amazon region, the model projects large rainfall reductions (up to 40 %) mainly during the rainy season (austral summer). During the dry season (austral winter), the changes projected are small. In both seasons, the

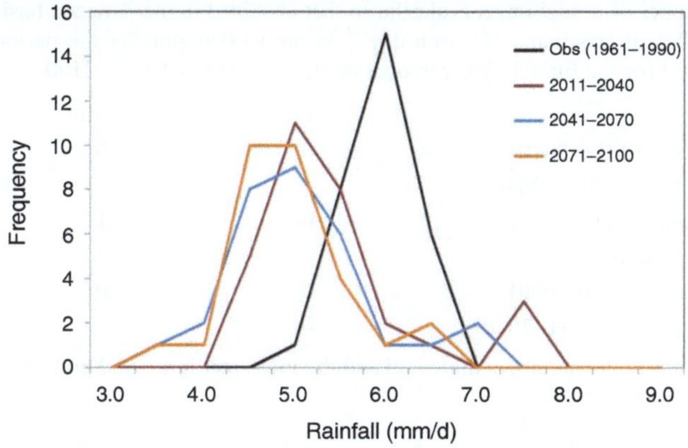

Fig. 4.5 Frequency distribution of projected mean annual rainfall in Southern Amazonia (75° W–50° W, 15° S–5° S) for the period 2011–2040 (*red line*), 2041–2070 (*blue line*), and 2071–2100 (*orange line*), based on Eta CPTEC model run with the HadCM3 global model for the A1B emission scenario. *Black line* shows the frequency for the observation from 1961 to 1990. Data are from GPCC-monitoring product available at 1.0° lat–lon gridbox area

model demonstrates that all changes are more intense after 2040, when CO_2 and temperatures are predicted to increase very rapidly (IPCC 2013).

A frequency distribution was fitted to the annual rainfall in southern Amazonia (75° W–50° W, 15° S–5° S) for both present and future (Fig. 4.5), to identify any long-term changes in its distribution patterns, as well as its changes relative to the present (1961–1990). The analysis of the frequency over various time slices for the present and future indicates some noticeable changes in rainfall distribution. These changes have some important implications regarding the critical values in the outer tails of the distributions and, consequently, the frequency of extreme rainfall events. As can be seen, the projected rainfall distribution exhibits a negative shift of the location (towards lower rainfall values) and an increase in the width of the distribution with respect to observations. Additionally, the thickening of the upper tail is indicative of a decreased frequency of extreme rainfall events since 2011–2040.

Further analyses using the Eta-CPTEC climate change projections have been carried out to analyse precipitation values and anomalies in four regions of the Amazon basin: northwest (75° W–62.5° W; 5° N–2.5° S), northeast (62.5° W–50° W; 5° N–2.5° S), west (75° W–62.5° W; 2.5° S–10° S), and east (62.5° W–50° W; 2.5° S–10° S). To identify the possibility of drought events similar to that in 2010 in the future, we use the GPCC rainfall data to determine rainfall anomalies for the June–October season for that year relative to 1961–1990, and this value is 1.4 mm day^{-1} for the west Amazon region. This is the largest negative rainfall anomaly identified for 2010, and considering this value as a reference we have identified the occurrence of droughts similar or worse than that in 2010 that could occur in the future until 2071–2100, as derived from the Eta-CPTEC regional model projections.

Table 4.2 Number of drought events similar to that observed in the Amazon basin in 2010, with rainfall anomalies of less than -1.4 mm day^{-1} in the west region for the period July–October (JASO) projected for the Eta-CPTEC regional model between 2010 and 2100

	Low	Mid	High
Number of events (2011–2040)	0	0	0
Return period (year) (2011–2040)	–	–	–
Number of events (2041–2070)	0	1	3
Return period (years) (2011–2040)	–	30	10
Number of events (2071–2099)	0	0	4
Return period (years) (2011–2040)	–	–	7.25

Projections represent the members that emulated the low-, medium-, and high-emission scenarios

Table 4.2 exhibits the number of events and return periods of droughts similar to that observed in 2010. For the period 2011–2040, no events were observed. During 2041–2070, one event is detected in the mid-emission scenario with a return period of 30 years, and three events were projected for the high-emission scenario with a return period of 10 years. In 2071–2099, four events were projected for the high-emission scenario, with a return period of 7.5 years.

From the projections shown in Table 4.2, it may be concluded that droughts similar to or worse than that of 2010 could occur more frequently in the future, especially after 2040, with shorter return periods, depending on the CO_2 emission scenario considered. Finally, it should be emphasised that these projections show uncertainties, arising from the intrinsic dynamics of the climate system. Such uncertainties are even more evident when considering factors, such as the assumptions concerning greenhouse gas emissions for the future and the coupling of the Eta-CPTEC regional model into the HadCM3 global model (Marengo et al. 2012b). In this regard, Cox et al. (2008) have suggested that the drought of 2005 was an approximately 1-in-20-year event, but will become a 1-in-2-year event by 2025 and a 9-in-10-year event by 2060 in a warmer future climate.

Projected impacts of such drying scenarios would result in a dieback effect on the forests of the Amazon basin, converting them from net absorbers to net emitters of carbon (Cox et al. 2004, 2008; Nobre and Borma 2009). However, the latest IPCC AR5 reports (2013, 2014) show that while some models have indicated drying in Amazonia and possible impacts on the forest, other models show that the Amazon forest would remain robust through the coming century. Therefore, the question remains unresolved and so does the possibility of an Amazon dieback.

4.6 Conclusions

Droughts and floods are part of the natural environmental cycles in Amazonia. While various intense droughts in the region have occurred during El Niño events (1926, 1983, 1998), we cannot generalise that El Niño is responsible for all droughts in Amazonia, nor point to La Niña as responsible for all floods in Amazonia. The tropical Atlantic also plays a major role in the rainfall regime in the region, such as in the droughts of 1963 and 2005, both occurring during non-El Niño years.

Droughts are extremely important due to the fact that they favour forest fires and can lead to tree mortality, leading to changes in the carbon release, which, in turn, switches the forest from sink to source of carbon. While the carbon release from the forest during the droughts of 2005 and 2010 was high, this release was lower during the wet years 2011–2012, as demonstrated by Lewis et al. (2011) and Gatti et al. (2014). The drought in 2010 was so severe that it affected the capacity of the forest to absorb CO_2, and analysis of the year after the rainy conditions shows that the vegetation was able absorb not only the CO_2 released by natural processes (e.g. respiration) but also the carbon emissions from human activities, including forest fires. Very low river levels during drought episodes also affect population living along the river banks and cities, as well as natural ecosystems.

In the long term, while there are no indicators of systematic increase or decrease in rainfall, we have noted inter-annual and decadal variations in rainfall towards drier or wetter conditions on shorter timescales in the recent decades. It is important also to consider rainfall variability during the wet and dry seasons, suggesting also that the onset of the rainy season and thus the length of the dry season are changing with time, with tendencies for a longer dry season in southern Amazonia since the middle 1970s (Marengo et al. 2011). New evidence on the possible role of human influences (deforestation, greenhouse gases, and aerosol release) on decadal rainfall and river variability and on droughts and floods has started to appear in the literature [See Marengo and Espinoza (2015) and references therein].

Modelling results suggest that the future climate in Amazonia may exhibit a risk of rainfall reduction in eastern Amazonia, as well as longer dry seasons in the region, while increased precipitation is projected for western Amazonia. However, uncertainties are large since some processes, such as dynamic vegetation and aerosol feedbacks, as well as the representation of El Niño and El Niño-related climate anomalies in the region, are still in development. In addition, large uncertainties in relation to the possibility of the dieback of the Amazon forest for the future still remain. Model improvements can help in assessing the risk of dangerous climate change, as well as its impacts on the functioning of the Amazon forest and its role in regional climate change.

Acknowledgements The research leading to these results has received funding from the Rede-CLIMA, the National Institute of Science and Technology for Climate Change funded by CNPq Grant Number 573797/2008-0 and FAPESP Grants 2008/57719-9, FAPESP-NERC 2011/51843-2, and FAPESP-GO AMAZON 2013/50538-7. ERW's contribution to this study was supported by the NASA TRMM program, under the direction of Dr. Ramesh Kakar.

References

Aceituno P (1988) On the functioning of the Southern Oscillation in the South American sector. Part I: surface climate. Mon Weather Rev 166:505–524

Ambrizzi T, Souza EB, Pulwarty RS (2004) The Hadley and Walker regional circulations and associated ENSO impacts on the South American seasonal rainfall. In: Diaz HF, Bradley RS (eds) The Hadley circulation: present, past and future, vol 21, 1st edn. Kluwer Academic Publishers, Netherlands, pp 203–235

Angelini IM, Coauthors (2011) On the coupling between vegetation and the atmosphere. Theor Appl Climatol 105:243–261. doi:10.1007/s00704-010-0377-5

Aragão LEOC, Malhi Y, Roman-Cuesta RM, Saatchi S, Anderson LO, Shimabukuro YE (2007) Spatial patterns and fire response of recent Amazonian droughts. Geophys Res Lett 34:L07701. doi:10.1029/2006GL028946

Arias PA, Fu R, Hoyos CD, Li W, Zhou L (2011) Changes in cloudiness over the Amazon rainforests during the last two decades: diagnostic and potential causes. Climate Dynam 37 (5–6):1151–1164

Arraut JM, Nobre CA, Barbosa H, Obregon G, Marengo J (2012) Aerial rivers and lakes: looking at large-scale moisture transport and its relation to Amazonia and to subtropical rainfall in South America. J Climate 25:543–556. doi:10.1175/2011JCLI4189.1

Betts RA, Cox PM, Collins M, Harris PP, Huntingford C, Jones CD (2004) The role of ecosystem–atmosphere interactions in simulated Amazonian precipitation decrease and forest die-back under global climate warming. Theor Appl Climatol 78:157–175. doi:10.1007/s00704-004-0050-y

Buarque DC, Clarke RT, Bulhoes Mendes CA (2010) Spatial correlation in precipitation trends in the Brazilian Amazon. J Geophys Res Atmos 115:D12108

Butt N, de Oliveira PA, Costa MH (2011) Evidence that deforestation affects the onset of the rainy season in Rondonia, Brazil. J Geophys Res 116:D11120. doi:10.1029/2010JD015174

Coelho C, Cavalcanti I, Ito R, Luz G, Santos L, Nobre CA, Marengo JA, Pezza AB (2013) As secas de 1998, 2005 e 2010. Analise climatologica. In: De Simone Borma L, Nobre CA (eds) Secas na Amazonia: Causas e Consequencias. Oficina de Textos Press, Sao Paulo, pp 89–116

Cox PM, Betts RA, Jones CD, Spall SA, Totterdell IJ (2000) Acceleration of global warming due to carbon-cycle feedbacks in a coupled climate model. Nature 408(6809):184–187

Cox PM, Betts RA, Collins M, Harris PP, Huntingford C, Jones CD (2004) Amazonian forest dieback under climate-carbon cycle projections for the 21st century. Theor Appl Climatol 78 (1–3):137–156

Cox PM, Harris PP, Huntingford C, Betts RA, Collins M, Jones CD, Jupp TE, Marengo JA, Nobre CA (2008) Increasing risk of Amazonian drought due to decreasing aerosol pollution. Nature 453:212–215. doi:10.1038/nature06960

Dai A (2006) Precipitation characteristics in eighteen coupled climate models. J Climate 19 (18):4605–4630

Espinoza JC, Guyot JL, Ronchail J, Cochonneau G, Filizola N, Fraizy P, Labat D, de Oliveira E, Julio Ordonez J, Vauchel P (2009a) Contrasting regional discharge evolutions in the Amazon basin (1974–2004). J Hydrol 375(3–4):297–311

Espinoza JC, Ronchail J, Guyot JL, Cochonneau G, Naziano F, Lavado W, de Oliveira E, Pombosa R, Vauchel P (2009b) Spatio-temporal rainfall variability in the Amazon basin countries (Brazil, Peru, Bolivia, Colombia, and Ecuador). Int J Climatol 29(11):1574–1594

Espinoza JC, Ronchail J, Guyot JL, Junquas C, Vauchel P, Lavado W, Drapeau G, Pombosa R (2011) Climate variability and extreme drought in the upper Solimões River (western Amazon Basin): understanding the exceptional 2010 drought. Geophys Res Lett 38:L13406. doi:10.1029/2011GL047862

Espinoza JC, Ronchail J, Guyot JL, Junquas C, Drapeau G, Martinez JM, Santini W, Vauchel P, Lavado W, Ordoñez JJ, Espinoza R (2012) From drought to flooding: understanding the abrupt

2010–11 hydrological annual cycle in the Amazonas River and tributaries. Environ Res Lett 7 (2):024008. doi:10.1088/1748-9326/7/2/024008

Espinoza JC, Ronchail J, Frappart F, Lavado W, Santini W, Guyot JL (2013) The major floods in the Amazonas River and tributaries (Western Amazon Basin) during the 1970–2012 period: a focus on the 2012 flood. J Hydrometeorol 14(3):1000–1008

Espinoza JC, Marengo JA, Ronchail J, Molina J, Noriega L, Guyot JL (2014) The extreme 2014 flood in South-Western Amazon basin: the role of tropical-subtropical South Atlantic SST gradient. Environ Res Lett 9:124007. doi:10.1088/1748-9326/9/12/124007

Fernandes K et al (2011) North tropical Atlantic influence on western Amazon fire season variability. Geophys Res Lett 38(12):L12701

Figueroa S, Nobre CA (1990) Precipitation distribution over central and western tropical South America. Climanalise 6:36–40

Filizola N, Latrubesse EM, Fraizy P, Souza R, Guimarães V, Guyot J-L (2014) Was the 2009 flood the most hazardous or the largest ever recorded in the Amazon? Geomorphology 215:99–105. doi:10.1016/j.geomorph.2013.05.028

Fu R, Coauthors (2013) Increased dry-season length over southern Amazonia in recent decades and its implication for future climate projection. doi:10.1073/pnas.1302584110/-/DCSupplemental. Online at www.pnas.org/lookup/suppl/doi:10

Fu R, Zhu B, Dickinson RE (1999) How do atmosphere and land surface influence seasonal changes of convection in the tropical Amazon? J Climate 12:1306–1321

Gatti LV et al (2014) Drought sensitivity of Amazonian carbon balance revealed by atmospheric measurements. Nature. doi:10.1038/nature12957

Gloor M, Brienen RJW, Galbraith D, Feldpausch TR, Schöngart J, Guyot J-L, Espinoza JC, Lloyd J, Phillips OL (2013) Intensification of the Amazon hydrological cycle over the last two decades. Geophys Res Lett 40. doi:10.1002/grl.50377

Huntingford C, Co-authors (2013) Simulated resilience of tropical rainforests to CO2-induced climate change. Nature. doi:10.1038/NGEO1741, Published online: 10 March 2014

IPCC (2013) Summary for policymakers. In: Field C, Barros VR, Dokken DJ, Mach KJ, Mastrandrea MD, Bilir TE, Chatterjhee M, Ebi KL, Estrada YO, Genova RC, Girma B, Kissel ES, Levy AN, MacCracken S, Mastrandrea PR, White LL (eds) Climate change 2013: impacts, vulnerability and adaptation. Part A: global and sectoral aspects. Contribution of working group 2 to the fifth assessment report of the Intergovernmental Panel on Climate Change. Cambridge University Press, Cambridge and New York, NY, pp 1–32

IPCC (2014) Climate change 2014: impacts, adaptation, and vulnerability. In: Field CB, Barros VR, Dokken DJ, Mach KJ, Mastrandrea MD, Bilir TE, Chatterjhee M, Ebi KL, Estrada YO, Genova RC, Girma B, Kissel ES, Levy AN, MacCracken S, Mastrandrea PR, White LL (eds) Part A: global and sectoral aspects. Contribution of working group II to the fifth assessment report of the Intergovernmental Panel on Climate Change. Cambridge University Press, Cambridge and New York, NY, p 1132

Jenkins HS (2009) Amazon climate reconstruction using growth rates and stable isotopes of tree ring cellulose from the Madre De Dios Basin, Peru. Dissertation, Department of Earth and Ocean Sciences, Duke University

Joetzjer EH, Douville C, Delire P, Ciais P (2013) Present-day and future Amazonian precipitation in global climate models: CMIP5 versus CMIP3. Clim Dyn 41:2921–2936. doi:10.1007/s00382-012-1644-1

Lewis SL, Brando PM, Phillips OL, van der Heijden GMF, Nepstad D (2011) The 2010 Amazon drought. Science 331:554

Li W, Fu R (2004) Transition of the large-scale atmospheric and land surface conditions from the dry to the wet season over Amazonia as diagnosed by the ECMWF reanalysis. J Climate 17:2637–2651

Li WH, Fu R, Juarez RIN, Fernandes K (2006) Causes of recent changes of rainfall variabilities and implications to the future climate in the Amazon region. Phil Trans R Soc B 363:1767–1772. doi:10.1098/rstb.2007.0022

Makaireva AM, Gorshkov VG, Sheil D, Nobre AD, Li BL (2013) Where do winds come from? A new theory on how water vapor condensation influences atmospheric pressure and dynamics. Atmos Chem Phys 13:1039–1056. doi:10.5194/acp-13-1039-2013

Malhi Y et al (2009) Exploring the likelihood and mechanism of a climate-change-induced dieback of the Amazon rainforest. Proc Natl Acad Sci USA 106(49):20610–20615

Marengo JA (2004) Inter-decadal variability and trends in rainfall in the Amazon basin. Theor Appl Climatol 78:79–96

Marengo JA (2006) On the hydrological cycle of the Amazon basin: a historical review. Rev Bras Meteorol 21:1–19

Marengo JA (2009) Long-term trends and cycles in the hydrometeorology of the Amazon basin since the late 1920s. Hydrol Processes 23(22):3236–3244. doi:10.1002/hyp.7396

Marengo JA, Espinoza JC (2015) Extreme seasonal droughts and floods in Amazonia: causes, trends and impacts. Int J Climatol doi:10.1002/joc.4420

Marengo JA, Tomasella J, Uvo C (1998) Long-term streamflow and rainfall fluctuations in tropical South America: Amazonia, eastern Brazil and northwest Peru. J Geophys Res 103:1775–1783 (American Geophysical Union-Was)

Marengo JA, Nobre CA, Tomasella J, Oyama MD, Oliveira GS, de Oliveira R, Camargo H, Alves LM, Brown IF (2008a) The drought of Amazonia in 2005. J Climate 21:495–516. doi:10.1175/2007JCLI1600.1

Marengo JA, Nobre CA, Tomasella J, Cardoso MF, Oyama MD (2008b) Hydro-climatic and ecological behaviour of the drought of Amazonia in 2005. Philos Trans R Soc Lond Ser B Biol Sci 363:1773–1778. doi:10.1098/rstb.2007.0015

Marengo JA, Tomasella J, Soares W, Alves LM, Nobre C (2010a) Extreme climatic events in the Amazon basin: climatological and hydrological context of previous floods. Theor Appl Climatol 85:1–13

Marengo JA, Ambrizzi T, da Rocha R, Alves L, Cuadra S, Valverde M, Torres R, Santos D, Ferraz S (2010b) Future change of climate in South America in the late twenty-first century: intercomparison of scenarios from three regional climate models. Climate Dynam 35(6):1073–1097

Marengo JA, Tomasella J, Alves LM, Soares W, Rodriguez DA (2011) The drought of 2010 in the context of historical droughts in the Amazon region. Geophys Res Lett 38:1–5

Marengo JA et al (2012a) Recent developments on the South American monsoon system. Int J Climatol. doi:10.1002/joc.2254

Marengo JA, Chou SC, Kay G, Alves LM, Pesquero JF, Soares WR, Santos DC, Lyra AA, Sueiro G, Betts R, Chagas DJ, Gomes JL, Bustamante JF, Tavares P (2012b) Development of regional future climate change scenarios in South America using the Eta CPTEC/HadCM3 climate change projections: climatology and regional analyses for the Amazon, São Francisco and the Paraná River basins. Climate Dynam 38(9–12):1829–1848

Marengo JA, Borma LS, Rodriguez DA, Pinho P, Soares WR, Alves LM (2013a) Recent extremes of drought and flooding in Amazonia: vulnerabilities and human adaptation. Am J Climate Change 2:87–96. doi:10.4236/ajcc.2013.22009, Published Online June 2013 (http://www.scirp.org/journal/ajcc)

Marengo JA, Alves LM, Soares WR, Rodriguez DA, Camargo H, Paredes M, Diaz Pablo A (2013b) Two contrasting seasonal extremes in tropical South America in 2012: flood in Amazonia and drought in Northeast Brazil. J Climate 26(22):9137–9154

Meade RH, Rayol JM, Conçeicão JM, Navidade JRG (1991) Backwater effects in the Amazon basin of Brazil. Environ Geol Water Sci 18(2):105–114

Meehl GA, Stocker TF, Collins WD, Friedlingstein P, Gaye AT, Gregory JM, Kitoh A, Knutti R, Murphy JM, Noda A, Raper SCB, Watterson IG, Weaver AJ, Zhao ZC (2007) Global climate projections. In: Solomon S, Qin D, Manning M, Chen Z, Marquis M, Avery KB et al (eds) Climate change 2007: the physical science basis. Contribution of working group I to the fourth assessment report of the Intergovernmental Panel on Climate Change. Cambridge University Press, Cambridge and New York, NY, pp 747–845

Meggers B (1994) Archeological evidence for the impact of Mega-El Niño events on Amazonia during the past two millennia. Clim Change 28:321–338. doi:10.1007/BF01104077

Mendes D, Marengo JA (2010) Temporal downscaling: a comparison between artificial neural network and autocorrelation techniques over the Amazon basin in present and future climate change scenarios. Theor Appl Climatol 100(3–4):413–421

Menendez CG, de Castro M, Boulanger J, D'Onofrio A, Sanchez E, Soerensson AA, Blazquez J, Elizalde A, Jacob D, Le Treut H, Li ZX, Nunez MN, Pessacg N, Pfeiffer S, Rojas M, Rolla A, Samuelsson P, Solman SA, Teichmann C (2010) Downscaling extreme month-long anomalies in southern South America. Clim Change 98(3–4):379–403

Minvielle M, Garreaud RD (2011) Projecting rainfall changes over the South American altiplano. J Climate 24(17):4577–4583

Moran EF, Adams R, Bakoyéma B, Fiorini S, Boucek B (2006) Human strategies for coping with El Niño related drought in Amazônia. Clim Change 77:343–361. doi:10.1007/s10584-005-9035-9

Myneni RB et al (2007) Large seasonal swings in leaf area of Amazon rainforests. Proc Natl Acad Sci 104(12):4820–4823

Nepstad DC, Veríssimo A, Alencar A, Nobre C, Lima E, Lefebvre P, Schlesinger P, Potterk P, Moutinho P, Mendoza E, Cochrane M, Brooks V (1999) Large-scale impoverishment of Amazonian forests by logging and fire. Nature 398:505–508

Nobre CA, Borma LS (2009) Tipping points' for the Amazon Forest. Curr Opin Environ Sustain 1:28–36. doi:10.1016/j.cosust.2009.07.003

Nuñez MN, Solman SA, Fernanda Cabré M (2009) Regional climate change experiments over southern South America. II: climate change scenarios in the late twenty-first century. Climate Dynam 32(7–8):1081–1095

Pinho PF, Marengo JA, Stafford-Smith M (2014) Complex socio-economical dynamics driven by extreme events in the Amazon. Reg Environ Change. doi:10.1007/s10113-014-0659-z

Rammig A, Jupp TE, Thonicke K, Tietjen B, Heinke J, Ostberg S, Lucht W, Cramer W, Cox P (2010) Estimating the risk of Amazonian forest dieback. New Phytol 187:694–706. doi:10.1111/j.1469-8137.2010.03318.x

Rodriguez DA, Tomasella J, Linhares C (2010) Is the forest conversion to pasture affecting the hydrological response of Amazonian catchments? Signals in the Ji-Paran'a Basin. Hydrol Processes 24:1254–1269

Rojas M, Seth A, Rauscher SA (2006) Relationship between precipitation and moisture flux changes in the SRES A2 scenario for the South American monsoon region. CLARIS News 3:14–20

Ronchail J, Cochonneau G, Molinier M, Guyot JL, de Miranda Chaves AG, Guimarães V, de Oliveira E (2002) Interannual rainfall variability in the Amazon basin and sea-surface temperatures in the equatorial Pacific and tropical Atlantic Oceans. Int J Climatol 22:1663–1686. doi:10.1002/joc.815

Ronchail J, Bourrel L, Cochonneau G, Vauchel P, Phillips L, Castro A, Guyot J-L, de Oliveira E (2005) Climate and inundations in the Mamoré basin (South-Western Amazon—Bolivia). J Hydrol 302:223–238

Ronchail J, Guyot J-L, Espinoza JC, Fraizy P, Cochonnieau G, de Oliveira E, Filizola N, Ordoñez J (2006) Impact of the Amazon tributaries on flooding in Obidos. IAHS Publ 308:220–225

Rudolf B, Beck C, Grieser J, Schneider U (2005) Global precipitation analysis products. Global Precipitation Climatology Centre (GPCC), DWD, Germany, pp 1–8

Salati E, Marquez J, Molion LC (1978) Origem e distribuiçao das chuvas na Amazônia. Interciencia 3:200–205

Saleska SR, Didan K, Huete AR, da Rocha HR (2007) Amazon forests green-up during 2005 drought. Science 318:612. doi:10.1126/science.1146663

Sampaio G, Nobre C, Costa MH, Satyamurty P, Soares-Filho BS, Cardoso M (2007) Regional climate change over eastern Amazonia caused by pasture and soybean cropland expansion. Geophys Res Lett 34(17):L17709

Satyamurty P, de Castro AA, Tota J, da Silva Gularte LE, Manzi AO (2010) Rainfall trends in the Brazilian Amazon in the past eight decades. Theor Appl Climatol 99:139–148. doi:10.1007/s00704-009-0133-x

Satyamurty P, Wanzeler da Costa CP, Manzi AO (2013) Moisture sources for the Amazon Basin: a study of contrasting years. Theor Appl Climatol 111:195–209. doi:10.1007/s00704-012-0637-7

Sena JA, Beser de Deus LA, Fretas MAV, Costa L (2012) Extreme events of droughts and floods in Amazonia: 2005 and 2009. Water Resour Manage 26:1665–1676. doi:10.1007/s11269-012-9978-3

Seth A, Rojas M, Rauscher SA (2010) CMIP3 projected changes in the annual cycle of the South American monsoon. Clim Change 98(3–4):331–357

Sheffield J, Wood EF (2011) Drought: past problems and future scenarios. Earthscan Ltd, London, Washington DC, p 211

Sombroek W (2001) Spatial and temporal patterns of Amazon rainfall: consequences for the planning of agricultural occupation and the protection of primary forests. Ambio 30 (7):388–396

Sternberg H (1968) Man and environmental change in South America. In: Fittkau EJ, Illies J, Klinge H, Schwabe GH, Sioli H (eds) Biogeography and ecology in South America. Dr. W. Junk N.V. Publishers, The Hague, pp 413–445

Sternberg H (1987) Aggravation of floods in the Amazon River as a consequence of deforestation? Geogr Ann 69:201–219

Tebaldi JM, Arblaster K, Hayhoe T, Meehl GA (2006) Going to the extremes: an intercomparison of model-simulated historical and future changes in extreme events. Clim Change 79:185–211

Tomasella J, Borma LS, Marengo JA, Rodriguez DA, Cuartas LA, Nobre CA, Prado MCR (2011) The droughts of 1996–1997 and 2004–2005 in Amazonia: hydrological response in the river main-stem. Hydrol Processes 25:1228–1242. doi:10.1002/hyp.7889

Tomasella J, Borma LS, Marengo JA, Rodriguez DA, Cuartas LA, Nobre CA, Prado MCR, Coauthors (2013) The droughts of 1997 and 2005 in Amazonia: floodplain hydrology and its potential ecological and human impacts. Clim Change 116:723–746. doi:10.1007/s10584-012-0508-3

Urrutia R, Vuille M (2009) Climate change projections for the tropical Andes using a regional climate model: temperature and precipitation simulations for the end of the 21st century. J Geophys Res Atmos 114:D02108

Vale R, Filizola N, Souza R, Schongart J (2011) Acheia de 2009 na Amazonia Brasileira. Rev Bras Geocienc 41(4):577–586

Vera CS, Silvestri G, Liebmann B, González P (2006) Climate change scenarios for seasonal precipitation in South America from IPCC-AR4 models. Geophys Res Lett 33:2–5. doi:10.1029/2006GL025759

Wang G, Sun S, Mei R (2011) Vegetation dynamics contributes to the multi-decadal variability of precipitation in the Amazon region. Geophys Res Lett 38:L19703

Williams E, Dall'Antonia A, Dall'Antonia V, de Almeida J, Suarez F, Liebmann B, Malhado ACM (2005) The drought of the century in the Amazon basin: an analysis of the regional variation of rainfall in South America in 1926. Acta Amazon 35(2):231–238. doi:10.1590/S0044-59672005000200013

Yoon JH, Zeng N (2010) An Atlantic influence on Amazon rainfall. Climate Dynam 34:249–264. doi:10.1007/s00382-009-0551-6

Zemp DC, Schleussner CF, Barbosa HMJ, van der Ent RJ, Donge JF, Heink J, Sampaio G, Rammig A (2014) On the importance of cascading moisture recycling in South America. Atmos Chem Phys 14:13337–13359

Zeng N, Yoon JH, Marengo JA, Subramaniam A, Nobre CA, Mariotti A, Neelin D (2008) Causes and impacts of the 2005 Amazon drought. Environ Res Lett 3. doi:10.1088/1748-9326/3/1/014002

Zhang Y, Fu R, Yu H, Qian Y, Dickinson R, Silva Dias MAF, da Silva Dias PL, Fernandes K (2009) Impact of biomass burning aerosol on the monsoon circulation transition over Amazonia. Geophys Res Lett 36:L10814

Part III
Carbon Balance

Chapter 5
The Amazon Carbon Balance: An Evaluation of Methods and Results

John Grace

5.1 Introduction

Scientific interest in the carbon balance of tropical biomes remains high because of the crucial effect that carbon dioxide released from deforestation may be having on the climate system and the fact that deforestation is occurring before its implications are fully understood. This is a matter of scientific, economic, and ethical concern, especially when associated with the many related impacts of tropical deforestation, i.e. on biodiversity, on changes in the water and energy balance, and on atmospheric chemistry.

Development in the Amazon basin has been mostly driven by a perceived need to clear carbon-rich forests and woodlands to make way for cattle ranching and various other productive forms of agriculture, most of which have very low stocks of carbon (Fearnside 2002, and Chap. 16). Development of the Amazon by building highways led to an increase in population of non-indigenous people from 2 million in the 1960s to 20 million by the year 2000 (Laurence and Fearnside 2002), but the basin still has a very low population density, around 3.3 people km^{-2}. Agricultural production in Amazonian countries is a major component of the economy: exports include coffee, soybeans, wheat, rice, corn, sugar, cocoa, citrus, beef, and ethanol (Jarvis et al. 2011). In the case of Brazil, total agricultural production accounts for 5.2 % of a GDP equal to 2.2×10^{12} dollars, although not all of this is from the Amazon basin. The true value of the forest, its biodiversity, and the environmental services cannot be calculated (although some have tried).

The questions relating to carbon that have been addressed in the last 20 years of research may be succinctly stated as follows:

J. Grace (✉)
University of Edinburgh, School of GeoSciences, Edinburgh, EH9 3FF, UK
e-mail: jgrace@ed.ac.uk

© Springer-Verlag Berlin Heidelberg 2016

L. Nagy et al. (eds.), *Interactions Between Biosphere, Atmosphere and Human Land Use in the Amazon Basin*, Ecological Studies 227,
DOI 10.1007/978-3-662-49902-3_5

1. What is the carbon balance of the Amazon, and how does it change from year to year according to climatological variations?
2. What is the seasonality in the carbon flux between the land and atmosphere?
3. How do changes in land use affect the carbon balance?

These general questions are currently being addressed for all tropical regions (Ciais et al. 2011; Gloor et al. 2012; Valentini et al. 2013; Patra et al. 2013; Grace et al. 2014). However, the science is more mature in the case of the Amazon, and we are closer to 'answers' here, simply because of the strong institutional developments and international collaborations that were stimulated as a result of the LBA (Keller et al. 2004).

The approach to answering these questions is scale dependent. The carbon budget of the Amazon basin is influenced primarily by the balance between photosynthesis and respiration, which at one level can be investigated at the ecosystem scale using plot-based approaches. Typically, these plots are small, from 0.001 to 1 km^2. But focusing on a few small-scale plots fails to address regional-scale carbon budget and misses processes such as fire, cultivation, and evasion of CO_2 from wetlands and rivers. For broad integration of all processes, the carbon dioxide fluctuations in the planetary boundary layer over large spatial scales may be measured and interpreted.

The difference between these two approaches is crucial. At the ecosystem scale, a 'site' for study is chosen by researchers to represent a particular land use where intensive studies may be made. This is more likely to provide insights into the environmental controls and the physiological processes, but many such samples, which must be representative, are required before anything conclusive can be said about the bigger picture that the region as a whole presents. This requires a further step of integration by the use of assumptions, models, and remotely sensed data on land surface cover. The atmospheric approach on the other hand is capable of integrating directly over all the land and all the processes.

These two measurement strategies are complementary, as recognised explicitly in other research programmes where they have been called 'bottom-up' and 'top-down', respectively (Dolman et al. 2008). They represent two independent approaches, and one test of whether the correct answers have been achieved is whether they produce the same conclusions.

5.2 Evaluation of Methods

5.2.1 Bottom-Up Estimates

5.2.1.1 Biomass Plots

The mass of any cone-like structure is geometrically related to diameter, height, and density by an allometric relationship, and so the biomass of a tree (a collection of long cones) can be estimated in principle from simple field measurements (Brown and Lugo 1990; Chave et al. 2005). In practice, the tree is not composed of perfect cones and so calibrations have to be made from laborious gravimetric measurements made on destructively sampled trees. The biomass can then be roughly determined from the diameter of the stem at 1.3 m, called 'the diameter at breast height' and abbreviated DBH, but it is more accurately found if both biomass and height are known. For the best determinations, the stem density also needs to be taken into account (Chave et al. 2005). The carbon content of trees is close to 50 % (Thomas and Martin 2012 found it to vary from 42 to 52 %), and so most researchers assume that the carbon content is half the dry biomass. Thus, repeated measurements of trees at permanent sample plots, typically every few years, can be an effective method to track changes in carbon stored at any forested site. If the sites are well chosen, revisited over a long period (every few years), and grouped to form a network (Malhi et al. 2002), it becomes possible to infer an important part of the tropical carbon balance.

Using these methods, it was demonstrated that many old-growth rainforest sites (sometimes called primary, virgin, primeval, or late seral forest) are accumulating biomass, both in the tropics (Phillips et al. 1998; Lewis et al. 2009) and in the temperate and boreal regions (Luyssaert et al. 2008). This is quite contrary to the classical assumption that such forests should reach an equilibrium where the biomass is constant (Odum 1969) and prompts the general question of whether trees are benefiting from long-term fertilisation effect of elevated CO_2 (Körner and Arnone 1992; Lloyd and Farquhar 1996; Clark 2002) and the specific question of whether the hypothesis should be tested with a large-scale CO_2 fertiliser experiment in the Amazon basin (Tollefson 2013).

Further measurements at such plots, involving collection of litter and measurement of CO_2 fluxes using chambers attached to the soil, can provide a comprehensive picture of the main components of carbon flow in any ecosystem (Malhi et al. 2009). These components are the photosynthetic input, termed the gross ecosystem productivity (GPP); the autotrophic and heterotrophic respiration R_a and R_h; and the net primary productivity (NPP). The overall carbon balance of the site is termed the net ecosystem productivity (NEP), which is the difference between the photosynthetic inputs (GPP) and the respiratory 'losses', i.e. R_a plus R_h. Many such datasets from the Amazon now exist. A selection is represented in Fig. 5.1: the series of carbon-flow plots from east to west across the basin shows very clearly a decline in biomass across this east–west progression, although GPP declines only at the driest site (at the end of the series, Fig. 5.1f); NPP, intriguingly, is not affected.

Sample plots are also important in estimating the impact of land use change by making comparison of the carbon content before and after a transition or comparison of the carbon contents in sites under different management practices which happen to be close to each other (Don et al. 2011; Ziegler et al. 2012). These investigations are very important if we are to properly quantify the impact on carbon stocks of changing from one land cover to another. The process of equilibration following deforestation takes time and involves changes in the below-

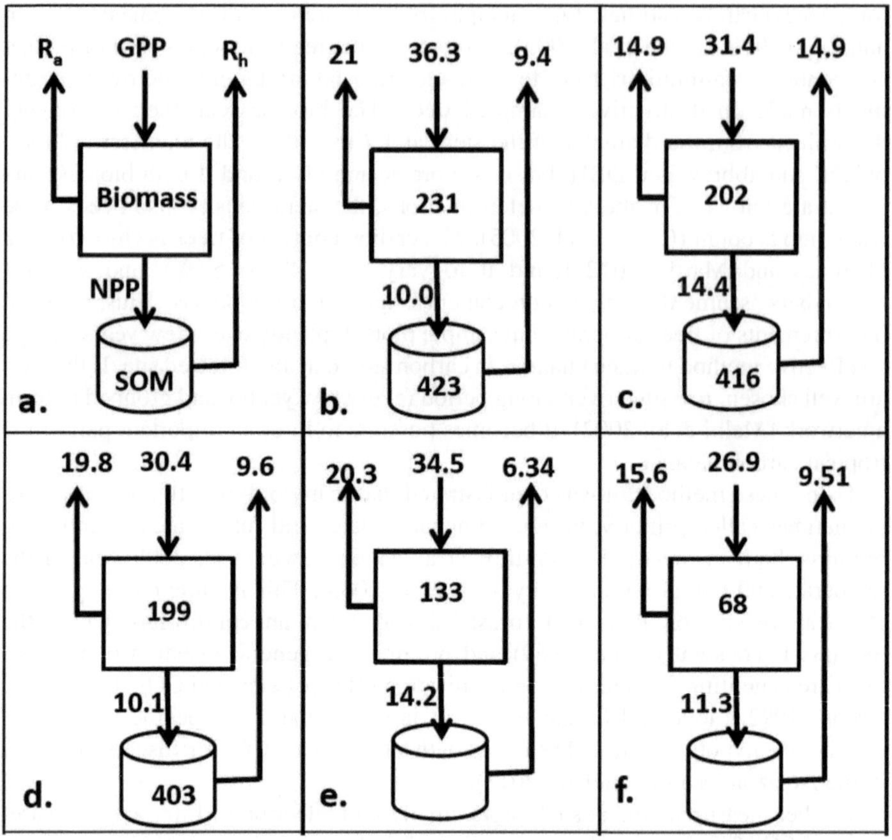

Fig. 5.1 Carbon flows inferred from plot-based methods as described by Malhi et al. (2009), arranged in order from east to west across the Amazon basin: (**a**) general scheme; (**b**) Caxiuanã; (**c**) Tapajos; (**d**) Manaus; (**e**) Tambopata; (**f**) dry forest near Santa Cruz in Bolivia. *Sources* of information: (**b, c, d**) are from Malhi et al. (2009), (**e**) is from Malhi et al. (2014), and (**f**) is from Araujo-Murakami et al. (2014). Fluxes (*arrows*) are in tC ha^{-1} annum^{-1}; stocks (*boxes* and *tubs*) are in tC ha^{-1}

ground carbon stocks which are hard to measure, and even when measured they are hard to upscale to larger areas (Houghton et al. 2012).

The main criticisms of estimating carbon fluxes from biomass plots are that (i) the sites may not have been representative, (ii) observer error can influence the measurement of stem diameter especially when different researchers are involved in data collection on different occasions, and when trees have rough surfaces and buttresses, (iii) changes in soil organic matter (SOM) are unlikely to be recorded as the sample size required to resolve small changes in SOM is larger than may practically be obtained, and (iv) sample plots of one hectare are too small and too much influenced by the death of large trees to be statistically robust. Most of these criticisms have been considered (Clark 2001; Phillips et al. 2002; Houghton et al. 2009; Gloor et al. 2009; Lloyd et al. 2009), but not all studies have been well enough designed to avoid error altogether (Bowman et al. 2014).

5.2.1.2 Eddy Covariance Sites

Eddy covariance measurements of CO_2 and H_2O fluxes were introduced in the 1990s as soon as suitably fast-responding CO_2 analysers became available. The technique is a relatively direct method to measure carbon fluxes over forests, integrating all gas exchange processes (photosynthesis, autotrophic and heterotrophic respiration, evasion from wetlands), and can in principle be extended to measurement of methane. Moreover, many thousands of trees are effectively sampled.

The technique relies on simultaneous measurement of the instantaneous vertical component of wind w' and CO_2 concentration c' over a large and uniform expanse of vegetation. In principle, the flux F can be estimated from a time series of the product of w' and c':

$$F = \overline{w'c'} + \frac{\partial}{\partial t} \int_0^h c(z)\mathrm{d}z$$

where over-bar denotes the mean of $w'c'$ over a sampling period (typically 0.5–2 h) and the second term in the equation adjusts the result to take into account the fact that carbon dioxide exchanged by plant and soil does not immediately appear at the top of the canopy, i.e. the second term represents storage which should be included if we wish to reveal more accurately the physiological activity of the vegetation.

When sensors are mounted well above the canopy on a tower, the system samples fluxes over the landscape in a variety of meteorological conditions at a scale of 0.05–1 km^2 and enables net CO_2 fluxes to be calculated directly and continuously on an hour-to-hour basis, thus giving important insights into environmental control of the net flux. With a few assumptions, the downward flux (photosynthesis) may be separated from the upward flux (respiration). Site requirements are however severe; for accurate eddy flux data the site should be homogenous and flat over a considerable distance upwind. A further requirement is that the air at sensor height is fully turbulent, a condition which is usually not satisfied for long periods at night when stably stratified meteorological conditions occur. When the requirements are not satisfied, a bias in F is very likely to be introduced, and when data are accumulated over an entire year a substantial error may appear in the annual estimate of carbon balance (Kruijt et al. 2004). The errors inherent in the eddy covariance technique have been widely discussed (Kruijt et al. 2004; Aubinet et al. 2000), and this discourse has resulted in standard protocols for filtering and correcting the data (Aubinet et al. 2012). However, critics of the eddy covariance technique have raised doubts about whether the annual flux from eddy covariance can ever be reliable, especially in tropical forests where site conditions are often less favourable than those in other ecosystems, and where regular maintenance is not always possible. Today, eddy covariance towers are generally established in association with biomass plots, thus providing the possibility of comparing the carbon fluxes calculated from the two independent methods. In cases where this has been tried, the agreement has been satisfactory (Malhi et al. 2009).

5.2.2 *Top-Down Estimates*

5.2.2.1 Measurements from Air-Borne Platforms

Fluxes over large areas of vegetation may be inferred from the analysis of vertical profiles of CO_2 concentration within the planetary boundary layer (PBL), as first illustrated with small datasets by Culf et al. (1999), Chou et al. (2002), Lloyd et al. (2007), and Gatti et al. (2010). More recently, Gatti et al. (2014) have utilised the technique over a large part of the entire Amazon basin, making use of profiles at different distances along the path of the air mass as it sweeps in from the Atlantic and travels across the basin:

$$F = \frac{\int_{h=100}^{h=5000} (C_1 - C_2)\mathrm{d}z}{t}$$

where F is the flux of CO_2; the profiles have been obtained (in the simple case) at two different points along the air flow (1,2); the concentrations C_1 and C_2 have been measured at heights from a few hundred to several thousand metres. The profiles are subtracted to reveal how much the air has become depleted or enriched in CO_2 during its journey; t is the time taken for the journey from point 1 to point 2. Gatti et al. (2014) worked with data from a small aircraft that descended from 4400 m to about 300 m a.s.l. and used four stations from east to west, essentially tracking the air from a 'control' site over the Atlantic Ocean until it reached the south-western part of the so-called Legal Amazon in Brazil.

The cost of such a large measurement programme at full Amazonian scale is high, and there can sometimes be difficulties with scheduling regular flights of research aircraft, partly due to vagaries of the weather. The weather dependency of flying introduces a possible systematic bias. In future, use may be made of commercial aircraft as regular 'ships of convenience' (Matsueda and Inoue 1996; Stephens et al. 2007; Machida et al. 2008).

5.2.2.2 Measurements from the Global Atmospheric Network

Since the early work of Bolin and Keeling (1963), the atmospheric research community has been establishing a global network of stations where concentrations of CO_2 are measured with maximum possible accuracy (usually better than

0.05 ppm) to yield data from which the distribution and strength of sources and sinks of carbon may be inferred through the 'atmospheric inversion' calculation developed by Enting and Mansbridge (1989) and Tans et al. (1989). Some authors have been bold enough to use this approach to suggest carbon balances of discrete regions including South America (Rödenbeck et al. 2003). However, in view of the poor coverage of the tropical region in the network, reporting is mostly confined to broad latitudinal zones (Gurney et al. 2002; Peylin et al. 2013).

Rödenbeck et al. (2006) have detailed the sources of error in this approach. They included uncertainties in the atmospheric transport model arising from vertical motion of the air and experimental error in measuring the very small concentration differences that arise in a well-mixed atmosphere. The main weakness of this approach is that there are too few stations, and the tropical regions are very poorly represented in the network. One possibility in the future is that the network will be used to calibrate worldwide observations on column-average CO_2 and CH_4 made from satellite, using variations in the short-wave infrared irradiance emitted from the Earth's surface. Satellite-based observations are currently much less accurate than those made from well-calibrated physical instrumentation near the ground, but recent data from GOSAT (http://www.gosat.nies.go.jp/en/) do seem to be well correlated with ground-based data (Cogan et al. 2012). However, unlike the case involving aircraft flights (see above) the concentrations seen are not only those of the well-mixed PBL, which carries the signal of interest, but also those aloft. Thus, distant signals, for example of biomass burning, may confuse the interpretation.

5.2.3 The Role of Satellite Remote Sensing

Satellite remote sensing is crucial for detecting the change in land surface cover, and especially deforestation. Satellite observations of land use change in the tropics were first used in the late 1970s for the detection of deforestation in Brazil using Landsat imagery (INPE 2003). The work continues to assist policy development in Brazil (see e.g. Chap. 15). Satellite missions widely used by the research community include the European SPOT (Système Pour l'Observation de la Terre) series of satellites, commencing with SPOT 1 (1986) and leading to SPOT 6 (launched 2012); ENVISAT which flew the moderate resolution imaging spectrometer MERIS from 2002 to 2012; the NOAA advanced very high resolution radiometer (AVHRR) from 1978 to the present; and NASA's moderate-resolution imaging spectroradiometer (MODIS) on board the Terra and Aqua satellites from 2000 to the present. Data from these satellites have provided valuable indications of the global rate of decline of forest cover (Achard et al. 2002; DeFries et al. 2002; Asner et al. 2010; Mayaux et al. 2013; Hansen et al. 2013). Knowledge of the deforestation rate can be converted into an estimate of carbon flux by making an assumption about the carbon stored per area and the extent to which it is oxidised when fired.

In recent years, satellite remote sensing has gone a stage further, from simply detecting forest versus non-forest towards the estimation of biomass. The most important technical development may be the deployment of active radar remote sensing from space. Because radar sensors can 'see' the land surface even at night and when there is cloud cover, much more data can be acquired than with optical sensors. Moreover, radar penetrates the forest canopy to an extent which depends on its wavelength, and the back-scattered signal provides information, which after calibration against ground-based observations can yield an estimate of the amount of biomass per area of land (Quegan et al. 2000; Le Toan et al. 2011). From 2006 to 2011, the Japanese *Advanced Land Observing Satellite* (ALOS) carried a synthetic aperture radar sensor (PALSAR, the phased array type L-band synthetic aperture radar) which has been used to map the biomass distribution of regions of Africa (Mitchard et al. 2009; Ryan et al. 2012).

The second very recent development is the use of space-borne LiDAR to measure the height of the vegetation and thus to estimate biomass from ground-based calibration data. The NASA satellite ICESat, designed primarily to measure the changing mass of polar ice sheets using LiDAR, flew from 2003 to 2009 and provided point estimates of tropical forest mass across the tropics, which were spatially extrapolated to prepare the first pantropical maps of above-ground carbon using ancillary full-cover datasets (Saatchi et al. 2011; Baccini et al. 2012). Both studies used the ICESat data in combination with remotely sensed information on forest cover to model and map the spatial distribution of biomass across three continents for the early 2000s.

It is clear that both Radar and LiDAR will be widely used for sensing forest biomass. In the future, there will be satellites designed specifically for monitoring biomass carbon, for example the European Space Agency's Biomass satellite mission, due to launch in 2020. This will solve some of the measurement problems, but estimation of biomass in very dense forest will still be hard to achieve because the back-scattered signal saturates when very high biomass is reached.

5.3 Evaluation of Results, with Emphasis on Uncertainties Arising from Methods

5.3.1 What Is the Carbon Balance of the Amazon Basin?

A major difficulty in speaking about the Amazon basin is that most of the available studies have been in Brazil, yet parts of the 6,915,000 km^2 basin are in other countries as follows: Peru (17 %), Bolivia (11 %), and Colombia (5.8 %) with small parts in Ecuador, Guyana, Venezuela, and Suriname. Although the part of the Amazon basin in Brazil covers some 5,016,136 km^2, i.e. 73 % of the total, the soils, climate, and government policy, the Amazonian countries are different from each other, and so it may be inadvisable to simply assume that data from Brazil applies to the entire Amazon basin.

To interpret the many studies, we postulate that the net carbon flux over the Amazon basin F_{Total} may be considered to be the sum of an array of processes, some of which have been investigated rather thoroughly, but several of which have been more or less neglected, and therefore coarse assumptions will have to be made to arrive at F_{Total}.

$$F_{Total} = F_{Deforestation} + F_{Degradation} + F_{old\text{-}growth} + F_{Secondary} + F_{Plantation} + F_{Harvest} + F_{Peat} + F_{Agriculture} + F_{Climate\ change}$$

In the following paragraphs, we consider the terms separately.

5.3.1.1 $F_{Deforestation}$ and $F_{Degradation}$ the Deforestation and Degradation Fluxes

Deforestation is achieved by means of fire. Burning is mostly in the dry season to achieve an efficient burn. Not all woody tissues are oxidised (Fearnside et al. 2007), but almost all are killed and the soil component decomposes over years and decades, with the exception of charcoal which may be 1 % of the biomass (Fearnside et al. 2007). When the land is converted into cattle ranching, the above-ground biomass may be only a few MgC ha^{-1} (McWilliam et al. 1993) compared with 100–200 MgC ha^{-1} for forest.

Data on deforestation rates may be obtained for each country from FAO official statistics (FAO 2011a), although these figures are often disputed (Grainger 2008). The Brazilian government pioneered the use of satellite remote sensing to track the transition from forest to pasture, a challenging operation when it started in the 1970s, but now routine. Their data show that deforestation in the Brazilian Legal Amazon (BLA) has fallen from 27,000 km^2 in 2004 to <5000 in 2012 (INPE 2012). In other parts of the Amazon basin, the rate of deforestation does not show this decline (Aragão et al. 2014). The most recent attempt to estimate forest cover is for the period 2000–2012 and covers the whole world (Hansen et al. 2013). For the case of Brazil, the figures from Hansen et al. (2013) agree reasonably well with government figures, given that the remote sensing data are for the whole of Brazil not just BLA. If the mean deforestation data for BLA (government figures, period 2000–2100, mean rate 16,000 km^2) are multiplied by the mean carbon stocks from ICESat (Saatchi et al. 2011, the mean carbon stock is 127 Mg ha^{-1}), we may conclude that Brazil's $F_{Deforestation}$ has been around 0.20 Pg C annum^{-1}, but that figure fell to only 0.06 in the year 2012. There is another dataset which provides deforestation data for the entire Amazon basin (FAO 2011b). According to this source, between the years 2000 and 2010 the forest cover fell by as much as 36,450 km^2 per year, implying a mean Amazon-wide flux for the period of 0.46 Pg C annum^{-1}.

The term 'Forest degradation' refers to a loss of biomass which is not visible by conventional remote sensing and which usually goes unreported. It arises mostly from selective logging, where the fraction of trees removed is not sufficient to change the land cover from 'forest' to 'non-forest' (Nepstad et al. 1999). High-resolution remote sensing has been used to detect degradation in the Amazon region: Asner et al. (2005) estimated the flux of carbon to be 25 % of that caused by deforestation. The DEGRAD System, developed by INPE in 2007, uses images of LANDSAT and CBERS satellites to map annually the areas where forest cover is not completely removed and therefore not counted as deforestation. In 2008, 27,417 km^2 was mapped as degraded forest. Later, much of this was cut and thus counted as deforestation. In 2010, degradation for the whole of Brazil was estimated to be 7508 km^2 (Serviço Florestal Brasileiro 2013). It is not possible to use these figures to compare with Asner et al. (2005) as they give no indication of the carbon content of the degraded component of forest, so we use Asner et al. (2005) and estimate $F_{\text{Degradation}}$ to be 0.11 Pg C annum^{-1}.

5.3.1.2 $F_{\text{old-growth}}$ and $F_{\text{Secondary}}$ Fluxes Over Forests

Measurements of trees at permanent sample plots (all old-growth forests) suggest that some plots have increased their carbon stocks whilst others have decreased or remain unchanged; however, on average one-hectare plots increased in biomass by 0.71 ± 0.34 Mg ha^{-1} annum^{-1}. (Phillips et al. 1998). More recently, Lewis et al. (2009) have upscaled the updated version of the dataset to the 7.9 million km^2 of Central and South America and estimated the carbon sink at 0.39–0.73 Pg C annum^{-1} for the whole region. A similar sink is estimated for other continents: the carbon sink for the entire tropics estimated by Lewis et al. (2009) is 0.79–1.56 Pg C annum^{-1} a figure considered by Wright (2013) to be too high.

The number of old-growth forest sites where eddy covariance has been carried out and published (5) is far too few to make any statistical average. All but one (Saleska et al. 2003) indicate a carbon sink. The sink has sometimes been reported to be very small and consistent with Phillips et al. (1998) and Lewis et al. (2009), for example see Grace et al. (1995), but in other locations a sink of several Mg ha^{-1} annum^{-1} is reported (Malhi et al. 1998; Araújo et al. 2002; Carswell et al. 2002; Saleska et al. 2009).

Plot-based estimates of carbon balances at specific old-growth sites, in which GPP has been estimated, often show that photosynthetic gains exceed respiratory losses by an average of several Mg per hectare (Fig. 5.1).

The remaining forest area in BLA is 3.37 million km^2, to which we add 1.73 million km^2 of other Amazonian countries, to give 5.10 million km^2 in all. Taking the plot-based figures from Lewis et al. (2009), tropical forests worldwide may be increasing by anything from 0.37 to 1.05 Mg ha^{-1}, giving an $F_{\text{old-growth}}$ for the basin of between 0.19 and 0.54 Pg C annum^{-1}, the most probable value being 0.35 Pg C annum^{-1}, somewhat higher than the value suggested by Chambers et al. (2001).

Much of all tropical forest is now secondary forest (*sensu* Brown and Lugo 1990). According to FAO (2011a), about 23 % of tropical forest is now secondary in South

America. This includes forest developing on abandoned farmland and forest regrowing from having been otherwise destroyed. Secondary forest often accumulates carbon rather rapidly when young and then more slowly (Brown and Lugo 1990; Helmer et al. 2009; Sierra et al. 2012). In a recent analysis of data from all three tropical continents, Bonner et al. (2013) found carbon uptake rates of secondary forest to vary from 0.25 to 6 Mg C ha^{-1} $annum^{-1}$, with a central tendency of about 3.0 Mg C ha^{-1} $annum^{-1}$, not very different from the value obtained from the much earlier (but smaller) dataset by Brown and Lugo (1990), and a more recent remote sensing study in the Amazon which suggested 4.0 Mg C ha^{-1} $annum^{-1}$ (Helmer et al. 2009). To estimate a sink from this value, we need to know the area of secondary forests. According to the remote sensing investigations of Neeff et al. (2006) and Almeida et al. (2010), Brazil had between 133,000 and 161,000 km^2 of secondary forest between 2000 and 2006, which implies that about 35 % of the area deforested since 1978 has been abandoned and is now secondary forest. This area may be an underestimate as the older secondary forests are hard to distinguish from old-growth forests by optical remote sensing, and so we assume 200,000 km^2 is the true figure and we take 4.0 Mg C ha^{-1} $annum^{-1}$ as the growth rate. This provides a sink of 0.08 Pg C $annum^{-1}$ for Brazil alone. For the entire Amazon, we estimate $F_{\mathrm{Secondary}}$ to be 0.12 Pg C $annum^{-1}$ with an uncertainty of ± 0.02.

5.3.1.3 $F_{\mathbf{Plantation}}$

Tropical plantations are capable of achieving very high growth rates and are sometimes heralded as a means of sequestering large quantities of CO_2. Laclau et al. (2000) found growth rates of 16 Mg biomass ha^{-1} $annum^{-1}$ for eucalyptus in the Congo, and much higher rates are possible with appropriate silviculture, especially when fast-growing clonal plantations are used (Stape et al. 2008). Brazil has 7.2 million hectares of plantations (Serviço Florestal Brasileiro 2013), mostly *Eucalyptus* and *Pinus*. If we assume these plantations have an average sequestration rate of 5 Mg C ha^{-1} $annum^{-1}$, we arrive at an estimate of 0.037 Pg C $annum^{-1}$ as the sink due to plantations.

5.3.1.4 $F_{\mathbf{Harvest}}$

Government data on quantities of harvested fuelwood and timber are compiled by FAO (2011a). Data are given in cubic metres, and here we have converted to an estimate of carbon by multiplying by 0.25. We may conclude that wood harvests in Brazil have amounted to rather small carbon fluxes, c. 0.035 Gt carbon per annum as fuelwood, and about the same as round-wood and sawn-timber combined. Only a small part is exported, so we may expect that 0.07 Gt C as CO_2 over Brazil derives from burning wood (assuming that the stock of long-lived products such as furniture is more or less at a steady value). Although these figures are from the whole of Brazil, and some is from plantations, we may consider that by far the largest part of the harvested wood is from the Amazon basin.

5.3.1.5 F_{Peat}

Attention has recently been drawn to the large stocks of tropical peat (Page et al. 2011). Until recently, most of this was thought to be in south-east Asia, but recent work suggests that peat accumulation is occurring in parts of the Amazon, in Peru (Lahteenoja et al. 2012, 2013). Accumulation rates of 28–108 g m^{-2} $annum^{-1}$ are reported over an area of 107,000 km^2, resulting in a carbon accumulation rate of a few million Mg of carbon annually. It is too early to estimate the extent of peat accumulation across the Amazon basin. It should be kept in mind that peat deposits can rapidly become a carbon source, as has happened in Indonesia (Page et al. 2011).

5.3.1.6 $F_{\text{Agriculture}}$

The Amazon basin continues to undergo rapid change, as forests are converted to agriculture for the production of soya bean, sugar cane, and beef (Davidson et al. 2012; Ortiz et al. 2013). This process involves a huge step-wise decline in above-ground biomass (McWilliam et al. 1993) and a slower decrease in the soil carbon as the woody root tissues gradually decompose (Fearnside and Barbosa 1998; Cusack et al. 2009). The grass, sugar cane, or soya bean that replaces the forest is thereafter cropped according to a seasonal cycle, but in this process some fraction of the production is added to the soil, as a crop residue, as root exudate, or as dead root material. In clay soils, the carbon may be partially protected from decay (Freibauer et al. 2004). These processes may result in a steady accumulation of carbon in the soil, i.e. a carbon sink. It was suggested 20 years ago that tropical grassland is indeed a significant global carbon sink (Neill et al. 1997) and some authors drew attention to deep-rooted tropical grasslands in particular (Fisher et al. 1994; Scurlock and Hall 1998). In temperate regions, the soil carbon sink has been measured and shown to depend on the management regime, especially the extent of tillage (Paustian et al. 2007; Freibauer et al. 2004; Poeplau and Don 2015). However, in the tropics there have been only a few long-term studies of soil carbon. In southern Brazil, four long-term experiments with no-tillage (7–19 years) showed accumulation rates that varied between 0.12 and 0.59 Mg C ha^{-1} $annum^{-1}$ (Amado et al. 2006). From these studies, and similar studies of soil carbon in croplands elsewhere (Poeplau and Don 2015), it appears that crops do indeed accumulate soil carbon at appreciable rates, but the rate depends on the tillage, the soil type, and especially the stabilising effect of clay particles. Changes from one form of agriculture to another introduce a further complication, and the overall picture is not yet clear (La Scala et al. 2012; Franco et al. 2015).

Gloor et al. (2012) presented Brazilian agricultural statistics on the areas of pasture (c. 0.55×10^6 km^2) and agriculture (c. 0.42×10^6 km^2) in the Amazon basin. Taking these areas and the sink strength from Amado et al. (2006), we may make a first tentative estimate of the potential agricultural sink of BLA between

0.01 and 0.05 Pg C annum^{-1}. This would apply if all the agriculture were no-till. We assume that the actual sink (in a scenario of mixed agricultural types) would be at the lower end of this range, and we here assume (for the purposes of estimation of F_{Total}), a sink of 0.02 ± 0.01 Pg C annum^{-1}.

Although the sequestration of carbon is significant, it should be thought of in relation to the 'carbon debt' that has accumulated since deforestation started in Brazil in the 1970s. If we take the figure for cumulative deforestation (750,000 km^2) and carbon loss (11.2 Pg) that Nogueira et al. (2015) have given, we may estimate how long it would take to repay the debt. To 'pay back' that loss by soil sequestration would take between 400 and 1100 years, even assuming that soil could store such huge amounts of carbon. Moreover, to sustain agricultural production (and sustained sequestration) would require intensification of cultivation and the application of fertilisers, which would reduce the effective sink strength.

5.3.1.7 $F_{\text{Climate change}}$

How far can plot-based measurements inform the debate on the impact of climate change on the forest carbon sink? There have been two major drought experiments in the Amazon basin, one near Santarém (Nepstad et al. 2002, 2007) and the other in the eastern part of the Amazon, Caxiuanã (Meir et al. 2009; Da Costa et al. 2010). The one-hectare plots are 'droughted' by excluding the throughfall—the rain is intercepted by covers near the ground and it is channelled away. Using this technique, a 50 % reduction in the supply of water to the rooting zone is achieved. In both cases, an increase in mortality of trees was observed; Da Costa et al. (2010) report a doubling of the tree mortality rate, with a loss of 37.8 ± 2.0 Mg ha^{-1} of carbon from 2002 to 2008 and a reduction in wood production. After a decade, the rate of loss increased, mostly due to mortality of the larger trees, and the droughted plot has now declined from 250 Mg C ha^{-1} to only 150 Mg C ha^{-1} (Rowland et al. 2015). If this were to be repeated in a natural drought on a large scale, the Amazon basin would become a substantial source of CO_2 to the atmosphere; however, droughts are not generally as prolonged as in this experiment. Information on the response to real droughts has been obtained from sample plots (Phillips et al. 2009, 2010). Data on mortality of tropical trees before, during, and after drought events show an extremely good relationship between soil water deficit and mortality, sometimes with a small delay, with mortality being most pronounced in larger trees.

We may conclude with Phillips et al. (2009) that drought certainly influences the carbon balance of Amazonian forest. It should be kept in mind that this is a direct effect, not the more obvious indirect effect associated with the fires which usually accompany drought (Chap. 13). Both the direct effect on mortality and indirect effect on fires presumably combine to account for the often-reported association between the CO_2 concentration in the global atmosphere (Heimann and Reichstein 2008). There may be additionally an effect of temperature on heterotrophic respiration, as often suggested by modellers (Tian et al. 1998; Cox et al. 2000), although this is not confirmed from plot-based studies where respiration has often been measured (e.g. Rowland et al. 2013).

5.3.1.8 $F_{\mathrm{Fossilfuel}}$

The population density in BLA is low, said to be 3.3 people km^2 (UNEP 2004). It is concentrated in urban centres, of which only three have populations exceeding 1 million (Manaus, Belém, La Santa Cruz). The total population of the top 11 urban centres is around 9 million, but there are many small settlements along roads and rivers as well as a population of forest-dwelling people, some of whom are 'uncontacted' and even undiscovered, who do not use fossil fuels. For the purposes of estimating the carbon emissions from fossil fuels, we assume a population of 20 million (Laurance et al. 2001), and we take the per capita emission as 0.50 Mg C annum^{-1}, representing emissions in Brazil since the year 2000 (Boden et al. 2010). This amounts to a very small basin-wide annual emission of only 10 million Mg of carbon.

5.3.2 *Total Estimate of Carbon Balance, Bottom-Up Evaluation*

Adding all the terms together provides an estimate of the Amazon carbon balance (Table 5.1). We find that the most likely result is a net all-basin carbon source of 0.11 Pg C annum^{-1}, but the result is highly dependent on the upscaled plot data (Phillips et al. 1998; Lewis et al. 2009). Taking the uncertainty bounds, and taking into account error propagation, we have 0.64 ± 0.04 minus 0.53 ± 0.16, which is a source of 0.11 ± 0.16, i.e. the balance is not distinguishable from zero.

Table 5.1 Annual carbon balance of the Amazon basin in an 'average year', obtained from plot measurements multiplied by measured and estimated areas, with the assumptions and caveats outlined in the text

	Loss (Pg C annum^{-1})	Gain (Pg C annum^{-1})
Deforestation	0.46 ± 0.04	
Degradation	0.11 ± 0.01	
Old growth forest		0.35 ± 0.16
Secondary forest		0.12 ± 0.02
Plantations		0.037 ± 0.004
Agriculture		0.02 ± 0.01
Harvested wood	0.07 ± 0.01	
Total	0.64 ± 0.04	0.53 ± 0.16

Units are Pg C annum^{-1}

5.3.3 *Total Estimate of Carbon Balance, Top-Down Evaluation*

A recent study provides basin-wide data on the carbon balance (Gatti et al. 2014), using the methods outlined above (Sect. 2.2.1). The vertical profiles of CO_2 showed marked differences in the burning season (July–October) when the surface concentrations were enriched (Fig. 5.2). In the rest of the year, the activity of the vegetation drew down the surface concentration below the background. By contrasting the year 2010 (a drought year) with 2011 (a more normal year), and by also measuring carbon monoxide (a marker for fire), the researchers were able to separate the impact of drought on the basic biological process and on the fire occurrence (Table 5.2). In the drought year, the uptake of carbon dioxide by the vegetation was reduced by 0.22 Pg C, and fires were increased by 0.21 Pg C of carbon. Overall, the Amazon basin was changed from being more or less carbon neutral in the normal year of 2011 to being a source of 0.48 Pg C of carbon in the drought year.

The results are in moderate agreement: the plot-based study shows a biological sink of some 0.53 ± 0.16 Pg C annum^{-1} whilst the aircraft study shows a sink that is somewhat smaller, of 0.25 ± 0.14 Pg C annum^{-1}. The deforestation figures based

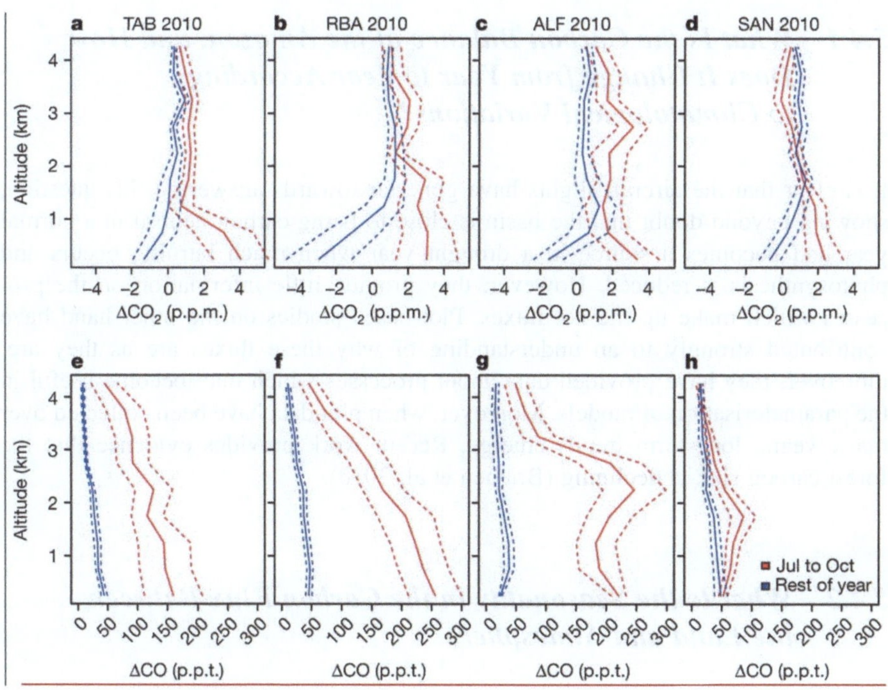

Fig. 5.2 Mean difference between profiles of CO_2 over the Amazon basin in 2010 arranged in order from west to east (Gatti et al. 2014). The difference is expressed relative to the background CO_2 from the Atlantic sites at Ascension and Barbados

Table 5.2 Annual carbon balance of the Amazon basin in a drought year versus a normal year, obtained from aircraft measurements in the planetary boundary layer (Gatti et al. 2014)

	Total	Fire	Net biome exchange
2010 (drought year)	0.48 ± 0.18	0.51 ± 0.12	-0.03 ± 0.22
2011 (normal year)	0.06 ± 0.01	0.30 ± 0.10	-0.25 ± 0.14

The net biome exchange is the exchange of carbon with the entire basin after removing the flux due to fire. Sign convention is as follows: negative means uptake by the land surface from the atmosphere. Units are Pg C $annum^{-1}$

on remote sensing and FAO data (Table 5.1) are in rough agreement with the fire estimate from Table 5.2.

5.4 Answers to the Science Questions

Which approach provides the best answers to the science questions posed on page 1 of this article? Here, we address them in turn.

5.4.1 *What Is the Carbon Balance of the Amazon, and How Does It Change from Year to Year According to Climatological Variations?*

It is clear that the aircraft flights have gone far towards answering this question, showing beyond doubt that the basin is close to being carbon neutral in a normal year and becomes a source in a drought year when much burning occurs and photosynthesis is reduced. However, they provide little information on the processes which make up the net fluxes. Plot-based studies on the other hand have contributed strongly to an understanding of why these fluxes are as they are; moreover, they have provided data about processes which may become useful in the parameterisation of models. Moreover, when plot data have been collected over many years, long-term trends emerge. Recent work provides evidence that the forest carbon sink is declining (Brienen et al. 2015).

5.4.2 *What Is the Seasonality in the Carbon Flux Between the Land and Atmosphere?*

This is answered by the aircraft measurements, which have picked up a major seasonality in the flux from burning. Seasonality is manifest as fire in a normal year and is essentially a measure of $F_{\mathrm{Deforestation}}$; the flux from burning is much intensified in a drought year. Seasonality in the biological fluxes is found by eddy covariance to be surprisingly low despite the definite seasonality in precipitation (Restrepo-Coupe et al. 2013; Baker et al. 2013), presumably because forests have deep roots and so they can access water at depth. Rowland et al. (2014) report relatively large seasonality in growth of trees in lowland Amazon, hinting at the capacity of tropical trees to vary allocation patterns according to the challenges of a fluctuating environment.

5.4.3 How Do the Changes in Land Use Affect the Fluxes?

This question is best answered by plot-based studies, such as those described in Don et al. (2011), Eclesia et al. (2012), and Ziegler et al. (2012). They provide insights into below-ground as well as above-ground processes and form the basis of the widely used 'bookkeeping' approach. Ziegler et al. (2012) showed how the carbon content of tropical ecosystems may vary from a few Mg ha^{-1} to over 400 Mg ha^{-1} and such data will in the future be used to model the process of transition. There are some surprises: the forest-to-pasture transition may even lead to an increase in soil carbon, especially in the surface layers of soil (Don et al. 2011; Eclesia et al. 2012; Smith et al. 2014) as suspected from earlier research on South American pastures made up of introduced grasses (Fisher et al. 1994). The review by Eclesia et al. (2012) shows the timing of the observed changes in soil organic carbon. When forest is replaced by tree plantations, the soil carbon content increases linearly, and after a century it far exceeds the levels found in the native forest, except in the wettest sites. When forest is replaced by pasture a somewhat similar picture emerges, except the largest increases are in the wetter sites. This analysis is intriguing and relevant to much of the land use change occurring in the tropics today, but the difference in behaviour between the two types of transition has not been explained.

Acknowledgements The author wishes to acknowledge support from the UK's Natural Environment Research Council through the Amazonica project and from the European FP7 project Geocarbon.

References

Achard F, Eva HG, Stigberg H-J et al (2002) Determination of deforestation rates of the world's Humid tropical forests. Science 297:999–1002

Almeida CA, Morisson VD, Sobral Escada MI et al (2010) Estimation of secondary vegetation area in the Brazilian Legal Amazon. Acta Amazon 40:289–301

Amado TJC, Bayer C, Conceicão et al (2006) Potential of carbon accumulation in no-till soils with intensive use and cover crops in southern Brasil. J Environ Qual. doi:10.2134/jeq2005.0233

Aragão LEOC, Poulter B, Barlow JB et al (2014) Environmental change and the carbon balance of Amazonian forests. Biol Rev 89:913–931

Araújo AC, Nobre A, Kruijt B et al (2002) Comparative measurements of carbon dioxide fluxes from two nearby towers in a central Amazonian rainforest: the Manaus LBA site. J Geophys Res 107(D20):8090. doi:10.1029/2001JD000676

Araujo-Murakami A, Doughty CE, Metcalke DB et al (2014) The productivity, allocation and cycling of carbon in a forest at the dry margin of the Amazon forest in Bolivia. Plant Ecol Divers 7:55–69

Asner GP, Knapp DE, Broadbent EN et al (2005) Selective logging in the Brazilian Amazon. Science 310:480–482

Asner GP, Powell GVN, Mascaro J et al (2010) High-resolution forest carbon stocks and emissions in the Amazon. Proc Natl Acad Sci 107:16738–16742

Aubinet M, Grelle A, Ibrom A et al (2000) Estimates of the annual net carbon and water exchange of forests: the EUROFLUX methodology. Adv Ecol Res 30:113–175

Aubinet M, Vesala T, Papale D (2012) Eddy covariance. A practical guide to measurement and data analysis. Springer, Dordrecht

Baccini A, Goetz SJ, Walker WS et al (2012) Estimated carbon dioxide emissions from tropical deforestation improved by carbon-density maps. Nat Climate Change 2:182–185

Baker IT, Harper AB, da Rocha HR et al (2013) Surface ecophysiological behavior across vegetation and moisture gradients in tropical South America. Agr Forest Meteorol 182:177–188

Boden TA, Marland G, Andres RJ (2010) Global, regional, and national fossil-fuel CO_2 emissions. Carbon Dioxide Information Analysis Center, Oak Ridge National Laboratory, U.S. Department of Energy, Oak Ridge, TN. doi:10.3334/CDIAC/00001_V2010

Bolin B, Keeling CD (1963) Large-scale atmospheric mixing as deduced from the seasonal and meridional variations of carbon dioxide. J Geophys Res 68:3899–3920

Bonner MTL, Schmidt S, Shoo LP (2013) A meta-analytical global comparison of aboveground biomass accumulation between tropical secondary forests and monoculture plantations. For Ecol Manage 291:73–86

Bowman DMJS, Brienen RJW, Gloor E et al (2014) Detecting trends in tree growth: not so simple. Trends Plant Sci 18:11–17

Brown S, Lugo AE (1990) Tropical secondary forests. J Trop Ecol 6:1–32

Brienen RJW, Phillips OL, Feldpausch TR et al (2015) Long-term decline of the Amazon carbon sink. Nature 519:344–348

Carswell FE, Costa AL, Palheta M (2002) Seasonality in CO_2 and H_2O flux at an eastern Amazonian rain forest. J Geophys Res 107. doi:10.1029/2000JD00284

Chambers JQ, Higuchi N, Tribuzy ES et al (2001) Carbon sink for a century. Nature 410:429

Chave J, Andalo C, Brown S et al (2005) Tree allometry and improved estimation of carbon stocks and balance in tropical forests. Oecologia 145:87–99

Chou WW, Wofsy SC, Harriss RC et al (2002) Net fluxes of CO_2 in Amazonia derived from aircraft observations. J Geophys Res 107. doi:10.1029/2001JD001295.

Ciais P, Bombelli A, Williams M et al (2011) The carbon balance of Africa: synthesis of recent research studies. Philos Trans R Soc A 369:2038–2057

Clark DA (2002) Are tropical forests an important carbon sink? Reanalysis of the long-term plot data. Ecol Appl 12:3–7

Clark DA, Brown S, Kicklighter DW et al (2001) Net primary production in tropical forests; an evaluation and synthesis of existing field data. Ecol Appl 11:371–384

Cogan AJ, Boesch H, Parker RJ (2012) Atmospheric carbon dioxide retrieved from the Greenhouse Gases Observing SATellite. J Geophys Res 117. doi:10.1029/2012JD018087

Cox PM, Betts RA, Jones CD et al (2000) Acceleration of global warming due to carbon cycle feedbacks in a coupled climate model. Nature 408:184–187

Culf AD, Fisch G, Malhi Y et al (1999) Carbon dioxide measurements in the nocturnal boundary layer over Amazonian forest. Hydrol Earth Syst Sci 3:39–53

Cusack DIF, Chou WW, Yang W et al (2009) Controls on long-term root and leaf litter decomposition in neotropical forests. Glob Chang Biol 15:1339–1355

Da Costa ACL, Galbraith D, Almeida S et al (2010) Effect of 7 yr of experimental drought on vegetation dynamics and biomass storage of an eastern Amazonian rainforest. New Phytol 187:579–591

Davidson EA, de Araújo AC, Artaxo P et al (2012) The Amazon basin in transition. Nature 481:321–328

DeFries RS, Houghton RA, Hansen MC et al (2002) Carbon emissions from tropical deforestation and regrowth based on satellite observations for the 1980s and 1990s. Proc Natl Acad Sci 99:256–261

Dolman AJ, Freibauer A, Valentini R (2008) The continental scale greenhouse gas balance of Europe, vol 203, Ecological studies. Springer, Heidelberg

Don A, Schumacher J, Freibauer A (2011) Impact of tropical land-use change on soil organic carbon stocks—a meta-analysis. Glob Chang Biol 17:1658–1670

Eclesia R, Jobbagy EG, Jackson RB et al (2012) Shifts in soil organic carbon for plantation and pasture establishment in native forests and grasslands of South America. Glob Chang Biol 18:3237–3251

Enting IG, Mansbridge JV (1989) Seasonal sources and sinks of atmospheric CO_2: direct inversion of filtered data. Tellus 41B:111–126

FAO (2011a) State of the world's forests 2011. Food and Agriculture Organization of the UN, Rome

FAO (2011b) The state of forests in the Amazon Basin, Congo Basin and Southeast Asia. A report prepared for the summit of the three rainforest basins Brazzaville, Republic of Congo, 31 May–3 June, 2011. FAO, Rome. http://www.fao.org/docrep/014/i2247e/i2247e00.pdf. Accessed 28 Jan 2014

Fearnside PM (2002) Environmental and social consequences of Brazil's planned infrastructure in Amazonia. Environ Manage 30:735–747

Fearnside PM, Barbosa RI (1998) Soil carbon changes from conversion of forest to pasture in Brazilian Amazon. For Ecol Manage 108:147–166

Fearnside PM, Barbosa RI, Alencastro Graça PML (2007) Burning of secondary forest in Amazonia: biomass, burning efficiency and charcoal formation during land preparation for agriculture in Apia, Roraima, Brazil. For Ecol Manage 242:678–687

Fisher MJ, Rao IM, Ayarza MA, Lascano CE, Sanz JI, Thomas RJ, Vera RR (1994) Carbon storage by introduced deep-rooted grasses in the South American savannas. Nature 371:236–238

Franco ALC, Cherubin MR, Pavinato PS et al (2015) Soil carbon, nitrogen and phosphorus changes under sugarcane expansion in Brazil. Sci Total Environ 515:30–38

Freibauer A, Roundsevell MDA, Smith P, Verhagen J (2004) Carbon sequestration in agricultural soils of Europe. Geoderma 122:1–23

Gatti LV, Miller JB, D'Amelio MTS et al (2010) Vertical profiles of CO2 above eastern Amazonia suggest a net carbon flux to the atmosphere and balanced biosphere between 2000 and 2009. Tellus Ser B 62:581–594

Gatti LV, Gloor M, Miller JB et al (2014) Atmospheric CO_2 measurements reveal strong drought sensitivity of Amazonian carbon balance. Nature 506:76–80

Gloor M, Phillips OL, Lloyd J et al (2009) Does the disturbance hypothesis explain the biomass increase in basin-wide Amazon forest plot data? Glob Chang Biol 15:2418–2430

Gloor M, Gatti L, Brienen R et al (2012) The carbon balance of South America: a review of the status, decadal trends and main determinants. Biogeosciences 9:5407–5430

Grace J, Mitchard E, Gloor E (2014) Perturbations in the carbon budget of the tropics. Glob Chang Biol 20:3238–3255

Grace J, Lloyd J, McIntyre J et al (1995) Carbon dioxide uptake by an undisturbed tropical rain forest in South-West Amazonia 1992–1993. Science 270:778–780

Grainger A (2008) Difficulties in tracking the long-term global trend in tropical forest area. Proc Natl Acad Sci 105:818–823

Gurney KR, Law RM, Denning AS et al (2002) Towards robust regional estimates of CO_2 sources and sinks using atmospheric transport models. Nature 415:626–630

Hansen MC, Potapov PV, Moore R et al (2013) High-resolution global maps of 21st-century forest cover change. Science 342:850–853

Heimann M, Reichstein M (2008) Terrestrial ecosystem carbon dynamics and climate feedbacks. Nature 451:289–292

Helmer EH, Lefsky MA, Roberts DA (2009) Biomass accumulations rates of Amazonian secondary forest and biomass of old-growth forests from Landsat time series and Geoscience Laser Altimeter System. J Appl Remote Sens 3:033505

Houghton RA, Gloor M, Lloyd J et al (2009) The regional carbon budget. In: Keller M, Bustamante M, Gash J, Dias PS (eds) Amazonia and global change. Geophysical monograph series. pp 409–428

Houghton R, Werf GVD, DeFries R et al (2012) Carbon emissions from land use and land-cover change. Biogeosciences 9:5125–5142

INPE (2003) Monitoring of the Brazilian Amazonian forest by satellite—2002. p 16

INPE (2012) Projeto PRODES: monitoramento da floresta amazônica brasileira por satélite. Disponível em http://www.obt.inpe.br/prodes/

Jarvis A, Ramirez J, Bonilla-Findji O, Zapata E (2011) Impacts of climate change on crop production in Latin America. In: Yadav SS, Redden RJ, Hatfield JL, Lotze-Campen H, Hall AE (eds) Crop adaptation to climate change. Wiley, Chichester, pp 44–56

Keller M, Alencar A, Asner GP (2004) Ecological research in the large-scale biosphere–atmosphere experiment in Amazonia: early results. Ecol Appl 14:S3–S16

Körner C, Arnone JA (1992) Responses to elevated carbon-dioxide in artificial tropical ecosystems. Science 257:1672–1675

Kruijt B, Elbers JA, von Randow C et al (2004) The robustness of eddy correlation fluxes for Amazon rain forest conditions. Ecol Appl 14:S101–S113

La Scala N, De Figueiredo EB, Panosso AR (2012) A review on soil carbon accumulation due to the management change of major Brazilian agricultural activities. Braz J Biol 72:775–785

Laclau J-P, Bouillet J-P, Ranger J (2000) Dynamics of biomass and nutrient accumulation in a clonal plantation of Eucalyptus in Congo. For Ecol Manage 128:181–196

Lahteenoja O, Reategui YR, Rasanen M et al (2012) The large Amazonian peatland carbon sink in the subsiding Pastaza-Maranon foreland basin, Peru. Glob Chang Biol 18:164–178

Lahteenoja O, Flores B, Nelson B (2013) Tropical peat accumulation in central Amazonia. Wetlands 33:495–503

Laurance WF, Cochrane MA, Bergen S et al (2001) The future of the Brazilian Amazon. Science 291:438–439

Laurence WF, Fearnside PM (2002) Issues in Amazonian development. Science 295:1643–1643

Le Toan T, Quegan S, Davidson MWJ et al (2011) The BIOMASS mission: mapping global forest biomass to better understand the terrestrial carbon cycle. Remote Sens Environ 115:2850–2860

Lewis SL, Lopez-Gonzalez G, Sonke B et al (2009) Increasing carbon storage in intact African tropical forests. Nature 457. doi:10.1038/nature07771

Lloyd J, Farquhar GD (1996) The CO_2 dependence of photosynthesis, plant growth responses to elevated atmospheric CO 2 concentrations and their interaction with soil nutrient status. I. General principles and forest ecosystems. Funct Ecol 10:4–32

Lloyd J, Kolle O, Fritsch H et al (2007) An airborne regional carbon balance for Central Amazonia. Biogeosciences 4:759–768

Lloyd J, Gloor EU, Lewis SL (2009) Are the dynamics of tropical forests dominated by large and rare disturbance events? Ecol Lett 12:19–21

Luyssaert S, Schulze ED, Börner A, Knohl A, Hessenmöller D, Law BE, Ciais P, Grace J (2008) Old-growth forests as global carbon sinks. Nature 455:213–215

Machida T, Matsueda H, Sawa Y et al (2008) Worldwide measurements of atmospheric CO_2 and other trace gas species using commercial airlines. J Atmos Ocean Technol 25:1744–1754

Malhi Y, Nobre AD, Grace J et al (1998) Carbon dioxide transfer over a Central Amazonian rain forest. J Geophys Res 103:31593–31612

Malhi Y, Phillips OL, Lloyd J et al (2002) An international network to monitor the structure, composition and dynamics of Amazonian forests (RAINFOR). J Veg Sci 13:439–450

Malhi Y, Eduardo LI, Aragao OC et al (2009) Comprehensive assessment of carbon productivity, allocation and storage in three Amazonian forests. Glob Chang Biol 15:1255–1274

Malhi Y, Farfán Amézquita F, Doughty CE et al (2014) The productivity, metabolism and carbon cycle of two lowland tropical forest plots in southwestern Amazonia, Peru. Plant Ecology and Diversity 7(1–2):85–105

Matsueda H, Inoue HY (1996) Measurements of atmospheric CO_2 and CH_4 using a commercial airliner from 1993 to 1994. Atmos Environ 30:1647–1655

Mayaux P, Pekel J-F, Desclée B et al (2013) State and evolution of the African rainforests between 1990 and 2010. Philos Trans R Soc B 368:20120300

McWilliam A-LC, Roberts JM, Cabral et al (1993) Leaf area index and above-ground biomass of terra firme rain forest and adjacent clearings in Amazonia. Funct Ecol 7:310–317

Meir P, Brando PM, Nepstad, D (2009) The effects of drought on Amazonian rain forests. In: Keller M, Bustamante M, Gash J et al (eds) Amazonia and global change. Geophysical monograph series, vol 186. pp 429–449

Mitchard ETA, Saatchi SS, Woodhouse IH (2009) Using satellite radar backscatter to predict above-ground woody biomass: a consistent relationship across four different African landscapes. Geophys Res Lett 36. doi:10.1029/2009GL040692

Neeff T, Lucas RM, dos Santos JR et al (2006) Area and age of secondary forests in Brazilian Amazonia 1978–2002: an empirical estimate. Ecosystems 9:609–623

Neill C, Melillo JM, Steudler PA et al (1997) Soil carbon and nitrogen stocks following forest clearing for pasture in the southwestern Brazilian Amazon. Ecol Appl 7:1216–1225

Nepstad DC, Verssimo A, Alencar A et al (1999) Large-scale impoverishment of Amazonian forests by logging and fire. Nature 398:505–508

Nepstad DC, Moutinho P, Dias-Filho MB, et al (2002) The effects of partial throughfall exclusion on canopy processes, aboveground production and biogeochemistry of an Amazon forest. J Geophys Res 107. doi:10.1029/2001JD000360

Nepstad DC, Tohver IM, Ray D, Moutinho P, Cardinot G (2007) Mortality of large trees and lianas following experimental drought in an Amazon forest. Ecology 88:2259–2269

Nogueira EM, Yanai AM, Fonseca FO, Fearnside PM (2015) Carbon stock loss through deforestation through 2013 in the Brazilian Amazon. Glob Chang Biol. doi:10.1111/gcb.12798

Odum EP (1969) The strategy of ecosystem development. Science 164:262–270

Ortiz R, Nowak A, Lavado A, Parker L (2013) Food security in Amazonia. Report for Global Canopy Programme and International Center for Tropical Agriculture as part of the Amazonia Security Agenda project

Page SE, Reiley, Banks CTJ (2011) Global and regional importance of the tropical peatland carbon pool. Glob Chang Biol 17:798–818

Patra PK, Canadell JG, Houghton RA et al (2013) The carbon budget of South Asia. Biogeosciences 10:513–527

Paustian K, Andrén O, Janzen HH et al (2007) Agricultural soils as a sink to mitigate CO_2 emissions. Soil Use Manag 13:230–244

Peylin P, Law RM, Gurney KR et al (2013) Global atmospheric carbon budget: results from an ensemble of atmospheric CO_2 inversions. Biogeosci Discuss 10:5301–5360

Phillips OL, Malhi Y, Higuchi N et al (1998) Changes in the carbon balance of tropical forests: evidence from long-term plots. Science 282:439–442

Phillips OL, Malhi Y, Vinceti B et al (2002) Changes in growth of tropical forests: evaluating potential biases. Ecol Appl 12:576–587

Phillips OL, Aragão LEOC, Lewis SL et al (2009) Drought sensitivity of the Amazon rainforest. Science 323:1344–1347

Phillips OL, van der Heijden G, Lewis SL et al (2010) Drought–mortality relationships for tropical forests. New Phytol 187:631–646

Poeplau C, Don A (2015) Carbon sequestration in agricultural soils via cultivation of cover crops—a meta-analysis. Agric Ecosyst Environ 200:33–41
Quegan ST, Le Toan JJ, Yu F et al (2000) Multitemporal ERS SAR analysis applied to forest monitoring. IEEE Trans Geosci Remote Sens 38:741–753
Restrepo-Coupe N, da Rocha HR, Hutyra LR et al (2013) What drives the seasonality of photosynthesis across the Amazon basin? A cross-site analysis of eddy flux tower measurements from the Brasil flux network. Agric Forest Meteorol 182(SI):128–144
Rödenbeck C, Houweling S, Gloor M, Heimann M (2003) CO_2 flux history 1982–2001 inferred from atmospheric data using a global inversion of atmospheric transport. Atmos Chem Phys 3:1919–1964
Rödenbeck C, Conway TJ, Langenfelds RL (2006) The effect of systematic measurement errors on atmospheric CO_2 inversions: a quantitative assessment. Atmos Chem Phys 6:149–161
Rowland L, Stahl C, Bonal D et al (2013) The response of tropical rainforest deadwood respiration to seasonal drought. Ecosystems 16:1294–1309
Rowland L, Malhi Y, Silva-Espejo JE (2014) The sensitivity of wood production to seasonal and interannual variations in climate in a lowland Amazonian rainforest. Oecologia 174:295–306
Rowland L, da Costa ACL, Galbraith DR et al (2015) Death from drought in tropical forests is triggered by hydraulics not carbon starvation. Nature 528:119–122
Ryan CM, Hill T, Woollen E et al (2012) Quantifying small-scale deforestation and forest degradation in African woodlands using radar imagery. Glob Chang Biol 18:243–257
Saatchi SS, Harris NL, Brown S et al (2011) Benchmark map of forest carbon stocks in tropical regions across three continents. Proc Natl Acad Sci 108:9899–9904
Saleska SR, Miller SD, Matross DM et al (2003) Carbon in Amazon forests: unexpected seasonal fluxes and disturbance-induced losses. Science 302:1554–1557
Saleska S, da Rocha H, Kruijt B, Nobre A (2009) Ecosystem carbon fluxes and Amazonian forest metabolism. In: Keller M, Bustamante M, Gash J, Dias PS (eds) Amazonia and global change, Geophysical monograph series 186. American Geophysical Union, Washington, DC, pp 389–407
Scurlock JMO, Hall DO (1998) The global carbon sink: a grassland perspective. Glob Chang Biol 4:229–233
Serviço Florestal Brasileiro (2013) Florestas do Brasil em resumo—2013: dados de 2007–2012. SFB, Brasília
Sierra CA, del Valle J, Restrepo H (2012) Total carbon accumulation in a tropical forest landscape. Carbon Balance Manage 7. doi:10.1186/1750-0680-7-12
Smith P, Bustamante H, Ahammad H et al (2014) Agriculture, forestry and other land use (AFOLU). In: Climate change 2014: mitigation of climate change. Contribution of working group III to the fifth assessment report of the Intergovernmental Panel on Climate Change. Cambridge University Press, Cambridge
Stape JL, Binkley D, Ryan MG (2008) Production and carbon allocation in a clonal Eucalyptus plantation with water and nutrient manipulations. For Ecol Manage 255:920–930
Stephens BB, Gurney KR, Tans PP et al (2007) Weak northern and strong tropical land carbon uptake from vertical profiles of atmospheric CO_2. Nature 316:1732–1743
Tans PP, Conway TJ, Nakazawa T (1989) Latitudinal distribution of the sources and sinks of atmospheric carbon dioxide from surface observations and an atmospheric transport model. J Geophys Res 94:5151–5172
Thomas SC, Martin AR (2012) Carbon content of tree tissues: a synthesis. Forests 3:332–352
Tian HQ, Melillo JM, Kicklighter DW et al (1998) Effect of interannual climate variability on carbon storage in Amazonian ecosystems. Nature 396:664–667. doi:10.1038/25328
Tollefson J (2013) Experiment aims to steep rainforests in carbon dioxide. Nature 496:405–406
UNEP (2004) Amazon Basin, GIWA regional assessment 40b. University of Kalmar, Kalmar
Valentini R, Arneth A, Bombelli A et al (2013) The full greenhouse gases budget of Africa: synthesis, uncertainties and vulnerabilities. Biogeosci Discuss 10:8343–8413
Wright SJ (2013) The carbon sink in intact tropical forests. Glob Chang Biol 19:337–339
Ziegler AD, Phelps J, Yuen JQ et al (2012) Carbon outcomes of major land-cover transitions in SE Asia: great uncertainties and REDD plus policy implications. Glob Chang Biol 18:3087–3099

Chapter 6
Climate and the Amazonian Carbon Balance

Emanuel Gloor

6.1 Introduction

The Amazon basin is an important component of the global carbon cycle and the earth climate system because of its vast size (ca. 6.5 mio. km^2) and its location in the tropics. The high annual precipitation resulting from lower troposphere wind and moisture convergence and subsequent upwelling and the high incoming solar radiation are the basis for extensive forests, the largest rainforests in the tropics. These forests store a large amount of organic carbon, in the order of 110 Pg C in above-ground biomass alone (Feldpausch et al. 2012), which may be released rapidly to the atmosphere by forest destruction and thus could potentially contribute substantially to greenhouse warming at short timescales (decades). Upwelling of air over the Basin is an important component of the Hadley and Walker circulations with the onset of the rainy season possibly being triggered by an increase in evapotranspiration at the end of the dry season (Fu et al. 1998; Silva Dias et al. 2002; Poeschl et al. 2010). Thus, the Amazon rainforests and the large-scale climate system are intertwined. The Amazon basin has over the last five decades and still is exposed to strong development pressure in an economically fast developing continent, with expanding agriculture having been an important force driving deforestation. Deforestation has over the last decades been particularly intense along the southern border of the Basin, the transition zone between forest and cerrado (or savanna) (e.g. Fearnside 2005). Replacement of forests by other vegetation types reduces evapotranspiration, i.e. reflux of water vapour to the atmosphere, and thus precipitation downwind (Salati et al. 1979; Martinelli et al. 1996;

E. Gloor (✉)
School of Geography, Faculty of Earth and Environment, University of Leeds, Woodhouse Lane, LS2 9JT Leeds, UK
e-mail: e.gloor@leeeds.ac.uk

© Springer-Verlag Berlin Heidelberg 2016

L. Nagy et al. (eds.), *Interactions Between Biosphere, Atmosphere and Human Land Use in the Amazon Basin*, Ecological Studies 227,
DOI 10.1007/978-3-662-49902-3_6

Silva Dias et al. 2002; Werth and Avissar 2002; van der Ent et al. 2010; Spracklen et al. 2012). In addition, the Basin, like the rest of the planet, is exposed to a warming climate in addition to elevated atmospheric CO_2. What will be the effect on vegetation and specifically rainforests? Will they thrive or decline in a warmer, probably wetter, elevated CO_2 world? Will there be an increased contribution to atmospheric CO_2 and thus greenhouse warming due to negative impacts on the rainforests (e.g. Huntingford et al. 2013) or will the forests instead have a moderating effect on greenhouse warming by forest growth stimulation? Finally, how much loss will be incurred due to forest destruction and induced fire feedbacks? The carbon balance of the Basin and its trend over time is a diagnostic for answering these questions and is the subject of this chapter.

We review the state of knowledge of carbon fluxes associated with specific processes, such as deforestation, and then put them into perspective with most recent carbon balances calculated at the whole basin scale for the years 2010 and 2011, based on regular lower troposphere greenhouse gas sampling along the main airstream over the Basin (Gatti et al. 2014). As a start and background of this review, we briefly document climatic variability and recent tendencies. Some of the material follows quite closely (Gloor et al. 2012, 2013; Gatti et al. 2014).

6.2 Climate Background

Documentation of climate of the Amazon basin is sparse; long-term records are few and thus emerging patterns need to be viewed with caution. Nonetheless, existing records from meteorological stations, here specifically those selected by the Climatic Research Unit (CRU) version 3.1 based on the criterion of continuity/length, as well as river stage records reveal the following picture. Firstly over the last two–three decades, total annual precipitation over the Basin has increased with the increase concentrated in the wet season (Fig. 6.1b, and Gloor et al. 2013) and the seasonal amplitude in precipitation has increased as well (Fig. 6.1c; see also Chap. 2). There have indeed been strong droughts in 1995, 1998, 2004/2005, and 2010 and unusually severe floods in 1999, 2009, and 2012 (e.g. Filizola et al. 2014). The south-western part of the Basin (Bolivia) has become slightly drier though and according to analyses of Marengo et al. (2011) and Fu et al. (2013) dry season length has increased in these southern regions (for possible consequences, see Chap. 3 by Coe et al.). Somewhat similar to precipitation records, temperatures have seen an upward swing between the mid-1950s and mid-1970s, followed by a downturn and, most significantly, a continuous increase thereafter paralleling the increase in precipitation and seasonality over the last three decades (Fig. 6.1a). An important feature for the vegetation in the Basin thus seems to be an increase in variability of the hydrological cycle paralleling an increase in temperatures. The compound effect will be an increase in periods of water deficit during comparably dry periods (e.g. Seneviratne et al. 2010) while increased very wet conditions will lead to more severe flooding with severe impacts on livelihoods (e.g. Filizola et al. 2014).

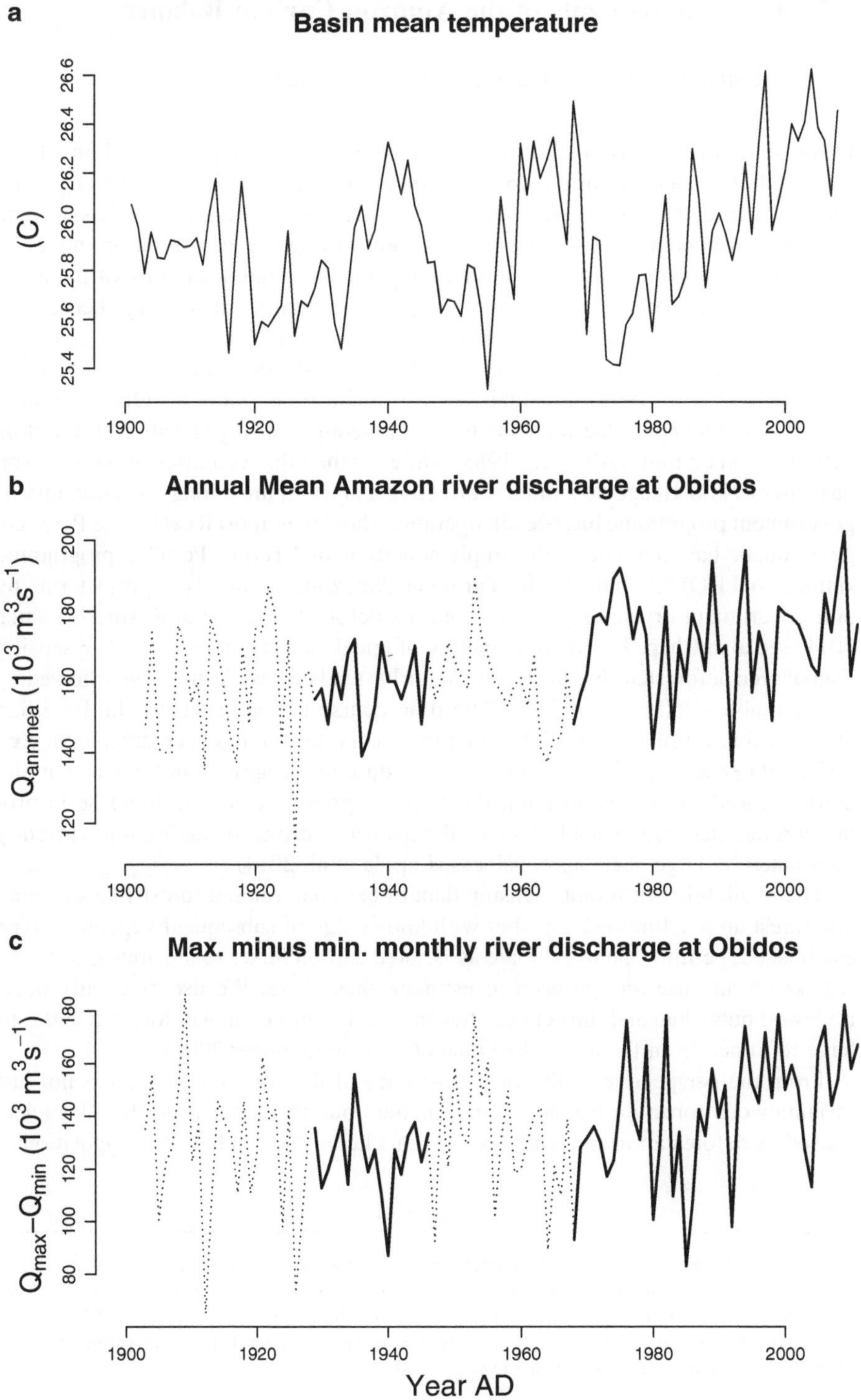

Fig. 6.1 (**a**) Annual mean Amazon-wide temperature estimated from the climatology Climate Research Unit (CRU) 3.1, (**b**) annual mean Amazon river discharge at Óbidos, draining ca. 80 % of

6.3 Flux Components of the Amazon Carbon Balance

6.3.1 Carbon Release Caused by Deforestation

Deforestation of Amazonian forests, which for the Brazilian part started on a large scale in 1970 as a result of government policy, has by today reduced the originally forested area by ca. 20 % (e.g. Fearnside 2005). With remote sensing it has become possible to obtain a clearer sense of the extent of ongoing deforestation; nonetheless, complete and accurate quantification of area cleared remains difficult for various reasons, including limitations in spatial and temporal coverage because of clouds but also the challenge of detecting partial deforestation, where trees are removed without any change in the forest/non-forest satellite classification (e.g. Nepstad et al. 1999). In Brazil, a government programme (PRODES, Projeto de Monitoramento do Desmatamento na Amazônia Legal por Satélite) monitors deforestation continuously since 1988, while for the other countries of the Amazon basin to our knowledge no similarly effective and transparent long-term continuous government programme has been in operation (however, most recently the Peruvian government has announced the implementation of "Terra-i-Peru" a programme similar to PRODES but for the Peruvian Amazon; Terra-i is a project run by non-government organisations and monitors deforestation of Latin America since 2011; see also Chap. 15. Keeping methodological caveats in mind, remote sensing data of area deforested (Fig. 6.2a) reveal fairly steady forest destruction followed by a remarkable decline since 2006. The time course is dominated by the Brazilian deforestation estimates (PRODES) and the recent decline has been attributed, e.g., by Lapola et al. (2014) to a temporary 'decoupling of agricultural expansion and deforestation' caused by an intensification of agriculture (i.e. an increase in production rate per area instead of an areal expansion) driven by agriculture currently dominated by large-scale agrobusiness (Lapola et al. 2014).

The availability of remote sensing data of cerradão (closed forest-like savanna) and forest area deforested, together with knowledge of subsequent vegetation type established, permits one to estimate associated carbon fluxes to the atmosphere and we sketch and use an approach to estimate these here. We use here only peer-reviewed published and thus clearly traceable estimates of area deforested and thus have refrained from the use of FAO data (see also Grainger 2008).

From the perspective of the carbon balance of the Amazon basin, it is not just instantaneous forest destruction which matters but also the carbon flux 'legacies' caused by deforestation and which occur years later. They include decomposition of

---

Fig. 6.1 (continued) the Basin, (**c**) annual maximum monthly mean minus annual minimum monthly mean Amazon river discharge at Óbidos. Óbidos is located ca. 800 km upstream from the Atlantic coast. Hydrological data are from the Brazilian Hydrological Service ANA, http://www2.ana.gov.br/, and, where measurements are missing (stippled), they are estimated from upstream river gauge stations by Callède et al. (2004)

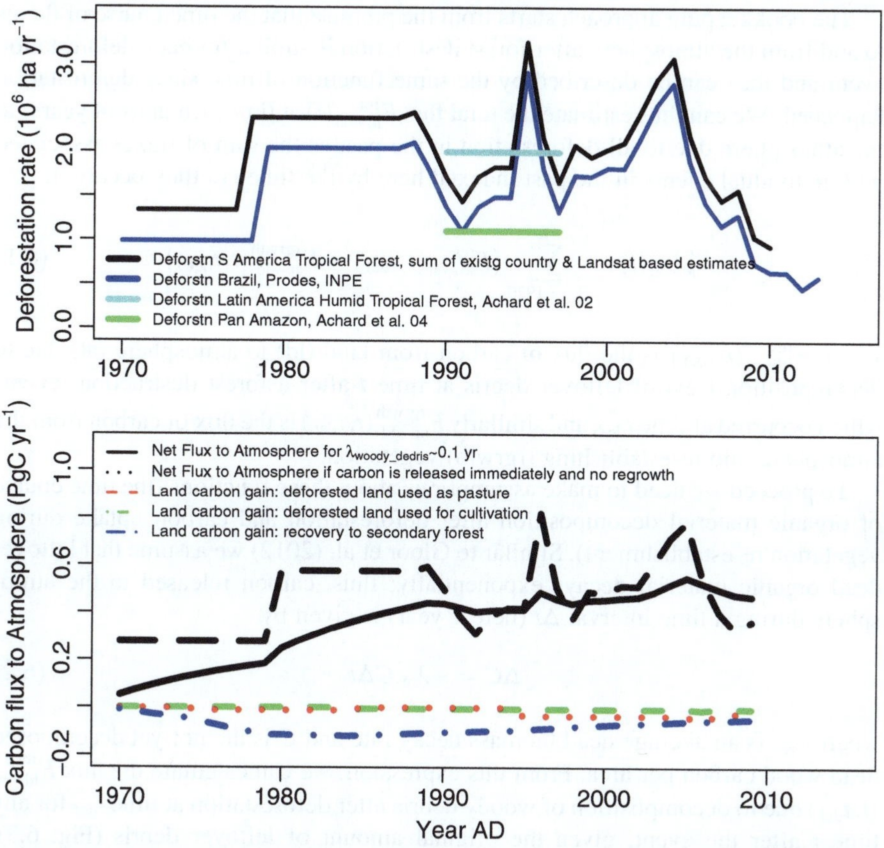

Fig. 6.2 (**a**) Comparison of deforestation rate estimates for the Amazon Basin/South American tropical forest area. The underlying numbers are given in Gloor et al. (2012) and complemented for this work with most recent numbers from PRODES. (**b**) Estimates of carbon release due to deforestation and carbon uptake by land vegetation due to regrowth

dead organic material which takes place over several years after forest destruction counteracted to some extent by carbon gains by regrowth, with the magnitude of those depending on the newly established (or old regrowing) vegetation type. The currently only available data-based approach to estimate the time course of carbon losses/gains after forest destruction is via model assumptions of the rates of decomposition and re-establishment of vegetation combined with data on areas deforested as, e.g., recorded from space. This method has been pioneered by Houghton et al. (1983) and is known as the 'bookkeeping' approach. We follow here in essence the same approach but use only deforestation area estimates based on remote sensing which have been published in the literature, i.e. are reproducible. As already mentioned unlike Houghton et al. (1983) we do not use estimates provided by the Food and Agriculture Organization of the United Nations (FAO).

The bookkeeping approach starts from the premise that the time course of fluxes to and from the atmosphere after forest destruction is similar for each deforestation event and thus can be described by the same function of time since deforestation happened. We can thus estimate the total flux $F^{tot}_{ld \to at}(t)$ at time t (in units of years) to the atmosphere due to all deforestation in the past as the sum of fluxes associated with individual events in the past indexed here by the time t_{def} they occurred:

$$F^{tot}_{ld \to at}(t) = \sum_{t_{def}=1970}^{t} \left(F^{res}_{ld \to at}(t, t_{def}) - F^{rgrwth}_{at \to ld}(t, t_{def})\right) \tag{6.1}$$

where $F^{res}_{ld \to at}(t, t_{def})$ is the flux of carbon from land (ld) to atmosphere (at) due to decomposition (res) of leftover debris at time t after a forest destruction 'event' which occurred at time t_{def}, and similarly $F^{rgrwth}_{at \to ld}(t, t_{def})$ is the flux of carbon from the atmosphere into re-establishing (rgrwth) vegetation.

To proceed we need to make assumptions about these functions (the time course of organic material decomposition after deforestation and carbon uptake during vegetation re-establishment). Similar to Gloor et al. (2012) we assume that leftover dead organic material decays exponentially; thus, carbon released to the atmosphere during a time interval Δt (here 1 year) is given by

$$\Delta C = -\lambda_{res} C \Delta t \tag{6.2}$$

where λ_{res} is an average dead biomass decay rate and C is the not yet decomposed dead wood carbon per area. From this expression, we can calculate the flux $F^{resp}_{ld \to at}$ (t, t_{def}) due to decomposition of woody debris after deforestation at time t_{def} for any time t after the event, given the original amount of leftover debris (Fig. 6.3). Re-establishment of vegetation after deforestation, in contrast, will cause uptake of carbon from the atmosphere. In this case, we assume that carbon content C of new vegetation will change over time following

$$C(t) = C_{steady}\left(1 - e^{-\lambda_{rgrwth} t}\right) \tag{6.3}$$

where C_{steady} is the new steady-state vegetation carbon content per area, i.e. it will first grow fast but then slow down and finally reach a steady state (Fig. 6.3). We may thus also calculate $F^{rgrwth}_{at \to ld}(t, t_{def})$, the flux from the atmosphere to vegetation at any time t after a deforestation event at time t_{def} (for explicit expressions, see Appendix) provided we know the newly establishing vegetation type and its steady-state carbon content. The simple choice for these functions permits to solve the problem analytically (see Appendix) without introducing much error.

We finally link the flux expressions to change in deforested area and new vegetation type estimated from remote sensing. We may for example estimate the amount of carbon which may maximally be released when an area ΔA is area deforested in year t_{def} as

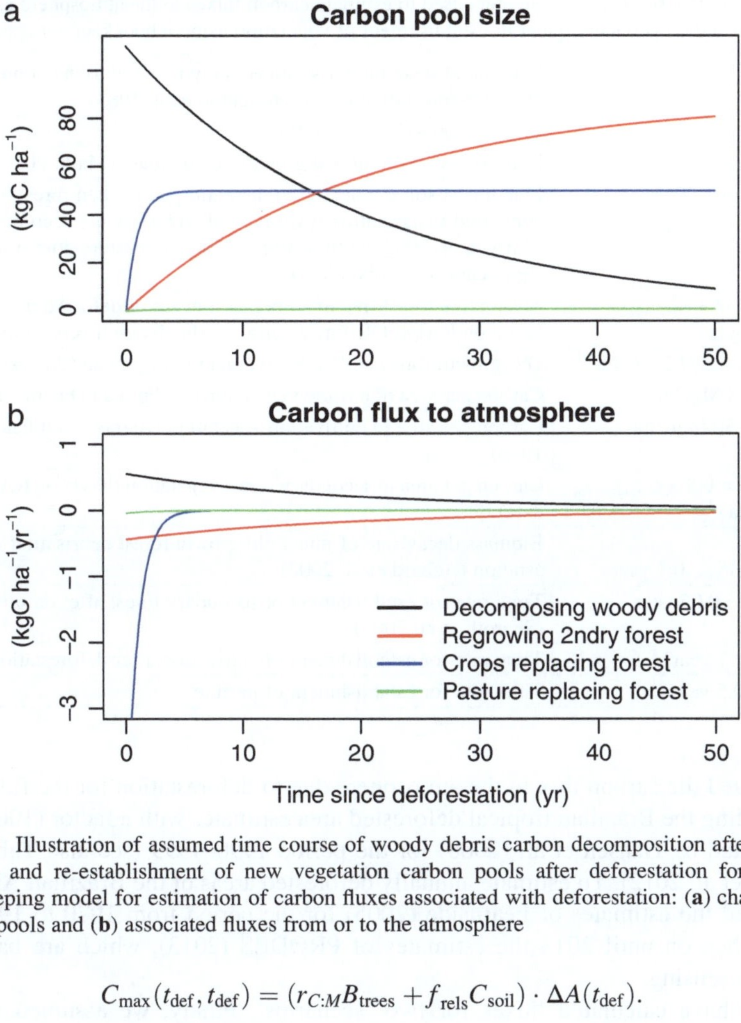

Fig. 6.3 Illustration of assumed time course of woody debris carbon decomposition after deforestation and re-establishment of new vegetation carbon pools after deforestation for use in bookkeeping model for estimation of carbon fluxes associated with deforestation: (**a**) changes of carbon pools and (**b**) associated fluxes from or to the atmosphere

$$C_{\max}(t_{\text{def}}, t_{\text{def}}) = (r_{C:M} B_{\text{trees}} + f_{\text{rels}} C_{\text{soil}}) \cdot \Delta A(t_{\text{def}}). \tag{6.4}$$

where $r_{C:M}$ is the carbon to biomass weight ratio of trees, B_{trees} the tree biomass per area, f_{rels} the fraction of soil organic carbon released to the atmosphere, and C_{soil} soil organic carbon content per area.

For our bookkeeping approach, we take into account that only a fraction α of woody carbon is immediately released to the atmosphere during destruction, while only the remaining fraction $(1 - \alpha)$ is decomposing over time following the time course described above and similar for soil organic carbon.

For carbon uptake by newly establishing vegetation after a deforestation event, we distinguished three vegetation types: pasture, crop cultivation, and secondary forest, the fraction of which we estimated from Brazilian government statistics (AGROPECUARIA), thus assuming the same land use time history after deforestation for all of the Amazon basin. Lacking data for the full Amazon basin, we

Table 6.1 Constants and parameters used to estimate carbon fluxes to the atmosphere as a result of deforestation and subsequent re-establishment of vegetation using a bookkeeping approach

$\alpha = 0.28$	Fraction of dead biomass immediately released to the atmosphere after a deforestation event (Houghton et al. 1983)
$r_{C:M} = 0.5$	Tree carbon to biomass ratio
$r_{\mathrm{BG:AG}} = 0.2$	Ratio of below- to above-ground tree biomass (Malhi 2010)
$r_{\mathrm{soil\ rel}} = 0.22$	Fraction of soil *C* released to the atmosphere when forest is converted to agriculture (Murty et al. 2002) (while according to Murty et al. 2002 the transition of forest to pasture does not lead to significant soil carbon loss)
$B_{\mathrm{trees}} = (1 + r_{\mathrm{BG:AG}}) \times$ 220 Mg ha^{-1}	Mean alive forest tree mass per area in arc of deforestation area based on RAINFOR forest censuses (Feldpausch pers. comm.)
$C_{\mathrm{for\ soil}} = 291$ MgC ha^{-1}	Old-growth forest soil carbon content (Jobaggy and Jackson 2000)
$C_{\mathrm{pas}} = 8$ MgC ha^{-1}	Carbon per area of pasture vegetation (Barbosa and Fearnside 1996)
$C_{\mathrm{cult}} = 50$ MgC ha^{-1}	Carbon per area in cultivation vegetation (Barbosa and Fearnside 1996)
$C_{\mathrm{sec\ for}} = 0.8 \times C_{\mathrm{oldgrw\ for}}$ MgC ha^{-1}	Carbon per area in secondary forest vegetation (based on RAINFOR data)
$\lambda_{\mathrm{oldgrw\ for}} = 0.05 \ldots 0.1$ year^{-1}	Biomass decay rate of intact old-growth forest debris after deforestation (Achard et al. 2002)
$\lambda_{\mathrm{sec\ for}} = 0.05$ year^{-1}	Time rate for establishment of secondary forest after deforestation (Schroth et al. 2002)
$\lambda_{\mathrm{cultiv}} = 1$ year^{-1}	Time rate for establishment of cultivation after deforestation
$\lambda_{\mathrm{pas}} = 0.5$ year^{-1}	Time rate for establishment of pasture

estimated the carbon flux to the atmosphere due to deforestation for the full Basin by scaling the Brazilian tropical deforested area estimates with a factor (100/79) as estimated by Hansen et al. (2008) for the period 1990–1999 (see also Table 7 in Gloor et al. 2012). To estimate annually deforested areas of the Brazilian Amazon, we used the estimates of Fearnside (2005) for the period from 1970 to 1987 and from then on until 2013 the estimates of PRODES (2013), which are based on remote sensing.

We have calculated fluxes for two ‘scenarios’. Firstly, we assumed that all biomass which eventually is released over time after forest destruction is released at once immediately after destruction. Secondly, we used our bookkeeping model to estimate lagged fluxes caused by decomposition which takes place over many years. The parameters used for the calculations are given in Table 6.1.

The results of the flux calculations for the two scenarios (Fig. 6.2b) illustrate two things. The net total flux to the atmosphere over recent years according to this calculation is 0.5 ± 0.15 Pg C year^{-1}, and the legacy of past deforestation extends over at least a decade or so past forest destruction. Uncertainties of our approach are mainly due to uncertainty in tree biomass per area and to area deforested, but this latter uncertainty is difficult to quantify (see also Gloor et al. 2012). Our estimates agree well with those of Achard et al. (2002, 2004) which cover the 1990–1995 period (Table 6.2).

Table 6.2 Summary of carbon flux estimates (Pg C $year^{-1}$) (sign convention: flux to atmosphere is positive)

	1975–1979	1980–1984	1985–1989	1990–1994
Deforestation				
Immediate release	0.37	0.63	0.62	0.47
Release if taking into account lags	0.20 ± 0.05	0.36 ± 0.09	0.48 ± 0.12	0.47 ± 0.12
Old-growth forest gains	–	−0.21 ± 0.23	−0.21 ± 0.23	0.57 ± 0.17
River carbon export	−0.07 ± 0.04	−0.07 ± 0.04	−0.07 ± 0.04	−0.07 ± 0.04
Total	–	+0.08	+0.20	−0.03
	1995–1999	2000–2004	2005–2009	2010–2013
Deforestation				
Immediate release	0.51	0.70	0.31	0.11
Release if taking into account lags	0.50 ± 0.12	0.57 ± 0.14	0.48 ± 0.12	–
Old-growth forest gains	−0.53 ± 0.14	−0.45 ± 0.25	−0.15 ± 0.23	–
River carbon export	−0.07 ± 0.04	−0.07 ± 0.04	−0.07 ± 0.04	−0.07 ± 0.04
Total	−0.10	+0.05	+0.26	–

6.3.2 *Intact Forest Carbon Balance*

Given that the atmospheric environment has been changing substantially over the last 50 years or so with CO_2 having increased from ca. 280 ppm during pre-industrial times to 400 ppm in 2014, and intact forests having also seen an increase in nitrogen deposition caused by fossil fuel burning, as well as a changing climate, the forests are expected to 'respond'. Such responses could for example be increased tree growth rates both below- and above-ground (Lloyd and Farquhar 1996) and/or shifts in species composition. One way to investigate whether such changes are indeed ongoing is to 'ask the trees themselves'. This is the approach taken by Phillips and co-workers (Phillips et al. 1998, 2009) who maintain a widespread network across the Amazon basin of permanent forest plots of 1 ha size where they undertake regularly detailed repeat censuses (Fig. 6.4, black dots). Thereby for each tree with a diameter larger than 10 cm, the diameter increment is being recorded continuously using girdles. Death and recruitment of tagged individuals is recorded as well and species of trees are identified. This dataset has revealed several very interesting results (see Chap. 10). Firstly, intact forests have gained carbon over about the last three decades, constituting an annual carbon sink of ca. 0.4 Pg C $year^{-1}$. Secondly, trees of the forest plots which were affected by the drought in 2005, which was located in the upper Solimões region and Southern fringes of the Basin (Espinoza et al. 2011), experienced higher than normal mortality. Mortality due to this event having been estimated to lead eventually to total carbon release to the atmosphere on the order of 1–1.5 Pg C, distributed over many years. Given the altered atmospheric composition, particularly increasing levels of CO_2, over the last decades the first of these two results seems not so

Fig. 6.4 Location of forest census plots of the RAINFOR network of Phillips et al. (2009) (*black dots*) and sites where regular vertical profiles of greenhouse gases are being measured (inside the Basin) using aircraft by Gatti and co-workers (see Gatti et al. 2014), as well as the National Oceanic and Atmospheric Administration (NOAA) background site at Ragged Point Barbados

surprising; nonetheless, it has been criticised on many grounds—the probably most important criticism being that rare severe disturbance events are not sufficiently represented in the dataset because of the limited number of plots (in the order of 150 by today) and the limited time span covered (the last 30 years) (e.g. Körner 2003). A related argument states that the limited size of the plots (1 ha) further biases low detected number of large disturbances because large disturbances are localised, i.e. spatially autocorrelated in contrast to randomly distributed 1 ha plots, which thus tend to miss them more often than would be expected just based on the ratio of area fraction affected by large disturbances (Chambers et al. 2013). Probably the best way to assess this issue is to determine the size frequency distribution of observed disturbances and then to explore its implications on limitations of conclusions based on forest plot censuses using a forest simulator. Such an analysis has recently been made by Espírito-Santo et al. (2014) who characterised the size frequency distribution across all spatial scales combining forest plot, LIDAR, and Landsat remote sensing data. The largest disturbances had nearly no effect on mean net gains although they strongly increased the variance of the mean, while intermediate-scale disturbances did affect slightly the mean but not much the variance. Most important for the net mean gains were small disturbances as captured by the forest census network by Phillips and co-workers. Altogether the Espírito-Santo et al.'s results give confidence in the forest census-based approach and confirm the results obtained of Phillips and co-workers.

6.3.3 Fossil Fuel Emissions

For completeness we mention that fossil fuel emissions in the Basin were negligibly small, <0.02 Pg C year^{-1} in 2010 and 2011 (Gatti et al. 2014, supplementary information online).

6.3.4 Riverine Carbon

Based on riverine CO_2 partial pressure measurements and estimates of gas exchange velocity across the river atmosphere interface, it is possible to estimate riverine efflux of CO_2 due to respiration of organic carbon in the waters. Furthermore, carbon transport to the Atlantic can be measured via measurements of dissolved inorganic carbon and river flow. The main results from such studies relevant for the whole Basin carbon balance are that there is a nearly closed loop of organic matter input into rivers and subsequent outgassing, while riverine export of carbon to the Atlantic is close to zero Pg C year^{-1} (Richey et al. 1990; Melack and Forsberg 2001; Melack et al. 2009). It is this latter quantity which is relevant for the whole-Basin carbon balance.

6.3.5 Amazon-Wide Carbon Balance Based on Lower Troposphere Greenhouse Gas Balances

We have described and attempted to quantify those component fluxes, which contribute to the Amazon carbon balance quantitatively at the Basin scale, and at the same time tried to make clear what the weaknesses of the approaches are as well as the nature of debates and critiques around them. Fairly recently new Basin-wide greenhouse gas concentration measurements have become available (Gatti et al. 2014) which provide a large-scale integrated picture of all land to atmosphere carbon fluxes and thus have the potential to resolve these debates and also test consistency of the component fluxes. The basic idea of the approach, building on the work of Chou et al. (2002), is to take advantage of the main lower-to-mid troposphere airstream over the Basin which enters in the equatorial region from the tropical Atlantic, travels over the full Basin towards the Andes where it is turned south-westwards, and then travels back to the Atlantic. This airstream can be compared to a virtual freight train with carbon being the load and the train being loaded by carbon fluxes coming from the land surface beneath the flow and offloaded by fluxes to the atmosphere. If the load far downstream increased compared to incoming air, then the land surface would be a carbon source for the atmosphere and vice versa. Air-parcel trajectories, calculated from meteorological data, are used to determine the air paths, and the carbon content of the air is estimated from vertical profiles at four locations distributed across the Basin along the main airstream (Fig. 6.3, red dots). For each measured profile and compound X measured ($X = CO_2$, CO, CH_4, ...), the flux is calculated as

$$F_X = \int_{z=0\,(\mathrm{agl})}^{4.4\,\mathrm{km}} \frac{\Delta X}{t(z)} \mathrm{d}z \tag{6.5}$$

where $\Delta X = X_{\mathrm{site}} - X_{\mathrm{bg}}$ is the difference between the concentration measured at the site and background (bg) air entering the Basin from the Atlantic, z the height above ground (agl), and $t(z)$ the air-mass trajectory travel time from the coast to height z of at the site. The greenhouse gas concentration of background air is estimated from SF_6 measured at the NOAA background stations Barbados (RGB) and Ascension (ASC) and in situ (site), using a linear mixing model:

$$X_{\mathrm{bg}} = f_{\mathrm{ASC}} \cdot X_{\mathrm{ASC}} + (1 - f_{\mathrm{ASC}}) \cdot X_{\mathrm{RPB}} \tag{6.6}$$

with weights f calculated as

$$f_{\mathrm{ASC}} = \frac{\mathrm{SF}_{6,\mathrm{site}} - \mathrm{SF}_{6,\mathrm{RPB}}}{\mathrm{SF}_{6,\mathrm{ASC}} - \mathrm{SF}_{6,\mathrm{RPB}}}. \tag{6.7}$$

SF_6 is suited for this purpose because it has virtually no sources in the Amazon basin, and atmospheric SF_6 concentration is substantially higher in the Northern compared to the Southern hemisphere. This is because most of its sources are located in the Northern hemisphere and sinks are negligible as it is destroyed only in the mesosphere.

As CO is also measured, it is possible to estimate the carbon flux component caused by burning using the same air path trajectory budget approach and using a $CO{:}CO_2$ mean emission ratio $r^{\mathrm{bb}}_{\mathrm{CO_2:CO}}$ to translate the CO flux to CO_2 flux as

$$F^{\mathrm{bb}}_{\mathrm{CO_2}} = r^{\mathrm{bb}}_{\mathrm{CO_2:CO}} \cdot \left(F_{\mathrm{CO}} - F^{\mathrm{bio}}_{\mathrm{CO}}\right). \tag{6.8}$$

where $F^{\mathrm{bb}}_{\mathrm{CO_2}}$ is CO_2 flux caused by biomass burning (bb), F_{CO} is net CO flux, and $F^{\mathrm{bio}}_{\mathrm{CO}}$ is a nearly constant small background carbon monoxide flux observed during the wet season attributed to plant and soil CO emissions (Gatti et al. 2010).

The study of Gatti et al. (2014) contrasted a dry and anomalously hot year (2010) and a wet year (2011). Regarding the full Basin carbon balance, the data suggest that during wet years the Basin is approximately in balance (0.06 ± 0.1 Pg C year^{-1}) with losses mediated by fires compensated by forest carbon gains of (0.25 ± 0.14 Pg C year^{-1}). They thus confirm the existence of an intact forest carbon sink during average conditions as has been found, based on forest census data. Since 2010 was a dry year, the results also give an assessment of the response of the land vegetation to such conditions—of particular interest given the current climate conditions of increasing variability of precipitation and increasing temperatures. During 2011, the land vegetation photosynthesis has temporally stalled, rendering the land vegetation neutral, while the Basin as a whole lost substantial amounts of carbon (ca. 0.48 ± 0.18 Pg C year^{-1}) via fires. Given that carbon release estimates based on remote sensing data and our bookkeeping approach are slightly larger than the estimates of carbon loss estimated from CO, and those include also fire-unrelated biomass decomposition components, the intact forest carbon sink may be even a bit larger than the numbers published by Phillips et al. (2009). Nonetheless, overall the numbers across different approaches and components are quite consistent.

6.4 Summary and Conclusions

Over the last decades, the two major processes determining the Amazon-wide carbon balance have been deforestation and intact forest carbon gains. Intact forest carbon gains tended to nearly cancel losses due to deforestation and thus the net balance of the Basin was close to zero within a range of ± 0.2–0.3 Pg C year^{-1}. Recent indications from both forest censuses and lower troposphere greenhouse gas budgets are that the carbon uptake of intact forests may be weakening in response to stress exerted by warmer and more variable climate. On the other hand, remote sensing data for the Brazilian Amazon suggest that deforestation activity has been decreasing over nearly the last 10 years, thus currently offsetting the weakening of the forest sink. Unfortunately, whether this recent trend of decreasing deforestation will continue and for how long is unclear.

Acknowledgements I thank John Grace, Laszlo Nagy, Jean Ometto, and an anonymous reviewer who have helped improve this chapter.

Appendix: Formalisation of Book Keeping Approach to Estimate Carbon Fluxes Caused by Deforestation Based on Data of Area Deforested and Simple Models of Woody Debris Decomposition and Vegetation Re-establishment

Carbon Release to the Atmosphere After Deforestation

Starting from Eq. (6.2) describing carbon release of woody debris after deforestation and using an annual time step $\Delta t = 1$ we obtain

$$C(t_{\mathrm{def}} + \Delta t, t_{\mathrm{def}}) = (1-\lambda)C(t, t_{\mathrm{def}}), C(t, t_{\mathrm{def}}) = (1-\lambda)^{t-t_{\mathrm{def}}} C(t_{\mathrm{def}}, t_{\mathrm{def}}) \quad (6.9)$$

where $C(t_{\mathrm{def}}, t_{\mathrm{def}})$ is the originally left over woody debris carbon not immediately released at the time t_{def} of deforestation and $C(t, t_{\mathrm{def}})$ the remaining not yet decomposed woody debris carbon. The flux to the atmosphere at time t caused by decomposition of leftover debris caused by deforestation at time t_{def} in the past is

$$\begin{aligned} F^{\mathrm{res}}_{\mathrm{ld}\to\mathrm{at}}(t, t_{\mathrm{def}}) &= -((C(t+\Delta t, t_{\mathrm{def}}) - C(t, t_{\mathrm{def}})) = \ldots = \\ &= \lambda_{\mathrm{res}}(1-\lambda_{\mathrm{res}})^{t-t_{\mathrm{def}}-1} C(t_{\mathrm{def}}, t_{\mathrm{def}}). \end{aligned} \quad (6.10)$$

The total flux to the atmosphere at time t due to deforestation in the past is then given by the sum of all contributions from the beginning of deforestation around 1970 until today or

$$\begin{aligned} F^{\mathrm{res,tot}}_{\mathrm{ld}\to\mathrm{at}}(t) = &\underbrace{\alpha \cdot r_{C:M} \cdot B_{\mathrm{res}} \cdot \Delta A(t)}_{\text{carbon immediately released during deforestation in year } t} \\ &\underbrace{+\lambda_{\mathrm{res}} \sum_{t_{\mathrm{def}}=1970}^{t} (1-\lambda_{\mathrm{res}})^{t-t_{\mathrm{def}}} \{(1-\alpha) \cdot r_{C:M} \cdot B_{\mathrm{trees}} + f_{\mathrm{rels}} \cdot C_{\mathrm{soil}}\} \cdot \Delta A(t_{\mathrm{def}})}_{\text{carbon released in year } t \text{ by decomposing leftover debris from deforestation in previous years}} \end{aligned} \quad (6.11)$$

with symbols explained in the main text and/or Table 6.1.

Carbon Uptake from the Atmosphere by Re-establishing land Vegetation

For the assumed time course of carbon uptake by regrowth after deforestation in year t_{def} (Eq. 6.3) the flux from atmosphere to land vegetation in year t is

$$F_{\text{at}\to\text{ld}}^{\text{rgrwth}}(t, t_{\text{def}}) = C(t + \Delta t, t_{\text{def}}) - C(t, t_{\text{def}}) = \ldots = \\ = \left(1 - e^{-\lambda_{\text{rgrwth}} \cdot \Delta t}\right) \cdot e^{-\lambda_{\text{rgrwth}}(t - t_{\text{def}})} \cdot C_{\text{steady}} \tag{6.12}$$

with $\Delta t = 1$ (in units of years, the time step we are using for summing contributions because satellite data of deforested area from Brazil are annual). The total flux of carbon at time t from the atmosphere to land due to regrowth in the wake of all deforestation events in the past is then

$$F_{\text{at}\to\text{ld}}^{\text{rgrwth, tot}}(t) = r_{C:M} \cdot B_{\text{veg}} \cdot \left(1 - e^{-\lambda_{\text{rgrwth}} \cdot \Delta t}\right) \cdot \sum_{t_{\text{def}}=1970}^{t-1} e^{-\lambda_{\text{rgrwth}} \cdot (t - t_{\text{def}})} \Delta A(t_{\text{def}}). \tag{6.13}$$

B_{veg} is the mass per area of the new vegetation type once fully established and as above $\Delta t = 1$.

References

Achard F, Eva HD, Stibig H-J, Mayaux P, Gallego J, Richards T, Malingreau JP (2002) Determination of deforestation rates of the world's humid tropical forests. Science 297:999–1002. doi:10.1126/science.1070656

Achard F, Eva HD, Mayaux P, Stibig HJ, Belward A (2004) Improved estimates of net carbon emissions from land cover change in the tropics for the 1990s. Global Biogeochem Cycles 18: GB2008. doi:10.1029/2003GB002142

Barbosa RI, Fearnside PM (1996) Pasture burning in Amazonia: dynamics of residual biomass and the storage and release of aboveground carbon. J Geophys Res 101:25847–25857

Callède J, Guyot JL, Ronchail J, L'Hôte Y, Niel H, De Oliveira E (2004) Evolution du débit de l'Amazone à Óbidos de 1903 à 1999. Hydrol Sci J 49:85–97

Chambers JQ, Negron-Juarez RI, Magnabosco Marra D, Di Vittorio A, Tews J, Roberts D, Ribeiro GHPM, Trumbore SE, Higuchi N (2013) The steady-state mosaic of disturbance and succession across an old-growth Central Amazon forest landscape. Proc Natl Acad Sci USA 110:3949–3954

Chou WW, Wofsy CS, Harriss RC, Lin JC, Gerbig C, Sachse GW (2002) Net fluxes of CO2 in Amazonia derived from aircraft observations. J Geophys Res 107(D22):4614. doi:10.1029/2001JD001295

Espinoza JC, Ronchail J, Guyot JL, Junquas C, Vauchel P, Lavado WS (2011) Climate variability and extreme drought in the upper Solimões River (Western Amazon basin): understanding the exceptional 2010 drought. Geophys Res Lett 38:L13406

Espírito-Santo FDB, Gloor M, Keller M, Malhi Y, Saatchi S, Nelson B, Oliveira Junior RC, Pereira C, Lloyd J, Frolking S, Palace M, Shimabukuro YE, Duarte V, Monteagudo Mendoza A, López-González G, Baker TR, Feldpausch TR, Brienen RJW, Asner GP, Boyd DS, Phillips OL (2014) Size and frequency of natural forest disturbances and the Amazon forest carbon balance. Nat Commun 5, 3434. doi:10.1038/ncomms4434

Fearnside PM (2005) Deforestation in Brazilian Amazonia: history, rates, and consequences. Conserv Biol 19:680–688

Feldpausch TR, Lloyd J, Lewis SL, Brienen RJW, Gloor M, Monteagudo Mendoza A, Lopez-Gonzalez G, Banin L, Phillips OL et al (2012) Tree height integrated into pan-tropical forest biomass estimates. Biogeosciences 9:3381–3403

Filizola N, Latrubesse EM, Fraizy P, Souza R, Guimarães V, Guyot JL (2014) Was the 2009 flood the most hazardous or the largest ever recorded in the Amazon? Geomorphology 215:99–105
Fu R, Zhu B, Dickinson RE (1998) How do atmosphere and land surface influence seasonal changes of convection in the tropical Amazon? J Climate 12:1306–1321
Fu R, Yin L, Lib W, Arias PA, Dickinson RE, Huanga L, Chakraborty S, Fernandes K, Liebmann B, Fisher R, Myneni RB (2013) Increased dry-season length over southern Amazonia in recent decades and its implication for future climate projection. Proc Natl Acad Sci USA 110:18110–18115
Gatti LV, Miller JB, D'Amelio MTS, Martinewski A, Basso LS, Gloor M, Wofsy S, Tans P (2010) Vertical profiles of CO2 above eastern Amazonia suggest a net carbon flux to the atmosphere and balanced biosphere between 2000 and 2009. Tellus Ser B 62:581–594. doi:10.1111/j.1600-0889.2010.00484.x
Gatti LV, Gloor M, Miller J, Doughty CE, Malhi Y, Domingues LG, Basso LS, Martinewski A, Correia CSC, Borges VF, Freitas S, Braz R, Anderson LO, Rocha H, Grace J, Phillips OL, Lloyd J (2014) Drought sensitivity of Amazonian carbon balance revealed by atmospheric measurements. Nature 506:76–80
Gloor M, Gatti LV, Brienen R, Feldpausch TR, Phillips OL, Miller J, Ometto JP, Rocha H, Baker T, de Jong B, Houghton R, Malhi Y, Aragão LEOC, Guyot JL, Zhao K, Jackson R, Peylin P, Sitch S, Poulter B, Lomas M, Zaehle S, Huntingford C, Lloyd J (2012) The carbon balance of South America: status, decadal trends and main determinants. Biogeosciences 9:5407–5430. doi:10.5194/bg-9-5407-2012
Gloor M, Brienen RJW, Galbraith D, Feldpausch TR, Schöngart J, Guyot JL, Espinoza JC, Lloyd J, Phillips OL (2013) Intensification of the Amazon hydrological cycle over the last two decades. Geophys Res Lett 2013(40):1–5. doi:10.1002/grl.50377
Grainger A (2008) Difficulties in tracking the long-term global trend in tropical forest area. Proc Natl Acad Sci USA 104:818
Hansen MC, Stehman SV, Potapov PV, Loveland TR, Townshend JRG, DeFries RS, Pittman KW, Arunarwati B, Stolle F, Steininger MK, Carroll M, DiMiceli C (2008) Humid tropical forest clearing from 2000 to 2005 quantified by using multi-temporal and multi-resolution remotely sensed data. Proc Natl Acad Sci USA 105:9439–9444
Houghton RA, Hobbie JE, Melillo JM, Moore B, Peterson BJ, Shaver GR, Woodwell GM (1983) Changes in the carbon content of terrestrial biota and soils between 1860 and 1980: a net release of CO_2 to the atmosphere. Ecol Monogr 53:235–262
Huntingford C, Zelazowski P, Galbraith D et al (2013) Simulated resilience of tropical rainforests to CO_2-induced climate change. Nat Geosci 6:268–273
Jobaggy E, Jackson R (2000) The vertical distribution of soil organic carbon and its relation to climate and vegetation. Ecol Appl 10:423–436
Körner C (2003) Slow in, rapid out—carbon flux studies and Kyoto targets. Science 300:1242–1243
Lapola DM, Martinelli LA, Peres CA, Ometto JPHB, Ferreira ME, Nobre CA, Aguiar APD, Bustamante MMC, Cardoso MF, Costa MH, Joly CA, Leite CC, Moutinho P, Sampaio G, Strassburg BBN, Vieira ICG (2014) Pervasive transition of the Brazilian land-use system. Nat Clim Change 4:27–35
Lloyd J, Farquhar GD (1996) The CO_2 dependence of photosynthesis, plant growth responses to elevated atmospheric CO_2 concentrations and their interaction with soil nutrient status. 1. General principles and forest ecosystems. Funct Ecol 10:4–32
Malhi Y (2010) The carbon balance of tropical forest regions, 1990–2005. Curr Opin Environ Sustain 2:237–244
Marengo JA, Tomasella J, Alves LM, Soares WR, Rodriguez DA (2011) The drought of 2010 in the context of historical droughts in the Amazon region. Geophys Res Lett 38. doi:10.1029/2011GL047436
Martinelli L, Victoria RL, Sternberg LSL, Moreira MZ (1996) Using stable isotopes to determine sources of evaporated water to the atmosphere in the Amazon basin. J Hydrol 183:191–204

Melack JM, Forsberg B (2001) Biogeochemistry of Amazon floodplain lakes and associated wetlands. In: McClain ME, Victoria RL, Richey JE (eds) The biogeochemistry of the Amazon Basin and its role in a changing world. Oxford University Press, New York, NY, pp 235–276

Melack JM, Novo EMLM, Forsberg BR, Piedade MTF, Maurice L (2009) Floodplain ecosystem processes. In: Keller M, Bustamante M, Gash J, Dias PS (eds) Amazonia and global change, 186, Geophysical monograph series. American Geophysical Union, Washington, DC, pp 525–542

Murty D, Kirschbaum MU, McMurtrie RE, McGilvray H (2002) Does the conversion of forest agricultural land change soil carbon and nitrogen? A review of the literature. Glob Chang Biol 8:105–123

Nepstad DC, Verissimo A, Alencar A, Nobre C, Lima E, Lefebvre P, Schlesinger P, Potter C, Moutinho P, Mendoza E, Cochrane M, Brooks V (1999) Large-scale impoverishment of Amazonian forests by logging and fire. Nature 398:505–508

Phillips OL, Malhi Y, Higuchi N, Laurance WF, Nunez PV, Vasquez RM, Laurance SG, Ferreira LV, Stern M, Brown S, Grace J (1998) Changes in the carbon balance of tropical forests: evidence from long-term plots. Science 282:439–442

Phillips OL, Aragão LEOC, Lewis SL, Fisher JB, Lloyd J et al (2009) Drought sensitivity of the Amazon rainforest. Science 323:1344–1347

Poeschl U et al (2010) Rainforest aerosols as biogenic nuclei of clouds and precipitation in the Amazon. Science 329:1513–1516

PRODES (2013) Brazilian government deforestation estimates based on remote sensing. http://www.obt.inpe.br/prodes/prodes_1988_2013.htm

Richey JE, Hedges JI, Devol AH, Quay P, Victoria RL, Martinelly L, Forsberg BR (1990) Biogeochemistry of carbon in the Amazon River. Limnol Oceanogr 35:352–371

Salati E, Dall'Olio A, Matsui E, Gat JR (1979) Recycling of water in the Amazon basin: an isotopic study. Water Resour Res 15:1250–1258

Schroth G, D'Angelo SA, Geraldes Teixeira W, Haag D, Lieberei R (2002) Conversion of secondary forest into agroforestry and monoculture plantations in Amazonia: consequences for biomass, litter and soil carbon stocks after 7 years. For Ecol Manage 163:131–150

Seneviratne SI, Corti T, Davin EL, Hirschi M, Jaeger EB, Lehner I, Orlowsky B, Teuling AJ (2010) Investigating soil moisture–climate interactions in a changing climate: a review. Earth Sci Rev 99:125–161

Silva Dias MAF, Rutledge S, Kabat P, Silva Dias PL, Nobre C, Fisch G, Dolman AJ, Zipser E, Garstang M, Manzi AO, Fuentes JD, Rocha HR, Marengo J, Plana-Fattori A, Sa LDÁ, Alvala ŔCS, Andreae MO, Artaxo P, Gielow R, Gatti L (2002) Cloud and rain processes in a biosphere–atmosphere interaction context in the Amazon Region. J Geophys Res 107 (D20):8072. doi:10.1029/2001JD000335

Spracklen DV, Arnold SR, Taylor CM (2012) Observations of increased tropical rainfall preceded by air passage over forests. Nature 489:282–285

Van der Ent RJ, Savenije HH, Schaefli B, Steele-Dunne SS (2010) Origin and fate of atmospheric moisture over continents. Water Resour Res 46:W09525. doi:10.1029/2010WR009127

Werth D, Avissar R (2002) The local and global effects of Amazon deforestation. J Geophys Res 107(D20):8087. doi:10.1029/2001JD000717

Chapter 7
Aquatic Ecosystems

John M. Melack

7.1 Introduction

The processing of carbon by aquatic ecosystems of inland waters is now recognised as a significant component of regional and global carbon dynamics. In particular, the high rates of sedimentation in lakes and reservoirs and considerable evasion of carbon dioxide and methane from many rivers, lakes, and wetlands lead to fluxes disproportionately large relative to the area of inland waters (Cole et al. 2007; Downing 2009; Battin et al. 2009; Aufdenkampe et al. 2011; Butman and Raymond 2011; Raymond et al. 2013; Stanley et al. 2015). Although the magnitude and variability of these fluxes remain uncertain, especially in tropical regions, recent studies are improving our understanding of carbon dynamics in the streams, rivers, lakes, reservoirs, and wetlands of the Amazon basin. The Large-scale Biosphere Atmosphere Experiment in Amazonia (LBA) and related activities have resulted in numerous relevant publications, many of which have been summarised in a recent monograph (Gash et al. 2009). In particular, Richey et al. (2009) described carbon processing from streams and rivers, and Melack et al. (2009) examined ecosystem processes in inundated areas. From these studies, it appears that evasion of carbon dioxide from Amazonian rivers, lakes, and temporally inundated aquatic habitats is of similar magnitude to net ecosystem exchanges in non-inundated upland forests (*terra firme* forests) derived from eddy covariance measurements. In addition, Amazonian aquatic ecosystems account for a significant proportion of the global methane flux from natural wetlands.

J.M. Melack (✉)
Bren School of Environmental Science and Management, University of California, Santa Barbara, CA, USA

Earth Research Institute, University of California, Santa Barbara, CA, USA
e-mail: melack@bren.ucsb.edu

© Springer-Verlag Berlin Heidelberg 2016

L. Nagy et al. (eds.), *Interactions Between Biosphere, Atmosphere and Human Land Use in the Amazon Basin*, Ecological Studies 227,
DOI 10.1007/978-3-662-49902-3_7

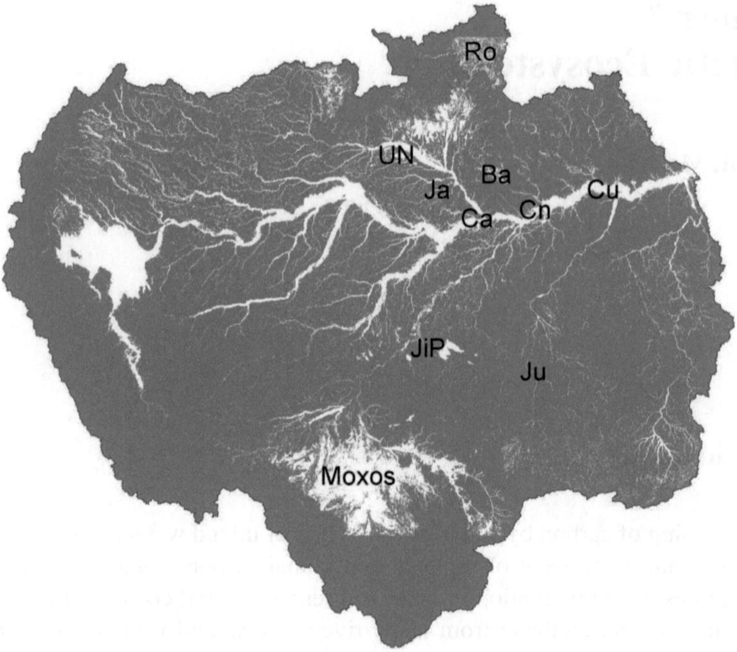

Fig. 7.1 Lowland Amazon basin (area ≤500 m; *grey*) with floodable areas (*white*), as described in Melack and Hess (2010) and Hess et al. (2003). Small areas of wetlands on the northern and southern edges are not shown because remote sensing data used to develop the wetland distribution were not available there. *Ba* Balbina Reservoir; *Ca* Lake Calado; *Cu* Lake Curuai; *Cn* Lake Canaçari; *Ja* Jau River; *JiP* Ji-Paraná River; *Ju* Juruena watershed; *Moxos* Llanos de Moxos; *Ro* Roraima; *UN* upper Negro interfluvial wetlands

With a focus on exchanges of carbon dioxide and methane between inland waters and the atmosphere, the purpose of this chapter is to extend the regional extent of previous analyses to the whole lowland Amazon basin (regions <500 m above sea level; 5.05 million km^2; Fig. 7.1; not including the Tocantins basin) by synthesis and critical evaluation of recent publications. Remote sensing analyses of inundation and wetland habitats, inundation modelling, measurements in rivers, reservoirs and other types of wetlands, and improved estimates of gas exchange coefficients for standing and flowing water are included. Regional estimates of gas fluxes are complemented with consideration of sources of the carbon being released, and remaining uncertainties and research needs are discussed. A discussion of climate trends and variability, exceptional events, and human impacts, as factors potentially altering carbon dynamics in the Amazon basin, is presented.

Several current projects are in the process of publishing their results, and will continue to contribute to our understanding of carbon dynamics in the Amazon, though only a portion is available for inclusion here. These projects include HYBAM (http://www.ore-hybam.org/index.php/eng/documents), CARBAMA (http://carbama.epoc.u-bordeaux1.fr/PUBLICATIONS.html), the 'Rede Beija Rio' network (http://boto.ocean.washington.edu/story/Amazon), A Biogeoquímica do Carbono e Mercúrio na Bacia Amazônica (Barbosa 2013) (http://www.biologia.ufrj.br/limnologia/projeto-carbono.php), and GEOMA (http://www.geoma.lncc.br/).

7.2 Inundation and the Variable Extent of Aquatic Habitats: Remote Sensing and Modelling

Carbon dynamics and carbon dioxide and methane exchanges in the Amazon basin vary among the diverse aquatic habitats. Junk et al. (2011) used information on climate, hydrology, water and sediment chemistry, and botany to delineate 14 major types of naturally occurring wetlands in the lowland Amazon. The amplitude, duration, frequency, and predictability of inundation are key criteria in this classification. While the classification has broad utility, data on all the criteria, including variations in inundation, are lacking for many parts of the Amazon. Hence, assigning spatial detail and areal extent to the various wetland types remains a challenge.

With a focus on gas exchange at the air–water interface, it is essential to have estimates of the surface area of the aquatic habitats, a daunting requirement given the large size of the Amazon basin and the wide range in dimensions of the habitats from headwater streams (<1 m across) to floodplains fringing major rivers (tens of km wide). While remote sensing can provide excellent information, currently available data on a regional scale have a spatial resolution of about 100 m, and sensors that allow seasonal and inter-annual variations in inundation to be recorded have spatial resolution of tens of km. Therefore, additional approaches are needed, especially for small rivers and streams, including geomorphology and modelling.

7.2.1 Remote Sensing

A variety of remote sensing approaches using passive and active microwave, laser, visible, and near-infrared and gravity anomaly detection systems are available and have been applied to tropical aquatic environments (Melack 2004). Melack and Hess (2010) applied the methodology of Hess et al. (2003), based primarily on mosaics of synthetic aperture radar (SAR) data obtained during a period of low and a period of high river stage, to the lowland Amazon basin to determine floodable area, inundated area, and areal extent of major habitats permanently or periodically inundated. Total floodable area within the lowland basin was estimated as 800,000 km^2. However, portions of southern Bolivia were not covered by available SAR data, and floodplains in the south-western Brazilian Amazon were not well delineated owing to the timing of data acquisitions. Open water, floating macrophytes or grasslands, and flooded or unflooded forests were distinguished, and the spatial resolution of the products was about 100 m. Areal extent of the aquatic habitats was reported for 31 river basins and five reaches of the Solimões-Amazonas River. While not so specific as the classification proposed by Junk et al. (2011), these habitats and divisions by river basin are relevant to the biogeochemical processes considered here. High-resolution, remotely sensed products are available for specific locations in the Amazon and provide information appropriate for validation of basin-scale products (e.g. Silva et al. 2010; Renó et al. 2011; Hawes et al. 2012; Arnesen et al. 2013).

Several remote sensing approaches have been used to estimate the areas of open water in the lowland Amazon basin. Open water area was determined by Melack and Hess (2010) as 64,800 km^2 from high water data. The Shuttle Radar Topography Mission (SRTM) offers digital elevations with 90 m horizontal postings (Jarvis et al. 2008) and a water body product (http://dds.cr.usgs.gov/srtm/version2_1/SWBD/). Open water area derived from the SRTM data is 72,000 km^2, with river channels wider than approximately 300 m covering 52,000 km^2. Hanson et al. (2013) provide a global composite of Landsat data at 30 m resolution from which an open water of 92,000 km^2 can be derived (Forsberg, personal communication). Global datasets of lakes, such as those in Lehner and Döll (2004), need refinement to be applicable to the dynamic and spatially complex conditions in the Amazon.

Remote sensing of seasonal variations in inundated area depends on systems with coarse spatial resolution, and time series products are available at a 25-km scale. Hamilton et al. (2002, 2004) used data from satellite-borne passive microwave sensors to determine monthly inundation on the mainstem Amazon River floodplain (Brazil), the Llanos de Moxos (Bolivia), the Bananal (Brazil), and Roraima savannas (Brazil and Guyana). Regressions between flooded area and stage heights in nearby rivers were used to extend the records of inundation for nearly a century for the Amazon mainstem and for several decades in the other floodplains. Prigent et al. (2007), Papa et al. (2008, 2010), and Prigent et al. (2012) provide inundation estimates at 0.25° resolution, derived from several satellite-borne sensors for the period from 1993 to 2007. Comparison of these products with SAR-based estimates of inundation areas in the Amazon basin indicated generally good correspondence for moderate to large inundated units. Aires et al. (2013) used the 25-km resolution products in conjunction with SAR-based images to develop a wetland dataset for the Amazon for a 15-year period with a resolution of about 500 m. Schröder and McDonald (pers. comm.) are producing a monthly inundation product for the period from 2002 to 2009 from AMSR-E (passive microwave) and QSCAT (radar scatterometry) data at 0.25° resolution.

Gravity anomalies detected by the GRACE satellites provide estimates of changes in water volumes partially associated with seasonal variations in inundation at a scale of 100,000 s of km^2, and have been used basin-wide in the Amazon (Alsdorf et al. 2010; Xavier et al. 2010). Frappart et al. (2005) combined satellite-derived SAR and altimetry data with in situ gauges to calculate water storage and inundation for the Negro River basin

7.2.2 *Geomorphological Approaches to River Areas*

Beighley and Gummadi (2011) combined relationships developed from hydraulic geometry with a high-resolution drainage network to estimate cumulative channel lengths and surface areas. Their analysis was done for drainage areas from 1 to 431,000 km^2. For channels >2 m in width, they estimated that the Amazon basin contains c. 4.4 million km of channels with a combined area of 59,700 km^2. Channels over 150 m in width were estimated to represent 29,500 km^2 of the combined area.

To determine the area of streams and rivers in the Ji-Paraná basin (Fig. 7.1), Rasera et al. (2008) developed empirical relationships between drainage area and channel width combined with river lengths derived from a digital river network (Mayorga et al. 2005a). Areas of rivers from third to sixth order covered on average 342 km^2 within the 75,400 km^2 Ji-Paraná basin, representing 0.45 % of the basin. Assuming that a similar relationship applied to the whole lowland Amazon basin, an area of 22,700 km^2 would result for rivers from third to sixth order. This value is similar to that estimated by Beighley and Gummadi (2011) for rivers in that size range.

Richey et al. (2002) estimated areas of moderate to small rivers and streams as a geometric series relating stream length and width to stream order. They calculated a total channel area of 21,000 km^2 at low water for a 1.77×10^6 km^2 area in the central Amazon, representing a fractional area of 1.2 %. If this fractional proportion is extrapolated to the whole lowland basin, a total channel area of 60,600 km^2 would result.

Downing et al. (2012) employed stream network theory combined with data on stream width to approximate the areal extent of streams and rivers, when within their channels, on continental scales, and estimated that rivers and streams were likely to cover 0.30–0.56 % of land surfaces. Stream and river areas in the conterminous United States represent 0.52 % of the land surface (Butman and Raymond 2011). If these percentages are applied to the lowland Amazon basin with an area of 5.05×10^6 km^2, areas for all rivers and streams range from 15,150 to 28,300 km^2.

These different areas for river and stream channels vary depending on approach and region considered, and given the basin-wide, explicit method used by Beighley and Gummadi (2011), their values are adopted here. Headwater streams are not represented by their values. For large rivers, the area derived from the SRTM data is also used. Further refinement of stream areas is required.

7.2.3 Modelling of Inundation

Several models of river discharge and associated inundation dynamics have recently been developed and applied to the Amazon basin (Coe et al. 2007; Beighley et al. 2009; Victoria 2010; Yamazaki et al. 2011; Getirana et al. 2012; Paiva et al. 2013). Miguez-Macho and Fan (2012) compiled a groundwater dataset for the Amazon and combined it with a hydrological model. While generally successful at calculating river discharges, the models' ability to represent inundation in the full range of floodplain and wetlands is only moderate, largely because of the lack of sufficiently accurate and detailed digital elevation models (DEMs). For example, Paiva et al. (2013) compared their modelled inundation in 3-month intervals with 0.25° remotely sensed estimates (Papa et al. 2010) averaged for the period from 1999 to 2004. The match was good in the central basin, but modelled areas significantly underestimated flooded area in the Bolivian Amazon and lower mainstem in Brazil. In the Peruvian Amazon, where the SAR analyses of Melack and Hess (2010) indicated a large area of wetland, the model also did so, but the coarse remotely sensed data did not. To better evaluate modelled results, it would be beneficial to examine SAR-derived products for specific periods and locations. Hydraulic models of flooding applied on a mesoscale have done well in locations with high-resolution DEMs (Wilson et al. 2007; Rudorff et al. 2014a, 2014b).

7.3 Gas Transfer Velocity Between Water and Atmosphere

Exchange of carbon dioxide and methane between surficial water and overlying atmosphere depends on the concentration gradient between air and water and on physical processes at the interface, usually parameterised as a gas transfer velocity (k), also called a piston velocity or gas exchange coefficient. Gas transfer velocities are a function of turbulence, kinematic viscosity of the water, and the molecular diffusion coefficient of the gas; the Schmidt number is the ratio of the latter two terms and is gas specific (MacIntyre et al. 1995). Schmidt numbers used here are normalised to carbon dioxide in freshwater at 20 °C and referred to as $k(600)$ (Engle and Melack 2000). Gas transfer velocities are influenced by atmospheric stability and, in water, are altered by currents, wind, and convection, as well as rain (Ho et al. 2007), temperature, organic surficial films, and changes in hydrostatic pressure. Methane can also exit via bubbles (ebullition; Crill et al. 1988) and pass through tissues of rooted aquatic plants, both herbaceous and woody (Brix et al. 1992; Rice et al. 2010).

In lakes, direct measurements of exchange can be made with floating chambers, as has been done in the Amazon since the 1980s (e.g. Crill et al. 1988; Guérin et al. 2007). Alternatively, measurements of gas concentrations can be combined with estimates of k to calculate diffusive fluxes. While collecting and assaying samples for carbon dioxide and methane are fairly straightforward, the selection of appropriate *k* values remains a challenge. In a study at Lake Calado (Fig. 7.1), Crill et al. (1988) used a surface renewal model, which has a sound theoretical basis (Banerjee and MacIntyre 2004), as the basis for determination of *k*. Empirical relations between wind speed and *k* have been applied as well (e.g. Engle and Melack 2000; Guérin et al. 2007). Rudorff et al. (2011) used three different models of *k*: a simple wind-based equation, a small eddy version of the surface renewal model, and a wind-based model that includes diel heating and cooling (as described by MacIntyre et al. 2010). In a short-term experiment in the open water of a floodplain lake in the central Amazon basin, Polsenaere et al. (2013) applied an eddy covariance technique to calculate fluxes and *k* values for CO_2.

Results from Rudorff et al. (2011) and Polsenaere et al. (2013) indicate that *k* values for standing waters in the Amazon are underestimated if based on simple wind-based relations commonly used. Rudorff et al. (2011) reported gas transfer coefficients that take into account wind as well as heating and cooling were on the order of 10 cm h^{-1}. Polsenaere et al. (2013) reported *k* values ranging from 1.3 to 31.6 cm h^{-1}, averaging 12.2 ± 6.7 cm h^{-1}. Under conditions with high sensible and latent heat fluxes, but low wind speeds (<2.7 m s^{-1}), *k* values near or above 20 cm h^{-1} were recorded. Based on floating chambers deployed in Balbina Reservoir (Fig. 7.1), Kemenes et al. (2011) calculated *k* values from 1.1 to 24.7 cm h^{-1}, with an average of about 12 cm h^{-1}. In contrast, Guérin et al. (2007) reported *k* values of about 2.5 cm h^{-1}, based on chambers and eddy covariance in Petit Saut Reservoir, but noted that rain led to *k* values for CO_2 from 0.8 to 13.4 cm h^{-1} at rain rates of 0.6–25 mm h^{-1}. These results indicate that *k* values in lakes are generally higher than those used in prior regional extrapolations; e.g. Richey et al. (2002) used *k* values of 2.7 ± 1 cm h^{-1} for floodplains and lakes.

In warm tropical waters, such as those in the Amazon basin, latent heat fluxes are especially important and lead to convective mixing and enhanced *k* values. Conversely, diurnal heating under strong insolation can cause stable stratification of the water column that may lead to low or high *k* values. Hence, given the pronounced diel cycle of heating and cooling often observed in shallow tropical lakes, it is important to measure stratification and mixing, and gas exchanges, over these diel cycles, although such measurements are very seldom done.

The lack of studies of *k* values and gas concentrations in vegetated habitats adds further uncertainty, especially because of the large areas of flooded forests and floating macrophytes throughout the Amazon basin (Melack and Hess 2010; Junk et al. 2011). Though winds and direct heating are lower in vegetated habitats than in open waters, convective mixing and horizontal exchanges driven by differential heating and cooling and associated eddies, as water moves through the vegetation (Ortiz et al. 2013), will likely increase *k* values. Release of hydrophobic organic molecules by aquatic plants may reduce gas exchange within flooded vegetation.

Spatial variations in CO_2 and CH_4 concentrations can be large, as reported for the Amazon basin (Rudorff et al. 2011; Polsenaere et al. 2013; Abril et al. 2014) and elsewhere (Roland et al. 2010; Hofmann 2013). The high variability in time and space of bubbling adds further variance to methane evasion rates. Though floating chambers can capture bubbles, to increase spatial and temporal coverage submerged funnels are usually used. Recent applications of hydroacoustic measurements have allowed significant improvements in estimation of ebullition (Del Sontro et al. 2011), though this method has yet to be applied in the Amazon basin.

In flowing waters, most calculations of gas exchange are based on *k* in combination with concentration measurements, though direct measurements with floating chambers have also been used. Values of *k* for Amazon waters have been derived from ^{222}Rn mass balances and floating chambers (Devol et al. 1987; Alin et al. 2011; Kemenes et al. 2011; Rasera et al. 2013). Alin et al. (2011) reported that values of *k* were significantly higher in small rivers and streams (channels <100 m wide), where current velocities and depth were found to be important, than in large rivers (channels >100 m wide), where wind was important. The range of *k* (600) values reported by Alin et al. (2011) for large rivers in the Amazon (1.2–31.1 cm h^{-1}) is quite similar to that reported by Rasera et al. (2013) from a multi-year study in six non-tidal rivers in the Amazon basin (1.3–31.6 cm h^{-1}). For the Uatumã River below Balbina Reservoir, Kemenes et al. (2011) reported average k values of 10.5 cm h^{-1}. As in the case for lakes, these results indicate that *k* values are generally higher in flowing waters of the Amazon basin than those used in prior regional extrapolations; e.g. Richey et al. (2002) used *k* values of 9.6 ± 3.8 cm h^{-1} (Amazon mainstem) and 5 ± 2 cm h^{-1} (major tributaries).

7.4 Carbon Dioxide and Methane Concentrations and Fluxes

Recent measurements of carbon dioxide and methane concentrations and fluxes have been made in streams, rivers, wetlands, and a few lakes and reservoirs. These results are summarised by region with the intent of providing the basis for improved basin-wide estimates. Hence, areal fluxes are calculated for each habitat as Mg C km^{-2} $year^{-1}$, kg C km^{-2} day^{-1}, or g C m^{-2} $year^{-1}$, as averages and/or ranges, if appropriate or possible.

7.4.1 *Streams and Rivers*

Working in remote headwater streams in the southern Amazon basin, Johnson et al. (2006, 2007, 2008) determined that most of the CO_2 in the streams had been terrestrially respired within soils and that almost all was evaded to the atmosphere within headwater reaches. Baseflow delivered groundwater highly supersaturated in CO_2, while during storms surface run-off and direct precipitation were relatively low in CO_2. Concentrations of CO_2 near the source of these headwater streams were 10,000–50,000 μatm. In a year-long study of a perennial, first-order stream in southern Mato Grosso draining a forested catchment, Neu et al. (2011) recorded pCO_2 concentrations from 6490 to 14,980 μatm and evasion rates from the stream surface of c. 6490 ± 680 g C m^{-2} $year^{-1}$; methane concentrations in the stream ranged from about 290–440 μatm and evasion averaged 990 ± 220 g C m^{-2} $year^{-1}$. Similarly, Davidson et al. (2010) measured high pCO_2 levels (average value of 19,000 μatm) in headwaters in remnant forests in northeastern Pará. Vihermaa et al. (2014) sampled two small streams and the La Torre and Tambopata rivers in the Madre de Dios region of the western Amazon; they reported CO_2 evasion rates from 1866 to 82,900 mg C m^{-2} day^{-1} for the Tambopata River and a perennial stream, respectively. Working on small to moderate-sized rivers in the Ji-Paraná basin (Rondônia), Rasera et al. (2008) measured CO_2 evasion rates per unit of river area from 695 to 13,095 mg C m^{-2} day^{-1} for third- and fourth-order rivers and from 622 to 4686 mg C m^{-2} day^{-1} for fifth- and sixth-order rivers in the basin.

Richey et al. (2009) reported pCO_2 concentrations ranging from 500 to 20,000 μatm and illustrated a positive correlation of pCO_2 with discharge over 4 years for the Solimões (at Manacapuru), the Madeira (at Porto Velho), and Ji-Paraná rivers. Borges et al. (2015) sampled along the mainstem Solimões and Amazon rivers and at the mouths of major tributaries and reported a range in pCO_2 concentrations from 70 to 16,880 ppm. Ellis et al. (2012) measured concentrations from 860 μatm in the Acre River to 12,900 μatm in a stream in *campina* vegetation. Based on a 5-year study with seasonal sampling of six non-tidal rivers (Negro, Solimões, Teles Pires, Cristalino, Araguaia, and Javaés) and one tidal river (Caxiuanã), Rasera et al. (2013) reported a range of pCO_2 concentrations from 259 to 7808 μatm and demonstrated a strong correlation between pCO_2 and discharge. They reported a range in CO_2 flux from uptake of 830 mg C m^{-2} day^{-1} to evasion of 15,860 mg C m^{-2} day^{-1}. Uptake occurred in clear water rivers at low water in conditions conducive to algal growth. Alin et al. (2011) reported a similar range of evasion rates (41–14,720 mg C m^{-2} day^{-1}), as did Ellis et al. (2012) with a range from 830 to 13,170 mg C m^{-2} day^{-1}. Abril et al. (2014) conducted eight 800-km cruises along the main channel of the Solimões-Amazon River and portions of its major tributaries in the central basin and measured pCO_2 every minute while underway. Values of pCO_2 were similar to previous studies and varied from approximately 1000–10,000 ppmv, except for those in the Tapajós River which were lower.

Raymond et al. (2013) included regional information in their global estimates of carbon dioxide emissions from streams and rivers. Carbon dioxide concentrations in streams and rivers of whole Amazon basin (including the Andes) averaged about 6890 μatm and their calculated efflux, using a $k(600)$ of 29 cm h^{-1}, averaged 19,100 mg C m^{-2} day^{-1}, expressed in relation to surface area of rivers and streams, not in relation to land area, as given in Raymond et al. (2013). The $k(600)$ value used is at the high end of those reported by others for the rivers of the lowland Amazon as it includes streams and rivers in highland portions of the basin. Hence, the areal efflux is probably too high as a basin-wide average.

Sawakuchi et al. (2014) reported measurements of methane concentrations and fluxes from the mainstem Solimões-Amazon River and five tributaries (Negro,

Madeira, Tapajós, Xingu, and Para) based on floating chambers. Sixteen of the 34 sites were sampled during low and high water. Dissolved, near-surface methane concentrations ranged from 0.02 to 0.5 μM. Overall, average riverine flux was 16.8 kg C km^{-2} day^{-1}.

7.4.2 Lakes

In Lago Grande de Curuai (Fig. 7.1), a floodplain composed of interconnected lakes with a flooded area ranging seasonally from 850 to 2274 km^2, Rudorff et al. (2011) noted gradients in CO_2 concentration with higher concentrations near littoral regions with floating macrophytes than farther off shore as well as seasonal differences in concentrations. Polsenaere et al. (2013) used an equilibrator connected to an infrared gas analyser to measure pCO_2 on transects in Lake Canaçari (Fig. 7.1), 450 km^2 in area, during a 4-day measurement period. Incorporation of extensive spatial sampling in these two studies permitted recognition of spatial patterns not possible from other studies having far fewer samples. Based on transects of pCO_2 and eddy covariance-based k values, Polsenaere et al. (2013) calculated mean evasion of 612 kg C km^{-2} day^{-1} during a 4-day low water period. Rudorff et al. (2011) estimated mean fluxes of CO_2 from open water in L. Curuai of 2930, 4180, 4450, and 4370 kg C km^{-2} day^{-1} during receding, low, rising, and high water levels, respectively. Abril et al. (2014) include five large lakes as part of their continuous transects of pCO_2 in the central Amazon and reported variations from approximately 20–20,000 ppmv. They found that the carbon dioxide efflux increased as the percentage of floating, emergent aquatic vegetation in the lakes increased.

Raymond et al. (2013) calculated carbon dioxide concentrations in lakes of whole Amazon basin (including the Andes) as averaging about 1906 μatm and an efflux, using a $k(600)$ of 5.8 cm h^{-1}, of 1230 kg C km^{-2} day^{-1}, expressed in relation to surface area of rivers and streams, not in relation to land area, as given in Raymond et al. (2013).

7.4.3 Wetlands

The Negro River basin includes extensive flooded forests (locally called *igapó*) and large areas of interfluvial wetlands (Hess et al. 2003). Based on multi-temporal synthetic aperture radar data and field measurements with floating chambers of methane flux in the Jau River basin (Fig. 7.1) made by Rosenqvist et al. (2002), mean annual emission of methane from *igapó* was 23 Mg C km^{-2} $year^{-1}$. Upper Negro interfluvial wetlands are a mosaic of emergent grasses, sedges, shrubs, and palms with shallow permanent water or seasonal flooding. Belger et al. (2011) measured methane uptake on unflooded lands, evasion from flooded areas as diffusive and ebullitive fluxes with chambers and funnels, and as transport through rooted plants. Carbon dioxide fluxes were calculated from measurements of CO_2 concentrations in air and water and a k value of 2.7 cm h^{-1}. Based on annual emission from two interfluvial wetlands representative of the region (Fig. 7.1), Belger et al. (2011) estimated average areal emission from wetland areas as 770 Mg C km^{-2} $year^{-1}$ for CO_2 and 21 Mg C km^{-2} $year^{-1}$ for CH_4.

Large savanna floodplains occur in the Llanos de Moxos (Bolivia) and in Roraima (Brazil) (Fig. 7.1) (Hamilton et al. 2002; Ferreira et al. 2007). Based on measurements in similar systems elsewhere, Melack et al. (2004) approximated mean annual methane emission from these two areas as 70 Mg C km^{-2} $year^{-1}$. Jati (2013) made monthly measurements of carbon dioxide and methane flux with floating chambers in 80 wetlands near Boa Vista (Roraima); mean values from his results were about 9670 kg C km^{-2} day^{-1} and 9.6 kg C km^{-2} day^{-1}, for CO_2 and CH_4, respectively. These are quite high CO_2 fluxes and rather low CH_4 fluxes compared to other Amazonian habitats. Emissions from cultivated rice in Roraima are not available.

7.4.4 Reservoirs

Five hydroelectric reservoirs (Tucuruí, Balbina, Samuel, Curuá-Una, Serra da Mesa), covering about 6300 km^2, currently exist in the lowland Amazon in Brazil. All were constructed decades ago and continue to release both carbon dioxide and methane from their surfaces and through their turbines and to enhance releases in downstream rivers. Only Balbina has data collected from multiple upstream and downstream stations over a full year as well as measurements of fluxes associated with turbines (Kemenes et al. 2007, 2011). Though not in the Amazon basin, multiyear studies at Petit Saut (French Guiana), located in tropical forest, include measurements in the reservoir and downstream (Abril et al. 2005). Data scattered through the years at other Amazonian reservoirs are also available (see citations in Melack et al. 2004; Guérin et al. 2006). Kemenes et al. (2016) report carbon dioxide and methane evasion via degassing through turbines and downstream for Tucuruí, Samuel, and Curuá-Una reservoirs. Barros et al. (2011) summarised much of the data from Amazonian reservoirs, though methane emission from Balbina is listed as 10 mg C m^{-2} day^{-1} rather than 47 mg C m^{-2} day^{-1}, as reported in Kemenes et al. (2007). Moreover, degassing through turbines and downstream is not included for Balbina or other Amazonian reservoirs.

Carbon dioxide emissions from the surface of Balbina and Petit Saut reservoirs averaged 1296 Mg C km^{-2} $year^{-1}$ and 473 Mg C km^{-2} $year^{-1}$, respectively. In the case of CO_2 total annual emission from the Balbina hydroelectric system, including the reservoir, turbine outflow, and river channel extending 30 km downstream, was 1340 Mg C km^{-2} $year^{-1}$, when expressed relative to the average reservoir area. Average CH_4 emissions from the reservoir surface were 18 Mg C km^{-2} $year^{-1}$, and total emissions were 39 Mg C km^{-2} $year^{-1}$, indicating the importance of degassing of methane through the turbines and downstream. Though Barros et al. (2011) present mean fluxes from Amazonian reservoirs for CO_2 of 400 Mg C km^{-2} year^{-1} and for CH_4 of 50 Mg C km^{-2} year^{-1}, these values should be used with caution since the data from which they were obtained are based on only a few samples from most of the reservoirs without adequate seasonal sampling and without data on downstream fluxes.

7.4.5 Airborne Surveys

Airborne campaigns have provided integrated coverage of subregions of the Amazon basin and permitted calculation of carbon balance (Lloyd et al. 2007) and methane emission (Beck et al. 2012; Miller et al. 2007). Miller et al. (2007) collected vertical profiles of methane over 4 years at sites near Santarém and Manaus and calculated average emissions of 20 kg C km^{-2} day^{-1}. Wetlands are likely the major source of methane, at least during seasons with extensive inundated areas. Other sources include fires, urban areas, termites, and, perhaps emissions associated with *terra firme* forests. Beck et al. (2012) described results of airborne campaigns in November and May, periods representing generally low and large inundation, during which continuous, in-flight measurements of CH_4 and sampling for isotopic analyses were conducted. The flights extended over much of the lowland Amazon in Brazil and were concentrated in the central basin. Isotopic measurements indicated that biogenic methane predominates, and wetlands are likely the major source though near Manaus anthropogenic sources, such as waste decomposition, contribute. A signature of biomass burning was detected in samples collected during the dry season, but this source appeared to be minor for CH_4. Beck et al. (2012) estimated a CH_4 flux for the lowland Amazon during November as 27 ± 9 kg C km^{-2} day^{-1} and during May as 32 ± 14 kg C km^{-2} day^{-1}.

7.5 Regionalisation of Fluxes

7.5.1 Prior Estimates

Though various estimates of regional carbon dioxide and methane fluxes have been made through the years, only recent estimates that used data available through the beginning of the twenty-first century are summarised here. Richey et al. (2002) and Melack et al. (2004) were the first to use regional analyses of microwave remote sensing data to establish inundated areas and habitats. Both applied Monte Carlo error propagation to establish uncertainties.

Richey et al. (2002) used carbon dioxide measurements primarily from the Solimões-Amazon River, its fringing floodplain and mouths of major tributaries, and conservatively low piston velocities to calculate an outgassing rate of 830 ± 240 Mg C km^{-2} $year^{-1}$ for the annual mean flooded area. The areal flux was combined with remote sensing-derived inundated areas at high and low stages for rivers and floodplains over 100 m across (Hess et al. 2003) to determine total outgassing in a 1.77 million km^2 quadrant in the central basin. The seasonal variation in inundated area was assumed to track river stage, as measured by the sparse network of gauges. A comparison of inundated area-derived passive microwave data versus stage for large floodplains supports this assumption (Hamilton et al. 2002). Areas of moderate to small rivers and streams in this quadrant were approximated as a geometric series relating stream length and width to stream order. To extrapolate to the whole Amazon basin (6.07 million km^2), an areal flux of half that used for the central quadrant was applied to the 4.3 million km^2 outside the central quadrant to yield a value of 470 Tg C $year^{-1}$ and that applied to the area of the lowland basin yields an evasion of 390 Tg C $year^{-1}$.

Rasera et al. (2013) extrapolated their results to a central Amazon quadrant (1.47 million km^2). To do so, they combined (1) areas of streams (<100 m wide) and the areas of rivers and floodplains (>100 m wide) for high and low stage as reported in Richey et al. (2002); (2) $k(600)$ values based on recent work for rivers and streams; and (3) pCO_2 values from measurements and from a relationship between measured values and average soil cation exchange capacity. If the annual total evasion of CO_2 calculated in this manner were increased in proportion to the slightly larger central basin area used by Richey et al. (2002), 432 ± 78 Tg C $year^{-1}$ would result. This annual rate is about twice that reported by Richey et al. (2002) and reflects higher k (600) values and improved data for streams. It is important to recognise that the fluxes calculated for the mainstem Amazon and tributaries in Rasera et al. (2013) include floodplain areas, not just river channels; if extrapolated to the area of the lowland basin a flux of 1240 ± 206 Tg C $year^{-1}$ results.

Regional extrapolation of fluxes of carbon dioxide from streams and small rivers is especially difficult because of the few measurements among the millions of kilometres of these systems and the large spatial and temporal variations observed. Johnson et al. (2008) approximated potential evasion of CO_2 from headwater streams basin-wide (an area of 6.07 million km^2, which included the Andean highlands) as 114 Tg C $year^{-1}$ or 19.5 Mg C km^{-2} $year^{-1}$, excluding inundated areas and accounting for human modified land uses, and where the areal rate is expressed per km^2 of total land area, not stream areas. Variations in annual water balances for the period from 1976 to 1996 would introduce about a 10 % increase or decrease between wet or dry years. Their regionalisation approach is based on groundwater fluxes, determined as the difference between average annual precipitation and evapotranspiration, and estimates of soil pCO_2 from carbonate equilibrium reactions at a spatial scale of 0.1°. Though an interesting approach, it requires validation based on actual measurements for a variety of headwaters, such as Andean, blackwater, or savanna, streams that are different from those examined by Johnson et al. (2006). Rasera et al. (2008) extrapolated from the Ji-Paraná River basin to the Amazon basin and arrived at a value substantially higher than can be calculated from the data presented. Based on the total CO_2 evasion from the

Ji-Paraná River basin divided by the area of the basin and multiplied by the area of third- to sixth-order rivers (derived from the fractional area of these rivers in the Ji-Paraná River basin; see above), the annual evasion is about 21 Tg C for these rivers in the lowland Amazon basin.

Melack et al. (2004) used available measurements of habitat-specific methane fluxes; all the data were from the central Amazon basin and based mainly on methane captured in floating chambers. By combining these values with remote sensing-based estimates of areal extent of aquatic habitats (open water in lakes, rivers, flooded forests, and aquatic macrophytes), they calculated regional estimates for the mainstem Solimões-Amazon in Brazil and for the same 1.77 million km^2 quadrant used by Richey et al. (2002). Seasonal changes in the areal extent of the aquatic habitats were approximated by interpolating between remote sensing-derived areas obtained for low and high water periods. A time series of inundation extent along the mainstem Solimões-Amazon floodplain, based on passive microwave data, was used to calculate inter-annual variations in methane flux for 3 years. To extrapolate to the whole lowland basin (5.05 million km^2), a single, habitat-averaged value of 30 Mg C km^{-2} $year^{-1}$, calculated from the mean annual emission estimated for the mainstem Solimões-Amazon in Brazil and the mean annual flooded area of this reach, was used, resulting in a flux of c. 22 Tg C $year^{-1}$. If expressed as the greenhouse gas warming potential equivalence of CO_2, this mean flux amounts to about 0.2 Pg C $year^{-1}$ (not 0.5 Pg C $year^{-1}$, as noted in the original paper).

Sawakuchi et al. (2014) subtracted estimates of Landsat-based lake areas from the water body category in an AVHRR 1-km regional land cover product (Brown et al. 2003) to obtain a large river channel area of approximately 91,000 km^2, a value considerably larger than others discussed earlier. When combined with their methane fluxes, they calculated an annual average flux of 0.37 Tg C as methane from the large rivers of the Amazon basin. If the areal estimate of Beighley and Gummadi (2011) is used, the flux is 0.12 Tg C $year^{-1}$, and if the SRTM areal estimate is used, the flux is 0.21 Tg C $year^{-1}$. These methane fluxes represent 0.5–1.6 % of the basin-wide emission calculated by Melack et al. (2004) and indicate a minor role for the large river channels. Methane emission from the mainstem Solimões-Amazon River in Brazil was estimated by Melack et al. (2004) to represent only 0.06 % of the total flux for that reach including the fringing floodplains.

Barros et al. (2011) estimated emission from extant tropical Amazonian reservoirs, based on an area of 20,000 km^2, of 8 Tg C $year^{-1}$ as CO_2 and 1.0 Tg C $year^{-1}$ as CH_4. We question these values based on the area of Amazonian reservoirs being 6300 km^2 and based on areal emission estimates available from Balbina and Petit Saut, as described above.

Global or continental scale calculations derived from satellite retrievals of atmospheric concentrations and inverse modelling provide coarse spatial resolution methane emission values for the Amazon basin (e.g. Frankenberg et al. 2008). Beck et al. (2012) examined the performance of CH_4 inversion models, constrained by observations from surface stations and SCIAMACHY retrievals, in comparison to their airborne campaigns for the Amazon basin. In this comparison, the global models used the same transport model but different prior CH_4 inputs from wetlands, none of which were well suited to the Amazon basin. One result derived from the transport modelling suggested that the Amazon basin is influenced by an atmospheric region larger than the basin. Furthermore, Beck et al. (2012) concluded that a reliable annual methane budget will require regional airborne campaigns over a full year.

7.5.2 New Estimates

In principle, total basin-wide fluxes of CO_2 and CH_4 (F, in units of Tg C year^{-1}) could be calculated using a general expression similar to that in Melack et al. (2004):

$$F = {}^{6}\Sigma_{j=1}{}^{12}\Sigma_{i=1} t_i A_{ij} f_{ij} \tag{7.1}$$

where F is the flux of CO_2 or CH_4 for each habitat (expressed as kg C km^{-2} day^{-1}); j is each habitat: (1) headwaters, (2) streams and moderate-sized rivers, (3) large rivers, (4) floodplains, (5) wetlands, and (6) hydroelectric reservoirs; A is an estimate of average flooded area of each habitat per month; t is the number of days per month; and i is each month incremented from 1 to 12. Depending on available data, several habitat categories could be subdivided: e.g. large rivers (e.g. Solimões-Amazon and white water tributaries, Negro and black water tributaries, Tapajós, and other clear waters), floodplains (e.g. open water lakes, flooded forests, aquatic macrophytes), and wetlands (e.g. upper Negro interfluvial, Roraima savanna, Llanos de Moxos). If data from multiple years were available, such as the inundation time series of Papa et al. (2010) or Paiva et al. (2013), or calculated from an empirical or mechanistic model of carbon dioxide and methane dynamics, Eq. (7.1) could be evaluated repeatedly to determine inter-annual variability.

Several challenges make it difficult to apply Eq. (7.1). In particular, sufficient information about the spatial and temporal variations of inundated areas, habitat characteristics, and associated fluxes on a basin-wide scale for multiple years is lacking. A modelling system that combines climatic and hydrological processes with biogeochemical and ecological processes is required.

As an alternative, a combination of the published values, summarised in prior sections, for specific regions and calculations based on averaged measurements of carbon dioxide from (1) moderate-sized rivers, (2) large rivers, (3) floodplains and wetlands, and (4) hydroelectric reservoirs as noted in Table 7.1 are used to estimate carbon dioxide emissions. These fluxes include those measured with floating chambers and those calculated with new estimates of gas exchange velocities and in situ gas concentrations. Remote sensing-based estimates of inundated areas at low and high water levels and modelled variations in inundation are used to

Table 7.1 Spatial components used in calculation of lowland Amazon basin carbon dioxide evasion

Headwaters: 95 Tg C $year^{-1}$ (Johnson et al. 2008 as lowland proportion based on area)
Moderate-sized rivers: Using an area of 31,000 km^2 (Beighley and Gummadi 2011) for river channels <150 m wide, and the annual average reported in Rasera et al. (2013) of 1880 Mg C km^{-2} $year^{-1}$ = 58 Tg C $year^{-1}$
Large rivers: Using an area of 29,500 km^2 (Beighley and Gummadi 2011) for river channels >150 m wide, an average of pCO_2 values (140 μM) from Richey et al. (2002) and Rasera et al. (2013) for the mainstem Amazon and large tributaries, and *k*(600) of 10 cm h^{-1} (mid-range of values reported in Devol et al. 1987, Alin et al. 2011, Kemenes et al. 2011 and Rasera et al. 2013) results in a flux of 1500 Mg C km^{-2} $year^{-1}$, which totals 44 Tg C $year^{-1}$. Using the SRTM value for the area of large river channels (52,000 km^2) would result in a total flux of 77 Tg C $year^{-1}$
Floodplain and wetland habitats: Using average of upstream and downstream pCO_2 values (335 μM) from Richey et al. (2002) and *k*(600) of 12 cm h^{-1} (mid-range of values reported by Rudorff et al. 2011 and Polsenaere et al. 2013) results in a flux of 4200 Mg C km^{-2} $year^{-1}$ with floodplain areas of 370,000 km^2 based on combination of monthly variations in modelled inundated fractions (Coe et al. 2007) and high and low water SAR analyses for lowland basin in 1995 and 1996 (Melack and Hess 2010) (not including upper Negro interfluvial wetlands, Roraima wetlands and large river areas, which are counted separately) = 1550 Tg C $year^{-1}$. If a lower *k*(600) is used to account for the large areas with vegetation where *k* values are likely to be lower, the flux would be proportionately lower
Interfluvial wetlands in upper Negro basin: 21 Tg C $year^{-1}$, based on an area of 50,000 km^2 (Belger et al. 2011)
Roraima wetlands: 13 Tg C $year^{-1}$, based on an area of 13,300 km^2 (Jati 2013)
Reservoirs: Using an area of 6300 km^2, the area of existing reservoirs in the lowland Amazon basin, and an average of fluxes from Balbina and Petit Saut reservoirs (about 890 Mg C km^{-2} $year^{-1}$) results in a total flux of 5.6 Tg C $year^{-1}$. Though not within the Amazon basin, Petit Saut Reservoir (French Guiana) has a watershed of tropical lowland forest and the longest set of measurements of a tropical South American reservoir including sampling in the reservoir, below the turbines and in the downstream river

approximate annual values for floodplain and wetland environments. River channel areal estimates from Beighley and Gummadi (2011) and SRTM are used. The estimates are standardised to the lowland basin below 500 m as delineated from JERS-1 synthetic aperture radar data (5.05 million km^2). Insufficient new measurements for methane are available to improve upon Melack et al. (2004).

Carbon dioxide evasion from rivers and streams is estimated to be c. 200 Tg C $year^{-1}$. Though fluxes per unit area for rivers and streams are based on updated *k*(600) and recent data from a range of stream and river sizes, a smaller surface area for these habitats than previous basin-wide estimates is being used. The annual flux from the category called floodplain and wetland habitats, which includes lakes, plus those from other wetlands and reservoirs is estimated to be approximately 1600 Tg C $year^{-1}$. These fluxes incorporate updated *k*(600) values; however, the lack of information about *k* values in vegetated areas is a concern because Melack and Hess (2010) estimate that about 79 % of the lowland basin is characterised by woody vegetation with another 13 % predominately herbaceous vegetation. Since *k* values are likely to be less in these areas in comparison to those in lakes, the annual fluxes would be lower. Furthermore, basin-wide modelled inundation fractions and remote sensing-based areas do not represent well the seasonality in areas or fluxes. With these and other issues noted above acknowledged, total evasion of carbon dioxide for the lowland Amazon basin is estimated to be c. 1.8 Pg C $year^{-1}$. Though a formal analysis of uncertainty cannot be done based on the heterogeneous information used for these estimates, based on spatial and temporal variability and uncertainty in measurements, an overall uncertainty of at least ±50 % is reasonable.

Raymond et al. (2013) calculated global annual evasion of carbon dioxide from streams and rivers ranging from 1.5 to 2.1 Pg C and evasion from lakes and reservoirs ranging from 0.06 to 0.84 Pg C. The ranges represent 5th and 95th confidence intervals derived from a Monte Carlo analysis. Wetlands were not included. That the value reported here for the lowland Amazon basin is similar to the sum of these global estimates indicates the importance of including tropical floodplains and other wetlands in calculations of carbon dioxide evasion from inland waters.

7.6 Sources and Decomposition of Organic Carbon

Floodplains and other wetlands are productive aquatic environments in which most of the production and evasion of CO_2 and CH_4 is likely derived from metabolic processing of the carbon fixed by aquatic plants. These environments also export considerable amounts of carbon to rivers and accumulate sediments. Estimates of carbon balances for floodplains at several spatial scales provide supporting evidence: Calado (Melack and Engle 2009), Curuai (Rudorff 2013), the mainstem Solimões-Amazon floodplain in Brazil (Melack and Forsberg 2001), and a 1.77 million-km^2 quadrant in the central Amazon (Melack et al. 2009; Abril et al. 2014). In particular, Abril et al. (2014) suggest that Amazonian wetlands export about half of their primary productivity to neighbouring waters where it is metabolised and much is released to the atmosphere. Further evidence is provided by estimates of root respiration by herbaceous and woody aquatic plants (Hamilton et al. 1995; Worbes 1997), isotopic studies of microbial respiration (Waichman 1996), calculation of aquatic macrophyte growth and decay (Engle et al. 2008; Silva et al. 2009, 2013), rates of methane oxidation in exposed wetland sediments (Koschorreck 2000), and the enrichment of $\delta^{13}C$ of CO_2 in lowland rivers, expected if C4 grasses are significant sources of respired carbon (Mayorga et al. 2005b).

Understanding the relevance of evasion of carbon dioxide from aquatic environments to the carbon balance of *terra firme* forests requires (1) determination of the proportion of the carbon fixed within aquatic ecosystems versus that imported from uplands as inorganic or organic carbon and (2) measurements of aquatic respiration and of decomposition of these sources of organic carbon. Though further work is needed, recent results indicate that the carbon inputs to aquatic systems that are eventually emitted as CO_2 or CH_4 vary among habitats and are related to hydrological conditions and proximity of carbon sources.

Headwater streams in the Juruena watershed (Fig. 7.1), and presumably elsewhere, receive most of the CO_2 degassed in their uppermost reaches directly from terrestrially derived respiration in soils with small contributions of organic C from riparian and upland litter that is gradually processed downstream (Johnson et al. 2006, 2007). In slightly larger streams and small rivers, relatively labile dissolved and particulate organic carbon and lateral inputs of dissolved CO_2 support evasion (earlier work summarised in Richey et al. 2009; Davidson et al. 2010). There is regional heterogeneity in the carbon sources, with ^{13}C-depleted CO_2 in streams draining sandy soils in the Negro basin indicative of C3 plants, while the ^{13}C-enriched CO_2 in streams passing through pastures in Rondônia indicating C4 grasses. Based on direct assays of $^{14}CO_2$ outgassed from small streams and rivers in the western Amazon, Vihermaa et al. (2014) report that a portion of the carbon dioxide is derived from sedimentary rock and carbonate weathering.

Carbon dioxide evasion from large rivers appears to be supported by a wide variety of carbon compounds from a combination of nearby and distant organic carbon sources. Especially relevant results are reported by Ellis et al. (2012), who determined the $\delta^{13}C$ of the CO_2 evaded from the Amazonian rivers and found that organic carbon from C3 and C4 plants and phytoplankton was evident and spatially and temporally variable. Another valuable result from Ward et al. (2013) demonstrated that the degradation of lignin and associated macromolecules in water from the lower Amazon River was sufficient to support considerable respiration and presumably CO_2 evasion. Since these compounds have been thought to be refractory and are often of terrestrial origin, this finding supports the notion that metabolism in the large rivers is supported by diverse carbon sources. Fatty acid and stable isotope analyses by Mortillaro et al. (2011, 2012) as well as studies by Kim et al. (2012), using the branched and isoprenoid tetraether index, offer further evidence of multiple carbon inputs to the lower Amazon River. Oxidation of petrogenic organic carbon has also been documented (Bouchez et al. 2010). Results by Mayorga et al. (2005b) indicated that the main source of respired carbon was <5 years old and that the dissolved CO_2 was isotopically different from organic carbon in the rivers sampled at the same time. These results imply that inputs of labile carbon, which is rapidly oxidised, support the generation of the high pCO_2 values observed. However, Ellis et al. (2012) found that the respired carbon was isotopically similar to that in the water. Further, as noted by Rasera et al. (2013), the high concentrations of CO_2 during high river stages may reflect export of labile organic carbon from fringing floodplains. In the Amazon, Madeira and Solimões rivers Ellis et al. (2012) found that riverine respiration could account for most or all of the CO_2 evaded, while in the Negro River it could account for only 15–34 %. Photo-oxidation of organic carbon appears to make small contributions to CO_2 in large rivers (Amaral et al. 2013; Remington et al. 2011).

In summary, the relative importance of carbon sources originating in *terra firme* forests versus aquatic habitats varies as a function of position in the continuum from headwaters to large rivers with fringing floodplains or associated wetlands. Hence, a general conclusion regarding the proportion of *terra firme* net productivity that is emitted to the atmosphere after lateral transport to aquatic environments is difficult to state. However, it is clear that CO_2 evasion, in particular, is supported by carbon fixed by *terra firme* plants in many streams and that this carbon contributes to metabolic processes in the largest rivers. In contrast, floodplains and wetlands likely represent environments where the CO_2 and CH_4 emitted to the atmosphere are derived largely from carbon fixed by aquatic plants and, to a lesser extent, algae, and that a portion of organic carbon metabolised in rivers is supplied by their floodplains. Based on this logic, the values in Table 7.1 lead to the conclusion that the majority of the basin-wide evasion of carbon dioxide is derived from plants in aquatic environments.

7.7 Uncertainties and Research Needs

7.7.1 Field Measurements

The largest uncertainties stem from the sparseness of measurements in time and space. For example, based on a Monte Carlo error analysis, Melack et al. (2004) noted that the uncertainty in actual methane fluxes, largely because of the high spatial and temporal variability, compounded by the episodic nature of ebullition, was larger than the uncertainty associated with remote sensing-based habitat extent. This result applied to the best sampled floodplains of the central basin. In many wetlands of the basin, few or no data are available. Particularly large gaps with no data exist in the Llanos de Moxos, Bananal, peatlands in the western Amazon (Lahteenoja et al. 2011), Peruvian lowlands, and habitats above 500 m. The extensive network of streams and medium-sized rivers is significantly under-sampled, and Richey et al. (2009) made a plea for collecting many spatially distributed measurements given the large variability observed among the data gathered. While reservoirs are receiving increased attention, time series data and measurements above and below the dams are required to better guide the management of greenhouse gas evasion from hydroelectric projects.

Within lakes, reservoirs, and wetlands, diel and seasonal variations in vertical stratification and horizontal advection influence the concentration gradients of gases and the transfer velocities. With the increasing availability of in situ sensors and automatic recording systems, it is possible to incorporate temporal and spatial variations in physical and chemical conditions in calculations of gas exchanges. For example, deep convective mixing at night, common in tropical waters, is likely to increase both transfer velocities and concentrations of carbon dioxide and methane in surficial waters, thus increasing fluxes.

Other components of the carbon balance in aquatic habitats can be challenging to measure, and data are lacking for many areas or are without sufficient spatial and temporal coverage even at better studied sites. Key processes that need attention are primary productivity by algae and higher plants, sedimentation in floodplains and other wetlands, transport of dissolved and particulate inorganic and organic carbon from uplands to inland waters, and rates of respiration and decomposition.

7.7.2 *Modelling*

While models of river discharge and inundation are improving, the current basin-scale models do not properly flood some important and large habitats, such as the largely rain-fed interfluvial wetlands in the upper Negro basin, the Roraima wetlands, or the Llanos de Moxos. One key to improvement is higher spatial resolution DEMs for floodplains. Biogeochemical models of carbon dioxide and methane production and evasion appropriate for conditions in the Amazon are not available (Riley et al. 2011), though relevant models are under development (Potter et al. 2016). While several models have potentially useful components or formulations (Ringeval et al. 2010; Bloom et al. 2010; Cao et al. 1996; Potter 1997; Walter and Heimann 2000), no spatially explicit model exists that incorporates the inundation dynamics and biogeochemical processes of aquatic environments in the Amazon.

7.8 Climate Change, Exceptional Events, and Human Impacts

Climate warming, climate variability, including exceptional droughts and floods and severe wind storms, and fires are influencing the Amazon basin (Davidson et al. 2012). Additionally, human alterations include agricultural expansion with associated deforestation, construction of dams, and fossil fuel exploration (Melack 2005; Costa et al. 2009; Pires and Costa 2013; Renó et al. 2011; Castello et al. 2013; Finer and Orta-Martinez 2010). A review of current climatic conditions in the context of aquatic environments in the Amazon basin by Melack and Coe (2013) noted the paucity of meteorological records in floodplains and other wetlands. Basin-wide, long-term warming trends in air temperature are generally 0.1–0.3 °C per decade (1960–2009; Burrows et al. 2011). Rainfall has negative trends in the northern Amazon and positive trends in the southern Amazon (1949–1999; Marengo 2004). Severe droughts occurred in 2005 and 2010 and were associated with tropical Atlantic warming and ENSO events (Marengo et al. 2008, 2009, 2011; Zeng et al. 2008; Villar et al. 2011). In contrast, an exceptional flood occurred in 2009 (Chen et al. 2010). Furthermore, increased variability in climatic conditions with increased frequency and severity of droughts and storms is projected by global models (Malhi et al. 2008; Gloor et al. 2013; Huntingford et al. 2013; Lau et al. 2013; Liu et al. 2013; Reichstein et al. 2013).

Another consequence of severe weather can be substantial disturbance of forests as trees are blown down (Negrón-Juárez et al. 2010; Chambers et al. 2007). Droughts can lead to increased incidence of fires (Aragão et al. 2008; Longo et al. 2009), even in *igapó* forests (Flores et al. 2014).

Melack and Coe (2013) ran simulations of inundation under altered climate and land uses specifically for the Amazon basin using a basin-wide hydrological model forced with observed climate data from 1950 to 2000. Simulations of inundation with 10 and 25 % decreases in rainfall (without seasonal and spatial variations included) resulted in reductions in inundation similar to reductions in rainfall: −5 to −20 % (10 %) and −12 to −30 % (25 %). Based on 35 % deforestation coupled to a global climate model, rainfall decreased and evapotranspiration decreased more; hence average maximum flooded area along mainstem Amazon increased slightly compared to current land cover. Others have considered actual or potential climate change or land use impacts on hydrological conditions in the Amazon basin (Coe et al. 2009, 2011; Casimiro et al. 2011, 2012; Langerwisch et al. 2012; Cox et al. 2013). In addition to the existing hydroelectric reservoirs, others are under construction (Belo Monte on the Xingu River; Santo Antonio and Jirau, run-of-river dams on the Madeira River), and many more are planned throughout the Amazon basin including the Tapajós hydroelectric complex which could inundate about 2000 km^2 (Finer and Jenkins 2012). To estimate the emissions of CO_2 and CH_4 from these new and planned reservoirs is difficult because of differences in construction and operation and because economic and environmental issues will likely play roles. Furthermore, road construction and agriculture create numerous small impoundments (Macedo et al. 2013), of unknown total area with unmeasured fluxes.

7.9 Conclusion

The updated value for annual carbon dioxide emission from aquatic habitats in the lowland Amazon basin of 1800 Tg C is larger than previous estimates. Almost 90 % of this flux is likely associated with lakes, floodplains, and other wetlands. Carbon fixed by *terra firme* plants contributes most of the carbon dioxide emitted from streams and adds organic carbon to rivers. Lakes, floodplains, and other wetlands represent environments where the CO_2 emitted to the atmosphere is derived largely from carbon fixed by aquatic plants with lesser contributions from algae, and a portion of organic carbon metabolised in rivers is supplied by their floodplains.

To further improve the updated estimates of carbon dioxide and methane evasion from aquatic habitats in the Amazon basin requires several activities. The largest uncertainties stem from the sparseness of measurements in time and space. Deployment of eddy covariance systems, continuous measurements along transects, and regional airborne campaigns would help considerably. Inclusion of habitats not well characterised, such as flooded forests, savannas, intermittently flooded regions along streams, and depressions within *terra firme* forests is needed. Hydroacoustic techniques and sampling releases via turbines in hydroelectric dams will improve estimation of ebullition. Seasonal and inter-annual variability of inundated areas derived from remote sensing and modelling should be incorporated. The lack of high-resolution digital elevation data are a serious limitation throughout the basin. Given the pronounced diel cycle of heating and cooling often observed in shallow tropical waters, it is important to measure stratification and mixing, and gas concentrations and exchanges, over these diel cycles. Biogeochemical models of carbon dioxide and methane production and evasion appropriate for conditions in the Amazon require further development. Climate warming, climate variability, including exceptional droughts and floods and severe wind storms, fires, agricultural expansion, and construction of dams are all influencing the Amazon basin with consequences for the role of aquatic environments in the carbon cycle and release of carbon dioxide and methane.

Acknowledgements S. MacIntyre, L. Hess, G. Abril, J. Richey, B.R. Forsberg, and L. Nagy provided helpful comments on the manuscript and its development. The writing and related research was supported by NASA's Terrestrial Ecology Program.

References

Abril G, Guérin F, Richard S, Delmas R, Galy-Lacaux C, Gosse P, Tremblay A, Varfalvy L, Santos MA, Matvienko B (2005) Carbon dioxide and methane emissions and the carbon budget of a 10-year old tropical reservoir (Petit Saut, French Guiana). Global Biogeochem Cycles 19:1–16

Abril G, Martinez J-M, Artigas LF, Moreira-Turcq P, Benedetti MF, Vidal L, Meziane T, Kim J-H, Bernardes MC, Savoye N, Deborde J, Souza EL, Albéric P, de Souza MFL, Roland F (2014) Amazon River carbon outgassing fuelled by wetlands. Nature 505:395–398

Aires F, Papa F, Prigent C (2013) A long-term high-resolution wetland dataset over the Amazon basin, downscaled from multiwavelength retrieval using SAR data. J Hydrometeorol 14:594–607

Alin SR, Rasera MFFL, Salimon CI, Richey JE, Holtgrieve GW, Krusche AV, Snidvongs A (2011) Physical controls on carbon dioxide transfer velocity and flux in low-gradient river systems and implications for regional carbon budgets. J Geophys Res Biogeosci 116:G01009. doi:10.1029/2010jg001398

Alsdorf D, Han S-C, Bates P, Melack J (2010) Seasonal water storage on the Amazon floodplain measured from satellites. Remote Sens Environ 114:2448–2456

Amaral JHF, Suhett AL, Melo S, Farjalla VF (2013) Seasonal variation and interaction of photodegradation and microbial metabolism of DOC in black water Amazonian ecosystems. Aquat Microb Ecol 70:157–168

Aragão LEOC, Malhi Y, Barbier N, Lima A, Shimabukuro Y, Anderson L, Saatchi S (2008) Interactions between rainfall, deforestation and fires during recent years in the Brazilian Amazonia. Philos Trans R Soc B 363:1779–1785

Arnesen AS, Silva TSF, Hess LL, Novo EMLM, Rudorff CM, Chapman BD, McDonald KC (2013) Monitoring flood extent in the lower Amazon River floodplain using ALOS/PALSAR ScanSAR images. Remote Sens Environ 130:51–61. doi:10.1016/j.rse.2012.10.035

Aufdenkampe AK, Mayorga E, Raymond PA, Melack JM, Doney SC, Alin SR, Aalto RE, Yoo K (2011) Rivers key to coupling biogeochemical cycles between land, oceans and atmosphere. Front Ecol Environ 9:53–60

Banerjee S, MacIntyre S (2004) The air-water interface: turbulence and scalar exchange. In: Grue J (ed) Advances in coastal and ocean engineering, vol 9. World Scientific, Hackensack, NJ, pp 181–237

Barbosa PM (2013) Avaliação do fluxo difusivo de metano (CH_4) em ambientes do médio-baixo Solimões. M.Sc. Thesis, Universidade Federal do Rio de Janeiro

Barros N, Cole JJ, Tranvik LJ, Prairie YT, Bastviken D, Huszar VLM, del Giorgi P, Roland F (2011) Carbon emission from hydroelectric reservoirs linked to reservoir age and latitude. Nat Geosci 4:593–596

Battin TJ, Luyssaert S, Kaplan LA, Aufdenkampe AK, Richter A, Tranvik LJ (2009) The boundless carbon cycle. Nat Geosci 2:598–600

Beck V, Chen H, Gerbig C, Bergamaschi P, Bruhwiler L, Houweling S et al (2012) Methane airborne measurements and comparison to global models during BARCA. J Geophys Res 117: D15310. doi:10.1029/2011JD017345

Beighley RE, Gummadi V (2011) Developing channel and floodplain dimensions with limited data: a case study in the Amazon basin. Earth Surf Process Landf 36:1059–1071

Beighley RE, Eggert KG, Dunne T, He Y, Gummadi V, Verdin KL (2009) Simulating hydrologic and hydraulic processes throughout the Amazon River Basin. Hydrol Process 23:1221–1235. doi:10.1002/hyp.7252

Belger L, Forsberg B, Melack JM (2011) Factors influencing carbon dioxide and methane emissions from interfluvial wetlands of the upper Negro River basin, Brazil. Biogeochemistry 105:171–183. doi:10.1007/s10533-010-9536-0

Bloom AA, Palmer PI, Fraser A, Reay DS, Frankenberg C (2010) Large-scale controls on methanogenesis inferred from methane and gravity spaceborne data. Science 327:322–325

Borges AV, Abril G, Darchambeau F, Teodoru CR, others (2015) Divergent biophysical controls of aquatic CO_2 and CH_4 in the world's two largest rivers. Sci Rep. doi:10.1038/srep15614

Bouchez J, Beyssac O, Galy V, Gaillardet J, France-Lanord C, Laurence M, Moreira-Turcq PF (2010) Oxidation of petrogenic organic carbon in the Amazon floodplain as a source of atmospheric CO_2. Geology 38:255–258

Brix H, Sorrell BK, Orr PT (1992) Internal pressurization and convective gas flow in some emergent freshwater macrophytes. Limnol Oceanogr 37:1420–1433

Brown J, Loveland T, Ohlen D, Zhu Z (2003) LBA regional land cover from AVHRR 1-km version 1.2 (IGBP). Oak Ridge National Laboratory Distributed Active Archive Center, Oak Ridge, TN

Burrows MT, Schoeman DS, Buckley LB, Moore P, Poloczanska ES, Brander KM et al (2011) The pace of shifting climate in marine and terrestrial ecosystems. Science 334:652–655

Butman D, Raymond PA (2011) Significant efflux of carbon dioxide from streams and rivers in the United States. Nat Geosci 4:839–842

Cao M, Marshall S, Gregson K (1996) Global carbon exchange and methane emission from natural wetlands: application of a process-based model. J Geophys Res 101:14399–14414

Casimiro L, Waldo S, Labat D, Guyot JL, Ardoin-Bardin S (2011) Assessment of climate change impacts on the hydrology of the Peruvian Amazon-Andes basin. Hydrol Process 25:3721–3734

Casimiro L, Waldo S, Labat D, Guyot JL, Ronchail J, Espinoza Villar JC (2012) Trends in rainfall rates and temperature fluctuations in the Peruvian Amazonas-Andes basin over the last 40 years. Hydrol Process. doi:10.1002/hyp.9418

Castello L, McGrath DG, Hess LL, Coe MT, Lefebvre PA, Petry P, Macedo MN, Renó VF, Arantes CC (2013) The vulnerability of Amazon freshwater ecosystems. Conserv Lett 6:217–229

Chambers JQ, Fisher JI, Zeng H, Chapman EL, Baker DB, Hurtt GC (2007) Hurricane Katrina's carbon footprint on U.S. Gulf Coast forests. Science 318:1107

Chen JL, Wilson CR, Tapley BD (2010) The 2009 exceptional Amazon flood and interannual terrestrial water storage change observed by GRACE. Water Resour Res 46. doi:10.1029/2010WR009383

Coe MT, Costa MH, Howard E (2007) Simulating the surface waters of the Amazon River basin: impacts of new river geomorphic and dynamic flow parameterizations. Hydrol Process 22:2542–2553

Coe MT, Costa MH, Soares-Filho BS (2009) The influence of historical and potential future deforestation on the stream flow of the Amazon River—land surface processes and atmospheric feedback. J Hydrol. doi:10.1016/j.jhydrol.2009.02.043

Coe MT, Latrubesse EM, Ferreira ME, Amsler ML (2011) The effects of deforestation and climate variability on the streamflow of the Araguaia River, Brazil. Biogeochemistry 105:119–131. doi:10.1007/s10533-011-9582-2

Cole JJ, Praire YT, Caraco NF, McDowell WH, Tranvik LJ, Striegl RR, Duarte CM, Kortelainen P, Downing JA, Middleburg J, Melack JM (2007) Plumbing the global carbon cycle: integrating inland waters into the terrestrial carbon budget. Ecosystems. doi:10.1007/s10021-006-9013-8

Costa MH, Coe MT, Guyot JL (2009) Effects of climatic variability and deforestation on surface water regimes. In: Gash J, Keller M, Silva-Dias P (eds) Amazonia and global change, Geophysical monograph series 186. American Geophysical Union, Washington, DC, pp 543–553

Cox PM, Pearson D, Booth BB, Friedlingstein P, Huntingford C, Jones CD, Luke CM (2013) Sensitivity of tropical carbon to climate change constrained by carbon dioxide variability. Nature 494:341–344

Crill PM, Bartlett KB, Wilson J, Sebacher DI, Harriss RC, Melack JM, MacIntyre S, Lesack L, Smith Morrill L (1988) Tropospheric methane from an Amazon floodplain lake. J Geophys Res 93:1564–1570

Davidson EA, Figueiredo RDO, Markewitz D, Aufdenkampe AK (2010) Dissolved CO_2 in small catchment streams of eastern Amazonia: a minor pathway of terrestrial carbon loss. J Geophys Res 115. doi:10.1029/2009JG001202

Davidson EA, de Araújo AC, Artaxo P, Balch JK, Brown IF, Bustamante MMC, Coe MT, DeFries RS, Keller M, Longo M, Munger JW, Schroeder W, Soares-Filho BS, Souza CM Jr, Wofsy SC (2012) The Amazon basin in transition. Nature 481:321–328

Del Sontro T, Kunz MJ, Kempter T, Wuest A, Wehrli B, Senn DB (2011) Spatial heterogeneity of methane ebullition in a large tropical reservoir. Environ Sci Technol 45:9866–9873

Devol AH, Quay PD, Richey JE, Martinelli LA (1987) The role of gas exchange in the inorganic carbon, oxygen and 222 radon budgets of the Amazon River. Limnol Oceanogr 32:235–248

Downing J (2009) Global limnology: up-scaling aquatic services and processes to planet Earth. Verh Int Verein Limnol 30:1149–1166

Downing JA, Cole JJ, Duarte CM, Middelburg J, Melack JM, Prairie YT, Kortelainen P, Striegl RG, McDowell WH, Tranvik LJ (2012) Global abundance and size distribution of streams and rivers. Inland Waters 2:229–236

Ellis EE, Richey JE, Aufdenkampe AK, Krusche AV, Quay PD, Salimon C, da Cunha HB (2012) Factors controlling water-column respiration in rivers of the central and southwestern Amazon Basin. Limnol Oceanogr 57:527–540

Engle D, Melack JM (2000) Methane emissions from the Amazon floodplain: enhanced release during episodic mixing of lakes. Biogeochemistry 51:71–90

Engle DL, Melack JM, Doyle RD, Fisher TR (2008) High rates of net primary productivity and turnover for floating grasses on the Amazon floodplain: implications for aquatic respiration and regional CO_2 flux. Glob Chang Biol 14:369–381

Ferreira E, Zuanon J, Forsberg B, Goulding M, Briglia-Ferreira SR (2007) Rio Branco: ecologia e conservação de Roraima. INPA/ACCA, Lima, p 201

Finer M, Jenkins CN (2012) Proliferation of hydroelectric dams in the Andean Amazon and implications for Andes-Amazon connectivity. PLoS One 7:e35126. doi:10.1371/journal.pone.0035126

Finer M, Orta-Martinez M (2010) A second hydrocarbon boom threatens the Peruvian Amazon: trends, projections, and policy implications. Environ Res Lett 5. doi:10.1088/1748-9326/5/1/014012

Flores BM, Piedade MTF, Nelson BW (2014) Fire disturbance in Amazonian blackwater floodplain forests. Plant Ecol Divers 7:319–327

Frankenberg C, Bergamaschi P, Butz A, Houweling S, Meirink JF, Notholt J, Petersen AK, Schrijver H, Warneke T, Aben I (2008) Tropical methane emissions: a revised view from SCIAMACHY onboard ENVISAT. Geophys Res Lett 35. doi:10.1029/2008GL034300

Frappart F, Seyler F, Martinez J-M, León JG, Cazenave A (2005) Floodplain water storage in the Negro River basin estimated from microwave remote sensing of inundation and water levels. Remote Sens Environ 99:387–399

Gash J, Keller M, Silva-Dias P (eds) (2009) Amazonia and global change, Geophysical monograph series 186. American Geophysical Union, Washington, DC

Getirana ACV, Boone A, Yamazaki D, Decharme B, Papa F, Mognard N (2012) The hydrological modeling and analysis platform (HyMAP): evaluation in the Amazon Basin. J Hydrometeorol 13:1641–1665

Gloor M, Brienen RJW, Galbraith D, Feldpausch TR, Schöngart J, Guyot JL, Espinoza JC, Lloyd J, Phillips OL (2013) Intensification of the Amazon hydrological cycle over the last two decades. Geophys Res Lett 40. doi:10.1002/grl.50377

Guérin F, Abril G, Richard S, Burban B, Reynouard C, Seyler P, Delmas R (2006) Methane and carbon dioxide emissions from tropical reservoirs: significance of downstream rivers. Geophys Res Lett 33. doi:10.1029/2006GL027929

Guérin F, Abril G, Serca D, Delon C, Richard S, Delmas R, Tremblay A, Varfalvy L (2007) Gas transfer velocities of CO_2 and CH_4 in a tropical reservoir and its river downstream. J Mar Syst 66:161–172

Hamilton S, Sippel S, Melack JM (1995) Oxygen depletion and carbon dioxide production in waters of the Pantanal wetland of Brazil. Biogeochemistry 30:115–141

Hamilton SK, Sippel SJ, Melack JM (2002) Comparison of inundation patterns among major South American floodplains. J Geophys Res 107. doi:10.1029/2000JD000306

Hamilton SK, Sippel SJ, Melack JM (2004) Seasonal inundation patterns in two large savanna floodplains of South America: the Llanos de Moxos (Bolivia) and the Llanos del Orinoco (Venezuela and Colombia). Hydrol Process 18:2103–2116

Hanson MC, Potapov PV, Moore R, Hancher M, Turubanov SA, Tyukavina A et al (2013) High-resolution global maps of 21st century forest cover change. Science 342:850–853

Hawes JE, Peres CA, Riley LB, Hess LL (2012) Landscape-scale variation in structure and biomass of Amazonian seasonally flooded and unflooded forests. For Ecol Manage 281:163–176

Hess LL, Melack JM, Novo EMLM, Barbosa CCF, Gastil M (2003) Dual-season mapping of wetland inundation and vegetation for the central Amazon basin. Remote Sens Environ 87:404–428

Ho DT, Veron F, Harrison E, Bliven LF, Scott N, McGillis WR (2007) The combined effect of rain and wind on air-water gas exchange: a feasibility study. J Mar Syst 66:150–160

Hofmann H (2013) Spatiotemporal distribution patterns of dissolved methane in lakes: how accurate are the current estimations of the diffusive flux path? Geophys Res Lett 40:2779–2784

Huntingford C, Jones PD, Livina VN, Lenton TM, Cox PM (2013) No increase in global temperature variability despite changing regional patterns. Nature 500:327–330

Jarvis A, Reuter HI, Nelson A, Guevara E (2008) Hoe-filled seamless SRTM data v4. International Center for Tropical Agriculture, Cali. Available at http://srtm.csi.cgiar.org

Jati SR (2013) Emissão de CO_2 and CH_4 nas savanas úmidas de Roraima. M.Sc. Thesis, Instituto Nacional de Pesquisas da Amazonia

Johnson MS, Lehmann J, Couto EG, Novaes JP, Riha SJ (2006) DOC and DIC in flowpaths of Amazonian headwater catchments with hydrologically contrasting soils. Biogeochemistry 81:45–57

Johnson MS, Weiler M, Couto EG, Riha SJ, Lehmann J (2007) Storm pulses of dissolved CO_2 in a forested headwater Amazonian stream explored using hydrograph separation. Water Resour Res 43:W11201. doi:10.1029/2007WR006359

Johnson MS, Lehmann J, Riha SJ, Krusche AV, Richey JE, Ometto J, Couto EG (2008) CO_2 efflux from Amazonian headwater streams represents a significant fate for deep soil respiration. Geophys Res Lett 35:L17401. doi:10.1029/2008GL034619

Junk WJ, Piedade MTF, Schöngart J, Cohn-Haft M, Adeney JM, Wittmann F (2011) A classification of major naturally-occurring Amazonian lowland wetlands. Wetlands 31:623–640

Kemenes A, Forsberg BR, Melack JM (2007) Methane release below a hydroelectric dam. Geophys Res Lett 34:L12809. doi:10.1029/2007GL029479

Kemenes A, Forsberg BR, Melack JM (2011) CO_2 emissions from a tropical hydroelectric reservoir (Balbina, Brazil). J Geophys Res Biogeosci 116:G03004. doi:10.1029/2010JG001465

Kemenes A, Forsberg BR, Melack JM (2016) Downstream emissions of CH_4 and CO_2 from Amazon hydroelectric reservoirs (Tucuruí, Samuel and Curuá-Una). Inland Waters (in press)

Kim JH, Zell C, Moreira-Turcq P, Pérez MAP, Abril G, Mortillaro JM, Weijers JWH, Meziane T, Sinninghe Damsté JS (2012) Tracing soil organic carbon in the lower Amazon River and its tributaries using GDGT distributions and bulk organic matter properties. Geochim Cosmochim Acta 90:163–180

Koschorreck M (2000) Methane turnover in exposed sediments of an Amazon floodplain lake. Biogeochemistry 50:195–206

Lahteenoja O, Reátegui YR, Räsänen M, Torres DC, Oinonen M, Page S (2011) The large Amazonian peatland carbon sink in the subsiding Pastaza-Maranon foreland basin, Peru. Glob Chang Biol 18:164–178

Langerwisch F, Rost S, Gerten D, Poulter B, Rammig A, Cramer W (2012) Potential effects of climate change on inundation patterns in the Amazon basin. Hydrol Earth Syst Sci Discuss 9:261–300

Lau KM, Wu HT, Kim KM (2013) A canonical response of precipitation characteristics to global warming from CMIP5 models. Geophys Res Lett 40:3163–3169

Lehner B, Döll P (2004) Development and validation of a global database of lakes, reservoirs and wetlands. J Hydrol 296:1–22

Liu J, Wang B, Cane MA, Yim S-Y, Leem J-Y (2013) Divergent global precipitation changes induced by natural versus anthropogenic forcing. Nature 493:656–659

Lloyd J, Kolle O, Fritsch H, Freitas SR, Dias MAFS, Artaxo P, Nobre AD, Araújo AC, Kruijt B, Sogocheva L, Fisch G, Thielmann A, Kuhn U, Andreae MO (2007) An airborne regional carbon balance for Central Amazonia. Biogeosciences 4:759–768

Longo KM, Freitas SR, Andreae MO, Yokelson R, Artaxo P (2009) Biomass burning in Amazonia: emissions, long-range transport of smoke and its regional and remote impacts. In: Gash J, Keller M, Silva-Dias P (eds) Amazonia and global change, Geophysical monograph series 186. American Geophysical Union, Washington, DC, pp 207–232

Macedo MN, Coe MT, DeFries R, Uriarte M, Brando PM, Neill C, Walker WS (2013) Land-use-driven stream warming in southeastern Amazonia. Philos Trans R Soc B 368:20120153, 10.1098/rstb.2012.0153

MacIntyre S, Wanninkhof R, Chanton J (1995) Trace gas exchange across the air-water interface in freshwater and coastal marine environments. In: Matson P, Harriss R (eds) Biogenic trace gases: measuring emissions from soil and water. Blackwell, Oxford, pp 52–97

MacIntyre S, Jonsson A, Jansson M, Åberg J, Turney DE, Miller S (2010) Buoyancy flux, turbulence, and the gas transfer coefficient in a stratified lake. Geophys Res Lett 37:L24604. doi:10.1029/2010GL044164

Malhi Y, Roberts JT, Betts RA, Killen TJ, Li W, Nobre CA (2008) Climate change, deforestation, and the fate of the Amazon. Science 319:169–172

Marengo JA (2004) Interdecadal variability and trends of rainfall across the Amazon basin. Theor Appl Climatol 78:79–96

Marengo JA, Nobre CA, Tomasella J, Oyama MD, de Oliveira GS, de Oliveira R, Camargo H, Alves LM, Brown IF (2008) The drought of Amazonia in 2005. J Clim 21:495–516

Marengo J, Nobre CA, Betts RA, Cox PM, Sampaio G, Salazar L (2009) Global warming and climate change in Amazonia: climate-vegetation feedback and impacts on water resources. In: Gash J, Keller M, Silva-Dias P (eds) Amazonia and global change, Geophysical monograph series 186. American Geophysical Union, Washington, DC, pp 273–292

Marengo JA, Tomasella J, Alves LM, Soares WR, Rodriguez DA (2011) The drought of 2010 in the context of historical droughts in the Amazon region. Geophys Res Lett 38:L12703. doi:10.1029/2011GL047436

Mayorga E, Logsdon MG, Ballester MVR, Richey JE (2005a) Estimating cell-to-cell land surface drainage paths from digital channel networks, with an application to the Amazon basin. J Hydrol 315:167–182

Mayorga E, Aufdenkampe AK, Masiello CA, Krusche AV, Hedges JI, Quay PD, Richey JE, Brown TA (2005b) Young organic matter as a source of carbon dioxide outgassing from Amazonian rivers. Nature 436:538–541. doi:10.1038/nature03880

Melack JM (2004) Remote sensing of tropical wetlands. In: Ustin S (ed) Manual of remote sensing, vol 4, Remote sensing for natural resources management and environmental monitoring. Wiley, New York, NY, pp 319–343

Melack JM (2005) Floodplain lakes and reservoirs in tropical and subtropical South America: limnology and human impacts. In: O'Sullivan P, Reynolds C (eds) Lakes handbook, vol 2. pp 241–257

Melack JM, Coe MT (2013) Climate change and the floodplain lakes of the Amazon basin. In: Goldman CR, Kumagai M, Robarts R (eds) Global impact of climate change on inland water systems. Wiley, New York, NY, pp 295–310

Melack JM, Engle D (2009) An organic carbon budget for an Amazon floodplain lake. Verh Int Verein Limnol 30:1179–1182

Melack JM, Forsberg B (2001) Biogeochemistry of Amazon floodplain lakes and associated wetlands. In: McClain ME, Victoria RL, Richey RE (eds) Biogeochemistry of the Amazon Basin and its role in a changing world. Oxford University Press, Oxford, pp 235–276

Melack JM, Hess LL (2010) Remote sensing of the distribution and extent of wetlands in the Amazon basin. In: Junk WJ, Piedade M, Wittmann F, Schöngart J, Parolin P (eds) Amazonian floodplain forests: ecophysiology, ecology, biodiversity and sustainable management, Ecological studies. Springer, Berlin, pp 43–59

Melack JM, Hess LL, Gastil M, Forsberg BR, Hamilton SK, Lima IBT, Novo EMLM (2004) Regionalization of methane emissions in the Amazon basin with microwave remote sensing. Glob Chang Biol 10:530–544

Melack JM, Novo EMLM, Forsberg BR, Piedade MTF, Maurice L (2009) Floodplain ecosystem processes. In: Gash J, Keller M, Silva-Dias P (eds) Amazonia and global change, Geophysical monograph series 186. American Geophysical Union, Washington, DC, pp 525–541

Miguez-Macho G, Fan Y (2012) The role of groundwater in the Amazon water cycle: 1. Influence on seasonal streamflow, flooding and wetlands. J Geophys Res 117. doi:10.1029/2012JD017539

Miller JB, Gatti LV, d'Amelio MTS, Crotwell AM, Dlugokencky EJ, Bakwin P, Artaxo P, Tans PP (2007) Airborne measurements indicate large methane emissions from the eastern Amazon basin. Geophys Res Lett 34:L10809. doi:10.1029/2006GL029213

Mortillaro JM, Abril G, Moreira-Turcq P, Sobrinh R, Perez M, Meziane T (2011) Fatty acid and stables isotopes (δ13C, δ15N) signatures of particulate organic matter in the Lower Amazon River: seasonal contrasts and connectivity between floodplain lakes and the mainstem. Org Geochem 42:1159–1168

Mortillaro JM, Rigal F, Rybarczyk H, Bernardes M, Abril G, Meziane T (2012) Particulate organic matter distribution along the lower Amazon River: addressing aquatic ecology concepts using fatty acids. PLoS One 7:e46141. doi:10.1371/journal.pone.0046141

Negrón-Juárez RI, Chambers JQ, Guimaraes G, Zeng H, Raupp CFM, Marra DM, Ribeiro GH, Higuchi N, Nelson B (2010) Widespread Amazon forest tree mortality from a single cross-basin squall line event. Geophys Res Lett 37:L16701. doi:10.1029/2010GL043733

Neu V, Neill C, Krusche AV (2011) Gaseous and fluvial carbon export from an Amazon forest watershed. Biogeochemistry 105:133–147

Ortiz AC, Ashton A, Nepf H (2013) Mean and turbulent velocity fields near rigid and flexible plants and the implications for deposition. J Geophys Res Earth Surf 118. doi:10.1002/2013JF002858

Paiva RCD, Buarque DC, Collischonn W, Bonnet MP, Frappart F, Calmant S, Mendes CAB (2013) Large-scale hydrologic and hydrodynamic modeling of the Amazon River basin. Water Resour Res 49:1226–1243

Papa F, Guntner A, Frappart F, Prigent C, Rossow WB (2008) Variations of surface water extent and water storage in large river basins: a comparison of different global data sources. Geophys Res Lett 35:L11401. doi:10.1029/2008GL033857

Papa F, Prigent C, Jimenez C, Aires F, Rossow WB (2010) Interannual variability of surface water extent at global scale, 1993–2004. J Geophys Res 115:D12111. doi:10.1029/2009JD012674

Pires GF, Costa MH (2013) Deforestation causes different subregional effects on the Amazon bioclimatic equilibrium. Geophys Res Lett 40:3618–3623

Polsenaere P, Deborde J, Detandt G, Vidal LO, Pérez MAP, Marieu V, Abril G (2013) Thermal enhancement of gas transfer velocity of CO_2 in an Amazon floodplain lake revealed by eddy covariance measurements. Geophys Res Lett 40:1734–1740

Potter CS (1997) An ecosystem simulation model for methane production and emission from wetlands. Global Biogeochem Cycles 11:495–506

Potter CS, Melack JM, Engle D (2016) Modeling carbon dynamics and methane emissions from Amazon floodplain ecosystems. Wetlands 11:501–511

Prigent C, Papa F, Aires F, Rossow WB, Matthews E (2007) Global inundation dynamics inferred from multiple satellite observations, 1993–2000. J Geophys Res 112:D12107. doi:10.1029/2006JD007847

Prigent C, Papa F, Jimenez C, Rossow WB, Matthews E (2012) Changes in land surface water dynamics since the 1990s and relation to population pressures. Geophys Res Lett 39:L08403. doi:10.1029/2012GL051276

Rasera MFFL, Ballester MVR, Krusche AV, Salimon C, Montebelo LA, Alin SR, Victoria RL, Richey JE (2008) Estimating the surface area of small rivers in the southwestern Amazon and their role in CO2 outgassing. Earth Interact 12:1–16. doi:10.1175/2008EI257.1

Rasera MFFL, Krusche AV, Richey JE, Ballester MVR, Victória RL (2013) Spatial and temporal variability of pCO_2 and CO_2 efflux in seven Amazonian rivers. Biogeochemistry 116:241–259

Raymond PA, Hartmann J, Lauerwald R, Sobek S, McDonald C, Hoover M, Butman D, Striegl R, Mayorga E, Humborg C, Kortelainen P, Dürr H, Meybeck M, Ciais P, Guth P (2013) Global carbon dioxide emissions from inland waters. Nature 503:355–359

Reichstein M, Bahn M, Ciais P, Frank D, Mahecha MD, Seneviratne SI et al (2013) Climate extremes and the carbon cycle. Nature 500:287–295

Remington S, Krusche AV, Richey JM (2011) Effects of DOM photochemistry on bacterial metabolism and CO_2 evasion during falling water in a humic and a whitewater river in the Brazilian Amazon. Biogeochemistry 105:185–200

Renó VF, Novo EMLM, Suemitsu C, Rennó CD, Silva TSF (2011) Assessment of deforestation in the lower Amazon floodplain using historical landsat MSS/TM imagery. Remote Sens Environ 115:3446–3456

Rice AL, Butenhoff CL, Shearer MJ, Teama D, Rosenstiel TN, Khalil AK (2010) Emissions of anaerobically produced methane by trees. Geophys Res Lett 37:L03807. doi:10.1029/2009GL041565

Richey JE, Melack JM, Aufdenkampe AK, Ballester VM, Hess L (2002) Outgassing from Amazonian rivers and wetlands as a large tropical source of atmospheric carbon dioxide. Nature 416:617–620

Richey JE, Krusche AV, Johnson MS, da Cunha HB, Ballester MV (2009) The role of rivers in the regional carbon balance. In: Gash J, Keller M, Silva-Dias P (eds) Amazonia and global change, Geophysical monograph series 186. American Geophysical Union, Washington, DC, pp 489–504

Riley WJ, Subin AM, Lawrence DM, Swenson SC, Torn MS, Meng L et al (2011) Barriers to predicting changes in global terrestrial methane fluxes: analyses using CLM4Me, a methane biogeochemistry model integrated in CESM. Biogeosciences 8:1925–1953. doi:10.5194/bg-8-1925-2011

Ringeval B, Noblet-Ducoudré N, Ciais P, Bousquet P, Prigent C, Papa F, Rossow WB (2010) An attempt to quantify the impact of changes in wetland extent on methane emissions on the seasonal and interannual time scales. Global Biogeochem Cycles 24:GB2003. doi:10.1029/2008GB003354

Roland F, Vidal LO, Pacheco FS, Barros NO, Assireu A, Ometto HPHB, Cimbleris ACP, Cole JJ (2010) Variability of carbon dioxide flux from tropical (Cerrado) hydroelectric reservoirs. Aquat Sci 72:283–293

Rosenqvist A, Forsberg BR, Pimentel T, Rauste YA, Richey JE (2002) The use of spaceborne radar data to model inundation patterns and trace gas emissions in the central Amazon floodplain. Int J Remote Sens 23:1303–1328

Rudorff CM (2013) Flooding and carbon dynamics on the lower Amazon floodplain. Ph.D. Thesis, University of California, Santa Barbara

Rudorff CM, Melack JM, MacIntyre S, Barbosa CCF, Novo EMLM (2011) Seasonal and spatial variability in CO_2 emissions from a large floodplain lake in the lower Amazon. J Geophys Res Biogeosci 116:G04007. doi:10.1029/2011JG001699

Rudorff CM, Melack JM, Bates PD (2014a) Flooding dynamics on the lower Amazon floodplain: 1. Hydraulic controls on water elevation, inundation extent and river-floodplain discharge. Water Resour Res 50. doi:10.1002/2013WR014091

Rudorff CM, Melack JM, Bates PD (2014b) Flooding dynamics on the lower Amazon floodplain: 2. Seasonal and interanual hydrological variability. Water Resour Res 50. doi:10.1002/2013WR014714

Sawakuchi HO, Bastiviken D, Sawakuchi AEO, Krusche AV, Ballester MVR, Richey JE (2014) Methane emissions from Amazonian rivers and their contribution to the global methane budget. Glob Chang Biol. doi:10.1111/gcb.12646

Silva TSF, Costa M, Melack JM (2009) Annual net primary production of macrophytes in the eastern Amazon floodplain. Wetlands 29:747–758

Silva TSF, Costa MPF, Melack JM (2010) Spatio-temporal variability of macrophyte cover and productivity in the eastern Amazon floodplain: a remote sensing approach. Remote Sens Environ 114:1998–2010

Silva TSF, Melack JM, Novo EML (2013) Responses of aquatic macrophyte cover and productivity to flooding variability on the Amazon floodplain. Glob Chang Biol 19:3379–3389

Stanley EH, Casson NJ, Christel ST, Others (2015) The ecology of methane in streams and rivers: patterns, controls, and global significance. Ecol Monogr. doi:10.1890/15-1027.1

Victoria DC (2010) Simulacao hidrologica de bacias amazonicas utilizando o modelo de Capacidade de Infiltracao Variavel (VIC). PhD. Thesis, University of Sao Paulo, Piracicaba

Vihermaa LE, Waldron S, Garnett MH, Newton J (2014) Old carbon contributes to aquatic emissions of carbon dioxide in the Amazon. Biogeosciences 11:3635–3645

Villar E, Jhan C, Ronchail J, Guyot JL, Junquas C, Vauchel P, Casimiro L, Waldo S, Drapeau G, Pombosa Loza LR (2011) Climate variability and extreme drought in the upper Solimões River (Western Amazon Basin): understanding the exceptional 2010 drought. Geophys Res Lett 38: L13406. doi:10.1029/2011GL047862

Waichman AV (1996) Autotrophic carbon sources for heterotrophic bacterioplankton in a floodplain lake of Central Amazon. Hydrobiologia 341:27–36

Walter BP, Heimann M (2000) A process-based, climate-sensitive model to derive methane emissions from natural wetlands: application to five wetland sites, sensitivity to model parameters, and climate. Global Biogeochem Cycles 14:745–765

Ward ND, Keil RG, Medeiros PM, Brito DC, Cunha AC, Dittmar T, Yager PL, Krusche AV, Richey JE (2013) Degradation of terrestrially derived macromolecules in the Amazon River. Nat Geosci. doi:10.1038/ngeo1817

Wilson MD, Bates PD, Alsdorf A, Forsberg B, Horritt M, Melack J, Frappart F, Famglietti JS (2007) Modeling large-scale inundation of Amazonian seasonally flooded wetlands. Geophys Res Lett 34:L15404. doi:10.1029/2007GL030156

Worbes M (1997) The forest ecosystem of the floodplains. In: Junk WJ (ed) The Central Amazon floodplain, Ecological studies. Springer, Berlin, pp 223–260

Xavier L, Becker M, Cazenave A, Longuevergne L, Llovel W, Rotunno Filho OC (2010) Interannual variability in water storage over 2003–2008 in the Amazon Basin from GRACE space gravimetry, *in situ* river level and precipitation data. Remote Sens Environ 114:1626–1637

Yamazaki D, Kanae S, Kim H, Oki T (2011) A physically based description of floodplain inundation dynamics in a global river routing model. Water Resour Res 47:W04501. doi:10.1029/2010WR009726

Zeng N, Yoon JH, Marengo JA, Subramaniam A, Nobre CA, Mariotti A, Neelin JD (2008) Causes and impacts of the 2005 Amazon drought. Environ Res Lett 3. doi:10.1088/1748-9326/3/1/014002

Chapter 8
Ecosystem–Atmosphere Exchanges of CO_2 in Dense and Open 'Terra Firme' Rainforests in Brazilian Amazonia

Alessandro C. Araújo, Celso von Randow, and Natalia Restrepo-Coupe

8.1 Introduction

The rainforests of the Amazon basin constitute the single largest expanse of tropical forest on the planet, covering an area of about 5.8×10^6 km^2 (Eva and Huber 2005). Its total above-ground live biomass is estimated to contain about 93 ± 23 Pg C (Malhi et al. 2006), about 74 % of it being in non-inundated (*terra firme*) forests (Vieira et al. 2004). In addition to the large carbon stock, the Amazon basin has important implications for climate, locally and regionally (Nobre et al. 2009).

Concerns about global climate change have stimulated renewed interest in, and created an unprecedented demand for, information about the least understood parts of the climate system, including the boundary between the land surface of the earth and the lowest part of the planetary boundary layer (PBL), the internal carbon fluxes within ecosystems, and trace gas exchange between the terrestrial biosphere and the atmosphere. To address these, several projects have been implemented in the Amazon basin since the early 1990s within the scope of the Large-scale Biosphere–Atmosphere Experiment/Programme in Amazonia (LBA). Additionally, LBA projects have been addressing the climatological, ecological, biogeochemical, and hydrological functioning in the Amazon basin, the impact of land use change on

A.C. Araújo (✉)
Brazilian Agricultural Research Corporation—Embrapa Eastern Amazon, Belém, Brazil
e-mail: alessandro.araujo@embrapa.br

C. von Randow
National Institute for Space Research, Earth System Science Center, Cachoeira Paulista, Brazil

N. Restrepo-Coupe
Climate Cluster, University of Technology Sydney, Sydney, NSW 2007, Australia

© Springer-Verlag Berlin Heidelberg 2016
L. Nagy et al. (eds.), *Interactions Between Biosphere, Atmosphere and Human Land Use in the Amazon Basin*, Ecological Studies 227,
DOI 10.1007/978-3-662-49902-3_8

these functions, and the interactions between Amazonia and the Earth system (Avissar et al. 2002; Gash et al. 2004; Keller et al. 2004).

LBA has undertaken the first comprehensive assessment of the spatial and temporal variability of CO_2 fluxes between the biosphere and atmosphere in the Amazon forest 'biome'. Estimates based on eddy covariance (EC) measurements have suggested that the forest areas of the basin might be either a carbon sink (c. 0.7–4.2 Pg C $year^{-1}$) (Grace et al. 1995a, b; Malhi et al. 1998; Araújo et al. 2002; Carswell et al. 2002; Kruijt et al. 2004; Gloor 2016; Grace 2016) or a carbon source (c. 0.2–0.8 Pg C $year^{-1}$) (Saleska et al. 2003; Miller et al. 2004; Hutyra et al. 2007, 2008). Moreover, recent estimates based on biomass inventory plot measurements have suggested that the Amazon forest is absorbing carbon at a rate of about 0.7 Pg C $year^{-1}$ during normal or wet years (Phillips et al. 1998; Baker et al. 2004; see also Chap. 10), but turns to neutral or sources of carbon during drought years (Gatti et al. 2014; Chaps. 5 and 6). The reported spatio-temporal variability is largely related to soil fertility and precipitation gradients and to climate-driven effects, such as the El Niño Southern Oscillation (ENSO). The soils are more fertile to the south-west of the basin, and there is more precipitation in the north-west. In addition, the above-ground live biomass is larger in the north than in the south (Malhi et al. 2006; Saatchi et al. 2007).

In this chapter, we present analyses of the relations between climatic variables and fluxes measured over the LBA towers in two types of non-inundated ('terra firme') evergreen lowland rainforest—(1) dense rainforest and (2) open rainforest in the Brazilian Legal Amazon (BLA) (IBGE 2012), hereinafter referred to as BLA (Table 8.1). We address the uncertainty in the responses of the vegetation (ecosystem–atmosphere exchanges of CO_2) to natural climate variations and possible future extreme conditions. We present analyses, combining information from flux tower observations and terrestrial process-based models.

8.2 Sites: Rainfall Patterns and Precipitation Seasonality

Climatic seasonality in BLA is related to annual rainfall patterns, in particular to the dry season, the period when monthly rainfall values are below 100 mm $month^{-1}$ (Sombroek 2001), or evapotranspiration (ET) is greater than precipitation (used interchangeably as mean monthly forest ET is c. 100 mm). The timing of the onset and length of the dry season and the amount of dry season precipitation are important variables for ecosystem functioning. Using precipitation averages for 1998–2013 from the Tropical Rainfall Measuring Mission (TRMM) database (NASA 2014), the following pattern emerges.

Mean annual (2552 mm) and minimum monthly (140 mm) precipitation across the equatorial zone (0–5°S) gradually decreases from the north-west of the basin (80° W–75° W) with no dry season to 2263 mm $year^{-1}$ and 46 mm $month^{-1}$ at the confluence of the Amazon and Tapajós rivers (57° W–54° W) and then increases again near to the Amazonas delta (0–2° S and 53° W–48° W) (2626 mm and

Table 8.1 Characteristics of LBA flux tower sites dense and open 'terra firme' rainforest of Brazilian Legal Amazon that were used in this study

ID	Site name	Lon. (° W)	Lat. (° S)	Measurement period	Total AG biomass (kg m^{-2})	Above-ground living biomass (kg m^{-2})	LAI (m^2 m^{-2})	Canopy height (m)	Soil depth (m)	Water table depth (m)
K34	Manaus	60.2	2.61	14-Jun-99 to	40.61	30.2	4.7	30	15	35
				30-Sep-06	Malhi et al. (2009)	Malhi et al. (2009)	Malhi et al. (2009)	Andreae et al. (2002)	Tomasella et al. (2008)	Tomasella et al. (2008)
K67	Tapajós	55	2.85	2-Jan-02 to	39.9 ± 1.6[a]	29.3 ± 1.1	6.4–5.7[a]	9-Feb-00	12 (free of hardpan or iron oxide concretions)	>100
				23-Jan-06	Rice et al. (2004)	Rice et al. (2004)	Nepstad et al. (1994)	Rice et al. (2004)	Oliveira et al. (2005)	Nepstad et al. (1994)
K83	Tapajós	54.6	3.01	29-Jun-00 to	37.20	33.5 (2001 pre-logging)	4.36–5.56	35–40 (Miller et al. 2004)	12 (free of hardpan or iron oxide concretions) Oliveira et al. (2005)	>100
				12-Mar-04	Keller et al. (2005)	30.0 (2004 after-logging)	Negrón Juárez et al. (2009)			Nepstad et al. (1994)
						Figueira et al. (2008)				
CAX	Caxiuanã	51.5	1.72	1-Jan-99 to	47.14	42.65	5.14	35	3–4 (free of hardpan)	~10 wet season
				30-Jul-03	Malhi et al. (2009)	Malhi et al. (2009)	Malhi et al. (2009)	Fisher et al. (2007)	Malhi et al. (2009)	Malhi et al. (2009)
RJA	Reserva Jarú	61.9	10.08	23-Mar-99 to	–	22.00	5.5	35.00	1.2–4	–
				14-Nov-02		Meir et al. (1996)	Andreae et al. (2002)	von Randow et al. (2004)	Andreae et al. (2002)	

[a]At a nearby forest, the control site at partial exclusion of precipitation experiment

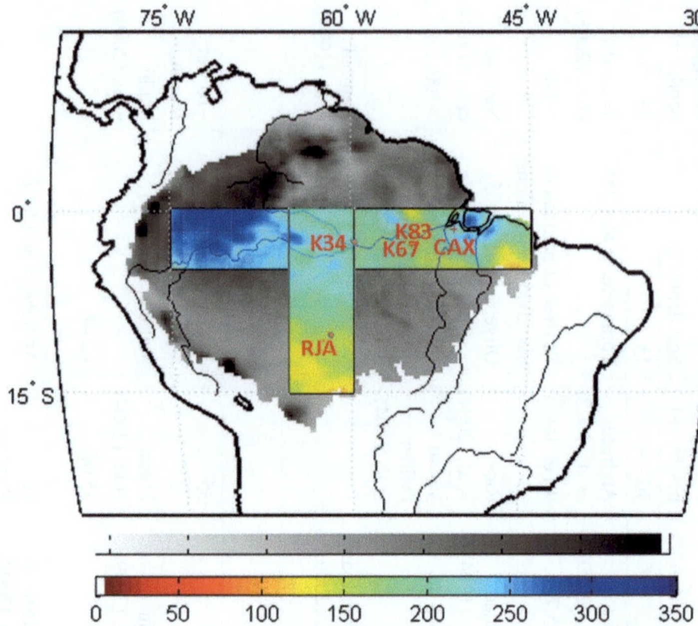

Fig. 8.1 Average monthly precipitation (mm month^{-1}) in the Amazon region (1998–2013), according to the values of the Tropical Rainfall Measuring Mission (TRMM). Location of eddy flux sites: Manaus dense 'terra firme' forest (K34), Santarém dense 'terra forme' forest (K67 and K83), Caxiuanã dense 'terra firme' forest (CAX), Reserva Jarú open 'terra firme' forest (RJA). *Colour* bands indicate equatorial Amazon region (latitude 0–5°S) and north–south transect (0–15°S) along longitude 65–60°W

58 mm month^{-1}) (Fig. 8.1). The flux towers of Manaus (K34), Tapajós National Forest (K67 and K83), and Caxiuanã National Forest (CAX) constitute a small sample along the above-mentioned gradient (Table 8.1), which are referred here as dense terra firme forest sites. At the K34 site (2.61°S; 60.2°W), dry season lasts 1 month and mean annual precipitation reaches 2520 mm (Araújo et al. 2002). To the east, K67 (2.85°S; 55°W) and K83 (3.01°S; 54.6°W) sites are located in a region with a mean annual precipitation of 1823 mm and which technically has a c. 5-month-long dry season (minimum monthly precipitation is on average 54–98 mm) (Goulden et al. 2004; Hutyra et al. 2007; Saleska et al. 2003). Finally, the most easterly site at CAX (1.72°S; 51.5°W) has the highest seasonality, as despite receiving 2551 mm annual rainfall and having the highest monthly precipitation among the three sites in the wet season, there is a dry season where during 4 months the average monthly precipitation is 77 mm (Carswell et al. 2002) (Figs. 8.1 and 8.2, lower rows). A similar precipitation gradient is observed from north to south, where at the open 'terra firme' rainforest of Jaru (RJA) (10.08°S; 61.9°W) mean precipitation is 2031 mm year^{-1} and there is a 5-month-long dry season with average monthly rainfall of 23 mm (von Randow et al. 2004) (Fig. 8.1). In addition, particularly in this part of BLA (Latitude > 7°S), there is an observed

tendency for an increase in dry and very dry events during the dry season, which is concomitant with an increase in the length of the low precipitation period (Marengo et al. 2011).

8.3 Ecosystem–Atmosphere CO_2 Fluxes in BLA

In the late 1980s, EC, a technique that quantifies the exchange rate of gases between the terrestrial ecosystems and atmosphere, emerged as an important tool (Baldocchi 2003). The technique measures the covariance between the fluctuations in vertical wind velocity and CO_2 mixing ratio (see also Chap. 5). This technique was first used in Amazonia for short-term measurements of CO_2 fluxes in the late 1980s by Fan et al. (1990) and in the early 1990s by Grace et al. (1995a, b). From middle 1990s onwards, long-term measurements became possible (Malhi et al. 1998; Araújo et al. 2002; Carswell et al. 2002; Kruijt et al. 2004; Saleska et al. 2003; Miller et al. 2004; Hutyra et al. 2007, 2008) after the introduction of fast-response instrumentation (see more detailed information in Saleska et al. 2009).

Except for atmospheric–oceanic phenomena such as ENSO, which caused a reduction in both net ecosystem–atmosphere CO_2 exchange (NEE) and ET during the dry season in a forest near to K34 site in Manaus (Malhi et al. 1998, 2002), results from flux towers indicate that NEE decreases (by convention carbon uptake has a negative sign; hence the more negative it becomes the higher is the carbon assimilation) during the dry season as compared to the wet season at K34, K67, and CAX sites (Araújo et al. 2002; Carswell et al. 2002; Kruijt et al. 2004; Saleska et al. 2003; Miller et al. 2004; Hutyra et al. 2007, 2008). This pattern, until recently, has not been incorporated into many modelling studies (Saleska et al. 2003, 2009), which, as a consequence, have continued to predict a reduction in NEE and ET in dry season. Thus, only at the RJA open 'terra firme' forest site was a good agreement between observations and model predictions (von Randow et al. 2004; Kruijt et al. 2004).

Although there has been a large improvement in our understanding of the environmental (net radiation) and biotic (rooting depth) controls of ecosystem–atmosphere water fluxes in Amazonia (Négron-Juárez et al. 2007; Hasler and Avissar 2007; da Rocha et al. 2009; Costa et al. 2010), the same does not apply to ecosystem–atmosphere CO_2 fluxes, which only recently started to be addressed (Restrepo-Coupe et al. 2013; von Randow et al. 2013). This is because CO_2 fluxes are inherently more complex, involving nuances of ecosystem metabolism, than the water fluxes (Saleska et al. 2009).

8.4 Seasonality of C-Fluxes

Across the central part of BLA, the timing of the dry season and the associated decrease in cloud cover coincides with the equinox, the timing of highest top of the atmospheric radiation (TOA). As a consequence, radiation at the canopy level, also measured as incoming photosynthetically active radiation (PAR), increases to its annual maximum in July–October (Fig. 8.2). By contrast, in the south (Lat. >7°S), as the TOA and precipitation are synchronous (July minimum and January maximum), PAR is higher and less seasonal than at the equatorial sites and its maximum coincides with the end of the dry season, which extends from May to September.

At all EC sites, the available energy values showed a direct relationship with ET as reported in previous studies (Costa et al. 2010; da Rocha et al. 2004; Hasler and Avissar 2007; Négron-Juárez et al. 2007). However, gross ecosystem productivity (GEP), understood as a measure of ecosystem photosynthetic flux, does not show a simple relationship with PAR when viewed on a month-by-month basis (Restrepo-

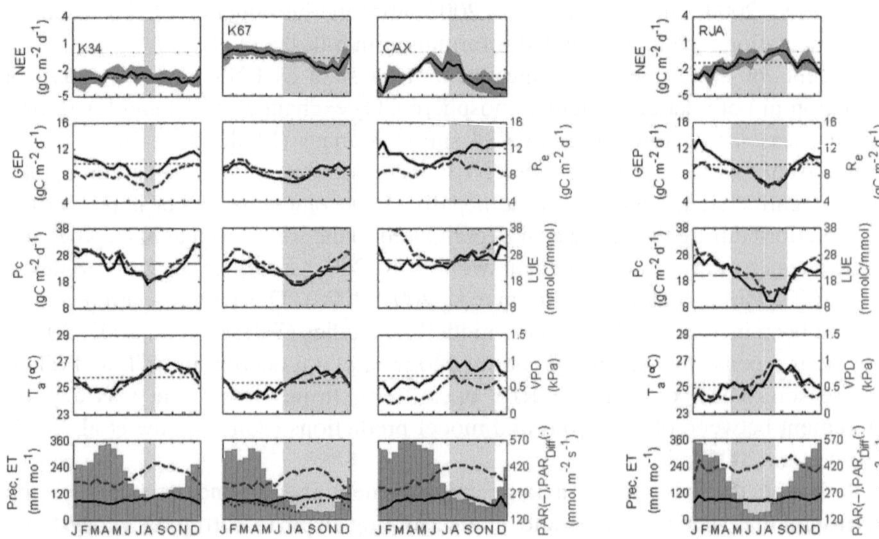

Fig. 8.2 Annual cycles, 16-day composites of the mean of all available years (Table 8.1) of net ecosystem exchange (NEE, *top row* for each site) in Brazilian Legal Amazonia. *Dark grey* areas indicate 95 % CI determined by the selection of u_* (see Restrepo-Coupe et al. 2013); gross ecosystem productivity (GEP, *second row top to bottom black continuous line*); ecosystem respiration (R_e, *second row to bottom, grey dashed line*); photosynthetic capacity (Pc, *third row* for each site, *black continuous line*); light use efficiency (LUE, *third panel, grey dashed line*); air temperature (T_a, *fourth row black continuous line*); vapour pressure deficit (VPD, *fourth row grey dashed line*); and precipitation (Prec, *lower row grey bars* at *lower row*), evapotranspiration (ET, *continuous black line*), and incoming photosynthetic active radiation (PAR, *fifth row dashed grey line*) and diffuse fractions (PAR_{Diff}, *fifth row dotted grey line*). Diffuse radiation only available at K67. *Grey shaded* areas indicate dry seasons (precipitation <100 mm month^{-1}). Dense 'terra firme' forest sites: Manaus (K34), Santarém (K67), Caxiuanã (CAX); open 'terra firme' site: Jarú (RJA)

Coupe et al. 2013). At the three central BLA sites (K34, K67, and CAX), we observed a gradual increase in GEP at the beginning of the dry season and it reached its maximum by the middle or end of the dry season (Fig. 8.2). Interestingly, the two eastern sites, K67 and CAX, where the dry season is longer compared to K34, have a second peak during the wet season. This peak in productivity is highly correlated to wood allocation as presented in the section on the seasonality of photosynthesis.

On an hour-by-hour basis GEP is driven by PAR, controlled by vapour pressure deficit (VPD), soil moisture, and air temperature (T_a), as previously reported by different authors (Law et al. 2002; Mercado et al. 2009; Oliveira et al. 2007). We also observed that at some sites, GEP was linearly related to the absolute value of available diffuse radiation ($R^2 = 0.82$, $p < 0.001$ at K67). However, on a seasonal basis (16-day time series) linear and multivariate regressions between GEP and PAR, and GEP and any meteorological variable, including T_a, VPD, precipitation, and PAR (Fig. 8.3, top rows), were not statistically significant (Table 8.2). An indication of a shared atmospheric and vegetation adaptation (e.g. leaf phenology, wood allocation) controls of ecosystem productivity.

The seasonality of C-fluxes presented here uses meteorological and carbon and water flux variables for the periods between 1999 and 2006 from five EC towers

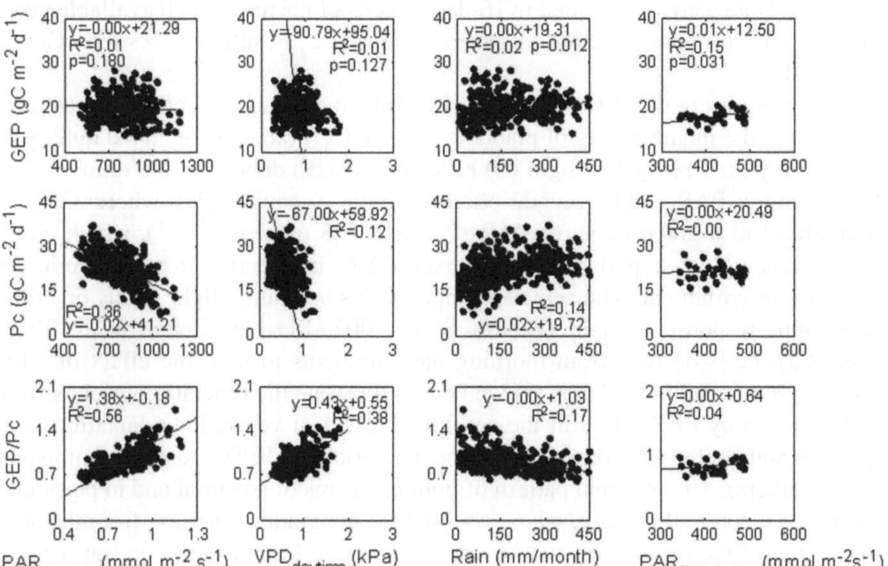

Fig. 8.3 From *right* to *left*: linear regression 16-day average gross ecosystem productivity, *GEP* (g C m^{-2} day^{-1}) (*top rows*), photosynthetic capacity, Pc (g C m^{-2} day^{-1}) (*middle rows*) and GEP/Pc (*bottom rows*) versus photosynthetic active radiation, PAR (mmol m^{-2} s^{-1}), vapour pressure deficit, VPD (kPa), precipitation from the Tropical Rainfall Measuring Mission (TRMM) (mm $month^{-1}$), and diffuse PAR, PAR_{Diff} (mmol m^{-2} s^{-1}) (only available at K67). Data from flux tower sites in dense 'terra firme' forests: Manaus (K34), Santarém (K67), Caxiuanã (CAX); and open 'terra firme': Jarú (RJA)

located at tropical forest ecosystems from the Brasil flux database (Restrepo-Coupe et al. 2013) (refer to Table 8.1 for site-specific measurement period). In 2005, large sections of southwestern Amazonia experienced one of the most intense droughts of the last 100 years (Marengo et al. 2008). A later analysis has shown that 1 year after the drought, the RJA open 'terra firme' forest site became a net source of carbon to the atmosphere (Zeri et al. 2014).

We derived GEP data from measured carbon fluxes, where gross ecosystem exchange (GEE) was obtained from EC measurements of daytime NEE by subtracting estimates of ecosystem respiration (R_e). R_e was calculated as the mean of filtered nighttime NEE (low turbulence threshold, u_* threshold values removed) within a 5-day wide window, expanded up to monthly average if not enough data were available. GEP was defined as minus GEE assuming negligible leaf re-assimilation of dark respiration and insignificant CO_2 recirculation within the canopy air space below the EC system (Stoy et al. 2006). Minimum missing data filling was carried out to the C-flux time series, as we wanted to base our analysis on observations rather than model results. GEP data was filled using a c. 10-day moving lookup table, informed by PAR and time of day (2-h-wide bins). The hourly bins account for morning and afternoon differences in GEP (generally driven by VPD) and the moving window accounts for sensitivity to seasonally varying soil moisture. Data were composited in 16-day periods of the mean of all available years (Table 8.1) for presentation purposes (see more details in Restrepo-Coupe et al. 2013).

Here we present two different measures of photosynthetic potential in an attempt to quantify the innate ability of plants to capture CO_2 (amount, age, and quality of leaves): first, the ecosystem light use efficiency (LUE) defined as the ratio of GEP to incoming PAR, and second, photosynthetic capacity (Pc) where GEP is normalised to a given range of PAR (725 and 925 μmol m^{-2} s^{-1}), which based on the rectangular hyperbola (GEP versus PAR) is understood as the optimum radiation environment. The selected range avoids saturation light levels or initial ecosystem response to light (Hutyra et al. 2007; Restrepo-Coupe et al. 2013) (Fig. 8.2). Pc is derived from morning measurements to limit the effect of VPD on ecosystem response. LUE is shown here to illustrate the vegetation response that is not driven by PAR, but still the combined effect of vegetation adaptation (leaf quantity and quality) and other environmental variables (VPD, T_a, and soil moisture among others). The seasonal pattern of both measures of potential and in particular Pc makes a compelling case, where we observe the gradual increase from the start to the end of the dry season and an opposite pattern (high to low) during the wet season (Fig. 8.2). This indicates that under the current precipitation regime the pattern of photosynthesis, based on EC data, shows little evidence of seasonal water limitation, as forests sustain high or increasing levels of Pc as the dry season progresses. Thus, in agreement with leaf-level stomatal conductance at saturating light (a proxy for measuring photosynthetic rate) measurements reported for K67 showed aseasonal or significantly higher values during the dry season (Domingues et al. 2014). Exploring the idea that the pattern of Pc may be consistent with light limitation (Goulden et al. 2004; Huete et al. 2006), we explored the relationships

Table 8.2 Pearson product–moment correlation coefficient between gross ecosystem productivity (GEP), ecosystem respiration (R_e), net ecosystem exchange (NEE), and different meteorological variables, air temperature (T_a), vapour pressure deficit (VPD), photosynthetic active radiation (PAR), and precipitation (Precipitation)

Correlation coefficient, R	C-flux (μmol m^2 day^{-1})	T_{air} (C)	Precipitation (mm day^{-1})	$PAR_{daytime}$ (μmol m^2 s^{-1})	$VPD_{daytime}$ (kPa)
K34	NEE	−0.15	0.15	−0.29	−0.14
	$GEP_{daytime}$	−0.26	0.13	−0.12	−0.24
	R_e	−0.17	0.18	*−0.55*	−0.32
K67	NEE	*−0.66*	*0.61*	−0.34	*−0.71*
	$GEP_{daytime}$	0.05	0.10	−0.25	−0.01
	R_e	−0.50	*0.56*	*−0.58*	*−0.55*
CAX	NEE	−0.07	0.29	0.49	0.29
	$GEP_{daytime}$	*0.59*	*−0.55*	0.26	0.26
	R_e	0.45	−0.33	0.38	0.30
RJA	NEE	0.34	−0.48	0.34	0.49
	$GEP_{daytime}$	−0.45	0.47	−0.37	*−0.75*
	R_e	−0.20	0.53	−0.45	−0.53
All sites	NEE	−0.32	−0.01	0.03	−0.20
	$GEP_{daytime}$	0.12	0.14	−0.08	−0.10
	R_e	−0.18	0.22	−0.39	−0.44

Data from flux tower sites in dense 'terra firme' forests: Manaus (K34), Santarém (K67, K83); and open 'terra firme': Jarú (RJA). Numbers in italic denote $R^2 > 0.3$

between Pc and direct (PAR) and diffuse radiation (PAR_{diff}), temperature and VPD (Fig. 8.2). In contrast to GEP, Pc showed good correlation with PAR ($R^2 = 0.53$ at K34, 0.56 at K67, 0.04 at CAX, 0.38 at RJA). However, the fraction of productivity used to build capacity (GEP/Pc) and PAR was the strongest of all ($R^2 = 0.57$ at K34, 0.64 at K67, 0.35 at CAX, 0.43 at RJA) (Fig. 8.3), thus contrasting with the low correlation coefficients found for key meteorological variables and GEP (Table 8.2). At the seasonal forest site (RJA), VPD was highly correlated with GEP ($R^2 > 0.5$, $p < 0.01$), a possible sign of water limitation as more information would be necessary to establish causality.

Even though there is no correlation between GEP and precipitation as a proxy for water availability at individual sites (Table 8.2), across the central Amazon the east to west differences in rainfall patterns are reflected in absolute values of GEP, R_e, and NEE (see Restrepo-Coupe et al. 2013). GEP at K34 ranges from a minimum of 7.9 ± 0.9 to a maximum of 11.5 ± 1.4 g C m^{-2} day^{-1}; lower values are calculated at K67 (6.9 ± 0.68–9.7 ± 0.5 g C m^{-2} day^{-1}) and higher values at CAX (9.4 ± 1.19–13.1 ± 0.0 g C m^{-2} day^{-1}). With lowest minimum and highest maximum values (6.5 ± 0.74–13.3 ± 0.38 g C m^{-2} day^{-1}), RJA is the most seasonal site regarding GEP. The annual cycle of R_e is similar at all sites (6.8 ± 0.9–10.3 ± 0.4 g C m^{-2} day^{-1}), where values of R_e follow seasonality in GEP. An exception to this pattern is CAX where max GEP lags max end of the wet season R_e; however, there is no statistically significant relationship between R_e and meteorological drivers ($R^2 > 0.3$) for this site.

The balance between photosynthetic activity and respiration (NEE = GEP − R_e) shows greater seasonality at CAX and RJA, similarly to that in precipitation. Carbon uptake by vegetation (negative values of NEE) is observed at all times at K34 and CAX, the two sites with highest rainfall values and shortest dry season. Interestingly, in the three dense 'terra firme' sites highest C-uptake is reached during the October–December period, the end of the dry season (K34, K67, and CAX). During the wet season, the rainforest at K67 and during the dry season the seasonal forest at RJA are both sources of CO_2 (Fig. 8.2). Thus, we can infer a negative effect on NEE (reduction of C-uptake) due to a decrease in rainfall amounts at the wet equatorial forests, a common future climatic scenario (Costa and Pires 2010). It is unclear the effect of a changing seasonal rainfall pattern as vegetation adaptations are designed to optimise photosynthesis at the current balance between light and water.

8.5 Seasonality of Photosynthesis

In dense 'terra firme' forests at central BLA, GEP shows no signs of water limitation, in contrast to the open 'terra firme' at RJA forest. At all sites, GEP responds to a combination of adaptive mechanisms and climate: (1) Biology determines photosynthetic capacity through leaf flush and litterfall seasonality; and (2) Meteorology determines available sunlight energy (which drives the fraction of photosynthetic capacity utilised). The observed increase in Pc during the dry season can be explained by the increase in the quantity of leaves (leaf area index, LAI) or by the production of new more productive leaves (leaf-level photosynthetic assimilation per unit area, A_{max}), or the shedding of old leaves, or any combination of the three (e.g. production of new leaves, coincident with the loss of old leaves, such that the foliage becomes younger, even though LAI may actually decrease) (Restrepo-Coupe et al. 2013). There is a c. 16-day lag between Pc and LAI, where LAI leads. The observed interval can be explained by the c. 10–30 days that take for new leaves to reach maximum A_{max} values as reported for tropical evergreen species by Sobrado (1994). This relationship between LAI and Pc ($R^2 = 0.22$, $P < 0.001$ at K67) denotes that Pc is an indicator of both quantity and quality of leaves and could be interpreted as a proxy for chlorophyll content.

Looking at the patterns of change in Pc and LAI along with leaf fall, leaf carbon assimilation at saturating light, and specific leaf area (SLA), one can derive a very simple leaf flush model (see Restrepo-Coupe et al. 2013). At K34 and K67, Pc and LAI start to increase as the dry season begins. Leaf-fall patterns show a similar peak, with a maximum in the dry season, indicating that leaf flush and leaf fall are somewhat synchronous processes in these forests (Fig. 8.4). At K67 and K34, the dry season production of leaves is out of phase with the production of wood, where the observed wet season wood allocation peak can be correlated to a second peak in

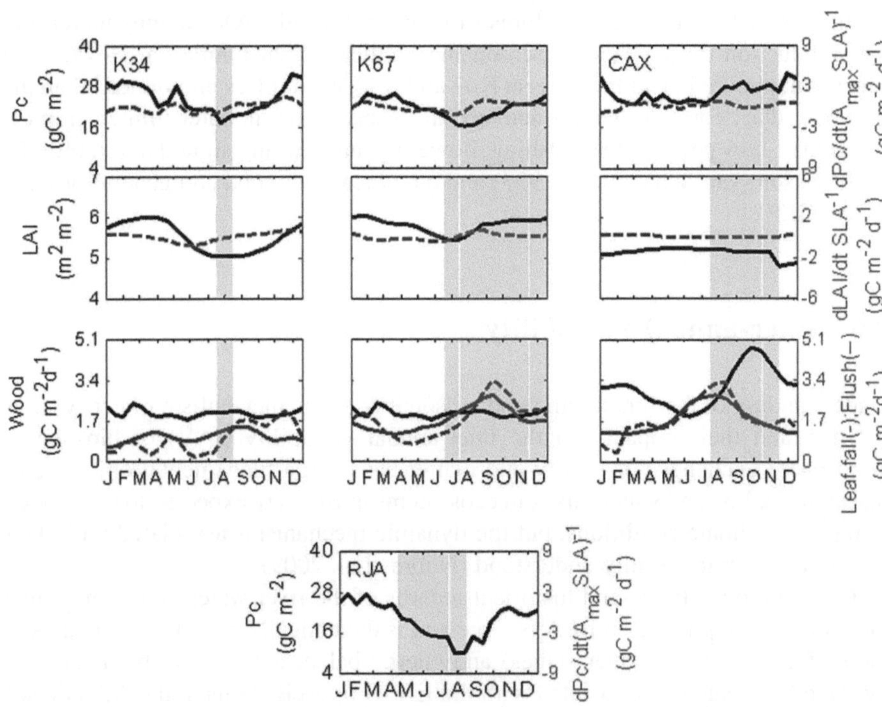

Fig. 8.4 Annual cycles, 16-day composites of the mean of all available years (Table 8.1) of photosynthetic capacity (Pc, g C m^{-2} day^{-1}) (*black continuous line*), change in Pc over time divided by the Specific Leaf Area and leaf carbon assimilation at saturating light (dPc/dt SLA^{-1} $A_{max}{}^{-1}$) (*grey dashed line*) (*top row*). *Second row* (for equatorial sites only): Leaf Area Index (LAI) (*black continuous line*) and change in LAI divided by the Specific Leaf Area (dLAI/dt SLA $^{-1}$) (*grey dashed line*). *Bottom row* (for equatorial sites only): measurements of wood increment (Wood, g C m^{-2} day^{-1}) (*black continuous line*), annual cycle of modelled leaf flush (Flush, g C m $^{-2}$ day^{-1}) (*grey dashed line*), and leaf fall (Leaf fall, g C m^{-2} day^{-1}) (*grey continuous line*). Areas in *grey* represent precipitation <100 mm $month^{-1}$ based on tower and Tropical Rainfall Measuring Mission (TRMM) data. Data from flux tower sites in dense 'terra firme' forests: Manaus (K34), Santarém (K67), Caxiuanã (CAX); and open 'terra firme': Jarú (RJA)

GEP. At equatorial BLA forests, periods of high irradiance are highly correlated with periods of leaf production, which appears to be prioritised over wood allocation (Fig. 8.4). This pattern could be altered when other resource constraints become more pressing than that of light. Trees at CAX exhibit a similar pattern to that in seasonally flooded forests (Dezzeo et al. 2003): leaf flush and leaf fall occur at the beginning of the dry season and allocation to wood peaks in October and November, in the middle of the dry season when rain is at its minimum.

Dry season leaf flush corroborates remotely sensed Enhanced Vegetation Index (EVI) from Moderate Resolution Imaging Spectroradiometer (MODIS) values that indicate maximum greenness at intact rainforests in the dry season (Huete et al. 2006). If light-limited trees are predicted to produce new leaves and flowers during the season of maximal irradiance (Wright and van Schaik 1994), a leaf-flushing model at the equatorial forests of K67, K34, and CAX showing increasing leaf production during the dry season may indicate light limitation. However, a second maximum in leaf flushing, at K67 and K34, even if less pronounced than the first one, does correlate with phenological observations in water-limited tropical forest that show peaks in leaf flushing, flowering, and fruiting around the start of the rainy season (van Schaik et al. 1993) and an increase in wood and general productivity (GEP).

8.6 Inter-annual Variability

Understanding of the functioning and seasonality of the metabolism of ecosystems in BLA and their impacts on the inter-annual variability (IAV) of biosphere–atmosphere exchanges of energy and matter has been a fundamental objective of research. Carbon and water fluxes in ecosystems in BLA are expected to be coupled to regional climate conditions, but the dynamic mechanisms associated with their IAV remain yet to be fully understood (Nobre et al. 2009).

Precipitation datasets and historical records of the river water yields show that the IAV of precipitation in BLA is large and is dynamically linked to anomalies in the surface water (e.g. river stages) and energy balances (e.g. sensible heat flux) over the basin and are associated with the ENSO or oscillations in the Atlantic sea surface temperature (SST) (Fu et al. 2001; Marengo 1992; Marengo et al. 1998; Poveda et al. 2006; Richey et al. 1989; Chap. 4). However, Nobre et al. (2009) have pointed out that the combined tropical Pacific and Atlantic SST variability explains little more than 50 % of inter-annual precipitation variance over Amazonia and not much is known about other mechanisms, internal or external to the region, responsible for the remaining unexplained IAV.

The proportions of IAV in net carbon or water fluxes directly related to variability in climate drivers then remain as an open question, and detailed assessment of the relative roles of changes in climate versus changes in vegetation response on the variability of fluxes is still needed. Some studies have found that the IAV of carbon exchange correlated poorly with climatic drivers (Richardson et al. 2007; Polley et al. 2010), but others have found strong correlations (Yuan et al. 2009; Desai 2010).

In a recent effort, the LBA Data Model Inter-comparison Project (LBA-DMIP) (http://www.climatemodeling.org/lba-mip/) run a suite of 21 land surface models (some of them including dynamic vegetation models), driven by standardised meteorological forcing data from the flux towers (for details, see Gonçalves et al. 2013; Von Randow et al. 2013). Here, we present the general relationships between the annual values of carbon and water exchange with climate variables, as measured at the LBA forest towers (K34, K67, K83, and RJA) and as an ensemble mean of the LBA-DMIP model simulations at each site and year (Figs. 8.5 and 8.6).

The four forest sites were grouped in the figures, resulting in a total of 13 site-years available in the dataset, but they are presented showing each site in different

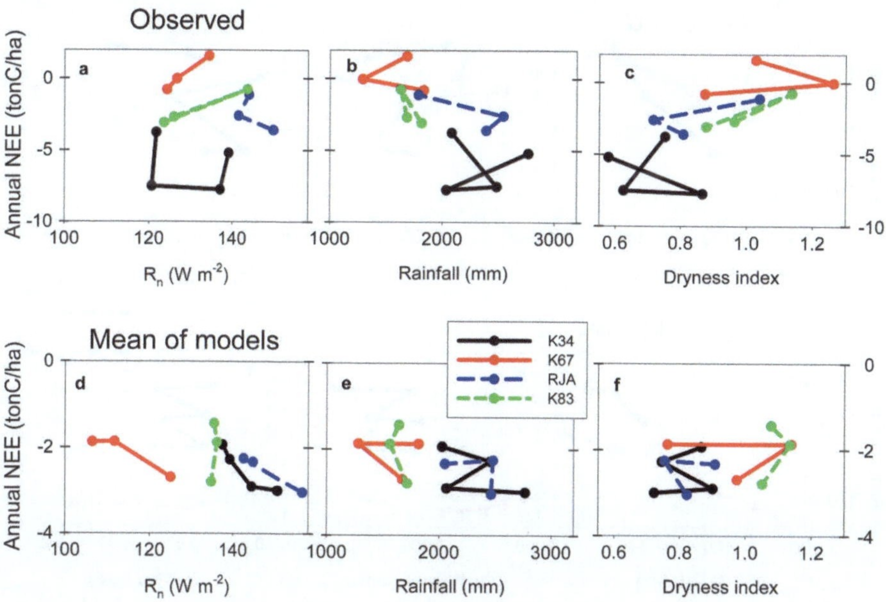

Fig. 8.5 The relationship between annual totals of NEE and climate drivers. *Top rows* observed data; *bottom rows* output averaged over a suite of 21 land surface models. Data from flux tower sites in dense 'terra firme' forests: Manaus (K34), Santarém (K67, K83); and open 'terra firme': Jarú (RJA).The four sites were grouped in the figures, resulting in a total of 13 site-years available in the dataset, but they are presented showing each site in different *colours* (adapted from von Randow et al. 2013)

colours. Although the grouping of the sites into broad categories may augment the spurious variability in each group, this classification is necessary because the dataset is too small for a more detailed analysis. Figure 8.5a–f shows NEE against annual average net radiation (R_n), total precipitation (P), and the dryness index ($D = R_n/\lambda P$, $\lambda = 2.45$ MJ kg^{-1}, latent heat of vaporisation).

There were similarities in the general responses in the annual values of carbon and water exchange to climate variables between the model simulations and the observed data; however, some differences were obvious. The magnitude of variability among sites was larger than the variability of model simulations (note the scale of the *y*-axis in the top rows is larger than in the lower rows). There appeared to be little relationship of the observed fluxes with R_n (Fig. 8.5a); contrary to this, model simulations were more radiation related (Fig. 8.5d). It appeared that the sites with low annual rainfall had lower C uptake than sites with high precipitation; some low precipitation sites in some years, in fact, were a source of carbon to the atmosphere (Saleska et al. 2003; Zeri et al. 2014; Gatti et al. 2014). This resulted in a pattern of higher net uptake at sites/years with higher annual rainfall or lesser D (Fig. 8.5b, c), which was also captured by the models (Fig. 8.5e, f).

Annual ET was largely related to R_n (Fig. 8.6a), and this pattern was well represented by the models for individual sites (Fig. 8.6d), but without a significant

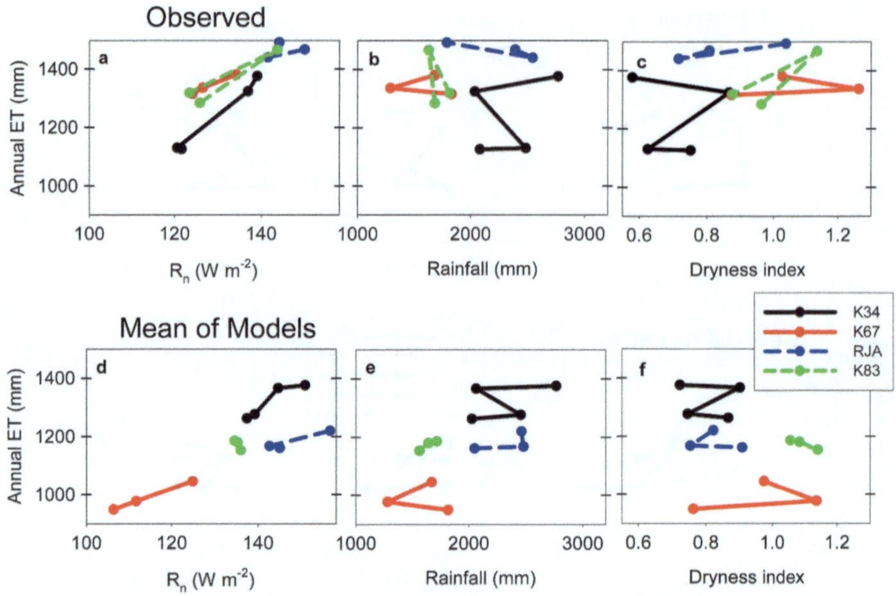

Fig. 8.6 The relationship between annual total evapotranspiration (ET) and observed climate drivers (*top rows*) or averaged over a suite of 21 land surface models (*bottom rows*). Data from flux tower sites in dense 'terra firme' forests: Manaus (K34), Santarém (K67, K83); and open 'terra firme': Jarú (RJA). Sites are grouped in the figures as in figure 6 (adapted from von Randow et al. 2013)

correlation (Von Randow et al. 2013). Previous studies have also shown a strong control of R_n on ET at seasonal scales (Costa et al. 2010; Fisher et al. 2009; da Rocha et al. 2009). Measured data showed a weak relationship between ET, P, and D, while the models were sensitive to these variables (Fig. 8.6b, c, e, and f).

In a study where worldwide tower flux measurements were scaled up, using a machine learning technique to provide a global grid of energy fluxes and NEE and its components gross primary productivity (GPP) and terrestrial ecosystem respiration (TER), it was found that the IAV of NEE was dominated by variability in GPP for the majority of the land surface, but not for the Amazonian region, where the dominant variability was derived from ecosystem respiration (Jung et al. 2011). Analysing the IAV of TER, the authors found that it was more strongly correlated with precipitation than with temperature, which also corroborates our results. This may be related to soil respiration in tropical forests being more limited by the moisture content of the soil/litter than by temperature.

Von Randow et al. (2013) further developed the analyses, by calculating the correlation coefficients of environmental variables and fluxes, finding that the negative correlation between NEE and annual rainfall was significant in the BLA dataset. While the average of the models also indicated a similar correlation, only three of the individual models had statistically significant values. For the ET values, the situation was reversed: measurements did not show any significant correlation with precipitation, but the majority of the simulations of ET were correlated with precipitation.

Should the general characteristics of the interactions between tropical forests and climate variables be maintained in the future, the findings of the LBA-DMIP suggest that forecast decreases in precipitation in some parts of the Amazonia basin could weaken CO_2 uptake by vegetation. The surface models are able to reproduce, to some extent, these general relationships; however, improvements are needed to better capture the characteristics of IAV.

8.7 Conclusions

At the dense 'terra firme' forests near the equator (K34, K67, and CAX), GEP increased gradually during the dry season, reaching a maximum from the middle to the end of this season, whereas at the southern open 'terra firme' site (RJA) it decreased. In absolute terms of GEP, RJA is the most seasonal site, followed by CAX.

Although GEP is strongly driven by PAR and controlled by VPD, soil moisture, and T_a on an hourly basis, the same is not observed on a seasonal basis. This mismatch is an indication that GEP responds to a combination of vegetation phenology (which biologically determines photosynthetic capacity through leaf flush and litterfall seasonality) and climate (available sunlight energy and water).

Net carbon uptake by vegetation is observed at all times at K34 and CAX. However, during the wet season the dense 'terra firme' forest at K67 and during the dry season the open 'terra firme' forest at RJA are sources of CO_2.

The Pc curves were similar to that of GEP throughout the year, most likely as a response of increase in LAI, or the production of new photosynthetically more efficient leaves, or the shedding of old leaves, or even any combination of the three. The dense 'terra firme' sites seem to be light limited, although among all four sites the weakest correlation with PAR was observed at CAX. Wood increment appears to be prioritised over leaf production by the forest of CAX. At the open 'terra firme' site RJA, Pc was much reduced in the dry season, corroborating the observed reduction in GEP. The forest at RJA seems to be water limited in the dry season as it takes little or no advantage of increased PAR.

Among the dense 'terra firme' sites, the relationship between LAI and Pc indicates that Pc is an indicator of both quantity and quality of leaves and could be interpreted as a proxy for chlorophyll content.

Assessment of the relative importance of particular climatic factors that affect carbon and water exchange at different site locations requires the analysis of the relationships between annual NEE and ET with climatic drivers (R_n, P, and D) as measured in LBA flux towers or computed by a suite of land surface model simulations.

Annual measurements of NEE are negatively correlated with annual rainfall. Although the average of land surface models yield a similar result, few individual models reproduce a significant negative correlation of simulated NEE with rainfall.

For the IAV of ET, data from tower measurements are related to annual variations in radiation which is partly captured by the average of the models, at individual sites, but no significant correlation between modelled ET and net radiation is found when all forest sites are grouped. Instead, simulated ET values are correlated with the amount of precipitation.

This work sheds some light on the mechanistic understanding of the dynamics of biosphere–atmosphere exchange of CO_2 in both dense and open 'terra firme' forests in BLA. The incorporation of this knowledge in the parameterization of the new generation of surface models is paramount for predicting the future of BLA under climate change.

Acknowledgements The authors thank the LBA Programme and the scientific coordination for support, discussion, and incentive to write this work. Two reviewers made insightful suggestions that much improved the quality of this chapter. N. R.-C. acknowledges the National Aeronautics and Space Administration (NASA) (LBA investigation CD-32 and the LBA-DMIP project, award #NNX09AL52G), the National Science Foundation (Amazon-PIRE, NSF award #OISE-0730305), the Gordon and Betty Moore Foundation Andes-Amazon Initiative, and the University Technology of Sydney and especially the Climate Change Cluster Remote Sensing Lab (Prof Alfredo Huete) for their kind support.

References

Andreae MO, Artaxo P, Brandão C, Carswell FE, Ciccioli P, Da Costa AL, Culf AD, Esteves JL, Gash JHC, Grace J, Kabat P, Lelieveld J, Malhi Y, Manzi AO, Meixner FX, Nobre AD, Nobre C, Ruivo MDLP, Silva-Dias MA, Stefani P, Valentini R, Von Jouanne J, Waterloo MJ (2002) Biogeochemical cycling of carbon, water, energy, trace gases, and aerosols in Amazonia: the LBA-EUSTACH experiments. J Geophys Res D: Atmos 107:33-31–33-25

Araújo AC, Nobre AD, Kruijt B, Elbers JA, Dallarosa R, Stefani P, von Randow C, Manzi AO, Culf AD, Gash JHC, Valentini R, Kabat P (2002) Comparative measurements of carbon dioxide fluxes from two nearby towers in a central Amazonian rainforest: the Manaus LBA site. J Geophys Res 107(D20):8090. doi:10.1029/2001JD000676

Avissar R, Dias PLS, Dias M, Nobre C (2002) The large-scale biosphere–atmosphere experiment in Amazonia (LBA): insights and future research needs. J Geophys Res Atmos 107(D20):8086. doi:10.1029/2002JD002704

Baker TR, Phillips OL, Malhi Y, Almeida S, Arroyo L, Di Fiore A, Erwin T, Higuchi N, Killeen TJ, Laurance SG, Laurance WF, Lewis SL, Monteagudo A, Neill DA, Vargas PN, Pitman NCA, Silva JNM, Martinez RV (2004) Increasing biomass in Amazonian forest plots. Philos Trans R Soc Lond Ser B Biol Sci 359:353–365

Baldocchi DD (2003) Assessing the eddy covariance technique for evaluating carbon dioxide exchange rates of ecosystems: past, present and future. Glob Chang Biol 9:479–492

Carswell FE, Costa AL, Palheta M, Malhi Y, Meir P, Costa JPR, Ruivo ML, Leal LSM, Costa JMN, Clement RJ, Grace J (2002) Seasonality in CO_2 and H_2O flux at an eastern Amazonian rain forest. J Geophys Res 107(D20):8076. doi:10.1029/2000JD000284

Costa MH, Pires GF (2010) Effects of Amazon and Central Brazil deforestation scenarios on the duration of the dry season in the arc of deforestation. Int J Climatol 30:1970–1979. doi:10.1002/joc.2048

Costa MH, Biajoli MC, Sanches L, Malhado ACM, Hutyra LR, da Rocha HR, Aguiar RG, de Araújo AC (2010) Atmospheric versus vegetation controls of Amazonian tropical rain forest evapotranspiration: are the wet and seasonally dry rain forests any different? J Geophys Res 115:G04021. doi:10.1029/2009JG001179

da Rocha HR, Goulden ML, Miller SD, Menton MC, Pinto LDVO, de Freitas HC, Silva Figueira AM (2004) Seasonality of water and heat fluxes over a tropical forest in eastern Amazonia. Ecol Appl 14:22–32. doi:10.1890/02-6001

da Rocha HR, Manzi AO, Cabral OM, Miller SD, Goulden ML, Saleska SR, Coupe NR, Wofsy SC, Borma LS, Artaxo P, Vourlitis G, Nogueira JS, Cardoso FL, Nobre AD, Kruijt B, Freitas HC, von Randow C, Aguiar RG, Maia JF (2009) Patterns of water and heat flux across a biome gradient from tropical forest to savanna in Brazil. J Geophys Res Biogeosci 114:G00B12. doi:10.1029/2007JG000640

Desai AR (2010) Climatic and phenological controls on coherent regional interannual variability of carbon dioxide flux in a heterogeneous landscape. J Geophys Res Biogeosci 115(G3): G00J02. doi:10.1029/2010jg001423

Dezzeo N, Worbes M, Ishii I, Herrera R (2003) Annual tree rings revealed by radiocarbon dating in seasonally flooded forest of the Mapire River, a tributary of the lower Orinoco River, Venezuela. Plant Ecol 168:165–175. doi:10.1023/A:1024417610776

Domingues TF, Martinelli LA, Ehleringer JR (2014) Seasonal patterns of leaf-level photosynthetic gas exchange in an eastern Amazonian rain forest. Plant Ecol Divers 7:189–203

Eva HD, Huber O (2005) A proposal for defining the geographical boundaries of Amazonia: synthesis of the results from an expert consultation workshop organized by the European Commission in collaboration with the Amazon Cooperation Treaty Organization—JRC Ispra, 7–8 June 2005

Fan SM, Wofsy SC, Bakwin PS, Jacob DJ, Fitjarrald DR (1990) Atmosphere–biosphere exchange of CO_2 and O_3 in a central Amazon forest. J Geophys Res 95(D10):16851–16864

Figueira AMS, Miller SD, de Sousa CAD, Menton MC, Maia AR, da Rocha HR, Goulden ML (2008) Effects of selective logging on tropical forest tree growth. J Geophys Res 113:G00B05. doi:10.1029/2007JG000577

Fisher RA, Williams M, da Costa AL, Malhi Y, da Costa RF, Almeida S, Meir PW (2007) The response of an Eastern Amazonian rainforest to drought stress: results and modelling analyses from a through-fall exclusion experiment. Glob Chang Biol 13:1–8

Fisher JB, Malhi Y, Bonal D, Rocha HR, Araújo AC, Gamo M, Goulden ML, Hirano T, Huete AR, Kondo H, Kumagai T, Loescher HW, Miller SD, Nobre AD, Nouvellon Y, Oberbauer SF, Panuthai S, Roupsard O, Saleska SR, Tanaka K, Tanaka N, Tu KP, von Randow C (2009) The land-atmosphere water flux in the tropics. Glob Chang Biol 15:2694–2714. doi:10.1111/j.1365-2486200801813x

Fu R, Chen M, Li W, Dickinson RE (2001) How do tropical sea surface temperatures influence the seasonal distribution of precipitation in the equatorial Amazon? J Climate 14:4003–4026

Gash JHC, Huntingford C, Marengo JA, Betts RA, Cox PM, Fisch G, Fu R, Gandu AW, Harris PP, Machado LAT, von Randow C, Dias MAS (2004) Amazonian climate: results and future research. Theor Appl Climatol 78:187–193

Gatti LV, Gloor M, Miller JB, Doughty CE, Malhi Y, Domingues LG, Basso LS, Martinewski A, Correia CSC, Borges VF, Freitas S, Braz R, Anderson LO, Rocha H, Grace J, Phillips OL, Lloyd J (2014) Drought sensitivity of Amazonian carbon balance revealed by atmospheric measurements. Nature 506(7486):76–80

Gloor E (2016) Climate and the Amazonian carbon balance. In: Nagy L, Forsberg B, Artaxo P (eds) Interactions between biosphere, atmosphere and human land use in the Amazon basin. Springer, Heidelberg, pp 99–115

Grace J (2016) The Amazon carbon balance: an evaluation of methods and results. In: Nagy L, Forsberg B, Artaxo P (eds) Interactions between biosphere, atmosphere and human land use in the Amazon basin. Springer, Heidelberg, pp 77–97

Gonçalves LGG, Borak JS, Costa MH, Saleska SR, Baker I, Restrepo-Coupe N, Muza MN, Poulter B, Verbeeck H, Fisher JB, Arain MA, Arkin P, Cestaro BP, Christoffersen B, Galbraith D, Guan X, van den Hurk BJJM, Ichii K, Imbuzeiro HMA, Jain AK, Levine N, Lu C, Miguez-Macho G, Roberti DR, Sahoo A, Sakaguchi K, Schaefer K, Shi M, Shuttleworth WJ, Tian H, Yang Z-L, Zeng X (2013) Overview of the large-scale biosphere–atmosphere experiment in Amazonia data model intercomparison project (LBA-DMIP). Agric For Meteorol 182–183:111–127

Goulden ML, Miller SD, da Rocha H, Menton MC, de Freitas HC, Figueira AMES, de Sousa CAD (2004) Diel and seasonal patterns of tropical forest CO_2 exchange. Ecol Appl 14:S42–S55

Grace J, Lloyd J, McIntyre J, Miranda A, Meir P, Miranda H, Moncrieff J, Massheder J, Wright I, Gash J (1995a) Fluxes of carbon dioxide and water vapor over an undisturbed tropical forest in Southwest Amazonia. Glob Chang Biol 1(1):1–12

Grace J, Lloyd J, McIntyre J, Miranda AC, Meir P, Miranda HS, Nobre C, Moncrieff J, Massheder J, Malhi Y, Wright I, Gash J (1995b) Carbon dioxide uptake by an undisturbed tropical forest in Southwest Amazonia. Science 270:778–780. doi:10.1126/science.270.5237.778

Hasler N, Avissar R (2007) What controls evapotranspiration in the Amazon Basin? J Hydrometeorol 8:380–395

Huete AR, Didan K, Shimabukuro YE, Ratana P, Saleska SR, Hutyra LR, Yang W, Nemani RR, Myneni R (2006) Amazon rainforests green-up with sunlight in dry season. Geophys Res Lett. doi:10.1029/2005GL025583

Hutyra LR, Munger JW, Saleska SR, Gottlieb E, Daube BC, Dunn AL, Amaral DF, de Camargo PB, Wofsy SC (2007) Seasonal controls on the exchange of carbon and water in an Amazonian rain forest. J Geophys Res 112:G03008. doi:10.1029/2006JG000365

Hutyra LR, Munger JW, Hammond-Pyle E, Saleska SR, Restrepo-Coupe N, Daube BC, de Camargo PB, Wofsy SC (2008) Resolving systematic errors in estimates of net ecosystem exchange of CO_2 and ecosystem respiration in a tropical forest biome. Agric For Meteorol 148:1266–1279. doi:10.1016/j.agrformet.2008.03.007

IBGE (2012) Manual Técnico da Vegetação Brasileira. Sistema fitogeográfico. Inventário das formações florestais e campestres. Técnicas e manejo de coleções botânicas. Procedimentos para mapeamentos Manuais Técnicos em Geociências vol. número 1. Instituto Brasileiro de Geografia e Estatística - IBGE, Diretoria de Geociências, Coordenação de Recursos Naturais e Estudos Ambientais, Rio de Janeiro

Jung M, Reichstein M, Margolis HA, Cescatti A, Richardson AD, Arain MA, Arneth A, Bernhofer C, Bonal D, Chen J, Gianelle D, Gobron N, Kiely G, Kutsch W, Lasslop G, Law BE, Lindroth A, Merbold L, Montagnani L, Moors EJ, Papale D, Sottocornola M, Vaccari F, Williams C (2011) Global patterns of land-atmosphere fluxes of carbon dioxide, latent heat, and sensible heat derived from eddy covariance, satellite, and meteorological observations. J Geophys Res 116:G00J07. doi:10.1029/2010JG001566

Keller M, Alencar A, Asner GP, Braswell B, Bustamante M, Davidson E, Feldpausch T, Fernandes E, Goulden M, Kabat P, Kruijt B, Luizão F, Miller S, Markewitz D, Nobre AD, Nobre CA, Filho NP, da Rocha H, Dias PS, von Randow C, Vourlitis GL (2004) Ecological research in the large-scale biosphere-atmosphere experiment in Amazonia: early results. Ecol Appl 14(4):S3–S16. doi:10.1890/03-6003

Keller M, Varner R, Dias J, Silva H (2005) Soil-atmosphere exchange of nitrous oxide, methane, and carbon dioxide in logged and undisturbed forest in the Tapajos National Forest, Brazil. Earth Interact 9(23):1–28

Kruijt B, Elbers JA, von Randow C, Araujo AC, Oliveira PJ, Culf A, Manzi AO, Nobre AD, Kabat P, Moors EJ (2004) The robustness of eddy correlation fluxes for Amazon rain forest conditions. Ecol Appl 14:S101–S113

Law BE, Falge E, Gu L, Baldocchi DD, Bakwin P, Davis K, Dolman AJ, Falk M, Fuentes JD, Goldstein A, Granier A, Grelle A, Janssens IA, Jarvis P, Jensen NO, Katul G, Malhi Y, Matteucci G, Meyers T, Monson R, Munger W, Oechel W, Olson R, Pilegaard K, Paw UKT, Thorgeirsson H, Valentini R, Verma S, Vesala T, Wilson K, Wofsy S (2002) Environmental controls over carbon dioxide and water vapor exchange of terrestrial vegetation. Agric For Meteorol 113:97–120

Malhi Y, Nobre AD, Grace J, Kruijt B, Pereira MGP, Culf A, Scott S (1998) Carbon dioxide transfer over a Central Amazonian rain forest. J Geophys Res Atmos 103:31593–31612

Malhi Y, Pegoraro E, Nobre AD, Pereira MGP, Grace J, Culf AD, Clement R (2002) Energy and water dynamics of a central Amazonian rain forest. J Geophys Res 107(D20):8061. doi:10.1029/2001JD000623

Malhi Y, Wood D, Baker TR, Wright J, Phillips OL, Cochrane T, Meir P, Chave J, Almeida S, Arroyo L, Higuchi N, Killeen TJ, Laurance SG, Laurance WF, Lewis SL, Monteagudo A, Neill DA, Vargas PN, Pitman NCA, Quesada CA, Salomao R, Silva JNM, Lezama AT, Terborgh J, Martinez RV, Vinceti B (2006) The regional variation of aboveground live biomass in old-growth Amazonian forests. Glob Chang Biol 12:1107–1138

Malhi Y, Aragão LEOC, Metcalfe DB, Paiva R, Quesada CA, Almeida S, Anderson L, Brando P, Chambers JQ, da Costa ACL, Hutyra LR, Oliveira P, Patiño S, Pyle EH, Robertson AL, Teixeira LM (2009) Comprehensive assessment of carbon productivity, allocation and storage in three Amazonian forests. Glob Chang Biol 15:1255–1274

Marengo JA (1992) Interannual variability of surface climate in the Amazon Basin. Int J Climatol 12:853–863

Marengo JA, Tomasella J, Uvo CR (1998) Trends in streamflow and rainfall in tropical South America: Amazonia, eastern Brazil and northwestern Peru. J Geophys Res 103 (D2):1775–1783

Marengo JA, Nobre CA, Tomasella J, Oyama MD, de Oliveira GS, de Oliveira R, Camargo H, Alves LM, Brown IF (2008) The drought of Amazonia in 2005. J Climate 21:495–516. doi:10.1175/2007JCLI1600.1

Marengo JA, Tomasella J, Alves LM, Soares WR, Rodriguez DA (2011) The drought of 2010 in the context of historical droughts in the Amazon region. Geophys Res Lett 38:L12703. doi:10.1029/2011GL047436

Meir P, Grace J, Miranda A, Lloyd J (1996) Soil respiration in a rainforest in Amazonia and in cerrado in Central Brazil. In: Gash JHC, Nobre CA, Roberts JM, Victoria RL (eds) Amazonian deforestation and climate. Wiley, Chichester, pp 319–329

Mercado LM, Bellouin N, Sitch S, Boucher O, Huntingford C, Wild M, Cox PM (2009) Impact of changes in diffuse radiation on the global land carbon sink. Nature 458:1014–1017. doi:10.1038/nature07949

Miller SD, Goulden ML, Menton MC, da Rocha HR, de Freitas HC, Figueira A, de Sousa CAD (2004) Biometric and micrometeorological measurements of tropical forest carbon balance. Ecol Appl 14:S114–S126

NASA (2014) Tropical rainfall measuring mission project (TRMM), 3B43: monthly 0.25 × 0.25 degree merged TRMM and other estimates v7 [WWW Document]. NASA Distrib Act. Arch Cent Goddard Space Flight Cent Earth Sci Greenbelt Md. http://mirador.gsfc.nasa.gov/cgi-bin/mirador/. Cited 27 Feb 2014

Negrón Juárez RI, Da Rocha HR, Figueira AMS, Goulden ML, Miller SD (2009) An improved estimate of leaf area index based on the histogram analysis of hemispherical photographs. Agric For Meteorol 149:920–928

Négron-Juárez RI, Hodnett MG, Fu R, Goulden ML, von Randow C (2007) Control of dry season evapotranspiration over the Amazonian forest as inferred from observations at a southern Amazon forest site. J Climate 20:2827–2839

Nepstad DC, De Carvalho CR, Davidson EA, Jipp PH, Lefebvre PA, Negreiros GH, Da Silva ED, Stone TA, Trumbore SE, Vieira S (1994) The role of deep roots in the hydrological and carbon cycles of Amazonian forests and pastures. Nature 372:666–669

Nobre CA, Obregon GO, Marengo JA, Fu R, Poveda G (2009) Characteristics of Amazonian climate: main features. In: Keller M, Bustamante M, Gash JHC, Silva Dias P (eds) Amazonia and global change, Geophysical monograph series. American Geophysical Union, Washington, DC, pp 149–162

Oliveira RS, Dawson TE, Burgess SSO, Nepstad DC (2005) Hydraulic redistribution in three Amazonian trees. Oecologia 145:354–363

Oliveira PHF, Artaxo P, Pires C, De Lucca S, Procópio A, Holben B, Schafer J, Cardoso LF, Wofsy SC, Rocha HR (2007) The effects of biomass burning aerosols and clouds on the CO_2 flux in Amazonia. Tellus B59:338–349. doi:10.1111/j.1600-0889.2007.00270.x

Phillips OL, Malhi Y, Higuchi N, Laurance WF, Nunez PV, Vasquez RM, Laurance SG, Ferreira LV, Stern M, Brown S, Grace J (1998) Changes in the carbon balance of tropical forests: evidence from long-term plots. Science 282:439–442

Polley HW, Emmerich W, Bradford JA, Sims PL, Johnson DA, Saliendra NZ, Svejcar T, Angell R, Frank AB, Phillips RL, Snyder KA, Morgan JA (2010) Physiological and environmental regulation of interannual variability in CO_2 exchange on rangelands in the western United States. Glob Chang Biol 16(3):990–1002. doi:10.1111/j.1365-2486.2009.01966.x

Poveda G, Waylen PR, Pulwarty R (2006) Annual and inter-annual variability of the present climate in northern South America and southern Mesoamerica. Palaleo 234:3–27

Restrepo-Coupe N, da Rocha HR, Christoffersen B, da Araujo AC, Borma LS, Cabral OMR, de Camargo PB, Cardoso FL, da Costa ACL, Fitzjarrald DR, Goulden ML, Hutyra LR, Kruijt B, Maia JMF, Malhi YS, Manzi AO, Miller SD, Nobre AD, von Randow C, Sá LDA, Sakai RK, Tota J, Wofsy SC, Zanchi FB, Saleska SR (2013) What drives the seasonality of productivity across the Amazon basin? A cross-site analysis of eddy flux tower measurements from the Brasil flux network. Agric For Meteorol 182–183:128–144

Rice AH, Pyle EH, Saleska SR, Hutyra LR, Palace M, Keller M, de Camargo PB, Portilho K, Marques DF, Wofsy SC (2004) Carbon balance and vegetation dynamics in an old-growth Amazonian forest. Ecol Appl 14:S55–S71

Richardson AD, Hollinger DY, Aber JD, Ollinger SV, Braswell BH (2007) Environmental variation is directly responsible for short- but not long-term variation in forest-atmosphere carbon exchange. Glob Chang Biol 13:788–803. doi:10.1111/j.1365-2486.2007.01330.x

Richey JE, Nobre CA, Deser C (1989) Amazon river discharge and climate variability: 1903–1985. Science 246:101–103

Saatchi SS, Houghton RA, Alvalá RCS, Soares JV, Yu Y (2007) Distribution of aboveground live biomass in the Amazon basin. Glob Chang Biol 13:816–837. doi:10.1111/j.1365-2486.2007.01323.x

Saleska SR, Miller SD, Matross DM, Goulden ML, Wofsy SC, Da Rocha HR, De Camargo PB, Crill P, Daube BC, De Freitas HC, Hutyra L, Keller M, Kirchhoff V, Menton M, Munger JW, Pyle EH, Rice AH, Silva H (2003) Carbon in Amazon forests: unexpected seasonal fluxes and disturbance-induced losses. Science 302:1554–1557

Saleska SR, da Rocha HR, Kruijt B, Nobre A (2009) Ecosystem carbon fluxes and Amazonian forest metabolism. In: Keller M, Bustamante M, Gash J, Dias PS (eds) Amazonia and global change, Geophysical monograph series. American Geophysical Union, Washington, DC, pp 389–407

Sobrado MA (1994) Leaf age effects on photosynthetic rate, transpiration rate and nitrogen content in a tropical dry forest. Physiol Plant 90:210–215. doi:10.1111/j.1399-3054.1994.tb02213.x

Sombroek W (2001) Spatial and temporal patterns of Amazon rainfall—consequences for the planning of agricultural occupation and the protection of primary forests. Ambio 30:388–396

Stoy PC, Katul GG, Siqueira MBS, Juang J-Y, Novick KA, Uebelherr JM, Oren R (2006) An evaluation of models for partitioning eddy covariance-measured net ecosystem exchange into photosynthesis and respiration. Agric For Meteorol 141:2–18

Tomasella J, Hodnett MG, Cuartas LA, Nobre AD, Waterloo MJ, Oliveira SM (2008) The water balance of an Amazonian micro-catchment: the effect of interannual variability of rainfall on hydrological behaviour. Hydrol Process 22:2133–2147. doi:10.1002/hyp.6813 van Schaik CP, Terborgh JW, Wright SJ (1993) The phenology of tropical forests: adaptive significance and consequences for primary consumers. Annu Rev Ecol Syst 24:353–377

Vieira S, de Camargo PB, Selhorst D, da Silva R, Hutyra L, Chambers JQ, Brown IF, Higuchi N, dos Santos J, Wofsy SC, Trumbore SE, Martinelli LA (2004) Forest structure and carbon dynamics in Amazonian tropical rain forests. Oecologia 140:468–479

von Randow C, Manzi AO, Kruijt B, de Oliveira PJ, Zanchi FB, Silva RL, Hodnett MG, Gash JHC, Elbers JA, Waterloo MJ, Cardoso FL, Kabat P (2004) Comparative measurements and seasonal variations in energy and carbon exchange over forest and pasture in South West Amazonia. Theor Appl Climatol 78:5–26

Von Randow C, Zeri M, Restrepo-Coupe N, Muza MN, de Gonçalves LGG, Costa MH, Araujo AC, Manzi AO, Rocha HR, Saleska SR, Alaf Arain M, Baker IT, Cestaro BP, Christoffersen B, Ciais P, Fisher JB, Galbraith D, Guan X, van den Hurk B, Ichii K, Imbuzeiro H, Jain A, Levine N, Miguez-Macho G, Poulter B, Roberti DR, Sahoo A, Schaefer K, Shio M, Tian H, Verbeeck H, Yang ZL (2013) Inter-annual variability of carbon and water fluxes in Amazonian forest, Cerrado and pasture sites, as simulated by terrestrial biosphere models. Agric For Meteorol 182:145–155

Wright SJ, Van Schaik CP (1994) Light and the phenology of tropical trees. Am Nat 143:192–199

Yuan W, Luo Y, Richardson AD, Oren RAM, Luyssaert S, Janssens IA, Ceulemans R, Zhou X, Grünwald T, Aubinet M, Berhofer C, Baldocchi DD, Chen J, Dunn AL, Deforest JL, Dragoni D, Goldstein AH, Moors E, William Munger J, Monson RK, Suyker AE, Starr G, Scott RL, Tenhunen J, Verma SB, Vesala T, Wofsy SC (2009) Latitudinal patterns of magnitude and interannual variability in net ecosystem exchange regulated by biological and environmental variables. Glob Chang Biol 15(12):2905–2920. doi:10.1111/j.1365-2486.2009.01870.x

Zeri M, Sá LDA, Manzi AO, Araújo AC, Aguiar RG, von Randow C, Sampaio G, Cardoso FL, Nobre CA (2014) Variability of carbon and water fluxes following climate extremes over a tropical forest in southwestern Amazonia. PLoS One 9(2):e88130. doi:10.1371/journal.pone.0088130

Chapter 9
Overview of Forest Carbon Stocks Study in Amazonas State, Brazil

Niro Higuchi, Rempei Suwa, Francisco G. Higuchi, Adriano J.N. Lima, Joaquim dos Santos, Hideyuki Noguchi, Takuya Kajimoto, and Moriyoshi Ishizuka

9.1 Background

Tropical forests are focal ecosystems in research on the effects of climate change on biological systems and their functioning, such as carbon fluxes. Amazonian forests account for a substantial portion of carbon stock and net primary production among global terrestrial ecosystems (e.g. Melillo et al. 1993; Malhi et al. 2006). However, large-scale estimates of carbon stock in the Amazon still include considerable uncertainty (Houghton et al. 2001; Eva et al. 2003; Fearnside and Laurance 2003; Chave et al. 2008) because of regional-scale variation in carbon stocks, likely caused by heterogeneity in vegetation structure among different Amazonian regions and methodological disagreements. For example, previous studies in Amazonian forests have reported that biomass, including roots, can range from 155 to 425 Mg ha^{-1} (Houghton et al. 2001) and 285–401 Mg ha^{-1} (Lima 2010). Saatchi et al. (2007) estimated the total carbon stock in the Amazonian forests at 86 Pg C including dead and below-ground biomass, based on remote sensing metrics and geoprocessing.

Amazonas State extends across the Central Amazon region and occupies over 1.5 million km^2, with geographical coordinates of the capital city, Manaus, of 03° 06′ S and 60° 01′ W. Based on the Deforestation Monitoring Program (PRODES) of the National Spatial Research Institute (INPE) of Brazil, the deforestation in

N. Higuchi (✉) • A.J.N. Lima • J. dos Santos
CADAF Project, National Institute for Research in the Amazon (INPA), Manaus, Brazil
e-mail: higuchi.niro@gmail.com; niro@inpa.gov.br

R. Suwa • H. Noguchi • T. Kajimoto • M. Ishizuka
Forestry and Forest Products Research Institute (FFPRI), Tsukuba, Japan

F.G. Higuchi
Federal University of Paraná, Curitiba, Brazil

© Springer-Verlag Berlin Heidelberg 2016

L. Nagy et al. (eds.), *Interactions Between Biosphere, Atmosphere and Human Land Use in the Amazon Basin*, Ecological Studies 227,
DOI 10.1007/978-3-662-49902-3_9

Amazonas has been modest (5 %) compared with that in neighbouring states, such as Pará (34 %), Rondônia (14 %), and Mato Grosso (34 %) (PRODES 2014). In light of the considerable uncertainty regarding carbon stocks, the introduction of forest conservation programmes, such as the programme for Reducing Emissions from Deforestation and Forest Degradation (REDD+) (see Box 9.1 for details), lacks a solid basis in the Amazonia. In particular, in a large part of Amazonas State there is a gap of basic information on baseline carbon stocks (Asner et al. 2005; Saatchi et al. 2007). This can partly be ascribed to the inaccessibility to extensive and remote forest areas.

Box 9.1: Reducing Emissions from Deforestation and Forest Degradation (REDD-plus)

Tropical deforestation is responsible for approximately 20 % of total human-caused greenhouse gas (GHG) emissions (IPCC 2007). REDD-plus is a proposed scheme to financially compensate countries to reduce such emissions by better managing forest resources. It is discussed extensively how to set up and incorporate a scheme for REDD-plus as a global warming mitigation option, into a post-2012 Kyoto Protocol framework. However, REDD-plus assumes that carbon emissions from deforestation and forest degradation are *M*easurable, *R*eportable, and *V*erifiable, namely MRV. In order for the REDD-plus scheme to be realised, it is necessary to develop a monitoring method for quantitative evaluation of carbon dynamics (emissions/uptakes) in the Amazonian forests.

To counter this, a pilot study on Continuous Forest Inventory (CFI) in forests areas of Amazonas State, including managed (logged) and primary forests, was launched by the Forest Management Laboratory (LMF) of the National Institute for Amazonian Research (INPA) in 2004. The project carries out field work across the whole of state. In total, 1272 plots (20 × 125 m and/or 100 × 100 m) were established at 16 different sites (including two managed forests). In 2009, a follow-up project (Carbon Dynamics of Amazonian Forests—CADAF, http://cse.ffpri.affrc.go.jp) was launched to use the plot data in ground-truthing and include remote sensing techniques for carbon estimation. An additional 6 sites, summing 564 plots, were added to the original dataset (total = 22 sites and 1836 plots) across Amazonas State.

At two of these sites, allometric equations for individual carbon stock estimation were developed based on a destructive method. The selected equations were single-entry models, considering exclusively diameter at breast height (DBH). For statistical analysis, models were tested based on DBH and height (stem and total height), but due to the unfeasibility (time-consuming, high uncertainty, and non-sampling errors in the field assessment) and the low benefit–cost ratio of the inclusion of a second independent variable (estimated error of the equation is not significantly lower), these models were discarded. In this chapter, we describe the recent development made in estimating the carbon stocks and show the plot-based estimates of carbon stocks in the Amazonas State on the basis of the results from these two projects.

9.2 Development of Allometric Equations

9.2.1 Allometric Equations for Old-Growth Amazonian Forests

Tree biomass can be estimated using a relationship between stem diameter at breast height (DBH) and tree mass, namely allometry. Past studies developed allometric models of Amazonian forests, primarily for above-ground biomass (Overman et al. 1994; Santos 1996; Higuchi et al. 1998; Araújo et al. 1999; Chambers et al. 2001), and there are few allometric models for below-ground biomass (Sierra et al. 2007; Silva 2007; Lima et al. 2012). Allometric models should be selected carefully by considering forest type and available tree and environmental variables (e.g. Brown 1997; Chave et al. 2005; Kenzo et al. 2009, 2010). Allometric models for estimating biomass can substantially differ between secondary and primary forests, as Silva (2007) showed in her experiment near Manaus in Amazonas State, Brazil.

Silva (2007) developed allometric equations for estimating above-ground and total (including coarse roots) biomass for *terra firme* forests in Manaus, based on 494 harvested trees (DBH > 5 cm) (see Box 9.2 for methodology) and the following model was developed:

$$\mathrm{PF_{abg}} = 2.274\,\mathrm{DBH}^{1.916}\ \left(n = 494, R^2 = 0.85,\ \text{uncertainty} = 8.4\%\right) \tag{9.1}$$

where $\mathrm{PF_{abg}}$ is individual fresh above-ground mass (kg) and the unit of DBH is cm. To evaluate the goodness of fit of model, the coefficient of determination and uncertainty were estimated (see Box 9.3 for uncertainty). For estimating total biomass, the following model was developed:

$$\mathrm{PF_{tot}} = 2.718\,\mathrm{DBH}^{1877}\left(n = 131, R^2 = 0.94, \text{uncertainty} = 7.8\%\right) \tag{9.2}$$

where $\mathrm{PF_{tot}}$ means whole individual weight of fresh mass (kg).

Box 9.2: Data Collection for Developing Allometric Models

The sample trees (DBH > 5 cm) were collected from a *terra firme* forest at ZF2 experimental forest of INPA, in Manaus. The DBH, tree heights (*H*), and fresh mass weights of each component (stem, branches, leaves, and roots) were measured in the field (see the left photograph). To facilitate mass measurements, stems of sample trees were cut into smaller pieces using a chainsaw. The mass of saw dust was also weighted. Large branches (diameter ≥10 cm) and small branches (diameter <10 cm) were weighed separately. All leaves were clipped completely and their fresh masses were measured. Below-ground parts were excavated manually to include most coarse roots (>~2 mm) (see the right photograph). Coarse roots were classified into large roots (diameter >5 cm) and small roots (diameter between 2 and 5 cm), which were weighed separately after carefully removing attached soil by hand. To determine the average water contents of stems, three discs (2–5 cm in thickness) were collected from the top, middle, and bottom positions of each individual. Three discs of large branch and coarse root were also cut to determine average water contents. We collected samples (~2 kg) of small branches, leaves, and small roots from each individual. The collected samples were transported to the laboratory and oven dried at 65 °C to constant dry mass.

(continued)

Box 9.2 (continued)

Box 9.3: Uncertainty

In this study, the uncertainty is considered as the confidence interval (CI), at a 95 % probability, of the estimated mean value, evaluated according to the IPCC guideline (IPCC 2006). The estimation process of the uncertainty is as below:

$$s_x = \frac{s}{\sqrt{n}}$$

where s_x, s, and n mean the standard deviation of the mean, standard deviation, and number of samples, respectively. The s_x is also known as the standard error of the mean. The uncertainty in percentage (U) can be defined as follows:

$$U = \frac{2s_x}{\overline{x}} \times 100$$

where $\overline{x}$ means the mean value of focal variable.

The U for the allometric models was estimated on the basis of the residual sum of squares.

For estimating dry mass above-ground and whole individual mass (PS_{abg} and PS_{tot}, respectively, in kg), the following conversion factors were determined (Silva 2007):

$$PS_{abg} = 0.592\,PF_{abg} \quad (9.3)$$

$$PS_{tot} = 0.584\,PF_{tot} \quad (9.4)$$

A conversion factor of 0.485 was applied to estimate the individual carbon stock, i.e. 48.5 % of the dry mass corresponds to its carbon amount, which was determined by measuring carbon contents of the tree samples used for developing the allometric equations (see Silva 2007 for details). Hence, all estimates presented in this chapter are in carbon units.

9.2.2 *Common Allometry with Dominant Height*

Allometric models vary by region (Malhi et al. 2006; Lima et al. 2012), regardless of the forest type. Lima et al. (2012) compared the allometric above-ground equations in three, very distinct, Amazonian regions, including Manaus (Silva 2007), Tomé Açu (Araújo et al. 1999), and São Gabriel da Cachoeira (Lima et al. 2012), and found that the regional difference in the allometric equations can

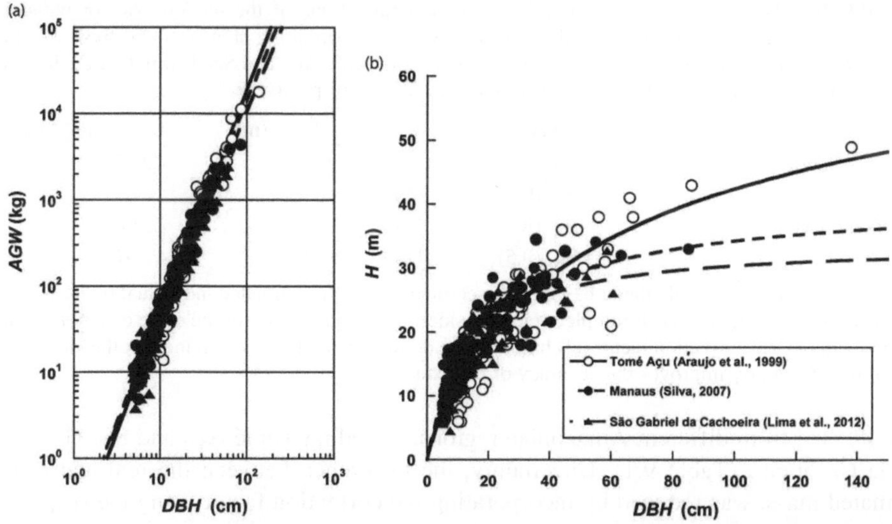

Fig. 9.1 (**a**) Relationships between stem diameter at breast height DBH and above-ground dry weight AGW at three Amazonian regions. (**b**) Relationships between DBH and tree height H at three Amazonian regions. *Open circles* and *solid line* represent data from Tomé Açu (Araújo et al. 1999) (R^2 0.80; $S_{yx\%}$ 1.69). *Closed circles* and *short dashed line*, data from Manaus (Silva 2007) (R^2 0.84; $S_{yx\%}$ 1.39). *Closed triangles* and *long dashed line*, data from São Gabriel da Cachoeira (Lima et al. 2012) (R^2 0.85; $S_{yx\%}$ 1.54). Those figures were drawn on the basis of Lima et al. (2012)

be partly ascribed to the regional differences in the DBH–height (H) relationships, especially for large trees (Fig. 9.1). In other words, the DBH–H relationship significantly differed among the three regions (F-test, $p < 0.001$), and tree height decreased in the sequence Tomé Açu > Manaus > São Gabriel da Cachoeira when DBH > 35 cm. Corroborating with the same descending order of the pattern of allometry for above-ground carbon stock.

To compensate for regional tree height differences in the allometric equations (of which considers only DBH), dominant height (Loetsch et al. 1973) can be used, as proposed by Higuchi et al. (2009). The dominant height (H_{dom}), according to Weise, cited by Loetsch et al. (1973), can be defined as the average height of the uppermost quintile for trees with a DBH > 10 cm. By incorporating H_{dom} into Eqs. (9.1) and (9.2), the following models can be obtained, respectively:

$$\mathrm{PF}_{\mathrm{abg}} = 2.274\,\mathrm{DBH}^{1.916}\frac{H_{\mathrm{dom}}}{28.6} \tag{9.5}$$

$$\mathrm{PF}_{\mathrm{tot}} = 2.718\,\mathrm{DBH}^{1877}\frac{H_{\mathrm{dom}}}{28.6} \tag{9.6}$$

where 28.6 is the dominant height (m) for Manaus site. The generality of Eqs. (9.5) and (9.6) was tested by applying them to the datasets of above-ground mass

Table 9.1 The results of the comparisons of the applications of the models with or without dominant height H_{dom} (Eqs. 9.1 and 9.5, respectively) for estimating above-ground fresh weight are shown. The datasets (DBH > 10 cm) collected from Tomé Açu and São Gabriel da Cachoeira were based on Araújo et al. (1999) and Lima et al. (2012), respectively

Site	Equation	N	H_{dom} (m)	Uncertainty[a] (%)
São Gabriel da Cachoeira	Eq. (9.1)	59	–	19.09
	Eq. (9.5)	59	25.7	14.43
Tomé Açu	Eq. (9.1)	127	–	13.90
	Eq. (9.5)	127	30.5	12.91

[a]The uncertainty was calculated based on the comparison of the estimated individual biomass and the observed weight of each sampled tree. Considering the application of the same equation, with and without the correction factor (cf) based on the dominant height, it was found that the inclusion of this cf actually improves the accuracy of the estimation

collected from different Amazonian regions, including Tomé Açu and São Gabriel da Cachoeira (Table 9.1). Uncertainty, the difference between the real and estimated mass, was reduced by incorporating the correction factor using the H_{dom}.

9.3 Plot-Based Carbon Stock Estimation

9.3.1 Structural Features of Inventory Plots

A grand total of 1836 plots were established in 22 different sites across Amazonas State, until June 2012 (Fig. 9.2 and Table 9.2). All sample units (plots) were placed based on a random and systematic sampling distribution, and stem diameter at breast height (DBH) of all alive and standing dead trees, with DBH equal or greater than 10.0 cm, was recorded (see Boxes 9.4 and 9.5 for details). For BIONTE and the two sites of EMBRAPA, the dimensions of each plot are 100 m by 100 m; for the remaining 1800 plots, they were 20 m × 125 m. Sampling was restricted to primary *terra firme* forests, with exception to two sites: Mil Madeiras and ST Manejo de Florestas, which are logging areas. In terms of area, the CFI system of the Forest Management Laboratory (LMF) of INPA has totaled 486 ha. The number of trees per hectare of CFI of LMF/INPA averaged 491 ± 47 (CI 95 %); and the minimum and maximum were 168 ha^{-1} at Manacapuru and 748 ha^{-1} at BIONTE, respectively. The mean basal area of the same plots was 25.7 ± 1.3 $m^2\ ha^{-1}$, and the minimum and maximum basal areas were 18 $m^2\ ha^{-1}$ at ST Manejo de Florestas Ltda in Lábrea and 31 $m^2\ ha^{-1}$ at BIONTE, respectively.

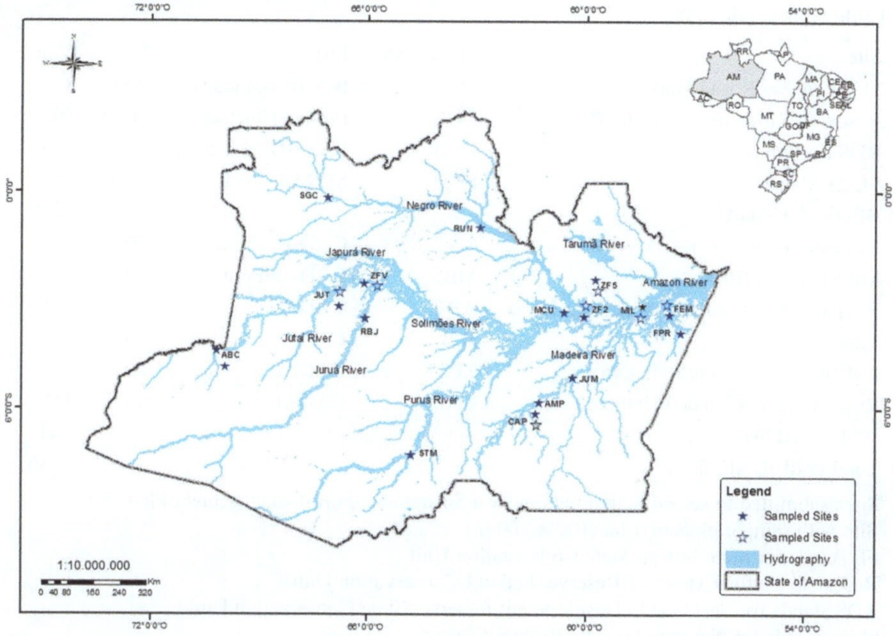

Fig. 9.2 Spatial distribution of study sites around Amazonas State. *Grey lines* indicate boundaries of municipalities

Table 9.2 General descriptions of the inventory sites: site name, project involved, year of installation, and number of plots (*n*)

Site	Site's code	Project	Year	*N* plots
Manacapuru	MCU	Chichuá	2004	32
Fonte Boa (ZFV)	ZFV	Chichuá	2004	72
Jutaí	JUT	Chichuá	2004	64
ST Manejo Florestal (Lábrea)[a]	STM	Chichuá	2005	87
EMBRAPA (Rio Preto da Eva)[b]	ZF5	Chichuá	2005	15
FE de Maués (Maués)[c]	FEM	PPOPE/Chichuá	2005	100
Resex do Baixo Juruá[d]	RBJ	ICMBio/Chichuá	2006	83
Mil Madeiras (Itacoatiara)[a]	MIL	Chichuá	2006	204
Resex Auati Paraná (Fonte Boa)[d]	AUP	ICMBio/Chichuá	2007	107
BIONTE—control (Manaus)	ZF2	Chichuá	2007	3
Resex Capanã Grande (Manicoré)[d]	CAP	ICMBio/Chichuá	2008	82
RDS do Amapá (Manicoré)[e]	AMP	SDS/Chichuá	2008	61

(continued)

Table 9.2 (continued)

Site	Site's code	Project	Year	*N* plots
Flona de Pau Rosa (Maués)[f]	FPR	ICMBio/Chichuá	2009	81
Resex do Rio Unini (Barcelos)[d]	RUN	Pronex/Chichuá	2009	90
RDS do Juma (Novo Aripuanã)[e]	JUM	FAZ/SDS/Chichuá	2009	115
FE de Maués (Maués)[c]	FEM	SDS/Chichuá	2010	76
BEFORE CADAF				1272
São Gabriel da Cachoeira	SGC	CADAF/Pronex	2010	100
Mil Madeiras (Itacoatiara)[a]	MIL	CADAF/Pronex	2010	119
Benjamin Constant and Atalaia do Norte	ABC	CADAF	2011	105
Jutaí	JUT	CADAF	2011	104
EMBRAPA (Rio Preto da Eva)[b]	ZF5	CADAF	2011	18
Resex Capanã Grande (Manicoré)[d]	CAP	CADAF	2012	118
WITH CADAF				564
Total until JUNE, 2012				1836

[a]Sites submitted to selective logging, under a Sustainable Forest Management Plan
[b]Site with sample plots of 1 ha (100 × 100 m)
[c]FE stands for State Forest, State Conservation Unit
[d]Resex stands for Extractive Reserve, Federal Conservation Unit
[e]RDS stands for Sustainable Development Reserve, State Conservation Unit
[f]Flona stands for National Foresta, Federal Conservation Unit

Box 9.4: Forest Inventory

The sampling of the forest inventory counted with the installation of fixed area units (sample plots), with ¼ ha each (25 × 125 m), with the exception of the EMBRAPA and BIONTE site, which counted with squared, 1 ha plots. The ¼ ha sample plots were established in conglomerates (cross and/or transects) as shown below. All sample points (the first sample plot of the conglomerate) were randomly selected, based on geographical coordinates and feasible accessibility, and the following plots were systematically installed.

Within each plot three main biomass categories were sampled: biomass (live trees), necromass (standing and/or fallen dead trees), and palm trees. All individuals were identified by its local common name, botany family, and genus name by a botanical team. The minimum diameter (DBH) measured was 10.0 cm; procedures of DBH measuring are described below. Additional information was registered, such as stem quality, decomposition status (of the necromass), and the presence of epiphytes (i.e. vines) among others. Sub-plots of 200 m^2 (10 × 20 m) were installed within the ¼ ha plots to sample the smaller individuals ($5.0 \leq \text{DBH} < 10.0$ cm).

(continued)

Box 9.4 (continued)

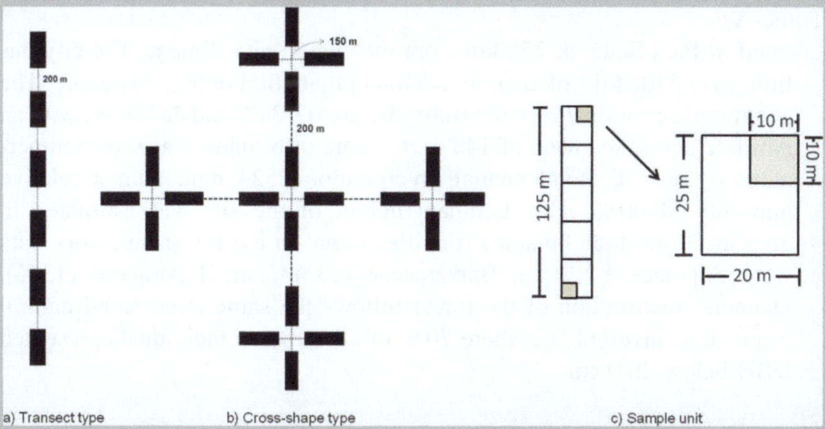

Measuring points (PDM) of the diameter of the sampled trees, in the forest inventories.

Box 9.5: Description of the Sites Where Biomass Destructive Allometry Were Conducted

Tomé-Açu

Located in Pará State, at 250 km from the capital city, Belém. The city has little over 5100 km^2 of area, with a total population of 56,500 people. The geographic coordinates of the study site are 02°30′S and 48°08′W, with an Altitude above sea level of 145 m. Climate information: average temperature is 26.2 °C, total annual precipitation 2524 mm, with a relative humidity of 80 %. The dominant height of the site was estimated in 30.5 m. Main three botanical families found in the forest inventory data were Sapotaceae (20 %), Burseraceae (13 %), and Piperaceae (13 %). Diameter distribution of the forest follows the same trend as all natural tropical, as inverted 'J', where 70 % of all sampled individuals presented DBH below 20.0 cm.

Manaus

Capital city of Amazonas State, with approximately 11,400 km^2, and 2 million people living within its limits. Average Altitude is 92 m above sea level. Climate, according to Köppen classification, is Awf. Mean temperature of 27.6 °C and annual precipitation of 2286 mm. The study site, where the destructive sampling took place, is located at the ZF2 Tropical Silviculture Experimental Station of INPA, at 90 km from the city. Coordinates: 02°35′S and 60°12′W. Dominant height estimated in 28.6 m. Main botanical families found in the study site, based on forest inventory data, are Euphorbiaceae (18 %), Fabaceae (15 %), and Lecythidaceae (10 %).

São Gabriel da Cachoeira

Located in Amazonas State, at 852 km from Manaus, with an area of approximately 109,000 km^2, 41,500 people, and an average Altitude of 90 m above sea level. The climate is classified as equatorial Af. Average annual rainfall is 2811 mm, with a mean temperature of 25.5 °C. Dominant height estimated is 27.3 m. Main botanical families found: Fabaceae (33 %), Euphorbiaceae (13 %), and Myristicaceae (11 %).

Considering the allometric model of Silva (2007) for carbon stock estimation, H_{dom} was estimated for six sites (Table 9.3). To overcome the difficulties of tree height measurement in the field (Chave et al. 2005), the heights of 1883 newly naturally fallen trees, whose canopy shape was entirely retained, were measured around the inventory plots (Lima 2010). This sampling was plot free, where each single fallen tree was considered a sample. From each fallen tree sampled, the DBH, stem height, and total height, with a metric tape were measured. The maximum and minimum of H_{dom} were recorded at Maués (30.2 m) and at Barcelos (25.3 m), respectively.

Table 9.3 Dominant height (H_{dom}) was estimated, with uncertainty, for six sites in Amazonas State

Site	n	H_{dom} (m)	Uncertainty (%)
Itacoatiara[a]	117	28.6	5.36
Fonte Boa	567	26.6	3.72
Manicoré	217	27.9	5.75
Maués	302	30.2	5.15
Barcelos	326	25.3	3.92
Novo Aripuanã	354	27.8	3.88

[a]H_{dom} at Itacoatiara was estimated using the datasets with DBH > 45 cm

Table 9.4 Carbon stocks at different sites in Amazonas State

Site	C_{abg} (Mg ha^{-1})	CI 95 % (±)	C_{tot} (Mg ha^{-1})	CI 95 % (±)
Manacapuru	122.8	9.6	124.2	9.6
Fonte Boa (ZFV)	163.1	6.9	167.6	7.0
Jutaí	167.2	7.2	172.7	7.4
ST Manejo Florestal (Lábrea)[a]	111.3	6.8	114.7	6.9
EMBRAPA (Rio Preto da Eva)	157.0	6.6	177.4	8.3
Maués	119.4	6.8	137.2	11.1
Resex do Baixo Juruá	150.2	4.6	170.5	5.5
Mil Madeiras (Itacoatiara) [a]	149.4	29.4	153.8	29.5
Resex Auati Paraná	148.9	5.3	170.7	6.3
BIONTE—control (Manaus)	188.6	12.6	194.7	13.3
Resex Capanã Grande	131.9	5.2	148.5	6.3
Flona de Pau Rosa	172.8	7.7	177.1	7.6
Resex do Rio Unini	147.1	5.0	151.5	5.0
RDS Juma (Manicoré)	146.8	6.1	151.1	6.0
FE de Maués (Maués)	145.5	6.2	168.0	7.7
São Gabriel da Cachoeira	109.0	5.5	125.1	6.1
Benj. Const. and Atalaia do Norte	143.9	4.1	175.4	5.3
Jutaí	139.2	4.7	169.7	6.0
EMBRAPA (Rio Preto da Eva)	158.1	7.6	178.1	8.9
Resex Capanã Grande	144.8	5.8	168.8	7.0
Mean	145.9	–	159.8	–
Confidence interval (95 %)	145.9 ± 8.8	–	159.8 ± 9.2	–

[a]Sites submitted to selective logging, under a Sustainable Forest Management Plan

9.3.2 *Carbon Stock Estimation*

As for the carbon stock estimation of the entire State of Amazonas, the average of each sampled site, except the selectively logged sites, was calculated. The above-ground and total (above-ground + coarse roots) oven-dry biomass averaged 300.7 ± 18.1 and 329.6 ± 19.0 Mg ha^{-1}, respectively (Table 9.4). In terms of

Table 9.5 Likelihood scale for describing quantified uncertainty (IPCC 2010)

Term	Likelihood of the outcome
Virtually certain	99–100 % probability
Extremely likely[a]	95–100 % probability
Very likely	90–100 % probability
Likely	66–100 % probability
About as likely as not	33–66 % probability
Unlikely	0–33 % probability
Very unlikely	0–10 % probability
Exceptionally unlikely	0–1 % probability

[a]Only for Assessment Report number 4

carbon stock, the above-ground and total averaged 145.9 ± 8.8 and 159.8 ± 9.2 Mg C ha^{-1}, respectively.

According to PRODES from INPE (2014), based solely on deforestation estimation, more than 1.5 million km^2 of old-growth forest of the Amazonas State still remain undisturbed. Based on this figure, the total C stock, considering above- and below-ground of live trees (DBH $\geq$ 10.0 cm), in Amazonas State should be from 22.6 to 25.4 Pg C (CI at 95 %). The estimated variation of the mean value is <6 %, which is '*extremely likely*' according to IPCC (2010) likelihood scale—see Table 9.5.

9.4 Perspective for Future Studies on the Carbon Stocks in Amazonas State

The availability of suitable allometric equations is one of the crucial requirements for accurate forest biomass/carbon assessments to reduce estimates uncertainties (Chave et al. 2004). Few allometric equations exist for Amazonian forests, and, as a result, sometimes generalised allometric equations (Brown 1997; Chave et al. 2005) are used as a default (Asner et al. 2010; Saatchi et al. 2011). In addition, since these generalised allometric equations involve wood density as an independent variable, systematic bias may exist in data inputs into the allometric equations: such as bias or errors in botanical identifications for assigning wood density information and the wood density data itself (Saatchi et al. 2011).

Saatchi et al. (2011) developed a benchmark map of forest carbon stocks in tropical regions across three continents using a combination of biomass data from in situ inventory plots and global forest height data measured by satellite LiDAR (GLAS). It has shown relatively low biomass density of the Rio Negro Basin due to low forest height measured by GLAS. However, the biomass carbon data of inventory plots in their paper was estimated from DBH, using generalised allometric equations, which might have introduced non-random, and potentially significant and systematic, errors as mentioned above. These non-random errors associated with ground-based estimation of forest biomass will remain uncertain until consistent allometric equations within forest types or regions are developed (Saatchi et al. 2011).

The common allometric equations with dominant height of canopy trees proposed by Higuchi et al. (2009), derived from destructive sampling in three distinct regions, in terms of geographical location, forest structure, DBH–*H* relationship, and tree species distribution (Tomé Açu, Manaus and São Gabriel da Cachoeira), are currently the most robust allometric equations applicable to areas of mature forests with different heights across the Brazilian Amazon. Dominant height of canopy trees can be measured directly from naturally recent fallen trees through field assessment, as for standing tree height, satellite, or airborne LiDAR are most commonly used.

There is an ongoing project (CADAF, http://cse.ffpri.affrc.go.jp) that aims to refine carbon stock distribution and its variations (over time and in different regions), which remains with large uncertainties (Saatchi et al. 2007, 2011), across central Amazonia, based on the common allometric equations newly developed and large-scale systematic forest inventories. We expect that these will improve the quality of tropical biomass estimates and enhance the awareness regarding the contribution of this biome to the global carbon cycle.

Acknowledgements The authors thank all members of the Forest Management Laboratory of the National Institute for Research in the Amazon (INPA) for their valuable suggestions and help in the fieldwork. This study was financially supported by the Amazonas State Science Foundation (FAPEAM), the National Council for Science and Technology Development (CNPq), the Science and Technology Research Partnership for Sustainable Development Program of Japan Science and Technology Agency (JST), and Japan International Cooperation Agency (JICA).

References

Araújo TM, Higuchi N, Carvalho JA (1999) Comparison of formulae for biomass content determination in a tropical rain forest site in the state of Pará, Brazil. For Ecol Manage 117:43–52

Asner GP, Knapp DE, Broadbent EN, Oliveira PJC, Keller M, Silva JN (2005) Selective logging in the Brazilian Amazon. Science 310:480–482

Asner GP, Powell GVN, Mascaro J, Knapp DE, Clark JK, Jacobson J, Kennedy-Bowdoin T, Balaji A, Paez-Acosta G, Victoria E, Secada L, Valqui M, Hughes RF (2010) High-resolution forest carbon stocks and emissions in the Amazon. Proc Natl Acad Sci USA 107:16738–16742

Brown S (1997) Estimating biomass and biomass change of tropical forests: a primer. FAO Forestry Paper 134, Rome, p 55

Chambers JQ, Santos J, Ribeiro RJ, Higuchi N (2001) Tree damage, allometric relationships, and above-ground net primary production in a tropical forest. For Ecol Manage 152:73–84

Chave J, Condit R, Aguilar S, Hernandez A, Lao S, Perez R (2004) Error propagation and scaling for tropical forest biomass estimates. Philos Trans R Soc Lond B 359:409–420

Chave J, Brown ACS, Cairns MA, Cahmbers JQ, Eamus D, Fölster H, Fromad F, Higuchi N, Kira T, Lescure JP, Nelson BW, Ogawa H, Puig H, Riéra B, Yamakura T (2005) Tree allometry and improved estimation of carbon stocks and balance in tropical forests. Oecologia 145:87–99

Chave J, Olivier J, Bongers F, Chatelet P, Forget P-M, Meer P, Norden N, Riera B, Charles-Dominique P (2008) Above-ground biomass and productivity in a rain forest of eastern South America. J Trop Ecol 24:355–366

IPCC (Intergovernmental Panel on Climate Change) (2007) IPCC fourth assessment report: climate change (AR4)

Deforestation Monitoring Program (PRODES) of the National Spatial Research Institute (INPE) (2014). http://www.obt.inpe.br/prodes/index.php

Eva HD, Achard F, Stibig H-J, Mayaux P (2003) Response to comment on 'determination of deforestation rates of the world's humid tropical forests'. Science 299:1015

Fearnside PM, Laurance WF (2003) Comment on "Determination of deforestation rates of the world's humid tropical forests". Science 229:1015

Higuchi N, Santos J, Ribeiro RJ, Minette L, Biot Y (1998) Biomassa da parte aérea da vegetação da floresta úmida de terra-firme da Amazônia brasileira. Acta Amazôn 28:153–166

Higuchi N, Pereira HS, Santos J, Lima AJN, Higuchi FG, Higuchi MIG, Ayres IGSS (2009) In: Higuchi N, Higuchi MIG (eds) Governos Locais Amazônicos e as Questões Climáticas Globais. Manaus

Houghton RA, Laurence KT, Hackers JL, Brown S (2001) The spatial distribution of forest biomass in Brazilian Amazon: a comparison of estimates. Glob Chang Biol 7:731–746

IPCC (2006) Annex 1: conceptual basis for uncertainty analysis. In: Penman J, Gytarsky M, Hiraishi T, Krug T, Kruger D, Pipatti R, Buendia L, Miwa K, Ngara T, Tanabe K, Wagner F (eds) Good practice guidance and uncertainty management in national greenhouse gas inventories. The institute for global environmental strategies for the IPCC, Kanagawa, p 22

IPCC (2010) Guidance note for lead authors of the IPCC fifth assessment report on consistent treatment of uncertainties. http://www.ipcc.ch/pdf/supporting-material/uncertainty-guidance-note.pdf

Kenzo T, Ichie T, Hattori D, Itioka T, Handa C, Ohkubo T, Kendawang JJ, Nakamura M, Sakaguchi M, Takahashi N, Okamoto M, Tanaka-Oda A, Sakurai K, Ninomiya I (2009) Development of allometric relationships for accurate estimation of above- and below-ground biomass in tropical secondary forests in Sarawak, Malaysia. J Trop Ecol 25:371–386

Kenzo T, Furutani R, Hattori D, Kendawang JJ, Tanaka S, Sakurai K, Ninomiya I (2010) Allometric equations for accurate estimation of above-ground biomass in logged-over tropical rainforests in Sarawak, Malaysia. J Forest Res 14:365–372

Lima AJN (2010) Avaliacão de um sistema de inventário florestal conti'nuo em áreas manejadas e não manejadas do estado do Amazonas (AM). Ph.D. Dissertation, Universidade Federal do Amazonas, Manaus, Brazil, p 183 (in Portuguese with English abstract)

Lima AJN, Suwa R, Ribeiro GHPM, Kajimoto T, Santos J, Silva RP, Souza CAS, Barros PC, Noguchi H, Ishizuka M, Higuchi N (2012) Allometric models for estimating above- and below-ground biomass in Amazonian forests at São Gabriel da Cachoeira in the upper Rio Negro, Brazil. For Ecol Manage 277:163–172

Loetsch F, Zohrer F, Haller KE (1973) Forest inventory, vol 2. BLV Verlagsgesellschaft, Munich

Malhi Y, Wood D, Baker TR, Wright J, Phillips O (2006) Regional variation of above-ground live biomass in old-growth Amazonian forests. Glob Chang Biol 12:1–32

Melillo JM, McGuire AD, Kicklighter DW, Moor IIIB, Vorosmarty CJ, Scholoss A (1993) Global climate change and terrestrial net primary production. Nature 363:234–240

Overman JPM, Witte HJL, Saldarriaga JG (1994) Evaluation of regression models for above-ground biomass determination in Amazon rain forest. J Trop Ecol 10:207–218

Saatchi SS, Houghton RA, Dos Santos Alvará RC, Soares JV, Yu Y (2007) Distribution of aboveground live biomass in the Amazon basin. Glob Chang Biol 13:816–837

Saatchi SS, Harris NL, Brown S, Lefsky M, Mitchard ETA, Salas W, Zutta BR, Buemann W, Lewis SL, Hagen S, Petrova S, White L, Silman M, Morel A (2011) Benchmark map of forest carbon stocks in tropical regions across three continents. Proc Natl Acad Sci USA 108 (24):9899–9904

Santos J (1996) Analize de modelos de regressao para estimar a fitomass de floresta tropical umida de terra-firme de Amazonia Brasiliera. Ph.D. Dissertation, Universidade Federal de Vicosa, Minas Gerais, Brazil, p 121 (in Portuguese with English abstract)
Sierra CA, Valle JI, Orrego SA, Moreno FH, Harmon MA, Zapata M, Colorado GJ, Herrera MA, Lara W, Restrepo DE, Berrouet LM, Loaiza LM, Benjumea JF (2007) Total carbon stocks in a tropical forest landscape of the Porce region, Colombia. For Ecol Manage 243:299–309
Silva RP (2007) Alometria, estoque e dinâmica da biomassa de florestas primárias e secundárias na regiaõ de Manaus (AM). Ph.D. Dissertation, Universidade Federal do Amazonas, Manaus, Brazil, p 152 (in Portuguese with English abstract)

Part IV
Environmental Variation and Global Change

Chapter 10
Recent Changes in Amazon Forest Biomass and Dynamics

Oliver L. Phillips, Simon L. Lewis, Niro Higuchi, and Tim Baker

10.1 Introduction

10.1.1 Overview

There is a major planet-wide experiment under way. Anthropogenic changes to the atmosphere–biosphere system mean that all ecosystems on Earth are now affected by human activities. While outright deforestation is physically obvious, other subtler processes, such as faunal imbalances and surface fires, impact forests in ways less evident to the casual observer (cf. Lewis et al. 2004a; Malhi and Phillips 2004; Estes et al. 2011). Similarly, anthropogenic atmospheric change is intensifying (Friedlingstein et al. 2006). By the end of the twenty-first century, carbon dioxide concentrations may reach levels unprecedented for at least 20 million years (e.g. Retallack 2001) and climates may move beyond Quaternary envelopes (Meehl et al. 2007). Moreover, the rate of change in these basic ecological drivers may be unprecedented in the evolutionary span of most species on Earth today. Additionally, these atmospheric changes are coinciding with probably the greatest changes in land cover and species' distributions since at least the last mass extinction at c. 65 million years ago (Ellis et al. 2011). The collective evidence points to conditions

O.L. Phillips (✉) • T. Baker
School of Geography, University of Leeds, Leeds LS2 9JT, UK
e-mail: o.phillips@leeds.ac.uk

S.L. Lewis
School of Geography, University of Leeds, Leeds LS2 9JT, UK

School of Geography, University College London, London WC1E 6BT, UK

N. Higuchi
Instituto Nacional de Pesquisas da Amazônia, Manaus, Brazil

© Springer-Verlag Berlin Heidelberg 2016
L. Nagy et al. (eds.), *Interactions Between Biosphere, Atmosphere and Human Land Use in the Amazon Basin*, Ecological Studies 227,
DOI 10.1007/978-3-662-49902-3_10

with no clear past analogue. We have entered the Anthropocene, a new geological epoch dominated by human action (Crutzen 2002; Steffen et al. 2011).

In this chapter, we focus on the changes occurring within remaining tropical forests, with an emphasis on Amazonia. Most forest vegetation carbon stocks lie within the tropics. Tropical forest ecosystems store 460 billion tonnes (Pg) of carbon in their biomass and soil (Pan et al. 2011), equivalent to more than half the total atmospheric stock, and annually process 40 Pg (Beer et al. 2010). They have other planetary influences via the hydrological cycle, and emit aerosols and trace gases, and are also characterised by their exceptional variety and diversity of life. Changes here therefore matter for several key reasons. First, the critical role that tropical forests play in the global carbon and hydrological cycles affects the rate and nature of climate change. Second, as tropical forests are home to at least half of all Earth's species, changes here impact on global biodiversity and the cultures, societies, and economies that are bound to this diversity (Groombridge and Jenkins 2003). Finally, as different plant species vary in their ability to store and process carbon, climate and biodiversity changes are linked by feedback mechanisms (e.g. Lewis 2006). The identities of the 'winner' species under environmental changes might exacerbate, or perhaps mitigate, human-driven climate change.

That remaining forests globally are now changing fast there is no doubt. Analysis of the global carbon cycle shows that after accounting for known atmospheric and oceanic fluxes there is a large, and increasing, carbon sink in the terrestrial biosphere, reaching >3 Pg by the middle part of the last decade (Le Quéré et al. 2009); independent analyses of atmospheric carbon dioxide concentration data to infer sources and sinks of carbon imply carbon uptake over the terrestrial land mass in both tropical and extra-tropical latitudes (Stephens et al. 2007); a ground-up, independent analysis by foresters suggests that forests in every vegetated continent are implicated in this terrestrial sink (Pan et al. 2011). One critical question is therefore: how should scientists go about documenting and monitoring the changing behaviour of tropical forests?

Of the many approaches and technologies available, it is careful, persistent, on-the-ground monitoring at fixed locations on Earth that can provide reliable long-term evidence of ecosystem behaviour, and this is the focus of this chapter. On-the-ground measurements can provide information on subtle changes in species composition, biomass, and carbon storage—none of which has been successfully done using satellites in mature lowland tropical forests, as signals saturate at high biomass (e.g. Mitchard et al. 2009) and cannot yet detect species composition and hence the density of each tree's wood, which substantially drives forest biomass (Mitchard et al. 2014). Yet, permanent sample plot work in the tropics has until quite recently been very sparse and mostly focused on a few well-known locations, leaving most of the ~10 million km^2 expanse of the world's richest ecosystems unstudied. This is particularly risky given that no one location, or small number of studied forests, can be taken as the mean conditions of all forests. Site-centric ecology is invariably skewed, since peculiar local features—such as fragmentation, unusual soil conditions, cyclones, or fires—strongly colour interpretations. In most fields, such as climate change, it would be an obvious folly to infer the presence or absence of global effects from records at a few sites, but long-term ecological monitoring is difficult and under-funded. As a result, attempts are still made to scale results from a few selected locations to draw conclusions about what the behaviour of the whole biome might be.

10.1.2 A Networked Approach

A robust approach to monitoring change needs to be much more synoptic. The first attempts to do this (Phillips and Gentry 1994; Phillips et al. 1994, 1998; Phillips 1996) drew inspiration from the macroecological work of Gentry, who had used intensive floristic inventories across hundreds of forest locations to reveal the major geographic gradients in diversity and composition (e.g. Gentry 1998a, b). But, unlike Gentry's floristic work, these first macroecological analyses of tropical forest dynamics lacked methodological standardisation. They relied heavily on published data from different teams worldwide and had limited sample sizes. To try to eliminate these weaknesses, since 2000 with many colleagues we have focused on developing standardised, international, long-term networks of permanent plots in mature forests across Amazonia and Africa. These first draw together the existing efforts of local foresters and ecologists, often working hitherto largely in isolation. Then, by analysing the gaps in geographical and environmental space, efforts can be made to extend the site network to fill the gaps and build support for long-term spatially extensive monitoring. The network of Amazonian-forest researchers, known as RAINFOR (*Red Amazónica de Inventarios Forestales*, http://www.geog.leeds.ac.uk/projects/rainfor/), now represents the long-term ecological monitoring efforts of 43 institutions worldwide including many from Amazonia itself. A parallel initiative in Africa, AfriTRON (*African Tropical Rainforests Observation Network*, http://www.geog.leeds.ac.uk/projects/afritron/), spans 11 countries across the African wet tropics. Here our aim is to synthesise some published results from RAINFOR to assess how Amazon forests have, on average, changed recently. Where appropriate, we also discuss results from the African network and from additional, individual sites where these shed light on the processes involved.

10.2 Methods

For these analyses, we define a monitoring plot as an area of old-growth, physiognomically mature forest where all trees $\geq$10 cm diameter at breast height (dbh, measured at 1.3 m height or above any buttress or other deformity) are tracked individually over time. 'Maturity' is inferred by the existence of a complex multi-age structure, multiple canopy tree species, large lianas, and large dead trees, together implying forests at least several centuries old. All trees are marked with a unique number, measured, mapped, and identified. Periodically (generally every 1–5 years) the plot is revisited, all surviving trees are remeasured, dead trees are noted, and trees recruited to 10 cm dbh are uniquely numbered, measured, mapped, and identified. This allows calculation of (i) the cross-sectional area that tree trunks occupy (basal area), which can be used with allometric equations to estimate tree biomass (Higuchi et al. 1998; Baker et al. 2004a; Chave et al. 2005); (ii) tree growth (the sum of all basal-area increments for surviving and newly recruited stems over a census interval); (iii) the total number of stems present; (iv) stem recruitment (number of stems added to a plot over time); and (v) mortality (either the number or basal area of stems lost from a plot over time). We present results from 50 to 123 plots, depending upon selection criteria for different analyses. The 'Amazon' plots span the forests of northern South America (Fig. 10.1), including Bolivia,

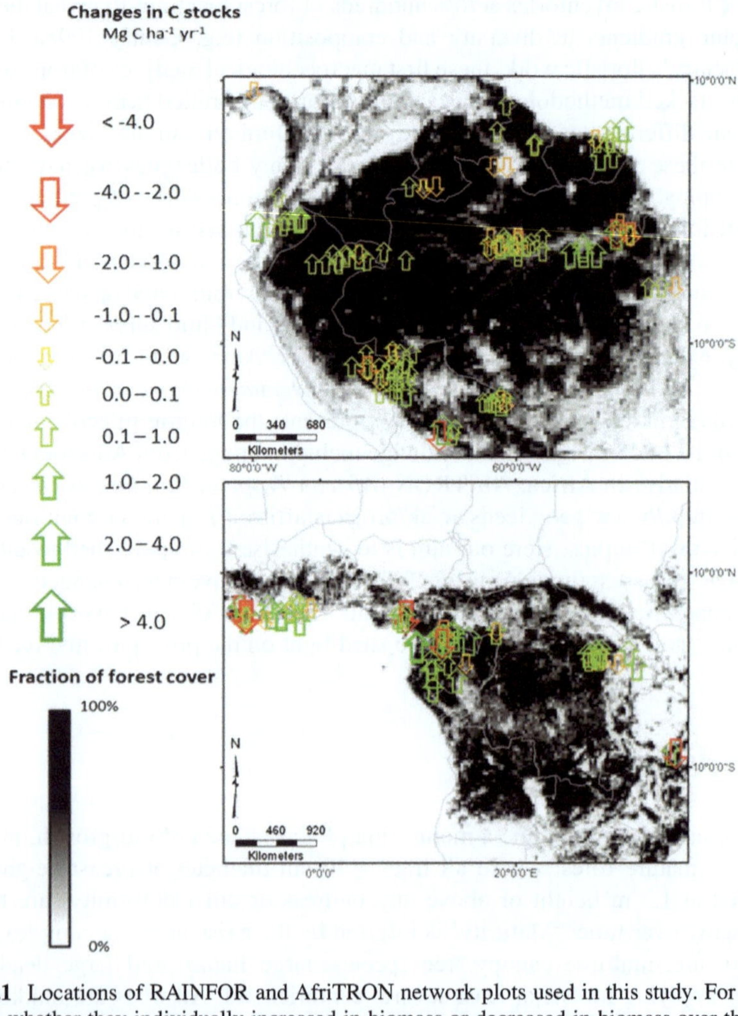

Fig. 10.1 Locations of RAINFOR and AfriTRON network plots used in this study. For each we indicate whether they individually increased in biomass or decreased in biomass over the period monitored (ending prior to the 2005 drought for Amazonia)

Brazil, Colombia, Ecuador, French Guiana, Peru, and Venezuela, from the driest to the wettest and the least to the most fertile soils. Most plots are 1 ha in size and comprise ~400–600 trees of ≥10 cm dbh, but the smallest is 0.25 ha and the largest 10 ha. The large majority result from applying locally randomised or systematic sampling protocols to locate plots in apparently old-growth forest landscape and then tracking their subsequent dynamics. Many plots have been monitored for more than a decade, although they range in age from 2 to 30 years (mean ~10 years). Here we analyse results of censuses completed up to 2007, but for Amazonia we first report results prior to the intense drought of 2005 (Aragão et al. 2007) and then also summarise the impact of the drought and briefly review the latest findings from RAINFOR (Brienen et al. 2015). Details of the exact plot locations, inventory and monitoring methods, and issues relating to collating and analysing plot data are discussed elsewhere in detail (Phillips et al. 2002a, b, 2004, 2009; Baker et al. 2004a, b; Malhi et al. 2002, 2004; Lewis et al. 2004b, 2009a; Lopez-Gonzalez et al. 2011). It is important to point out that the samples are not evenly distributed over Amazonia because they use historical plot data, where possible, and also because considerations of access limit where it is practical to work; nevertheless, a wide range of environmental space is captured by the samples (c.f. Fig. 10.2).

Scaling from individual tree to Amazon plot biomass is based on the diameter-based allometric equations detailed in Baker et al. (2004a). Thus, we used an equation developed for the Manaus area (Chambers et al. 2001a), modified by taking account of the taxon-specific wood density of each tree relative to the mean wood density of trees in the Manaus region. Alternatively, biomass can be estimated by universal, tropical forest equations such as those of Chave et al. (2005). The Manaus equation is based on a smaller sample size but has the advantage of being local. More allometric equations have been developed by the research community. For simplicity, we only show results using the Baker et al. (2004a) equation here, but note that while different methods certainly result in systematic differences in biomass estimates (e.g. Chave et al. 2003; Peacock et al. 2007; Feldpausch et al. 2012), the rates of net biomass change calculated across Amazonia are largely insensitive to the equation used (Baker et al. 2004a). For Africa, we use the Chave et al.'s (2005) moist forest equation including tree height and propagate the uncertainty in both the diameter and height parameters to obtain final biomass estimates (Lewis et al. 2009a). We summarise findings from mature forests in terms of (a) structural change, (b) dynamic-process change, and (c) functional and compositional change, over the past two–three decades, including taking account of recent droughts in Amazonia.

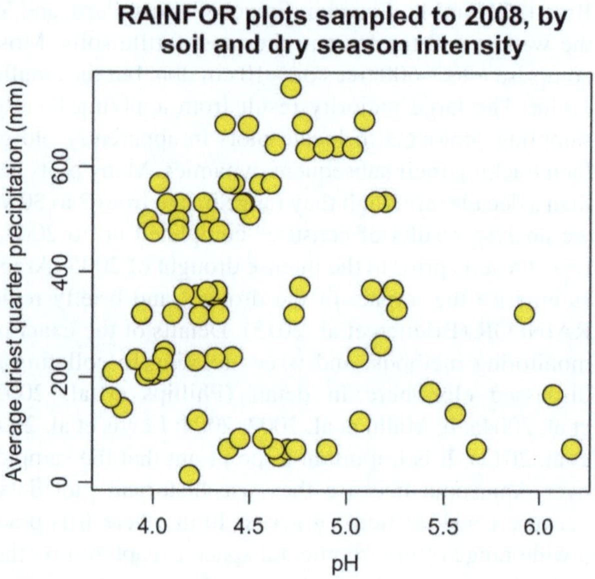

Fig. 10.2 Sampling in environmental space across Amazonia. Most plots sampled are on acid soils in weakly seasonal rainforest climates, reflecting the dominant conditions of the region, but span a broad environmental range that includes more than two orders of magnitude in soil acidity for example. Figure depicts the location of 87 Amazon plots whose soil has been characterised by the time of the post-2005 drought analysis

10.3 Results and Discussion

10.3.1 Structural Change

Among 123 long-term mature forest Amazonian plots with full tree-by-tree data, there was a significant increase in above-ground biomass between the first measurement (late twentieth century, median date 1991) and the last measurement before the 2005 drought (median date 2003). For trees $\geq$10 cm diameter, the increase has been 0.45 (0.33, 0.56) Mg C ha^{-1} year^{-1} (mean and 2.5, 97.5 % confidence limits; Phillips et al. 2009). Across all 123 plots, the above-ground change was approximately normally distributed and shifted to the right of zero (Fig. 10.3a). The overall net increase estimated is slightly lower than but statistically indistinguishable from the 0.54 ± 0.29 Mg C ha^{-1} year^{-1} estimated by Phillips et al. (1998) for the lowland Neotropics using 50 sites up to 1996, and the Baker et al.'s (2004a) estimate of 0.62 ± 0.23 Mg C ha^{-1} year^{-1} for 59 core RAINFOR Amazon plots up to 2000. In the large dataset now available, estimates of biomass carbon change are also rather insensitive to different weightings based on measurement interval and plot area (supplementary information in Phillips et al. 2009). Using the same approach, we also discovered a similar phenomenon in African forests (see Box: 'Changing African Forests').

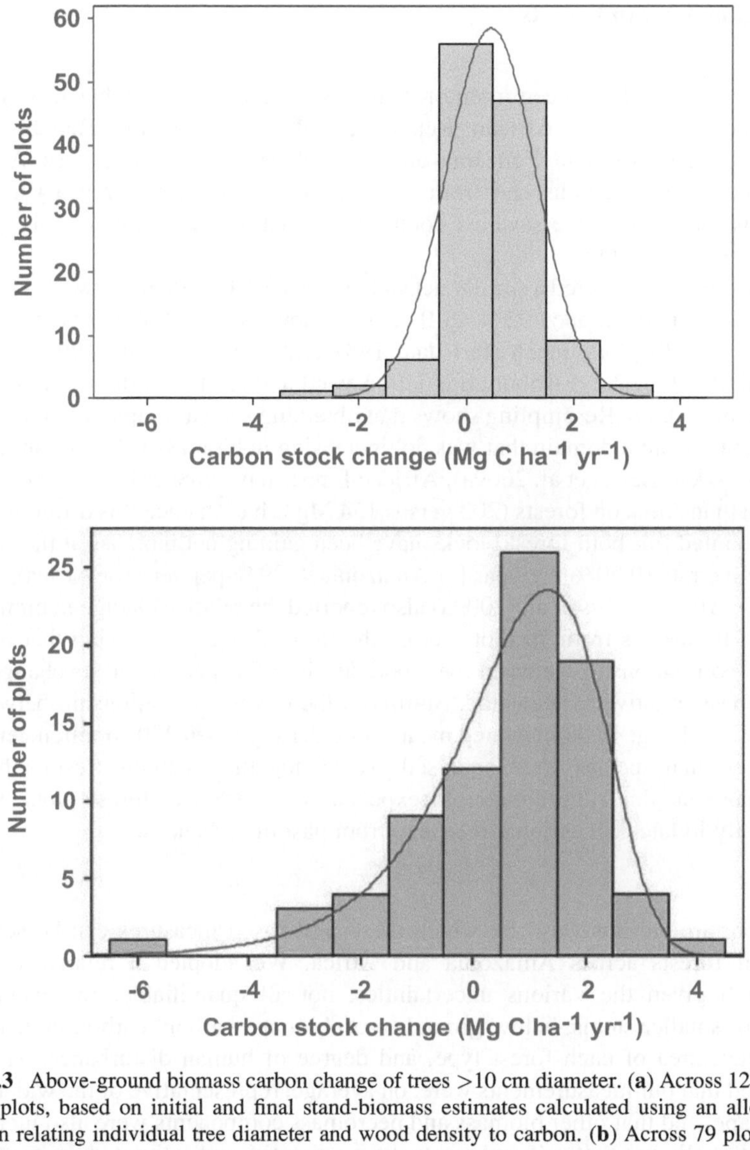

Fig. 10.3 Above-ground biomass carbon change of trees >10 cm diameter. (**a**) Across 123 Amazonian plots, based on initial and final stand-biomass estimates calculated using an allometric equation relating individual tree diameter and wood density to carbon. (**b**) Across 79 plots from Africa, but including estimated tree height for each stem, in addition to diameter and wood density, to estimate carbon, with uncertainty in the height and diameter measurement both propagated to final biomass carbon change estimates. As would be expected in a random sample of small plots measured for a finite period, some sites show a decline in biomass carbon during that period indicating that at that particular point in space and time tree mortality has exceeded tree growth. However, the mean and median are shifted significantly to the right for both datasets ($P < 0.001$)

Changing African Forests

African plots have been monitored in Liberia, Ivory Coast, Ghana, Nigeria, Cameroon, Central African Republic, Gabon, Democratic Republic of Congo, Uganda, and Tanzania and span West, Central, and East Africa biogeographic regions and from wet (~3000 mm rainfall per year) to dry (those adjacent to the savanna boundary) climates and a range of soil types (Lewis et al. 2009a).

Here, we measured a similar net sink in trees >10 cm diameter with a mean of +0.63 (bootstrapped 95 % CI 0.22–0.94) tonnes of carbon per hectare per year ($n = 79$ plots, mean start date 1987 and mean end date 1996; Lewis et al. 2009a). The distribution is left skewed and shifted to the right of zero (see Fig 10.3b). Resampling shows that obtaining such a sample of increasing biomass from a domain that was not increasing in biomass is highly unlikely ($P < 0.001$; Lewis et al. 2009a). African forests have greater biomass per unit area than Amazon forests (202 versus 154 Mg C ha^{-1}); once this difference is accounted for, both forest blocks have been gaining net biomass at the same relative rate (0.30 % per year for Amazonia, 0.29 % per year for Africa).

In Africa, Lewis et al. (2009a) also reported the relative change in biomass for 916 species from 79 plots across the tropical region, showing that there was no relationship between the wood density of a species and its change in biomass, relative to the stand. Similarly, there was no relationship between relative change in biomass and mean wood density when 200 common genera rather than species were analysed. Taken together with the results from Amazonia, this runs counter to expectations if tropical forest plots were mostly in late successional recovery from past disturbances.

There are various ways by which these plot-based measures can be scaled to tropical forests across Amazonia and Africa. We adopted a relatively simple approach given the various uncertainties, not all quantifiable, for example in terms of smaller stems, below-ground (root) biomass carbon, carbon in dead trees and litter, area of each forest type, and degree of human disturbance. Thus, we assumed that our measurements were, on average, representative of the wider forest landscape, and that other biomass and necromass components were also increasing proportionally but that soil carbon stocks were static, and estimated the magnitude of the sink in each continent by multiplying the plot-based net carbon gain rate by a series of correction factors to account for biomass of lianas, trees <10 cm diameter, necromass, and below-ground carbon, and a mid-range estimate of the surviving forest area for the year 2000 (Table 10.1). For the 1990s, this yielded a total estimated South American forest sink of 0.65 ± 0.17 Pg C $year^{-1}$ and a corresponding sink in African forests of 0.53 ± 0.30 and 0.14 ± 0.04 Pg C $year^{-1}$ in mature undisturbed Asian forests if these responded as other tropical forests did (Pan et al. 2011). Thus, the combined mature tropical forest sink in the 1990s is

Table 10.1 Estimated annual change in carbon stock (Tg C $year^{-1}$) in tropical intact forests by region for the periods of 1990–1999 and 2000–2007

	1990–1999								2000–2007							
	Estimated annual change in C stock (Tg C $year^{-1}$)							Change per area	Estimated annual change in C stock (Tg C $year^{-1}$)							Change per unit area
	Biomass	Dead wood	Litter	Soil	Wood product	Total change	Uncertainty	Mg C ha^{-1} $year^{-1}$	Biomass	Dead wood	Litter	Soil	Wood product	Total change	Uncertainty	Mg C ha^{-1} $year^{-1}$
Asia	125	13	2	ND	5	144	38	0.88	100	10	1	ND	6	117	30	0.90
Africa	469	48	7	ND	9	532	302	0.94	425	43	6	ND	8	482	274	0.94
Americas	573	48	9	ND	22	652	166	0.77	345	45	5	ND	23	418	386	0.53
Total	1167	108	17	ND	35	1328	347	0.84	870	98	13	0	36	1017	474	0.71

Table adapted from Pan et al. (2011), see that paper for details of calculations
ND data are not available

estimated to have been 1.3 ± 0.35 Pg C year^{-1}, before allowing for any possible net change in soil carbon stock. This is similar to the figure given by Lewis et al. (2009a), of 1.3 Pg C year^{-1} (bootstrapped CI 0.8–1.6) using plots with a mean interval of 1987–1997 and slightly differing methodology (Tropical America, 0.62; Tropical Africa, 0.44; and Tropical Asia 0.25 Pg C year^{-1}). In the decade of the 2000s, the American tropical sink apparently declined partly as a result of the 2005 Amazon drought; we discuss this later below.

The validity of these estimates depends on (i) measurement techniques; (ii) how representative the plots are of mature forests in South America, and the rest of the tropics; (iii) assumptions about the extent of mature forest remaining; and (iv) the extent to which we have sampled the regional-scale matrix of natural disturbance and recovery. Moreover, they represent average annual estimates for the period around the turn of the twenty-first century—forest plots are rarely measured sufficiently frequently in enough places to estimate biome carbon balance on a year-by-year basis. However, they are consistent with independent evidence from recent inversion-based studies, showing the tropics are either carbon neutral or sink regions, despite widespread deforestation (Denman et al. 2007, p. 522; Stephens et al. 2007), and the large net sink in the terrestrial biosphere after accounting for other sources and sinks (e.g. Le Quéré et al. 2009).

10.3.2 Recovery from Large Disturbances?

The finding of increasing forest biomass over recent decades has been remarkably controversial (cf. for example Clark 2002; Phillips et al. 2002a; Wright 2005; Lewis et al. 2006, 2009b), despite the fact that an uptake of >2 Pg C year^{-1} somewhere on Earth's land surface is evident from independent mass-balance observations of the global carbon cycle. The most persistent area of controversy has been prompted by the statement 'Slow in, rapid out' (Körner 2003). The 'Slow in, rapid out' argument stresses that forest growth is a slow process while mortality can potentially be singular in time, thereby causing rapid biomass loss and sometimes resetting forest stand structure. Consequently, limited sampling or sampling over short observation periods may tend to miss severe perturbation events over large scales. Inferences based on such sampling could therefore result in positively biased estimates of above-ground biomass trends in mature forests when results from plot networks are extrapolated to a large area. Given the still relatively small number of tropical plots—relative to the total biome area—this concern is understandable. However, it is unlikely to be a major source of uncertainty or bias in our calculations for three reasons.

Firstly, large and intense natural disturbances are rare in the lowland humid tropics, and probably much rarer than in temperate and boreal forests. Thus, even when accounting for Landsat-based measurements of large disturbances and conservatively using a disturbance frequency/magnitude model fit that overestimates the frequency of large magnitude disturbances, it is clear that disturbances capable of removing 100 Mg above-ground biomass at the 1-ha scale (i.e. about one-third of total standing biomass) have return times of 1000 years or more in Amazonia (Gloor et al. 2009: Table 1). Furthermore, Gloor et al. (2009) used a stochastic simulator to show for South American forests that any sampling biases resulting from such a disturbance regime, given the sample sizes available in the RAINFOR network, are too small to explain the gains detected by the plot network. Recent, independent analyses using satellite data from across the Amazon basin (Espírito-Santo et al. 2010, 2014) show just how rare large, intense disturbances in fact are. Thus, the return time of stand-initiating scale disturbances in western Amazonia is c. 27,000 years, while in eastern Amazonia it is $\approx$90,000 years. The basin-wide mean, c. 39,000 years, is so large that any impacts on our Amazon dataset are negligible (and in light of these values it is hardly surprising that RAINFOR plots have yet to sample a single stand-resetting disturbance). This accords with the first pioneering large-scale analysis, which also showed the rarity of large-scale disturbance events in the Amazon basin (Nelson et al. 1994). The 'Slow-in, rapid-out' debate was magnified by a theoretical paper which attempted to simulate its impacts (Fisher et al. 2008), but which parameterised the size-frequency distribution of disturbance events incorrectly—thus overestimating the frequency of large disturbance events (c.f. Lloyd et al. 2009)—, and then managed to compare single-year time-step simulations with actual RAINFOR results which averaged intervals of 10 years. These large errors both exaggerated the apparent magnitude of the 'Slow-in, rapid-out' effect. A recent analysis of the entire Amazon disturbance intensity/frequency spectrum using LiDAR, Landsat, and multiple on-the-ground measurements has shown quite clearly that the large majority of tree death in Amazonia occurs in very small events (one, two, or three trees) and that large disturbances are much too rare to affect the inference from the plot network of a sustained, widespread biomass carbon sink into mature Amazon forests (Espírito-Santo et al. 2014).

Secondly, the RAINFOR network was successfully utilised to detect the impact of a major disturbance (the 2005 Amazon drought; see below) and to differentiate its dynamic and floristic effects from the background state of long-term biomass accumulation. This biomass decline was in fact dominated by a clearly detectable increase in mortality (Phillips et al. 2009). Thus, if there was a dominating impact of very large, late-twentieth century disturbance events on Amazon forests, for example associated with intense El Niño or La Niña years, these should have been detected. It has been suggested that Amazonia may still be recovering from earlier potentially greater 'mega-disturbances', such as the 1926 drought, or even widespread deforestation pre-1492. While such ideas can be appealing, simple back-of-the-envelope calculations show that for the forest to still be recovering at the end of the twentieth century at the rate of hundreds of millions Mg C year^{-1}, then the carbon lost to the atmosphere at the time needs to be in the order of tens of billions of tonnes—or more—well beyond anything recorded in the global atmospheric record of CO_2. We note that while the 2005 impact was large in magnitude at the scale of 1–2 years, its impacts were in the order of one to two billion tonnes of carbon, much too weak to shut down the sink even for a decade.

Thirdly, our plots lack the basic ecological signatures of forests recovering from large disturbances. Thus, the biomass increase has been only one of several changes recorded in Amazonian forest plots. Across 91 RAINFOR plots where we tracked populations to 2002, there was also a small increase in the stand density between the first and last measurements, of 0.84 ± 0.77 stems ha^{-1} $year^{-1}$, an annual increase of $0.15 \pm 0.13\,\%$ (Phillips et al. 2004). The same test using a longer-term subset of plots (50 plots from Lewis et al. 2004b) showed a slightly larger increase ($0.18 \pm 0.12\,\%$ per year). These increases in stand density, while proportionally smaller than the biomass changes, are counter to expectations if the plots were in an advanced state of secondary succession (e.g. Coomes and Allen 2007), as do the simultaneous increases in tree growth rates (see below). In Africa stand density changes have yet to be evaluated, but in both Africa and Amazonia there has been no shift in species composition towards more shade-tolerant taxa that would occur in a domain that was recovering from past disturbance events (e.g. Lewis et al. 2009a; Phillips et al. 2009), nor has there been on average over the networks the clear decline in growth and slowing stem dynamics that late-recovery tropical forests exhibit (Chambers et al. 2004). In sum, analysis of other structural, dynamic, and floristic change in the same plots is not consistent with a widespread disturbance-recovery signature. These results argue against the notion that the generalised biomass increase observed across Amazon and African plots can be explained as a result of a combination of disturbance recovery and small sample sizes.

Lastly, independent evidence from a network of large plots (Chave et al. 2008a), from atmospheric CO_2 data (Denman et al. 2007; Stephens et al. 2007), and from carbon mass-balance approaches (Le Quéré et al. 2009) all imply a carbon sink in tropical forests (c.f. Phillips and Lewis 2014). Parsimony therefore suggests that the increase in biomass is not the result of a statistical artefact based on frequent forest disturbance episodes that have been poorly sampled.

The plots in the long-term monitoring networks are, however, not randomly distributed. It is possible to test whether this spatial bias might be driving the result by assessing whether we have oversampled unusually heavily in regions that happened to be gaining biomass, and under-sampled those that happened to lose biomass. At smaller scales this is unlikely, since the long-term mean net gain in Amazonia is almost identical whether the sampling unit is taken to be the ‘plot’ (as here), or a larger unit such as a ‘landscape cluster of plots’ in both Amazonia and Africa (Phillips et al. 2009; Lewis et al. 2004b, 2009a). At larger scales while the networks still leave large expanses of Brazilian Amazonia and the Central Congo Basin unmonitored (Fig. 10.1), the climate- and soil-environmental space is well covered (Fig. 10.2).

Looking forward, much greater monitoring efforts in the vast still difficult-to-access regions of Amazonia are clearly needed in the future to reduce the uncertainty due to incomplete spatial coverage. One strategy would be to purposefully try to fill the large spatial gaps. Alternatively/additionally, a systematic grid-based approach is also appealing and is being planned by some national-level inventory programmes. The access challenges in many tropical forests for once-off inventory, let alone for sustaining multiple high-quality remeasurements in remote locations, can hardly be overstated however. We expect that the most cost-effective strategy for monitoring the more remote remaining tropical forests would combine (1) - gap-filling the monitoring networks where possible—with locally randomised plots—with (2) extensive remote sensing (viz. LiDAR, radar). The potential power of remote sensing techniques for scaling-up is abundantly clear. The need for careful ground-based assessments to permit calibrating and validating forests' electromagnetic reflectance in terms of productivity, biomass, and biodiversity is equally obvious.

10.3.3 Dynamic Changes

An alternative way of examining forest change is to look for changes in the processes (growth, recruitment, death), as well as the structure: have these forests simply gained mass, or have they become more or less dynamic too? For Amazonia we have measured the dynamics of forests in two ways. Firstly, we examined changes in stem population dynamics. By convention we estimated stem turnover between any two censuses as the mean of annual mortality and recruitment rates for the population of trees $\geq$10 cm diameter (Phillips and Gentry 1994). Secondly, we examined changes in biomass fluxes of the forest—in terms of growth of trees and the biomass lost with mortality events. These stand-level rates of biomass growth and biomass loss should be approximately proportional to the rate at which surviving and recruiting trees gain basal area and the rate at which basal area is lost from the stand through tree death (Phillips et al. 1994).

Among 50 mature forest plots across tropical South America with at least three censuses to 2002 (and therefore at least two consecutive monitoring periods that can be compared), we found that all of these key ecosystem processes—stem recruitment, mortality, and turnover, and biomass growth, loss, and turnover—increased significantly (Fig. 10.4) when the first monitoring period is compared with the second (Lewis et al. 2004b). Thus, over the 1980s and 1990s these forests on average became faster growing and more dynamic, as well as bigger. The increases in the rate of the dynamic stem fluxes (growth, recruitment, and mortality) were about an order of magnitude greater than the increases in the structural pools (above-ground biomass and stem density; Lewis et al. 2004b).

These and similar results can be demonstrated graphically in a number of ways. In Fig. 10.5, we plot the across-site mean values for stem recruitment and mortality as a function of calendar year. The increase is evidently not the short-term result of a year with unusual weather: recruitment rates on average consistently exceeded mortality rates, and mortality appears to have lagged recruitment (Phillips et al. 2004; Lewis et al. 2004b).

For the 50 Amazon plots which have two consecutive census intervals, we can separate them into two groups, one fast-growing and more dynamic (mostly in western Amazonia), and one slow-growing and much less dynamic (mostly in eastern and central Amazonia), which reflects the dominant macroecological

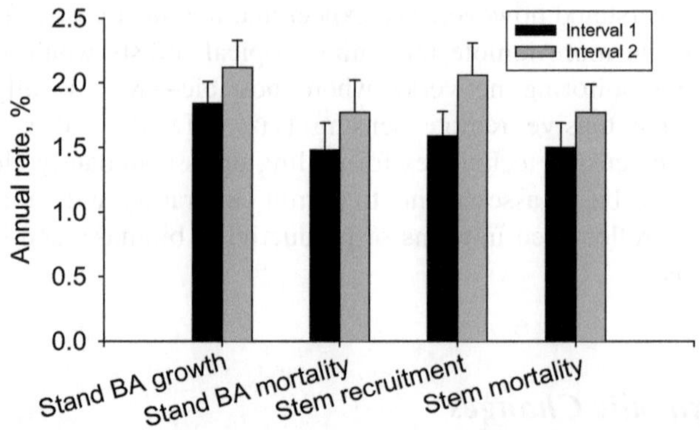

Fig. 10.4 Annualised rates of stand-level basal-area growth, basal-area mortality, stem recruitment, and stem mortality from plots with two consecutive census intervals, each giving the mean from 50 plots with 95 % confidence intervals. Paired *t*-tests show that all of the increases are significant. The average mid-year of the first and second censuses was 1989 and 1996, respectively (from Lewis et al. 2004b)

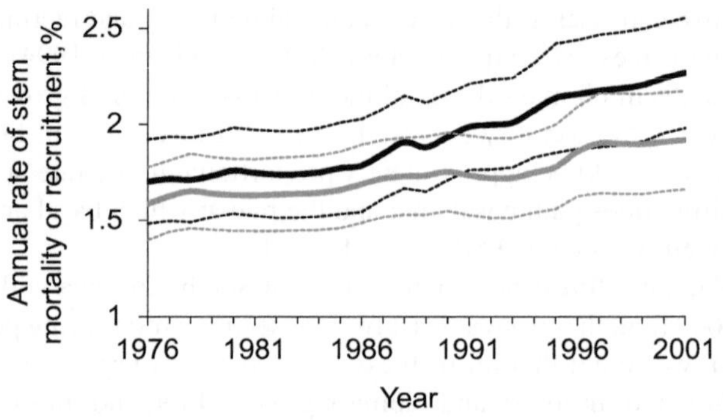

Fig. 10.5 Mean and 95 % confidence intervals for stem recruitment and mortality rates against calendar year, for plots arrayed across Amazonia. Rates for each plot were corrected for the effects of differing census-interval lengths, for 'site-switching' (changes through time in the plots being measured), and for 'majestic-forest bias' (potential avoiding of gaps when establishing plots). A detailed justification methodology for these corrections is given in Phillips et al. (2004); all trends hold if these corrections are not applied; *black* indicates recruitment, *grey* indicates mortality, *solid lines* are means, and *dots* are 95 % confidence intervals (from Phillips et al. 2004)

gradient across Amazonia (Phillips et al. 2004; ter Steege et al. 2006; Quesada et al. 2012). Both groups showed increased stem recruitment, stem mortality, stand basal-area growth, and stand basal-area mortality, with greater absolute increases in rates in the faster-growing and more dynamic sites than in the slower-growing and less dynamic sites (Fig. 10.6; Lewis et al. 2004b), but proportional increases in rates that were similar and statistically indistinguishable among forest types (Lewis et al. 2004b). It should be stressed that these results represent the mean response of all mature forests measured. Within the dataset naturally there are many

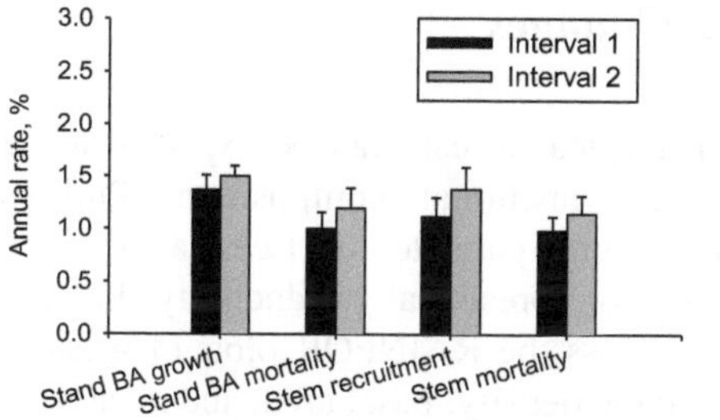

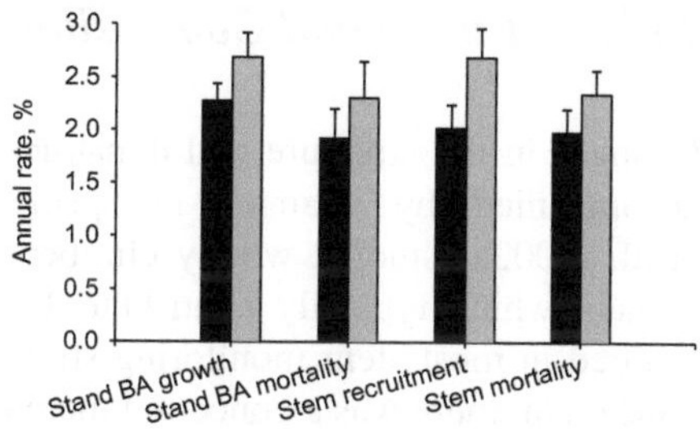

Fig. 10.6 Annualised rates of stand-level basal-area growth, basal-area mortality, stem recruitment, and stem mortality over consecutive census intervals for plots grouped into 'slower-growing less-dynamic' (*left*) and 'faster-growing more-dynamic' (*right*) forests. Of the slower-dynamics group, 20 of 24 plots are from eastern and central Amazonia, whereas just two are from western Amazonia. Of the faster-dynamics group, 24 of 26 plots are from western Amazonia, with just one from central Amazonia. The remaining three plots are from Venezuela and outside the Amazon basin. Changes have occurred across the South American continent and in both slower- and faster-dynamic forests (from Lewis et al. 2004b)

individual plots showing different, individual responses, just as within the whole literature there are some reports of individual sites showing similar or different patterns (e.g. Chave et al. 2008b; Feeley et al. 2007). Nevertheless, when viewed as whole the permanent plot record from Amazon and neotropical mature forests indicates that increasing growth, recruitment, and mortality occurred for at least two decades across different forest types and geographically widespread areas.

The simultaneous recent increases in plot dynamic rates, biomass, and stand density raise the question—for how long has this been going on? Only a handful of Amazon plots were monitored before the 1980s. To go further back in time requires alternative methods, for example annual dating of growth rates of a large sample of individual trees from different species, such as has been done in two locations in non-flooded mature forest (Vieira et al. 2005), using radiocarbon dating. Although the majority of trees tested did grow faster since 1960 than before 1960, the null hypothesis of no change in growth rate could not be rejected. This technique is complicated by potential ontogenetic variation in growth rates, partly related to changing light environments (e.g. Worbes 1999), and could overestimate stand-level growth rates in the past because individual trees with slow and declining growth are more susceptible to mortality (Chao et al. 2008) and therefore less likely to survive to the point at which they are dated. Similarly, as trees mature and increasingly allocate resources to flower, fruit, and seed production, ageing cohorts can exhibit slowing growth over time.

An alternative approach has been to analyse multiple dated herbarium samples stretching back to the nineteenth century for $\delta^{13}C$, ^{18}O, and stomatal density to assess possible changes in photosynthesis (and by implication, growth). For two species from the Guiana Shield, both showed implied increased photosynthesis over the past century (Bonal et al. 2011). A third approach—using tree rings to project growth of some tree species back in time—suffers from similar biases as the radiocarbon studies.

10.3.4 Functional Compositional Changes

Changes in the structure and dynamics of tropical forests can be expected to be accompanied by changes in species and functional composition. Phillips et al. (2002a) studied woody climbers (structural parasites on trees, also called lianas), which typically contribute 10–30 % of forest leaf productivity, but are ignored in most stem monitoring studies. Across the RAINFOR plots of western Amazonia, there was a concerted increase in the density, basal area, and mean size of lianas (Fig. 10.7; Phillips et al. 2002b). Over the last two decades of the twentieth century, the density of large lianas relative to trees increased here by 1.7–4.6 % per year—i.e. roughly doubling over the period, albeit from a low base. This was the first direct evidence that mature tropical forests are changing in terms of their life form composition. Other scattered reports since have suggested that the phenomenon of increasing lianas extended across other neotropical forests too (reviewed by Schnitzer and Bongers 2011). There is some limited experimental evidence (Granados and Körner 2002) for growth responses in tropical lianas to elevated atmospheric CO_2 concentrations to be stronger than those of trees.

Finally, a handful of studies have considered whether there have been consistent changes in tree species composition in forests over the past two decades (Laurance et al. 2004; Chave et al. 2008a; Lewis et al. 2009a; Butt et al. 2012). In the first, on a large cluster of plots in north of Manaus, many faster-growing genera of canopy and emergent stature trees increased in basal area or density, whereas some slower-growing genera of subcanopy or understory trees declined. Laurance et al. (2004) provided evidence of pervasive changes in this locality: growth, mortality, and recruitment all increased significantly over two decades (total basal area also increased, but not significantly so), with faster-growing genera showing larger increases in growth, relative to slower-growing genera. Further studies are needed to determine whether comparable shifts in tree communities are occurring throughout Amazonia.

Second, in Africa (see Box) wood density of a taxon has no predictive value of its change through time (Lewis et al. 2009a). Thirdly, Chave et al. (2008a) reported functional changes across ten forest plots across the tropics, by grouping species into quartiles based on growth rate, wood density, seed size, and maximum plant size. On an absolute basis, there were significant increases in biomass of the fastest- and slowest-growing quartiles of species, no significant change in biomass of the highest and lowest quartiles based on wood density, a significant increase in absolute biomass of the quartile with the smallest seed size, and no changes in biomass of quartiles based on maximum tree size. However, if these shifts are calculated relative to the changes in biomass of the stand, only one result is significant: the largest trees significantly decreased in biomass relative to the stand, whereas the smallest trees showed no relative change. Collectively, results from these three studies suggest that the increase in forest stand biomass is being caused by concurrent increases of many species with differing ecological habits.

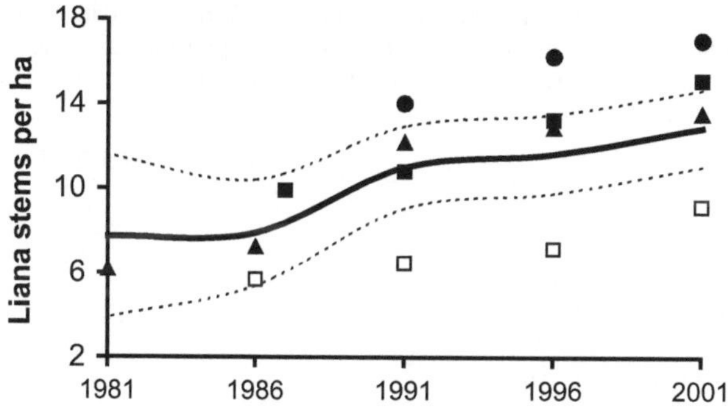

Fig. 10.7 Five-year running means (*solid line*) with 95 % confidence intervals (*dashed lines*) of liana stem density per hectare (≥10 cm diameter at breast height), with values plotted separately for northern Peru (*filled squares*), southern Peru (*filled triangles*), Bolivia (*filled circle*), and Ecuador (*unfilled squares*) (adapted from Phillips et al. 2002b; see that paper for full details of field and analytical methodology)

10.3.5 Recent Drought Impacts in Amazonia

The Amazon results discussed so far reflect forest changes up to the early part of the first decade of the twenty-first century. In 2005, the region was struck by a major drought, with the unusual characteristic of being driven by strong warming in the tropical north Atlantic, a feature that appears in some global circulation model outcomes of projected climate change (e.g. Cox et al. 2004). With the RAINFOR network largely in place and a good forest dynamics baseline established, we had an opportunity to use this 'natural experiment' to assay the sensitivity of the largest tropical forest to an intense, short-term drought, by rapidly re-censusing plots across the basin. Of 55 plots which we surveyed in 2005, the mean annual above-ground biomass change was −0.59 (−1.66, +0.35) Mg ha^{-1}, and among those plots that were actually impacted by drought the above-ground biomass change rate was clearly negative [−1.62 (−3.16, −0.54) Mg ha^{-1}]. Moreover, across the measured plots the magnitude of the biomass change anomaly was closely correlated to the magnitude of the moisture deficit anomaly experienced in the same period, with most of the difference being related to increased mortality, implying that it was the unusual moisture deficits that were responsible for the biomass loss by contributing to an enhanced mortality.

We estimated the basin-wide impact of the drought on biomass carbon, as compared to the baseline of a net biomass sink in pre-drought measurement period, as between −1.21 (−2.01, −0.57) Pg C and −1.60 Pg C (−2.63, −0.83) (Tables 10.2 and 10.3). The first value is based on scaling the per-plot impact by the total area impacted by drought; the second greater value is based on using remotely sensed rainfall data to scale from the relationship of biomass change data with relative drought intensity. The biomass dynamics/climate approach makes use of more of the information on forest response to drought than the simple scaling-up of the mean drought-impacted plot effect, but requires additional assumptions

Table 10.2 Estimated 2005 Amazon drought impact, from plot data, using two different approaches. (A) Scaling from the statistical distribution of plot biomass change data (sampling effort-corrected AGB change relative to pre-2005 for each plot) to the whole area of Amazon forest affected in 2005

			Mean	2.5 % CI	97.5 % CI
		Mg AGB ha^{-1} a^{-1}	*−2.39*	*−1.12*	*−3.97*
	Expansion factor				
Mean 2005 interval length, years	*1.97*				
Smaller trees and lianas	*1.099*				
Droughted forest area, ha (TRMM data)	*3.31 × 10^8*				
Additional fraction of Amazonia with unreliable TRMM data	*1.031*				
Sum AGB impact, Mg			-1.76×10^9	-8.25×10^8	-2.93×10^9
Below ground	*1.37*		-6.52×10^8	-3.05×10^8	-1.08×10^9
Sum biomass impact, Mg			-2.41×10^9	-1.13×10^9	-4.01×10^9
Sum carbon impact, Mg	***0.5***		**-1.21×10^9**	**-0.57×10^9**	**-2.01×10^9**

For details of the methodology, see online supplementary information published with Phillips et al. (2009)

*The text in italics and bold italics signify 95% confidence intervals

Table 10.3 Estimated 2005 Amazon drought impact, from plot data, using two different approaches. (B) Scaling from the plot biomass dynamics versus climate relationship to the whole Amazon forest area affected in 2005, using the relationship between change in mean annual maximum cumulative water deficit values and change in biomass dynamics

			Mean	2.5 % CI	97.5 % CI
		Mg AGB	*-1.05×10^{-9}*	*-5.46×10^{-8}*	*-1.72×10^{-9}*
	Expansion factor				
Mean 2005 interval length, years	*1.97*				
Additional proportion of Amazonia with unreliable TRMM data	*1.031*				
Smaller trees and lianas	*1.099*				
Sum AGB impact, Mg			-2.33×10^{-9}	-1.22×10^{-9}	-3.83×10^{-9}
Below ground, Mg	*1.37*		-8.63×10^{-8}	-4.51×10^{-8}	-1.42×10^{-9}
Sum biomass impact, Mg			-3.20×10^{-9}	-1.67×10^{-9}	-5.25×10^{-9}
Sum carbon impact, Mg	***0.5***		**-1.60×10^{-9}**	**-0.83×10^{-8}**	**-2.63×10^{-9}**

For details of the methodology, see online supplementary information published with Phillips et al. (2009)

*The text in italics and bold italics signify 95% confidence intervals

which may introduce error. The consistency of the results from the different analyses indicated a significant regional impact (confirmed now by new, independent analyses, e.g. Gatti et al. 2014; Grace et al. 2014), but much of this might not be 'seen' by the atmosphere until future years. Thus, the main impact was a temporary increase in dead wood production, implying losses to the atmosphere over future years as these dead trees decompose. In the drought year itself the sink may decline (reduction in growth) but not halt (as the new necromass only begins to decompose). Potentially, the sink could even increase temporarily if potential short-term reductions in soil respiration are allowed for, although aircraft measurements of atmospheric CO_2 concentrations show that in the 2010 drought the overall effect was indeed to completely suppress the long-term biomass sink for a year (Gatti et al. 2014).

Regardless of these details, the total committed carbon impact of the 2005 drought exceeds the annual net C emissions due to land use change across the Neotropics (0.5–0.7 Pg C) (Pan et al. 2011, Figure 1). By combining results from 2005 with published and unpublished information on tropical tree mortality from elsewhere, we have been able to extend the drought–mortality response relationship further. This second analysis (Phillips et al. 2010) suggested that across the biome, forest sensitivity to moisture anomalies may be predictable, and that even relatively weak drying compared to normal climatology can cause excess deaths once adequate sampling is in place to detect them. However, it should be noted that these are the impacts of short-term intense drought events. Recent analyses of 19 long-term plots from West Africa, which has seen a decades-long drying, saw a net increase in biomass coupled with strong increase in dry-adapted species over a c. 20-year period of monitoring (Fauset et al. 2012), similar to preliminary findings from western Amazonia (Butt et al. 2012). This might reflect the fact that the more marginal forest climate in West Africa and Pleistocene history of dry episodes may have already reduced the diversity and importance of wet forest taxa (Parmentier et al. 2007). Fuller understanding of the impacts of drought will require monitoring of forests through post-drought recovery and repeated droughts (such as occurred also in Amazonia in 2010) over the long term.

10.3.6 What is Driving these Changes?

What could have caused the continent-wide increases in tree growth, recruitment, mortality, stem density, and biomass? Many factors could be invoked, but there is only one parsimonious explanation for the pre-2005 pattern. The results appear to show a coherent fingerprint of increasing growth [i.e. increasing net primary productivity (NPP)] across tropical South America, probably caused by a long-term increase in resource availability (Lewis et al. 2004a, b; 2009a, b). According to this explanation, increasing resource availability increases NPP, which then increases stem growth rates. This accounts for the increase in stand basal-area growth and stem recruitment rates and the fact that these show the clearest, most highly significant changes (Lewis et al. 2004b). Because of increased growth, competition for limiting resources, such as light, water, and nutrients, increases. Over time some of the faster-growing, larger trees die, as do some of the 'extra' recruits, as the accelerated growth percolates through the system. This accounts for the increased losses from the system: biomass mortality and stem mortality rates increase. Thus, the system gains biomass and stems, while the losses lag some years behind, causing an increase in above-ground biomass and stems. Overall, this suite of changes may be qualitatively explained by a long-term increase in a limiting resource.

The changes in composition may also be related to increasing resource availability, as the rise in liana density may be either a direct response to rising resource supply rates or a response to greater disturbance caused by higher tree mortality rates. The changing tree composition in central-Amazonian plots (Laurance et al. 2004) is also consistent with increasing resource supply rates, as experiments show that faster-growing species are often the most responsive, in absolute terms, to increases in resource levels (Coomes and Grubb 2000). Others, however, have argued (e.g. Körner 2004; Lloyd pers. comm.) that the greatest proportional response should be in understory seedlings and saplings for whom a small increase in photosynthetic rate here could have a great proportional impact on carbon balance—and there is some experimental evidence to support this view (e.g. Kerstiens 2001; Aidar et al. 2002).

What environmental changes could increase the growth and productivity of tropical forests? While there have been widespread changes in the physical, chemical, and biological environment of tropical trees (Lewis et al. 2004a), only increasing atmospheric CO_2 concentrations, increasing solar radiation inputs (Wong et al. 2006), rising air temperatures, and changing precipitation patterns (Trenberth et al. 2007) have been documented across most or all of Amazonia over the relevant time period and could be responsible for increased growth and productivity. Additionally, it is conceivable that nutrient inputs have increased, first, from biomass burning that is occurring closer to once-remote tropical forest plots that are increasingly encroached upon by deforestation (Laurance 2004), and second, long-range inputs of Saharan dust to Amazonia (and west African forests) have increased over recent decades, possibly in response to climate change.

Yet for only one of these changes do we have clear evidence that the driver has both certainly changed over a large enough area and that such a change is likely to accelerate forest growth (Lewis et al. 2004a, 2009b). The increase in atmospheric CO_2 is the primary candidate, because of the undisputed long-term increase in CO_2 concentrations, the key role of CO_2 in photosynthesis, and the demonstrated positive effects of CO_2 fertilisation on plant growth rates, including experiments on forest stands, although not yet in the tropics (Norby et al. 2002; Hamilton et al. 2002; Lewis et al. 2004a, 2009b; Norby and Zak 2011). However, some role for increased insolation (e.g. Nemani et al. 2003; Ichii et al. 2005), or aerosol-induced increased diffuse fraction of radiation (e.g. Oliveira et al. 2007), or nutrient inputs, or rising temperatures increasing soil nutrient mineralisation rates, cannot be ruled out (Lewis et al. 2004a, 2006; 2009b; Malhi and Phillips 2004, 2005). Lastly, given the global nature of the CO_2 increase and ubiquitous biochemistry of the plant response involved, we may expect to see the same phenomenon in other biomes. Similarly to the tropics, increases in biomass and/or growth have recently been reported in maritime forests of western Canada (Hember et al. 2012), and across the temperate forests of the northern hemisphere (Luyssaert et al. 2008), and indeed on every continent where foresters are making measurements in multiple sites (Pan et al. 2011).

10.3.7 The Future: Potential Susceptibility of Amazon Forest to Environmental Stress and Compositional Changes

Long-term observations indicate that Amazonia, the world's largest tract of tropical forest, has shown concerted changes in forest dynamics over recent decades. Such unexpected and rapid alterations—regardless of the cause—were not anticipated by ecologists and raise concerns about other possible surprises that might arise as global changes accelerate in coming decades. On current evidence, tropical forests are sensitive to changes in resource levels and may show further structural and dynamic changes in the future, as resource levels alter further, temperatures continue to rise, and precipitation patterns shift. The implications of such rapid changes for the world's most biodiverse region could be substantial.

Mature Amazonian forests have evidently helped to slow the rate at which CO_2 has accumulated in the atmosphere, thereby acting as buffer to global climate change. The concentration of atmospheric CO_2 has risen recently at an annual rate equivalent to ~4 Pg C; this would have been significantly greater without the tropical South American biomass carbon sink of 0.4–0.7 Pg C $year^{-1}$ (and an African sink of 0.3–0.5 Pg C $year^{-1}$). This subsidy from nature could be a relatively short-lived phenomenon. Mature Amazonian forests may (i) continue to be a carbon sink for decades (e.g. Chambers et al. 2001b; Cramer et al. 2001; Rammig et al. 2010), (ii) soon become neutral or a small carbon source (Cramer et al. 2001; Phillips et al. 2002b; Körner 2004; Laurance et al. 2004; Lewis et al. 2011), or (iii) become a mega-carbon source (Cox et al. 2000; Cramer et al. 2001; Rammig et al. 2010; Galbraith et al. 2010), with modelling results reporting all three responses following a major model-inter-comparison project (Friedlingstein et al. 2006). Given that a 0.3 % annual increase in Amazonian forest biomass roughly compensates for the entire fossil-fuel emissions of Western Europe (or the deforestation in Amazonia), a switch of mature tropical forests from a moderate carbon sink to even a moderate carbon source would impact on global climate and human welfare. The ~0.3 % annual increase in carbon storage represents the difference between two much larger values: stand-level growth (averaging ~2 %) and mortality (averaging ~1.7 %), so a small decrease in growth or a sustained increase in mortality would be enough to shut the sink down. There are several mechanisms by which such a switch could occur, apart from the obvious and immediate threats posed by land use change and associated disturbances by fragmentation and fire, which we discuss below.

10.3.7.1 Moisture Stress

Anthropogenic climate change will include altered precipitation patterns. There are critical thresholds of water availability below which tropical forests cannot persist and are replaced by savanna systems; currently, as long as soil conditions are adequate, the threshold lies around 1000–1500 mm rainfall per annum (Salzmann and Hoelzmann 2005; Staver et al. 2011), but this level could increase with rising temperatures which increase evaporation, or it could decrease if rising atmospheric CO_2 concentrations reduce transpiration. The outcome of the interplay between these factors is therefore critical to determining transitions between carbon-dense tropical forests and carbon-light savanna systems. The degree to which tropical forests may be ecophysiologically resilient to extreme temperatures, particularly in the context of rising atmospheric CO_2 concentration, is a subject of active research, reviewed elsewhere (Lloyd and Farquhar 2008; Lewis et al. 2009b), and explored on a biome-wide scale by Zelazowski et al. (2011).

The 2005 drought provides direct evidence of the potential for intense dry periods to impact rainforest vegetation. However, while events such as the 2005 Amazon drought are clearly capable of at least temporarily disrupting some of the long-term trends in forest biomass, it remains to be seen whether they are powerful and frequent enough to permanently shift the dominant regime of biomass gains witnessed across mature tropical forests wherever they have been extensively monitored. The 1998 El Niño drought was equally strong in parts of Amazonia, but its impacts are not distinguishable from the signal of increased biomass and growth over the c. 5-year mean interval length available for plots at that time (fig. 1 in Phillips et al. 2009), implying a rapid recovery. We expect therefore that only frequent, multiple droughts would cause the sustained increases in necromass production needed to turn the long-term carbon sink in mature forest into a sustained source.

In 2010, a new drought affected the Amazon forest, again dropping some rivers to record lows. If the water deficit intensity–forest carbon loss relationship that we measured during 2005 also holds for the 2010 drought, then a total impact on mature forest biomass carbon in the region of c. 2 Pg can be anticipated (Lewis et al. 2011). It remains an open question as to how much the forest had recovered from the 2005 drought before the 2010 drought impacted the forest plots. Only coordinated monitoring with distributed networks of plots can determine whether the recent events represent one-off perturbations for forest carbon stocks from which a full recovery is made within 5 years, or the start of a longer-term climate-induced phase shift in which mature Amazonia becomes carbon neutral or a carbon source. A more recent, long-term analysis from a larger RAINFOR plot dataset (Brienen et al. 2015) has found evidence of a progressive decline in the net Amazon sink, in spite of the long-term growth gains. Recent droughts have played a part, but this enhanced dataset also indicates clearly that mortality has been increasing over a longer period, suggesting that other mechanisms are also constraining the carbon density of Amazon forests.

10.3.7.2 Photosynthesis/Respiration Changes

Forests remain a sink as long as carbon uptake associated with photosynthesis exceeds the losses from respiration. Under the simplest scenario of a steady rise in forest productivity over time, it is predicted that forests would remain a carbon sink for decades (e.g. Lloyd and Farquhar 1996). However, the recent increases in productivity, apparently caused by continuously improving conditions for tree growth, cannot continue indefinitely: if CO_2 is the cause, trees are likely to become CO_2 saturated (i.e. limited by another resource) at some point in the future. More generally, whatever the driver for recently accelerated growth, forest productivity will not increase indefinitely, as other factors, such as soil nutrients, will limit productivity.

Rising temperatures could also reduce the forest sink or cause forests to become a source in the future. Warmer temperatures increase the rates of virtually all chemical and biological processes in plants and soils, until temperatures reach inflection points where enzymes and membranes lose functionality. There is some evidence that the temperatures of leaves at the top of the canopy, on hot days, may be reaching such inflection points around midday at some locations (Lewis et al. 2004a; Doughty and Goulden 2008). Canopy-to-air vapour deficits and stomatal feedback effects may also be paramount in any response of tropical forest photosynthesis to future climate change (Lloyd et al. 1996). Simulations suggest that the indirect effect of rising temperatures on photosynthesis via stomatal closure is the dominant negative impact on tropical forest growth (Lloyd and Farquhar 2008), which is currently more than offset by increases in photosynthesis from increasing atmospheric CO_2. Alternatively, there is evidence that electron transport is the critical step in maximising photosynthesis at a given light level (Haxeltine and Prentice 1996), and the electron transport chain undergoes a reversible point of inflection as low as c. 37 °C (Lloyd and Farquhar 2008), thereby reducing photosynthesis at higher temperatures. Additionally, higher air temperature also means higher respiration costs which will also impact on the ability of plants to maintain a positive carbon balance in the future, which has been argued to be already affecting one forest in Costa Rica (Clark et al. 2010).

Understanding this complex relationship between temperature changes and their impacts on respiration and photosynthesis, plus the impact of rising atmospheric CO_2 on tree growth, is critical. The first global circulation model (GCM) to include dynamic vegetation and a carbon cycle that is responsive to these dynamic changes suggested that under the 'business as usual' scenario of emissions, IS92a, atmospheric CO_2 concentrations reach >900 ppmv (parts per million by volume) in 2100, compared to ~700 ppmv from previous GCMs (Cox et al. 2000, 2004). These concentrations depend on (1) dieback of the eastern Amazonian forests, caused by climate change-induced drought, and (2) the subsequent release of C from soils. The release of C from soils is critically dependent on the assumed response of respiration to temperature, coupled with the simplified representation of soil moisture and soil carbon. A decade on, both Galbraith et al. (2010) and Rammig et al. (2010) reanalysed the climate and dynamic vegetation models and found that rising air temperature was an important cause of dieback in most models. However, the dominant temperature-related mechanism differed among models, with, variously, increases in plant respiration, reduced photosynthesis, and increased vapour pressure deficit all resulting in loss of carbon. Yet, the level of the positive impact of CO_2 fertilisation had the largest single impact within the models, larger than the negative temperature and negative rainfall reduction impacts. In a recent review of free-air carbon dioxide enrichment (FACE) experiments, Norby and Zak (2011) conclude that photosynthetic C uptake increases in response to CO_2 under field conditions and that the enhancement is sustained over time. But they point out that it remains to be seen whether these growth-related responses also apply to tropical forests. Thus, overall, the uncertainties about (1) how much tropical plants will respond to CO_2, (2) how they will respond to long-term increases in high air temperatures, (3) how much rainfall may decline by in the dry season, and (4) how much plants will respond to the decline in rainfall still preclude robust statements about the timing and magnitude of any reversal of the tropical forest carbon sink.

To conclude, carbon losses from respiration will almost certainly increase as air temperatures continue to increase. The key question is what form this relationship takes. Carbon gains from photosynthesis cannot rise indefinitely and will almost certainly reach an asymptote. Thus, we conclude the sink in mature tropical forests is bound to diminish and possibly even reverse. The more catastrophic outcomes of large-scale biomass collapse indicated in some models appear very unlikely, but cannot be ruled out.

10.3.7.3 Compositional Change

Biodiversity change has inevitable consequences for climate change because different plant species vary in their ability to store and process carbon and different plant species will benefit and decline as global environmental changes unfold. Yet most models that project the future carbon balance in Amazonia (and future climate-change scenarios) make no allowance for changing forest composition. Representation of composition is challenging, both because of the computational complexities in integrating ecological processes into ecophysiology-driven models and because the ecological data themselves are sparse. Representing composition better, and its potential for change, is important. Lianas, for example, ignored in all forest models, often contribute little to forest biomass but heavily affect productivity (Schnitzer and Bongers 2002), while killing trees (Phillips et al. 2005) and preferentially infesting denser-wooded species (van der Heijden et al. 2008); their recent increase suggests that the tropical carbon sink might shut down sooner than models suggest. Large changes in tree communities could also lead to net losses of carbon from tropical forests (Phillips and Gentry 1994; Körner 2004). One way this could happen is a shift to faster-growing species, driven by increasing tree mortality rates and frequency of gap formation (Phillips and Gentry 1994; Phillips et al. 2004; Lewis 2006). Such fast-growing species generally have lower wood specific gravity, and hence less carbon (West et al. 1999), than shade-tolerant trees. The potential scope for such impacts of biodiversity changes on carbon storage is highlighted by Bunker et al. (2005), who explored various biodiversity scenarios based on the tree species at Barro Colorado Island: if slower-growing tree taxa were lost from an accelerated, liana-dominated forest, as much as one-third of the carbon storage capacity of the forest could be lost. In Amazonia a small and sustained basin-wide annual decrease in mean wood specific gravity could potentially cancel out the carbon sink effect. Currently, the more dynamic forests in the west of Amazonia have c. 20 % less dense wood than the slower-growing forests of the east (Baker et al. 2004b); because these faster-growing western forests also have lower basal area, the differences in terms of biomass carbon stored are somewhat greater still (Fig. 10.8). Concerted compositional changes driven by greater resource supply, increased mortality rates, and possible increases in the proportion

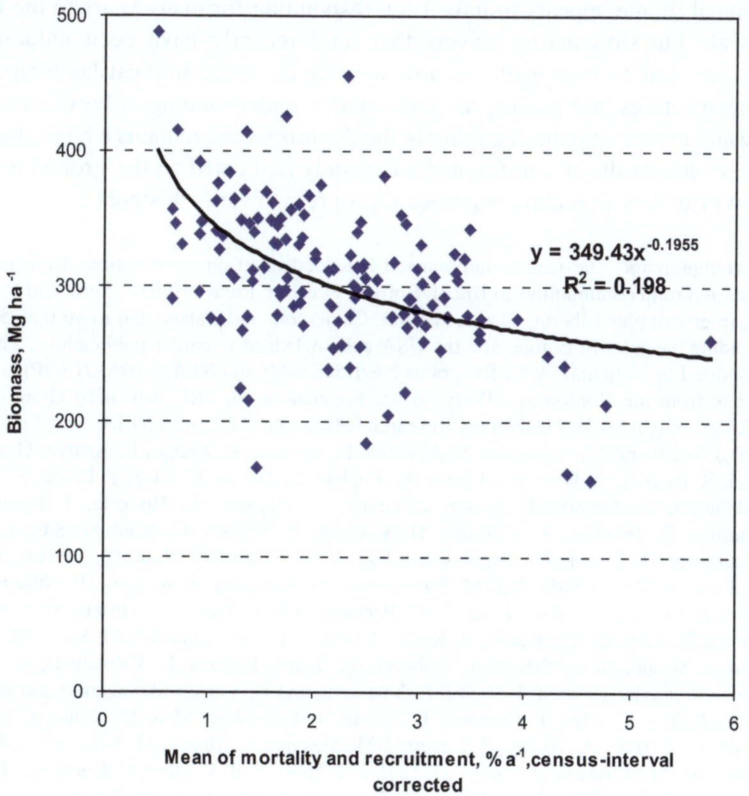

Fig. 10.8 Tree biomass as a function of mean stem turnover rates, for 127 lowland forest plots across South America monitored in the 1980s and 1990s. Note that the faster forests, typically in western Amazonia, have lower wood density and much lower biomass

of faster-growing trees which escape lianas could therefore shut down the carbon sink function of tropical forests earlier than ecophysiological analyses predict. While the initial moves towards individual-based models within Global Circulation Models provide a framework within which to evaluate these types of interaction and changing composition (Purves and Pacala 2008), any such analyses will need to be data driven.

10.4 Conclusion

By carefully tracking the lives, deaths, and identities of trees at hundreds of plots, it has been possible over the past three decades to build a preliminary understanding of how the world's mature tropical forests have been changing. The picture that emerges is at once both surprising and, perhaps for some, not so unexpected. Thus, in experiencing accelerated growth, mortality, and generally increasing biomass, the tropical biome appears to have been responding for many years to the kind of large-scale but slow-acting drivers that until recently have been unfamiliar to ecologists. New technologies are now helping us to see tropical landscapes with fresh perspectives, but gaining an authoritative understanding of how forest biodiversity and carbon may be changing in the Anthropocene remains a huge challenge. Repeated, standardised, careful, and adequately replicated on-the-ground measurements will be key to making significant progress towards this goal.

Acknowledgements The results summarised here depended on contributions from numerous assistants and rural communities in Brazil, Bolivia, Ecuador, French Guiana, Peru, and Venezuela (South America), plus Liberia, Ghana, Nigeria, Cameroon, and Gabon, and more than 50 grants from funding agencies in Europe and the USA acknowledged in earlier publications. This paper was supported in particular by NERC grants NE/B503384/1 and NE/D01025X/1 (OP), a grant to RAINFOR from the Gordon and Betty Moore Foundation, an ERC Advanced Grant (OP: 'T-FORCES'), a Royal Society University Research Fellowship (SL), and a NERC New Investigators grant (SL). We thank M. Alexiades, S. Almeida, E. Alvarez, A. Araujo, L. Arroyo, G. Aymard, D. Bonal, R. Brienen, S. Brown, J. Chave, K.-J, Chao, N. Dávila, C. Diaz, T. Erwin, E. Esparza, T. Feldpausch, A. Gentry, E. Gloor, J. Grace, N. Higuchi, E. Honorio, I. Huamantupa, N. Jaramillo, E. Jiménez, K. Johnson, H. Keeling, T. Killeen, C. Kuebler, S.G. Laurance, W.F. Laurance, J. Lloyd, G. Lopez-Gonzalez, A. Malhado, Y. Malhi, J. Martin, P. Meir, C. Mendoza, A. Monteagudo, H.E.M. Nascimento, D. Navarrete, D.A. Neill, P. Núñez Vargas, S. Patiño, J. Peacock, A. Peña Cruz, M.C. Peñuela, N.C.A. Pitman, A. Prieto, C.A. Quesada, F. Ramirez, H. Ramirez, J. Ricardo, R. Rojas, A. Rudas, L. Salorzano, J.N.M. Silva, M. Silveira, M. Stern, J. Stropp, H. ter Steege, J. Terborgh, A. Torres Lezama, L. Valenzuela, R. Vásquez Martínez, G. van der Heijden, I. Vieira, E. Vilanova, and B. Vinceti (Amazonia) and B. Sonké, K. Affum-Baffoe, L. Ojo, J. Reitsma, L. White, J. Comiskey, M-N Djuikouo, C. Ewango, A. Hamilton, T. Hart, A. Hladik, J. Lovett, J-M. Makana, F. Mbago, H. Ndangalasi, D. Sheil, T. Sunderland, M.D. Swaine, J. Taplin, D. Taylor, S. Thomas, R. Voytere, H Woell, A. Moungazi, S. Mbadinga, H. Bourobou, L. N. Banak, T. Nzebi, K. Jeffery, K. Ntim, K. Opoku, T. Tafoek, S. Grahame, R. Lowe, L. Banin, K. C. Nguembou, R. Condit, C. Chatelain, and L. Poorter, for contributing their effort, data, and/or discussions including to earlier papers on which this overview is partly based. We thank the editor and an anonymous reviewer for helpful comments which improved the manuscript.

References

Aidar MP, Martinez CA, Costa AC, Costa PMF, Dietrich SMC, Buckeridge MS (2002) Effect of atmospheric CO_2 enrichment on the establishment of seedlings of Jatobá, *Hymenaea courbaril* L. (Leguminosae, Caesalpinioideae). Biota Neotropica 2:BN01602012002

Aragão LEOC, Malhi Y, Roman-Cuesta RM, Saatchi S, Anderson LO, Shimabukuro YE (2007) Spatial patterns and fire response of recent Amazonian droughts. Geophys Res Lett 34. doi:10.1029/2006GL028946

Baker TR, Phillips OL, Malhi Y, Almeida S, Arroyo L, Di Fiore T, Higuchi N, Killeen T, Laurance SG, Laurance WF, Lewis SL, Monteagudo A, Neill DA, Pitman NCA, Silva JNM, Martínez RV (2004a) Increasing biomass in Amazonian forest plots. Philos Trans R Soc Ser B 359:353–365

Baker TR, Phillips OL, Malhi Y, Almeida S, Arroyo L, Di Fiore T, Killeen T, Laurance SG, Laurance WL, Lewis SL, Lloyd J, Monteagudo A, Neill D, Patiño S, Pitman N, Silva JN, Martínez RV (2004b) Variation in wood density determines spatial patterns in Amazonian forest biomass. Glob Chang Biol 10:545–562

Beer C, Reichstein M, Tomelleri E, Ciais P, Jung M, Carvalhais N, Rödenbeck C, Arain M, Baldocchi D, Bonan G, Bondeau A, Cescatti A, Lasslop G, Lindroth A, Lomas M, Luyssaert S, Margolis H, Oleson K, Roupsard O, Veenendaal E, Viovy N, Williams C, Woodward FI, Papale D (2010) Terrestrial gross carbon dioxide uptake: global distribution and covariation with climate. Science 329:834–838

Bonal D, Ponton S, Le Thiec D, Ningre N, Herault B, Ogee J, Gonzalez S, Pignal M, Sabatier D, Guehl J-M (2011) Leaf functional response to increasing atmospheric CO_2 concentrations over the last century in two northern Amazonian tree species: a historical delta ^{13}C and delta ^{18}O approach using herbarium samples. Plant Cell Environ 34:1332–1344

Brienen RJW, Phillips OL, Feldpausch TR, Gloor E, Baker TR, Lloyd J, Lopez-Gonzalez G, Monteagudo-Mendoza A, Malhi Y, Lewis SL, Vásquez Martinez R, Alexiades M, Álvarez Dávila E, Alvarez-Loayza P, Andrade A, Aragão LEOC, Araujo-Murakami A, Arets EJMM, Arroyo L, Bánki OS, Baraloto C, Barroso J, Bonal D, Boot RGA, Camargo JLC, Castilho CV, Chama V, Chao KJ, Chave J, Comiskey JA, Cornejo Valverde F, da Costa L, de Oliveira EA, Di Fiore A, Erwin TL, Fauset S, Forsthofer M, Galbraith DR, Grahame ES, Groot N, Hérault B, Higuchi N, Honorio Coronado EN, Keeling H, Killeen TJ, Laurance WF, Laurance S, Licona J, Magnussen WE, Marimon BS, Marimon-Junior BH, Mendoza C, Neill DA, Nogueira EM, Núñez P, Pallqui Camacho NC, Parada A, Pardo-Molina G, Peacock J, Peña-Claros M, Pickavance GC, Pitman NCA, Poorter L, Prieto A, Quesada CA, Ramírez F, Ramírez-Angulo H, Restrepo Z, Roopsind A, Rudas A, Salomão RP, Schwarz M, Silva N, Silva-Espejo JE, Silveira M, Stropp J, Talbot J, ter Steege H, Teran-Aguilar J, Terborgh J, Thomas-Caesar R, Toledo M, Torello-Raventos M, Umetsu RK, van der Heijden GMF, van der Hout P, Guimarães Vieira IC, Vieira SA, Vilanova E, Vos VA, Zagt RJ (2015) Long-term decline of the Amazon carbon sink. Nature 519:344–348

Bunker D, De Clerck F, Bradford J, Colwell R, Garden P, Perfecto I, Phillips OL, Sankaran M, Naeem S (2005) Carbon sequestration and biodiversity loss in a tropical forest. Science 310:1029–1031

Butt N, Malhi Y, New M, Maciá MJ, Lewis SL, Lopez-Gonzalez G, Laurance WF, Laurance S, Luizão R, Andrade A, Baker TR, Phillips OL (2012) Shifting dynamics of climate-functional groups in old-growth Amazonian forests. Plant Ecol Divers 7:267–279

Chambers JQ, Santos J, Ribeiro RJ, Higuchi N (2001a) Tree damage, allometric relationships, and above-ground net primary production in central Amazon forest. For Ecol Manage 152:73–84

Chambers JQ, Higuchi N, Tribuzy ES, Trumbore SE (2001b) Carbon sink for a century. Nature 410:429

Chambers JQ, Higuchi N, Teixeira LM, dos Santos J, Laurance SG, Trumbore SE (2004) Response of tree biomass and wood litter to disturbance in a Central Amazon forest. Oecologia 141:596–611

Chao K-J, Phillips OL, Gloor E, Monteagudo A, Torres-Lezama A, Vásquez Martínez R (2008) Growth and wood density predicts tree mortality in Amazon forests. J Ecol 96:281–292
Chave J et al (2003) Error propagation and scaling for tropical forest biomass estimates. Philos Trans R Soc Ser B 359:409–420
Chave J, Andalo C, Brown S, Cairns MA, Chambers JQ, Eamus D, Folster H, Fromard F, Higuchi N, Kira T, Lescure JP, Nelson BW, Ogawa H, Puig H, Riera B, Yamakura T (2005) Tree allometry and improved estimation of carbon stocks and balance in tropical forests. Oecologia 145:87–99
Chave J, Condit R, Muller-Landau HC, Thomas S, Ashton PS, Bunyavejchewin S, Co L, Dattaraja HS, Davies SJ, Esufali S, Ewango CEN, Feeley K, Foster RB, Gunatilleke N, Gunatilleke S, Hall P, Hart TB, Hernández C, Hubbell SP, Itoh A, Kiratiprayoon S, LaFrankie J, Loo de Lao S, Makana J-R, Supardi MNN, Rahman Kassim A, Samper C, Sukumar R, Suresh H, Tan S, Thompson J, Tongco M, Valencia R, Vallejo M, Villa G, Yamakura T, Zimmerman JK, Losos E (2008a) Assessing evidence for a pervasive alteration in tropical tree communities. PLoS Biol 6:e45
Chave J, Olivier J, Bongers F, Chatelet P, Forget P-M, van der Meer P, Norden N, Riera B, Charles-Dominique P (2008b) Above-ground biomass and productivity in a rain forest of eastern South America. J Trop Ecol 24:355–366
Clark DA (2002) Are tropical forests an important carbon sink? Reanalysis of the long-term plot data. Ecol Appl 12:3–7
Clark DB, Clark DA, Oberbauer SF (2010) Annual wood production in a tropical rain forest in NE Costa Rica linked to climatic variation but not to increasing CO_2. Glob Chang Biol 16:747–759
Coomes DA, Allen RB (2007) Mortality and tree-size distributions in natural mixed-age forests. J Ecol 95:27–40
Coomes DA, Grubb PJ (2000) Impacts of root competition in forests and woodlands: a theoretical framework and review of experiments. Ecol Monogr 200:171–207
Cox PM, Betts RA, Jones CD, Spall SA, Totterdell IJ (2000) Acceleration of global warming due to carbon-cycle feedbacks in a coupled climate model. Nature 408:184–187
Cox PM et al (2004) Amazonian forest dieback under climate-carbon cycle projections for the 21st century. Theor Appl Climatol 78:137–156
Cramer W, Bondeau A, Woodward FI, Prentice IC, Betts RA, Brovkin V, Cox PM, Fisher V, Foley JA, Friend AD, Kucharik C, Lomas MR, Ramankutty N, Sitch S, Smith B, White A, Young-Molling C (2001) Global response of terrestrial ecosystem structure and function to CO_2 and climate change: results from six dynamic global vegetation models. Glob Chang Biol 7:357–373
Crutzen PJ (2002) Geology of mankind. Nature 415:23
Denman KL, Brasseur G, Chidthaisong A, Ciais P, Cox PM, Dickinson RE, Hauglustaine D, Heinze C, Holland E, Jacob D, Lohmann U, Ramachandran S, da Silva Dias PL, Wofsy SC, Zhang X (2007) Couplings between changes in the climate system and biogeochemistry. In: Solomon S, Qin D, Manning M, Chen Z, Marquis M, Averyt KB, Tignor M, Miller HL (eds) Climate change 2007: the physical science basis. Contribution of working group I to the fourth assessment report of the Intergovernmental Panel on Climate Change. Cambridge University Press, Cambridge and New York, NY
Doughty CE, Goulden ML (2008) Are tropical forests near a high temperature threshold? J Geophys Res Biogeosci 113:G00B07
Ellis EC (2011) Anthropogenic transformation of the terrestrial biosphere. Philos Trans R Soc A Math Phys Eng Sci 369:1010–1035
Espírito-Santo FDB et al (2010) Storm intensity and old-growth forest disturbances in the Amazon region. Geophys Res Lett 37:L11403
Espírito-Santo FDB, Gloor M, Keller M, Malhi Y, Saatchi S, Nelson B, Oliveira Junior RC, Pereira C, Lloyd J, Frolking S, Palace M, Shimabukuro YE, Duarte V, Monteagudo Mendoza A, López-González G, Baker T, Feldpausch T, Brienen RJW, Asner GP, Boyd DS,

Phillips OL (2014) Size and frequency of natural forest disturbances and the Amazon forest carbon balance. Nat Commun 5. doi:10.1038/ncomms4434

Estes JA, Terborgh J, Brashares J, Power M, Berger J, Bond WJ, Carpenter SR, Essington T, Holt RD, Jackson JB, Marquis RJ, Oksanen L, Oksanen T, Paine RT, Pikitch E, Ripple W, Sandin S, Scheffer M, Schoener TW, Shurin J, Sinclair AR, Soule ME, Virtanen R, Wardle D (2011) Trophic downgrading of planet Earth. Science 333:301–306

Fauset S, Baker TR, Lewis SL, Feldpausch T, Affum-Baffoe K, Foli E, Hamer K, Swaine MD (2012) Drought induced shifts in the floristic and functional composition of tropical forests in Ghana. Ecol Lett 15:1120–1129

Feeley KJ, Wright SJ, Supardi MNN, Kassim AR, Davies SJ (2007) Decelerating growth in tropical forest trees. Ecol Lett 10:461–469

Feldpausch TR, Lloyd J, Lewis SL, Brienen R, Gloor E, Monteagudo A, Lopez-Gonzalez G, Banin L, Abu Salim K, Affum-Baffoe K, Alexiades M, Almeida S, Amaral I, Andrade A, Aragão LEOC, Araujo Murakami A, Arets E, Arroyo L, Aymard GA, Baker TR, Banki OS, Berry NJ, Cardozo N, Chave J, Comiskey JA, Davila EA, de Oliveira A, Di Fiore A, Djagbletey G, Domingues TF, Erwin TL, Fearnside PM, Franca M, Freitas MA, Higuchi N, Honorio E, Ishida Y, Jimenez E, Kassim A, Killeen TJ, Laurance WF, Lovett J, Malhi Y, Marimon BS, Marimon-Junior BH, Lenza E, Marshall AR, Mendoza C, Metcalfe DJ, Mitchard ETA, Neill DA, Nelson BW, Nilus R, Nogueira EM, Parada A, Peh KS, Peña Cruz A, Peñuela MC, Pitman NC, Prieto A, Quesada CA, Ramirez F, Ramirez-Angulo H, Reitsma J, Rudas A, Saiz G, Salomão R, Schwarz M, Silva N, Silva-Espejo JE, Silveira M, Sonké B, Stropp J, Taedoumg H, Tan S, ter Steege H, Terborgh J, Torello-Raventos M, van der Heijden GMF, Vásquez R, Vilanova E, Vos V, White L, Willcock S, Woell H, Phillips OL (2012) Tree height integrated into pantropical tree biomass estimates. Biogeosciences 9:3381–3403. doi:10.5194/bg-9-3381-2012

Fisher JI, Hurtt GC, Thomas RQ, Chambers JQ (2008) Clustered disturbances lead to bias in large-scale estimates based on forest sample plots. Ecol Lett 11:554–563

Friedlingstein P, Cox P, Betts R, Bopp L, Von Bloh W, Brovkin V, Cadule P, Doney S, Eby M, Fung I, Bala G, John J, Jones C, Joos F, Kato T, Kawamiya M, Knorr W, Lindsay K, Matthews H, Raddatz T, Rayner P, Reick C, Roeckner E, Schnitzler KG, Schnur R, Strassmann K, Weaver AJ, Yoshikawa C, Zeng N (2006) Climate-carbon cycle feedback analysis: results from the (CMIP)-M-4 model intercomparison. J Climate 19:3337–3353

Galbraith D, Levy PE, Sitch S, Huntingford C, Cox P, Williams M, Meir P (2010) Multiple mechanisms of Amazonian forest biomass losses in three dynamic global vegetation models under climate change. New Phytol 187:647–665

Gatti LV, Gloor M, Miller JB, Doughty CE, Malhi Y, Domingues LG, Basso LS, Martinewski A, Correia CSC, Borges VF, Freitas S, Braz R, Anderson LO, Rocha H, Grace J, Phillips OL, Lloyd J (2014) Drought sensitivity of Amazonian carbon balance revealed by atmospheric measurements. Nature 506:76–80

Gentry AH (1988a) Tree species richness of upper Amazonian forests. Proc Natl Acad Sci 85:156–9

Gentry AH (1988b) Changes in plant community diversity and floristic composition on environmental and geographical gradients. Ann Mo Bot Gard 75:1–34

Gloor M, Phillips OL, Lloyd J, Lewis SL, Malhi Y, Baker TR, Lopez-Gonzalez G, Peacock J, Almeida S, de Oliveira ACA, Amaral I, Arroyo L, Aymard G, Banki O, Blanc L, Bonal D, Brando P, Chao K-J, Chave J, Davila N, Erwin T, Espejo J, di Fiore A, Feldpausch T, Freitas A, Herrera R, Higuchi N, Honorio E, Jiménez E, Killeen T, Laurance W, Mendoza C, Monteagudo A, Andrade A, Neill D, Nepstad D, Núñez Vargas P, Peñuela MC, Peña Cruz A, Prieto A, Pitman N, Quesada C, Salamão R, Schwarz M, Stropp J, Ramírez F, Ramírez H, Rudas A, ter Steege H, Silva N, Torres A, Terborgh J, Vásquez R, van der Heijden G (2009) Does the disturbance hypothesis explain the biomass increase in basin-wide Amazon forest plot data? Glob Chang Biol 15:2418–2430

Grace J, Mitchard E, Gloor E (2014) Perturbations in the carbon budget of the tropics. Glob Chang Biol. doi:10.1111/gcb.12600

Granados J, Körner C (2002) In deep shade, elevated CO_2 increases the vigour of tropical climbing plants. Glob Chang Biol 8:1109–1117

Groombridge B, Jenkins MD (2003) World Atlas of biodiversity. University of California Press, Berkeley

Hamilton JG, DeLucia EH, George K, Naidu SL, Finzi AC, Schlesinger WH (2002) Forest carbon balance under elevated CO_2. Oecologia 131:250–260

Haxeltine A, Prentice IC (1996) A general model for the light-use efficiency of primary production. Funct Ecol 10:551–561

Hember RA, Kurz WA, Metsaranta JM, Black TA, Guy RD, Coop N (2012) Accelerating regrowth of temperate-maritime forests due to environmental change. Glob Chang Biol 18:2026–2040

Higuchi N, dos Santos J, Ribeiro JR, Minette L, Biot Y (1998) Biomassa da parte aérea da floresta tropical úmida de terra firme da Amazônia Brasileira. Acta Amazon 28:153–166

Ichii K, Hashimoto H, Nemani R, White M (2005) Modeling the interannual variability and trends in gross and net primary productivity of tropical forests from 1982 to 1999. Global Planet Change 48:274–286

Kerstiens G (2001) Meta-analysis of the interaction between shade-tolerance, light environment and growth response of woody species to elevated CO_2. Acta Oecol 22:61–69

Körner C (2003) Slow in, rapid out—carbon flux studies and Kyoto targets. Science 300:1242–1243

Körner C (2004) Through enhanced tree dynamics carbon dioxide enrichment may cause tropical forests to lose carbon. Philos Trans R Soc Ser B 359:493–498

Laurance WF (2004) Forest–climate interactions in fragmented tropical landscapes. Philos Trans R Soc Ser B 359:345–352

Laurance WF, Oliveira AA, Laurance SG, Condit R, Nascimento HEM, Sanchez-Thorin AC, Lovejoy TE, Andrade A, D'Angelo S, Ribeiro JE, Dick CW (2004) Pervasive alteration of tree communities in undisturbed Amazonian forests. Nature 428:171–174

Le Quéré C, Raupach MR, Canadell J, Marland G, Bopp L, Ciais P, Conway T, Doney S, Feely R, Foster P, Friedlingstein P, Gurney K, Houghton RA, House J, Huntingford C, Levy P, Lomas M, Majkut J, Metzl N, Ometto JP, Peters GP, Prentice IC, Randerson J, Running S, Sarmiento JL, Schuster U, Sitch S, Takahashi T, Viovy N, van der Werf GR, Woodward FI (2009) Trends in the sources and sinks of carbon dioxide. Nat Geosci 2:831–836

Lewis SL (2006) Tropical forests and the changing earth system. Philos Trans R Soc Lond Ser B Biol Sci 361:195–210

Lewis SL, Malhi Y, Phillips OL (2004a) Fingerprinting the impacts of global change on tropical forests. Philos Trans R Soc Ser B 359:437–462

Lewis SL, Phillips OL, Baker TR, Lloyd J, Malhi Y, Almeida S, Higuchi N, Laurance WF, Neill DA, Silva JNM, Terborgh J, Torres Lezama A, Vásquez Martinez R, Brown S, Chave J, Kuebler C, Núñez Vargas P, Vinceti B (2004b) Concerted changes in tropical forest structure and dynamics: evidence from 50 South American long-term plots. Philos Trans R Soc Ser B 359:421–436

Lewis SL, Phillips OL, Baker TR (2006) Impacts of global atmospheric change on tropical forests. Trends Ecol Evol 21:173–174

Lewis SL, Lopez-Gonzalez G, Sonké B, Affum-Baffo K, Baker TR, Ojo LO, Phillips OL, Reitsma J, White L, Comiskey JA, Ewango C, Feldpausch T, Hamilton AC, Gloor E, Hart T, Hladik A, Kamdem M-ND, Lloyd J, Lovett J, Makana J-R, Malhi Y, Mbago FM, Ndangalasi HJ, Peacock J, Peh KS, Sheil D, Sunderland T, Swaine MD, Taplin J, Taylor D, Thomas S, Votere R, Wöll H (2009a) Increasing carbon storage in intact African tropical forests. Nature 477:1003–1006

Lewis SL, Lloyd J, Sitch S, Mitchard ETA, Laurance WF (2009b) Changing ecology of tropical forests: evidence and drivers. Annu Rev Ecol Evol Syst 40:529–549

Lewis SL, Brando PM, Phillips OL, van der Heijden GMF, Nepstad D (2011) The 2010 Amazon drought. Science 331:554–555

Lloyd J, Farquhar GD (1996) The CO_2 dependence of photosynthesis, plant growth responses to elevated atmospheric CO_2 concentrations and their interaction with plant nutrient status. Funct Ecol 10:4–32

Lloyd J, Farquhar GD (2008) Effects of rising temperatures and (CO_2) on the physiology of tropical forest trees. Philos Trans R Soc B 363:1811–1817

Lloyd J, Grace J, Miranda AC, Meir P, Wong SC, Miranda HS, Wright IR, Gash JHC, MacIntyre JA (1996) A simple calibrated model of Amazon rainforest productivity based on leaf biochemical properties. Plant Cell Environ 18:1129–1145

Lloyd J, Gloor M, Lewis SL (2009) Are the dynamics of tropical forests dominated by large and rare disturbance events? Ecol Lett 12:E19–E21

Lopez-Gonzalez G, Lewis SL, Burkitt M, Phillips OL (2011) ForestPlotsnet: a web application and research tool to manage and analyse tropical forest plot data. J Veg Sci 22:610–613

Luyssaert S, Schulze E-D, Börner A, Knohl A, Hessenmöller D, Law BE, Ciais P, Grace J (2008) Old-growth forests as global carbon sinks. Nature 455:213–215

Malhi Y, Phillips OL (2004) Tropical forests and global atmospheric change: a synthesis. Philos Trans R Soc Ser B 359:549–555

Malhi Y, Phillips OL (2005) Tropical forests and global atmospheric change. Oxford University Press, Oxford, p 260

Malhi Y, Phillips OL, Baker TR, Almeida S, Frederiksen T, Grace J, Higuchi N, Killeen T, Laurance WF, Leaño C, Lewis S, Lloyd J, Meir P, Monteagudo A, Neill D, Núñez Vargas P, Panfil SN, Pitman N, Rudas-Ll A, Salomão R, Saleska S, Silva N, Silveira M, Sombroek WG, Valencia R, Vásquez Martínez R, Vieira I, Vinceti B (2002) An international network to understand the biomass and dynamics of Amazonian forests (RAINFOR). J Veg Sci 13:439–450

Malhi Y, Baker TR, Phillips OL, Almeida S, Alvarez E, Arroyo L, Chave J, Czimczik C, Di Fiore A, Higuchi N, Killeen T, Laurance SG, Laurance WF, Lewis SL, Mercado LM, Monteagudo A, Neill DA, Núñez Vargas P, Patiño S, Pitman NCA, Quesada A, Silva N, Torres Lezama A, Terborgh J, Vásquez Martínez R, Vinceti B, Lloyd J (2004) The above-ground coarse woody productivity of 104 neotropical forest plots. Glob Chang Biol 10:563–591

Meehl GA, Stocker TF, Collins WD, Friedlingstein P, Gaye AT, Gregory JM, Kitoh A, Knutti R, Murphy JM, Noda A, Raper S, Watterson IG, Weaver AJ, Zhao Z-C (2007) Global climate projections. In: Solomon S, Qin D, Manning M, Chen Z, Marquis M, Averyt K, Tignor M, Miller HL (eds) Climate change 2007: the physical science basis contribution of working group I to the fourth assessment report of the Intergovernmental Panel on Climate Change. Cambridge University Press, Cambridge and New York, NY

Mitchard ETA, Saatchi SS, Woodhouse IH, Nangendo G, Ribeiro NS, Williams M, Ryan CM, Lewis SL, Feldpausch TR, Meir P (2009) Using satellite radar backscatter to predict above-ground woody biomass: a consistent relationship across four different African landscapes. Geophys Res Lett 36:L23401

Mitchard ET, Feldpausch TR, Brienen RJ, Lopez-Gonzalez G, Monteagudo A, Baker TR et al (2014) Markedly divergent estimates of Amazon forest carbon density from ground plots and satellites. Glob Ecol Biogeogr. doi:10.1111/geb.12168

Nelson BW, Kapos V, Adams JB, Oliveira WJ, Braun OP (1994) Forest disturbance by large blowdowns in the Brazilian Amazon. Ecology 75:853–858

Nemani RR, Keeling CD, Hashimoto H, Jolly WM, Piper SC, Tucker CJ, Myneni RB, Running SW (2003) Climate-driven increases in global terrestrial net primary production from 1982 to 1999. Science 300:1560–1563

Norby RJ, Zak DR (2011) Ecological lessons from free-air CO2 enrichment (FACE) experiments. Annu Rev Ecol Evol Syst 42:181–203

Norby RJ, Hanson P, O'Neill E, Tschaplinski T, Weltzin J, Hansen R, Cheng W, Wullschleger S, Gunderson C, Edwards N, Johnson D (2002) Net primary productivity of a CO_2-enriched deciduous forest and the implications for carbon storage. Ecol Appl 12:1261–1266

Oliveira PHF, Artaxo P, Pires C, De Lucca S, Procópio A, Holben B, Schafer J, Cardoso LF, Wofsy SC, Rocha HR (2007) The effects of biomass burning aerosols and clouds on the CO_2 flux in Amazonia. Tellus 59B:338–349

Pan Y, Birdsey R, Fang J, Houghton R, Kauppi P, Kurz WA, Phillips OL, Shvidenko A, Canadell J, Ciais P, Jackson RB, Pacala S, Lewis SL, McGuire A, Piao S, Rautiainen A, Sitch S, Hayes D (2011) A large and persistent carbon sink in the world's forests. Science 333:988–993

Parmentier I, Malhi Y, Senterre B, Whittaker R, Alonso A, Balinga MPB, Bakayoko A, Bongers F, Chatelain C, Comiskey J, Corta R, Djuikouo Kamdem M-N, Doucet J-L, Gautier L, Hawthorne W, Issembe YA, Kouamé FN, Kouka LA, Leal ME, Lejoly J, Lewis SL, Newbery DM, Nusbaumer L, Parren M, Peh K, Phillips OL, Poorter L, Sheil D, Sonké B, Sosef MSM, Sunderland TCH, Stropp J, ter Steege H, Swaine MD, Tchouto MGP, van Gemerden BS, van Valkenburg J, Wöll H (2007) The odd man out? Might climate explain the lower tree α-diversity of African rain forests relative to Amazonian rain forests? J Ecol 95:1058–1071

Peacock J, Baker TR, Lewis SL, Lopez-Gonzalez G, Phillips OL (2007) The RAINFOR database: monitoring forest biomass and dynamics. J Vegn Sci 18:535–542

Phillips OL (1996) Long-term environmental change in tropical forests: increasing tree turnover. Environ Conserv 23:235–248

Phillips OL, Gentry AH (1994) Increasing turnover through time in tropical forests. Science 263:954–958

Phillips OL, Lewis SL (2014) Evaluating the tropical forest carbon sink. Glob Chang Biol 7:2039–2041

Phillips OL, Hall P, Gentry AH, Sawyer SA, Vásquez R (1994) Dynamics and species richness of tropical forests. Proc Natl Acad Sci USA 91:2805–2809

Phillips OL, Malhi Y, Higuchi N, Laurance WF, Nuñez PV, Vásquez Martínez R, Laurance SG, Ferriera LV, Stern M, Brown S, Grace J (1998) Changes in the carbon balance of tropical forest: evidence from long-term plots. Science 282:439–442

Phillips OL, Malhi Y, Vinceti B, Baker T, Lewis SL, Higuchi N, Laurance WF, Vargas PN, Martínez RV, Laurance SG, Ferreira LV, Stern M, Brown S, Grace J (2002a) Changes in the biomass of tropical forests: evaluating potential biases. Ecol Appl 12:576–587

Phillips OL, Martínez RV, Arroyo L, Baker TR, Killeen T, Lewis SL, Malhi Y, Mendoza AM, Neill D, Vargas PN, Alexiades M, Cerón C, Di Fiore A, Erwin T, Jardim A, Palacios W, Saldias M, Vinceti B (2002b) Increasing dominance of large lianas in Amazonian forests. Nature 418:770–774

Phillips OL, Baker TR, Arroyo L, Higuchi N, Killeen T, Laurance WF, Lewis SL, Lloyd J, Malhi Y, Monteagudo A, Neill D, Núñez Vargas P, Silva N, Terborgh J, Vásquez Martínez R, Alexiades M, Almeida S, Brown S, Chave J, Comiskey J, Czimczik C, Di Fiore A, Erwin T, Kuebler C, Laurance SG, Nascimento HEM, Palacios M, Patiño S, Pitman N, Olivier J, Quesada CA, Saldias M, Torres Lezama A, Vinceti B (2004) Pattern and process in Amazon tree turnover, 1976–2001. Philos Trans R Soc Ser B 359:381–407

Phillips OL, Vásquez Martínez R, Monteagudo A, Baker T, Núñez P (2005) Large lianas as hyperdynamic elements of the tropical forest canopy. Ecology 86:1250–1258

Phillips OL, Aragão LEOC, Lewis SL, Fisher JB, Lloyd J, López-González G, Malhi Y, Monteagudo A, Peacock J, Quesada CA, van der Heijden G, Almeida S, Amaral I, Arroyo L, Aymard G, Baker TR, Bánki O, Blanc L, Bonal D, Brando P, Chave J, Alves de Oliveira ÁC, Dávila Cardozo N, Czimczik CI, Feldpausch T, Freitas MA, Gloor E, Higuchi N, Jiménez E, Lloyd G, Meir P, Mendoza C, Morel A, Neill D, Nepstad D, Patiño S, Peñuela MC, Prieto A, Ramírez F, Schwarz M, Silva J, Silveira M, Sota Thomas A, ter Steege H, Stropp J, Vásquez R, Zelazowski P, Alvarez Dávila E, Andelman S, Andrade A, Chao K-J, Erwin T, Di Fiore A, Honorio E, Keeling H, Killeen TJ, Laurance WF, Peña Cruz A, Pitman NCA, Núñez Vargas P, Ramírez-Angulo H, Rudas A, Salamão R, Silva N, Terborgh J, Torres-Lezama A (2009) Drought sensitivity of the Amazon rainforest. Science 323:1344–1347

Phillips OL, van der Heijden G, López-González G, Aragão LEOC, Lewis SL, Lloyd J, Malhi Y, Monteagudo A, Almeida S, Alvarez E, Amaral I, Andelman S, Andrade A, Arroyo L, Aymard G, Baker TR, Blanc L, Bonal D, Alves de Oliveira AC, Chao K-J, Dávila C, da Costa L, Feldpausch TR, Fisher JB, Freitas MA, Galbraith D, Gloor E, Higuchi N, Honorio E, Jiménez E, Keeling H, Killeen TJ, Lovett JC, Meir P, Mendoza C, Morel A, Núñez P, Patiño S, Peh K, Peña Cruz A, Prieto A, Quesada CA, Ramírez F, Ramírez H, Rudas A, Salamão R, Schwarz M, Silva J, Silveira M, Sonké B, Sota Thomas A, Stropp J, Vásquez R, Taplin J, Vilanova E (2010) Drought mortality relationships for tropical forests. New Phytol 187:631–646

Purves D, Pacala S (2008) Predictive models of forest dynamics. Science 320:1452–1453

Quesada CA, Phillips OL, Schwarz M, Czimczik C, Baker TR, Patiño S, Fyllas NM, Hodnett MG, Herrera R, Almeida S, Alvarez Dávila E, Arneth A, Arroyo L, Chao KJ, Dezzeo N, Erwin T, di Fiore A, Higuchi N, Honorio Coronado E, Jimenez EM, Killeen T, Lezama AT, Lloyd G, López-González G, Luizão FJ, Malhi Y, Monteagudo A, Neill DA, Núñez Vargas P, Paiva R, Peacock J, Peñuela MC, Peña Cruz A, Pitman N, Priante Filho N, Prieto A, Ramírez H, Rudas A, Salomão R, Santos AJB, Schmerler J, Silva N, Silveira M, Vásquez IR, Vieira I, Terborgh J, Lloyd J (2012) Basin-wide variations in Amazon forest structure and function are mediated by both soils and climate. Biogeosciences 9:2203–2246

Rammig A, Jupp T, Thonicke K, Tietjen B, Heinke J, Ostberg S, Lucht W, Cramer W, Cox P (2010) Estimating the risk of Amazonian forest dieback. New Phytol 187:694–706

Retallack GJ (2001) A 300-million-year record of atmospheric carbon dioxide from fossil plant cuticles. Nature 411:287–290

Salzmann U, Hoelzmann P (2005) The Dahomey Gap: an abrupt climatically induced rain forest fragmentation in West Africa during the late Holocene. Holocene 15:190–199

Schnitzer SA, Bongers F (2002) The ecology of lianas and their role in forests. Trends Ecol Evol 17:223–230

Schnitzer SA, Bongers F (2011) Increasing liana abundance and biomass in tropical forests: emerging patterns and putative mechanisms. Ecol Lett 14:397–406

Staver AC, Archibald S, Levin SA (2011) The global extent and determinants of savanna and forest as alternative states. Science 334:230–232

Steffen W, Grinevald J, Crutzen P, McNeill J (2011) The Anthropocene: conceptual and historical perspectives. Philos Trans R Soc A Math Phys Eng Sci 369:842–867

Stephens BB, Gurney KR, Tans PP, Sweeney C, Peters W, Bruhwiler L, Ciais P, Ramonet M, Bousquet P, Nakazawa T, Aoki S, Machida T, Inoue G, Vinnichenko N, Lloyd J, Jordan A, Heimann M, Shibistova O, Langenfelds RL, Steele LP, Francey RJ, Denning AS (2007) Weak northern and strong tropical land carbon uptake from vertical profiles of atmospheric CO_2. Science 316:1732–1735

ter Steege H, Pitman NCA, Phillips OL, Chave J, Sabatier D, Duque A, Molino JF, Prévost MF, Spichiger R, Castellanos H, van Hildebrand P, Vásquez R (2006) Continental-scale patterns of canopy tree composition and function across Amazonia. Nature 443:444–447

Trenberth KE, Jones PD, Ambenje P, Bojariu R, Easterling D, Klein Tank A, Parker D, Rahimzadeh F, Renwick JA, Rusticucci M, Soden B, Zhai P (2007) Observations: surface and atmospheric climate change. In: Solomon S, Qin D, Manning M, Chen Z, Marquis M, Averyt K, Tignor M, Miller HL (eds) Climate change 2007: the physical science basis contribution of working group I to the fourth assessment report of the Intergovernmental Panel on Climate Change. Cambridge University Press, Cambridge and New York, NY

Van der Heijden G, Healey J, Phillips OL (2008) Infestation of trees by lianas in a tropical forest in Amazonian Peru. J Veg Sci 19:747–756

Vieira S, Trumbore S, Camargo PB, Selhorst D, Chambers J, Higuchi N, Martinelli LA (2005) Slow growth rates of Amazonian trees: consequences for carbon cycling. Proc Natl Acad Sci 102:18502–18507

West GB, Brown JH, Enquist BJ (1999) A general model for the structure and allometry of vascular plant systems. Nature 400:664–667

Wong T, Wielicki BA, Lee RB, Smith GL, Bush KA, Willis JK (2006) Reexamination of the observed decadal variability of the earth radiation budget using altitude-corrected ERBE/ERBS nonscanner WFOV data. J Climate 19:4028–4040
Worbes M (1999) Annual growth rings, rainfall-dependent growth and long-term growth patterns of tropical trees from the Caparo Forest Reserve in Venezuela. J Ecol 87:391–403
Wright SJ (2005) Tropical forests in a changing environment. Trends Ecol Evol 20:553–560
Zelazowski P, Malhi Y, Huntingford C, Sitch S, Fisher JB (2011) Changes in the potential distribution of humid tropical forests on a warmer planet. Philos Trans R Soc A 369:137–160

Chapter 11
The Biogeochemistry of the Main Forest Vegetation Types in Amazonia

Erika Buscardo, Gabriela Nardoto, Flávio Luizão, Maria T.F. Piedade, Jochen Schöngart, Florian Wittmann, Christopher E. Doughty, Carlos A. Quesada, and Laszlo Nagy

11.1 Introduction

Proponents of the ecosystem ecology approach have argued that integrative methods that ignore species composition/vegetation type are valid because most photosynthetic organisms operate the Calvin cycle; 'greenness' can be used to estimate photosynthesis; productivity is dependent on energy absorbed; and convergent vegetation types have comparable productivities. In the meantime, it is also recognised that individual species can have species-specific effects on resource capture and use and thus impact biogeochemistry (Mooney 2001). Thus, we may postulate that different habitats (vegetation–soil complexes or henceforth

E. Buscardo (✉)
Centro de Ecologia Funcional, Universidade de Coimbra, Coimbra, Portugal

Departamento de Biologia Vegetal, Universidade Estadual de Campinas, Campinas, Brazil

Instituto Nacional de Pesquisas da Amazônia, Manaus, Brazil
e-mail: erikatea@ci.uc.pt

G. Nardoto
Departamento de Ecologia, Universidade de Brasília, Brasília, Brazil

F. Luizão • M.T.F. Piedade • J. Schöngart • F. Wittmann • C.A. Quesada
Instituto Nacional de Pesquisas da Amazônia, Manaus, Brazil

C.E. Doughty
University of Oxford, Oxford, UK

L. Nagy
Departamento de Biologia Vegetal, Universidade Estadual de Campinas, Campinas, Brazil

© Springer-Verlag Berlin Heidelberg 2016

L. Nagy et al. (eds.), *Interactions Between Biosphere, Atmosphere and Human Land Use in the Amazon Basin*, Ecological Studies 227,
DOI 10.1007/978-3-662-49902-3_11

vegetation types) will have differences in their biogeochemical functioning. Vegetation types, especially those related to deficiency or overabundance of certain resources (e.g. heath forests vs. seasonally flooded forests), can differ in their elemental cycles and the characteristic environment–vegetation type correlations can be used to estimate differences in biogeochemical properties.

The prevailing focus of studies on biogeochemistry has centred on the carbon cycle. Carbon makes up about 50 % of vegetative matter and large quantities of it are stored in the vegetation–soil system; its atmospheric cycling has become of great interest for counterbalancing carbon emissions by human activities.

Patterns of the carbon cycle and balance estimated by various indirect methods are available elsewhere in this volume: Brazilian Legal Amazon (Gloor 2016), eddy covariance methods, using the network of flux towers in Amazonia (Araújo et al. 2016) secondary forests (Fearnside 2016), forest carbon stocks (Higuchi et al. 2016; Phillips et al. 2016), and an evaluation of the methods and a recap of the Amazon-wide C balance (Grace 2016). This chapter synthesises C, N, and P stocks and fluxes from plot-based studies in the major vegetation types within the Amazon basin—unflooded or *terra firme* lowland evergreen rainforests (both dense and open), seasonal unflooded forests, seasonally flooded forests (*várzea* and *igapó*), heath forests (tall and low-stature heath forests locally known as *campina(rana)*, *varillal*, *caatinga*, *bana*), and montane forests.

11.2 Physiography and the Atmosphere–Vegetation–Soil System in the Major Vegetation Formations of the Amazon Basin

11.2.1 Introduction: Nutrient Cycles

Mineral nutrients in forest ecosystems originate from a variety of sources from weathering of parent material to deposition from the atmosphere. The atmosphere supplies CO_2 for photosynthesis, N_2 for N fixation, and various organic and inorganic compounds as part of air- and precipitation-borne input. Some of the nutrients in precipitation can be absorbed by the canopy, while incoming precipitation can also leach elements and compounds from the vegetation.

Nutrients can become temporarily unavailable for plant uptake by 'immobilisation' in microbial biomass, and, depending on soil type and geological processes, they can be adsorbed to clay surfaces, organic matter, or fixed into mineral lattices. Part of the mineral nutrients that are taken up by plants is incorporated into organic compounds and some of it forms biomass. A fraction of these elements from biomass returns to the forest floor and through litter decomposition process is converted from organic into inorganic form and/or accumulates as decomposed organic matter. Leaching of elements from the soil (in inorganic or organic forms) occurs, and these eventually enter into aquatic ecosystems.

The major factors underlying patterns of nutrient cycling in ecosystems, including tropical forests, are represented by climate, species composition, soil fertility, and successional status (Vitousek and Sanford 1986). Various levels of interdependence among these factors and their correlation with physiography make it difficult to evaluate the contribution of each single component to the variation in nutrient cycling.

11.2.2 Physiography of the Amazon Basin

The physiography of the Amazon hydrographic basin includes the extensive humid and seasonal tropical lowlands (ca. 80 %) and it is characterised by a steep elevation gradient along the Andes, extending from tropical forest through montane forest to alpine biomes at over 5000 m a.s.l. The unflooded lowlands in the tropical forest biome are termed *terra firme*; substantial parts of this *terra firme* are characterised by a dissected topography, consisting of plateaux with clayey soil (referred in this chapter as clayey ferralsols) and valleys with sandy soil (referred in this chapter as sandy ferralsols), which may be occasionally affected by excess surface run-off. The lowland part of the Amazon basin is characterised by a gradient of decreasing rainfall and increasing seasonality from the perhumid, north-west, to the southern and eastern regions, where a strongly seasonal climate prevails with a dry season (precipitation < 100 mm per month) reaching > 5 months in the Cerrado biome (Fig. 11.1). Rainfall in Amazonia is intensely affected by El Niño/Southern Oscillation (ENSO) and by sea surface temperature anomalies from the tropical Atlantic Ocean (Marengo et al. 2016) and the central and eastern portions of the basin can be subject to episodic droughts (Davidson et al. 2012). There is a steep decrease in temperature with elevation on the Andean slopes; precipitation initially increases and then, above the cloud base, decreases. The decrease in precipitation is also apparent in north to south direction from páramo (humid alpine) to puna (xeric alpine).

Climate exerts a dominant control on the spatial distribution of biomes and their vegetation types at the hydrographic basin scale and on ecosystem biogeochemical properties at the macro scale. Within-biome scale soil physical and chemical properties and topography-related hydrological characteristics are among the most important environmental determinants of observed vegetation types and their productivity and nutrient cycling.

We consider here vegetation formations on a variety of soil types (see also Quesada and Lloyd (2016)) and observe case by case the patterns of nutrient biomass, cycling, and loss in different forest types: dense and open lowland evergreen rainforests on ferralsols and acrisols (*terra firme*), seasonal unflooded forests, heath or white sand forests/vegetation on heavily eluted 'white sand' or podzol, seasonally flooded forests (*igapó*, *várzea*) on fluvisol and partially on gleysol, and montane forests from tropical to alpine elevations. The selected vegetation types are the best studied in the Amazon basin and represent a range

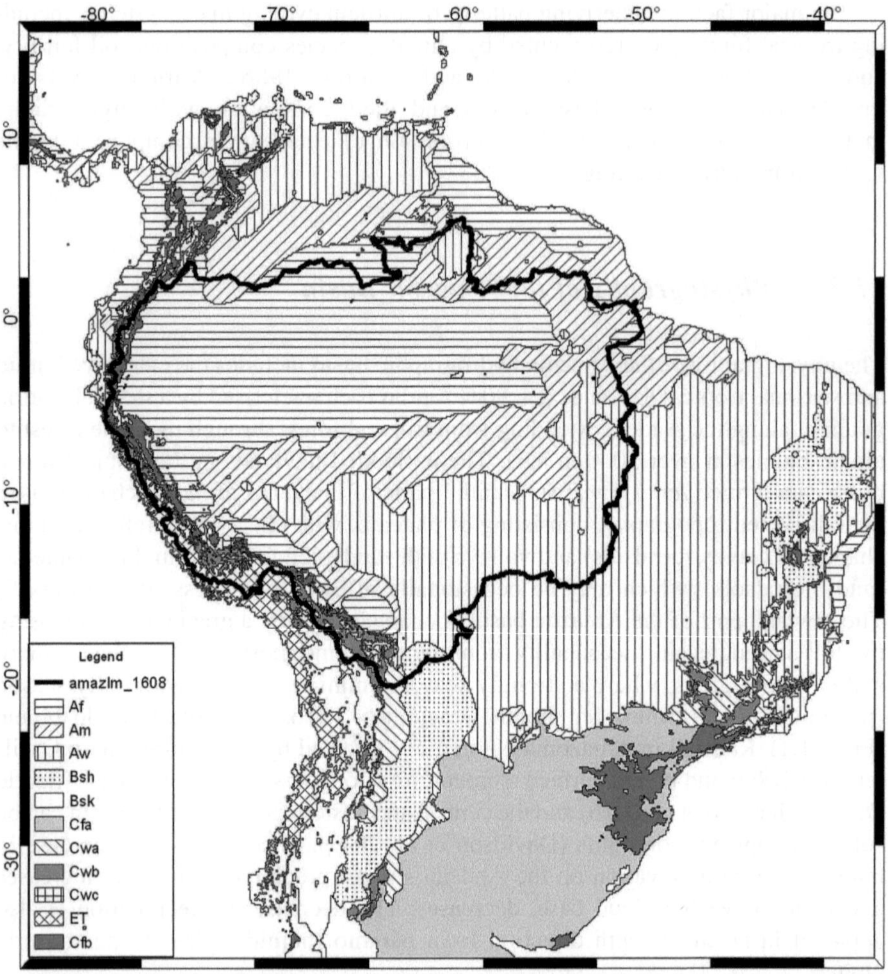

Fig. 11.1 The Amazon basin (*thick black line*) encompasses a range of climates, mostly aseasonal to seasonal tropical and various arid and temperate variants. Köppen–Geiger classifications, following the rules defined in (Hijmans et al. 2005) as applied to the 5′ resolution WorldClim global climatology (www.worldclim.org; Version 1.4, release 3; (Kriticos et al. 2012), were downloaded from the CliMond set of climate data products (www.climond.org; Kriticos et al. 2012). Climate classes: Af, tropical wet; Am, tropical monsoonal; Aw, tropical wet—dry (savanna); Bsh, subtropical dry semiarid; Bsk, mid-latitude dry semiarid; Cfa, humid subtropical; Cwa, Cwb, Cwc, tropical montane; ET, alpine climate above the treeline

of resource availabilities and habitat differences at the extremes of gradients: a topo-hydro gradient (seasonally flooded vs. unflooded), an elevation gradient (lowland vs. montane), and between major soil formations (non-podzolic to heavily podzolised). This is the basis for our treatment of biogeochemistry in the terrestrial part of the Amazon basin. The aquatic systems are treated in Melack (2016).

11.3 Carbon

11.3.1 Carbon Stocks, Allocation, and Fluxes

For C stocks of major importance is how much of the photosynthetically fixed C is retained (carbon use efficiency) in perennating organs, primarily in the stem, branches, and coarse root wood. Part of the fixed C in organs of shorter longevity contributes to annual fluxes in the internal ecosystem cycling of elements in the form of litterfall and fine root turnover. Most measurements of stocks and growth (total net annual primary production or NPP) report total stem volume and growth from repeated tree census data. Few studies have recorded simultaneously all components of stock and fluxes in vegetation (Table 11.1), and studies that also integrated soil carbon are rare (e.g. Malhi et al. 2009).

11.3.1.1 Carbon Stocks

Above-ground biomass (AGB) AGB carbon stocks in unflooded lowland vegetation are highly variable, with average values that decrease from *terra firme* forests (137 Mg C ha^{-1}) to tall heath forest formations, seasonal forests, open *terra firme*, and low-stature heath forest formations (37 Mg C ha^{-1}; Table 11.2; Fig. 11.2). *Várzea* and pre-montane forests show values comparable with those of *terra firme*, while AGB carbon stocks decrease with elevation from lower (92 Mg C ha^{-1}) to upper montane forests (55 Mg C ha^{-1}).

Wood debris (WD) Most of the available data on WD refer to dense *terra firme* forests (14 Mg C ha^{-1}) with values similar to those found in two studies for open *terra firme* forests (16 Mg C ha^{-1}), while heath forest formations show pronounced differences between tall (13 Mg C ha^{-1}) and low-stature heath forests (4 Mg C ha^{-1}), the latter being comparable with average values reported for seasonally flooded forests (Table 11.2; Fig. 11.2). In montane forests values of WD carbon stocks are intermediate between those found in *terra firme* and low-stature heath forests and are higher in upper (11 Mg C ha^{-1}) than in lower montane forests (7 Mg C ha^{-1}).

Coarse roots C stock values for coarse roots were estimated on total AGB data, considering a root:shoot ratio of 0.21 following Malhi et al. (2009) and they follow the pattern found for AGB (Table 11.2). This value by Malhi et al. (2009) is based on values reported for tropical forests by Jackson et al. (1996; 0.34 for tropical deciduous and 0.19 for tropical evergreen forests) and by Cairns et al. (1997; 0.24). Higuchi et al. (2016) based on allometric equations fitted to values obtained by destructive sampling reported root:shoot ratios between 0.01 and 0.22 in *terra firme* across 18 locations in the State of Amazonas, Brazil (in 10 locations the value was < 0.1, an unlikely low figure); Lima et al. (2012) reported a ratio of 0.14, using allometric equation, based on 101 trees in north-western Amazonas State, Brazil.

Table 11.1 Net primary production (NPP) and respiration (in Mg C ha^{-1} $year^{-1}$) as determined by measuring individual components in 16 paired 1-ha plots that were established by the GEM network. The plots are distributed across lowland dense evergreen and seasonal deciduous forests to various montane and upper montane formations across Brazil, Bolivia, and Peru

				Above-ground net primary productivity (NPP_{AG})		
Site name	Country	Forest type	Soil type	NPP_{ACW}	NPP_{BT}	$NPP_{litterfall}$
Caxiuanã National Forest Reserve, Pará	Brazil	*Terra firme*	Acrisol	2.55 ± 0.06	1.20 ± 0.12	2.94 ± 0.04
Caxiuanã National Forest Reserve, Pará	Brazil	*Terra firme* (secondary forest)	Terra preta	2.92 ± 0.29	1.10 ± 0.11	4.52 ± 0.18
Hacienda Kenia, Guarayos Province, Santa Cruz	Bolivia	*Terra firme* (drought-deciduous trees species)	Inceptisol (shallow)	2.56 ± 0.26	0.58 ± 0.06	4.23 ± 0.57
Hacienda Kenia, Guarayos Province, Santa Cruz	Bolivia	*Terra firme* (typical)	Inceptisol (deep)	4.12 ± 0.41	0.57 ± 0.06	5.65 ± 0.65
Allpahuayo-Mishana National Reserve, Maynas Province	Peru	*Terra firme*	Haplic Gleysol	3.45 ± 0.35	1.42 ± 0.14	4.20 ± 0.85
Allpahuayo-Mishana National Reserve, Maynas Province	Peru	*Terra firme*	Haplic Arenosol	2.73 ± 0.27	1.01 ± 0.10	5.66 ± 0.80
Tambopata Candamo Reserve, Madre de Dios Region	Peru	*Terra firme* (on Holocene floodplain terrace)	?	2.64 ± 0.24	0.95 ± 0.10	5.61 ± 0.36
Tambopata Candamo Reserve, Madre de Dios Region	Peru	*Terra firme* (on Pleistocene terrace)	?	2.64 ± 0.25	0.50 ± 0.05	7.75 ± 0.91
Parque Nacional del Manú, Cusco, San Pedro	Peru	Lower montane forest, 1500 m a.s.l.	Umbrisol	2.93 ± 1.47	0.52 ± 0.07	5.33 ± 0.22
Parque Nacional del Manú, Cusco, San Pedro	Peru	Lower montane forest, 1750 m a.s.l.	Umbrisol	1.87 ± 1.18	0.38 ± 0.04	3.52 ± 0.24
Parque Nacional del Manú, Cusco, Esperanza	Peru	Upper montane forest, 2825 m a.s.l.	Umbrisol	1.65 ± 0.15	0.75 ± 0.07	2.69 ± 0.28
Parque Nacional del Manú, Cusco, Wayqecha	Peru	Upper montane forest, 3025 m a.s.l.	Umbrisol (100 m below the tree line)	1.35 ± 0.12	0.54 ± 0.05	3.67 ± 0.28
Fazenda Tanguro, Mato Grosso	Brazil	Semi deciduous transitional forest	Ferralsol	2.63 ± 0.29	0.08 ± 0.01	5.00 ± 0.47

NPP_{ACW} above-ground coarse wood NPP; NPP_{BT} branch turnover NPP; $NPP_{litterfall}$ litterfall NPP; $NPP_{herb.}$ loss to leaf efflux; R_H heterotrophic soil CO_2 efflux; R_{leaves} canopy respiration; R_{stems} above-ground live wood respiration; $R_{c.roots}$ herbivory; $NPP_{c.roots}$ coarse root NPP; $NPP_{f.roots}$ fine root NPP; R_{soil} total soil CO_2 efflux; $R_{rhizosph.}$ autotrophic soil CO_2 coarse root respiration

	Below-ground net primary productivity (NPP_{BG})		Autotrophic and heterotrophic respiration							
$NPP_{herb.}$	$NPP_{c.roots}$	$NPP_{f.roots}$	R_{soil}	$R_{rhizosph.}$	R_{soilH}	R_{leaves}	R_{stems}	$R_{c.roots}$	References
0.09 ± 0.001	0.54 ± 0.84	3.89 ± 0.80	15.99 ± 1.69	9.93 ± 1.63	6.06 ± 0.47	5.69 ± 2.14	10.21 ± 4.49	2.14 ± 3.50	da Costa et al. (2014)
0.21 ± 0.01	0.53 ± 0.05	6.41 ± 1.08	16.75 ± 1.40	4.58 ± 0.67	12.17 ± 1.79	5.09 ± 1.67	8.46 ± 2.82	1.94 ± 1.03	Doughty et al. (2014)
0.42 ± 0.05	0.44 ± 0.04	3.04 ± 0.28	11.91 ± 1.90	2.40 ± 0.53	9.51 ± 1.37	4.43 ± 1.24	7.26 ± 3.16	1.52 ± 0.19	Araujo-Murakami et al. (2014)
0.55 ± 0.06	0.57 ± 0.06	4.04 ± 0.51	12.87 ± 1.97	4.40 ± 0.99	8.47 ± 0.98	5.23 ± 1.59	7.44 ± 3.14	1.56 ± 0.19	Araujo-Murakami et al. (2014)
0.50 ± 0.06	0.69 ± 0.07	3.02 ± 0.29	18.12 ± 2.39	4.44 ± 0.92	13.69 ± 2.64	8.92 ± 3.00	9.63 ± 3.05	1.89 ± 0.97	del Aguila-Pasquel et al. (2014)
0.76 ± 0.11	0.55 ± 0.06	3.50 ± 0.38	18.18 ± 1.71	6.38 ± 0.93	11.80 ± 1.97	11.35 ± 3.50	8.11 ± 2.55	1.77 ± 0.91	del Aguila-Pasquel et al. (2014)
0.76 ± 0.05	0.51 ± 0.05	4.54 ± 0.71	12.15 ± 0.82	5.07 ± 0.61	7.08 ± 8.86	8.86 ± 2.84	5.43 ± 1.77	1.14 ± 0.59	Malhi et al. (2014)
0.70 ± 0.06	0.52 ± 0.05	2.11 ± 0.31	10.97 ± 0.54	4.62 ± 0.57	6.34 ± 0.76	6.43 ± 2.07	7.62 ± 2.48	1.60 ± 0.82	Malhi et al. (2014)
0.66 ± 0.03	0.61 ± 0.06	1.89 ± 0.30	13.42 ± 1.16	8.79 ± 1.36	4.63 ± 0.70	7.06 ± 2.48	8.91 ± 2.82	1.87 ± 0.95	Huasco et al. (2014)
0.42 ± 0.03	0.36 ± 0.04	1.22 ± 0.23	10.47 ± 0.86	6.11 ± 0.96	4.37 ± 0.62	6.55 ± 2.17	9.70 ± 3.07	2.04 ± 1.02	Huasco et al. (2014)
0.25 ± 0.04	0.32 ± 0.03	1.41 ± 0.21	9.52 ± 0.59	2.71 ± 0.36	6.81 ± 0.87	6.10 ± 1.92	4.87 ± 1.54	1.02 ± 0.52	Girardin et al. (2014)
0.32 ± 0.02	0.25 ± 0.03	1.90 ± 0.35	10.63 ± 0.81	3.42 ± 0.50	7.22 ± 1.02	5.18 ± 1.63	7.69 ± 2.42	1.61 ± 0.81	Girardin et al. (2014)
0.32 ± 0.26	0.47 ± 0.14	1.86 ± 0.13	14.47 ± 1.19	3.30 ± 0.63	11.17 ± 1.01	7.48 ± 2.30	4.57 ± 1.82	0.96 ± 0.36	Rocha et al. (2014)

Table 11.2 Carbon cycling in different forest formations in the Amazon basin

		Terra firme	Open *Terra firme*	Seasonal forest	THF
Pools (Mg ha^{-1})					
AGB[a]		137 ± 2 (182)	62 ± 15 (3)	104 ± 9 (32)	116 ± 10 (7)
WD[d]		14.1 ± 1.3 (51)	15.6 ± 0.6 (2)	–	12.5 ± 4.3 (4)
Fine roots[c]	0–10 cm	1.4 ± 0.2 (4)	–	–	–
	0–20 cm	6.2 ± 4.5 (3)	–	–	5.5 (1)
	0–30 cm	6.5 ± 0.8 (3)	–	–	–
	0–40 cm	–	–	–	–
Coarse roots		29 ± 0 (182)	13 ± 3 (3)	22 ± 2 (32)	24 ± 2 (7)
Soil[c]					
TOTAL[c]					
Inputs (kg ha^{-1} year^{-1})					
C in precipitation		–	106 (1)[DOC]	–	–
Outputs (kg ha^{-1} year^{-1})					
BVOCs		190[b]	–	–	–
Surface flow		–	–	0.5 (1)[DIC]/5.5 (1)[DOC]	–
Leaching		–	–	–	–
Base flow		–	–	–	–
Deep groundwater flow		–	–	113 (1)[DIC]/	–
Stream export		1.5 ± 0.5 (2)[DOC]	4.4 (1)[DOC]	6.4(1)[DOC]/	–
				25 ± 24 (2)[TC]/	
				7.8 (1)[DIC]/	
				12.5 ± 9.5 (3)[DOC]	
Balance (INPUTS–OUTPUTS)[c]					
Internal cycling					
Litterfall (Mg ha^{-1} year^{-1})		4.4 ± 0.1 (36)	5 (1)	4.9 ± 0.7 (2)	2.8 ± 0.2 (3)
Fine root turnover[c] (Mg ha^{-1} year^{-1})					
	0–10 cm	–	–	–	–
	0–20 cm	1 ± 0.1 (2)	–	–	3.2 ± 0.2 (2)
	0–30 cm	1.3 ± 0.1 (3)	–	–	–
	0–40 cm	3 ± 0.4 (12)	–	4 ± 1 (2)	–
C in throughfall (kg ha^{-1} year^{-1})		166 ± 10 (4)[DOC]	302 (1)[DOC]	–	–
TOTAL[c]					

THF tall heath forest; *LSHF* low-stature heath forest; *DOC* dissolved organic C; *DIC* dissolved inorganic C; *TC* total C

[a]Trunks and branches

[b]Estimated value

[c]All underground components have been reported in an individualistic manner based on sampling to different depths. For this reason, no total estimates are given. We call for standardised protocols to allow comparative biogeochemical/earth system science studies

[d]The values include the undetermined proportion relative to annual decomposed wood, which should be considered as part of flux. The rate of decomposition, however, is highly variable (see values and references in text Section 'Carbon Fluxes')

LSHF	*Várzea*	*Igapó*	Pre-montane forest	Lower montane	Upper montane
37 ± 14 (6)	151 ± 8 (18)	111 ± 23 (4)	144 ± 16 (7)	92 ± 14 (10)	55 ± 5 (10)
4.4 ± 2.5 (5)	5.6 ± 1.3 (6)	7.5 ± 0.3 (2)	–	7.3 ± 1.5 (3)	10.7 ± 4.6 (4)
–	–	–	–	–	–
–	–	–	–	–	–
–	–	–	2.1 ± 0.7 (2)	3.1 ± 0 (2)	5.4 (1)
–	–	–	2.6 (1)	6.7 ± 0.1 (2)	4.5 ± 0.5 (2)
8 ± 3 (6)	32 ± 8 (18)	23 ± 5 (4)	30 ± 3 (7)	19 ± 3 (10)	12 ± 1 (8)
–	–	–	–	102–117 (3)^{TOC}	–
–	–	–	–	–	–
–	–	–	–	–	–
–	–	–	–	–	–
–	–	–	–	–	–
–	–	–	–	–	–
–	–	–	–	–	–
1.2 (1)	4 ± 0.3 (9)	4 ± 0.4 (2)	3.7 (1)	4.5 ± 0.4 (4)	2.9 ± 0.4 (3)
–	–	–	–	–	–
–	–	–	–	–	–
–	–	–	1.4 ± 0.2 (2)	1.6 ± 0.2 (4)	2.7 ± 1 (3)
–	–	–	2.4 (1)	2.5 ± 0.7 (2)	1.5 ± 0.2 (2)
–	–	–	–	–	–

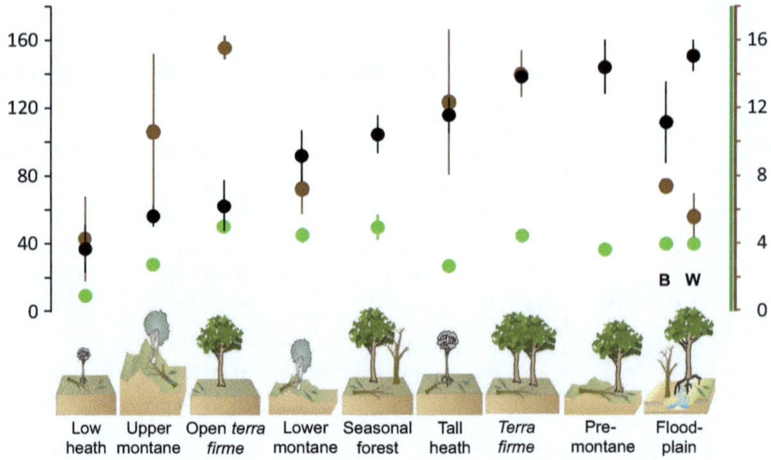

Fig. 11.2 Carbon stocks (in Mg ha^{-1}) in above-ground biomass (*left scale*, *black circles*), wood debris, and litterfall (*right scale*, *brown* and *green circles*) in different forest type formations within the Amazon basin. W, white water (locally known as *várzea*); B, black water (locally known as *igapó*). For number of replicates used for each value see Table 11.2

Fine roots Patterns in root biomass, growth, and decomposition are influenced by numerous physiological and environmental factors, including nutrient accumulation in the upper horizons of mineral soil and in litter, possibly stimulating the development of a mat of absorbing fine roots near the soil surface and even within the freshly fallen litter layer (Cuevas and Medina 1988).

While in *terra firme* forests the root mat develops near the soil surface as well as within the non-decomposed litter layer, heath forest formations are characterised by an accumulation of fine roots in a humus matrix near the soil surface and do not grow into the litter layer (Cuevas and Medina 1988; Coomes and Grubb 1996). The efficiency of the root mat to capture or to extract nutrients from decomposing litter appears to be associated with the presence of mycorrhizal fungi that colonise the majority of the roots present in the root mat (Herrera et al. 1978).

What can be considered as fine root stock depend on turnover rates: the lesser the rate of death and production of new roots the larger the stocks are (and the smaller the fluxes are). Data on carbon stocks in fine roots (diameter $\leq$ 2 mm) in forest formations within the Amazon basin are scant and they are difficult to express on a common basis as they have been obtained at different soil depths. Carbon stocks in fine roots in *terra firme* forests increase from 1.4 Mg C ha^{-1} at 0–10 cm to 6.2 Mg C ha^{-1} at 0–20 cm and 6.5 Mg C ha^{-1} at 0–30 cm (Table 11.2).

Major variation in root dynamics can be observed even within one soil type. Improved nutrient and carbon availability, aeration, and more penetrable soil concur to create a favourable environment for fine root growth in organic layers (Girardin et al. 2013). Studies in lowland *terra firme* forest showed that fine root biomass was higher in sandy soils than clayey soils (Silver et al. 2000). The authors suggested that patterns of root dynamics in sandy soils were possibly driven by water limitation for decomposer activity, poor litter quality, and low soil nutrient availability. Notably, the usual topography-related pattern present in large areas of undulating terrain in Amazonia, where waterlogged sandy soils are found in depressions, did not apply in this study, where the terrain was flat. Aragão et al. (2009) suggested that sandy soils might favour the allocation to below-ground structures due to their greater penetrability than clay-rich soils and to poorer water holding capacity of sandy soils that may induce the increase of root productivity to maintain soil water access during dry periods. Fine root biomass in *terra firme* forests was found to be negatively correlated with N mineralisation rates and positively correlated with extractable soil P concentrations (Silver et al. 2000), while changes in the availability of mineral N over time and the initial soil NO_3^- pool were positively correlated with fine root decay (Silver et al. 2005). The correlation could result from an increase in substrate availability for N mineralisation and nitrification, together with decreased plant uptake of mineral N because of root mortality.

Carbon stocks in the few available datasets for fine roots of pre-montane and montane forests at a soil depth of 0–30 cm and 0–40 cm appear to have a tendency to increase from pre- to upper montane at both soil depths (Table 11.2).

11.3.1.2 Carbon Fluxes

Litterfall Total annual litterfall production for *terra firme* forests averages 4.4 Mg C ha^{-1} $year^{-1}$ (Table 11.2; Fig. 11.2) and there is no or little difference in total fine litter production with soil type. Similar values were recorded for flooded, lower montane, and seasonal forests, while much lower values were found for tall (2.8 Mg C ha^{-1} $year^{-1}$) and low-stature heath forests (1.2 Mg C ha^{-1} $year^{-1}$) and for upper montane forests (2.9 Mg C ha^{-1} $year^{-1}$).

Release from branch turnover Part of the C from wood debris (WD) decomposition is lost into the atmosphere as CO_2 and part is redistributed via surface and sub-surface water flow. Chambers et al. (2001b) estimated 76 % C loss via microbial respiration (1.9 C Mg C ha^{-1} $year^{-1}$) and 24 % redistributed in a *terra firme* forest in central Brazilian Amazonia. Above-ground coarse WD accounts for 14–19 % of the annual above-ground C flux in tropical forests (Palace et al. 2008). However, decomposition rates are variable depending on climate, WD quality (e.g. chemical composition) and size, and decomposer communities (i.e. bacteria, fungi, and soil fauna), with values that vary across different forest types and within the same forest type. For example, decomposition constants in *terra firme* forests in eastern Amazonia decreased with WD size, being 0.12 $year^{-1}$ for large, 0.33 $year^{-1}$ for medium, and 0.47 $year^{-1}$ for small size class (Palace et al. 2008), and, on average, they have been estimated between 0.12 and 0.17 $year^{-1}$ in central and eastern Amazonian *terra firme* forests (Chambers et al. 2000, 2001b; Rice et al. 2004; Palace et al. 2007, 2008) and 0.09 $year^{-1}$ in a montane forest in Ecuador (Wilcke et al. 2005).

Fine root turnover Average C values for fine root production in *terra firme*, pre-montane, and montane forests, at different soil depths, appear to follow the same pattern as that for C stocks in fine roots (Table 11.2); in tall heath forest formations it averages 3.2 Mg C ha^{-1} $year^{-1}$; no data are available for flooded forests.

Nutrient fluxes produced by the release of nutrients after the death and decomposition of fine roots depend on the rate of annual turnover (calculated as the product of fine root production/fine root mass). Values found for turnover (range 0.13–0.92) indicate that fluxes vary between 13.5 % and 92.4 % of the stocks. In addition, there is a poorly quantified component of C flux from root exudates and carbohydrates transferred to mycorrhiza. This portion was estimated recently in eight locations across the Amazon basin as part of measured/estimated rhizosphere respiration (Table 11.1).

It has been shown that fine root dynamics is not limited by the same set of nutrients in different forest types (Cuevas and Medina 1988). Increased root growth was observed after annual P and Ca addition in a *terra firme* forest and N and P addition in an open low-stature heath forest stand, while fine root growth was significantly increased by N addition in a tall heath forest formation.

11.4 Nitrogen and Phosphorus

11.4.1 External Sources

Biological nitrogen fixation is the main pathway by which atmospheric nitrogen (N) is converted and introduced as ammonium into the pedosphere. Additional sources of N include atmospheric inputs in particulate, dissolved, and gaseous inorganic and organic forms. Phosphorus (P) enters ecosystems mainly by weathering of rocks, but can be deposited from aerosol/dust (largely unknown in amounts actually reaching soil surface), aerosol from biomass burning, and biogenic materials, such as pollen and spores.

11.4.1.1 Nitrogen Fixation

Nodulation Leguminous plants have evolved specialised symbioses with a group of bacteria called *Rhizobia* that exchange fixed N for C compounds. Leguminous tree species are abundant in Amazonian lowland forests, but only some of them nodulate and fix N symbiotically (Sprent 2009). Nodulation rates can range from low to abundant across forest types (Moreira et al. 1992; de Faria et al. 2010). In contrast to low rates of legume nodulation in mature lowland forests (Sylvester-Bradley et al. 1980), major nodulation rates were observed in seasonally flooded and in disturbed *terra firme* forests (Moreira et al. 1992). Such landscape-level pattern is consistent with the facultative fixation hypothesis in response to local variations in N availability (Hedin et al. 2009): low rates of nodulation point to downregulation of fixation as a consistent plant strategy and suggest that sufficient soil N is available to support demand, while high rates of nodulation point to upregulation in environments where soil N is scarce (Barron et al. 2011). For example, N leaching, low rates of organic matter mineralisation, and high denitrification rates appear to favour nodulation in flooded forests (Moreira et al. 1992; Barrios and Herrera 1994; Sprent 1999; James et al. 2001).

N_2-fixation rates N_2-fixation rates in natural ecosystems are most frequently assessed by using indirect approaches, such as the ^{15}N abundance method. This approach assumes that N_2-fixation can be calculated from a simplified isotope mixing equation between atmospheric N_2 and soil N (Hedin et al. 2009). Several such studies have indicated that putatively N_2-fixing leguminous species in undisturbed *terra firme* forests often do not fix N (Gehring et al. 2005; Ometto et al. 2006; Nardoto et al. 2008; Barron et al. 2011), despite having higher foliar N concentrations than non-leguminous plants (Thompson et al. 1992; Martinelli et al. 1999; Davidson et al. 2007; Nardoto et al. 2008; Fyllas et al. 2009).

Dense and open terra firme *forests*. Estimated values for N_2-fixation in undisturbed forests on *terra firme* are relatively low. Using a combination of model-generated and empirically derived data Cleveland et al. (2010) estimated rates of N_2-fixation for open *terra firme* forests on acrisols at the dry end of the rainfall spectrum for Amazonian tropical rainforests at between 4 and 7 kg N ha^{-1} $year^{-1}$. These values are comparable with the fixation rate of 3 kg N ha^{-1} $year^{-1}$ estimated for lowland tropical forests across the Amazon basin with the isotopic abundance method by Nardoto et al. (2014). The latter study has shown that ^{15}N variation was attributable to site-specific conditions, with ^{15}N being strongly related to extractable soil P and dry season precipitation (Nardoto et al. 2014). The authors suggested that low N availability is only likely to affect forest growth on immature or old weathered soils and/or where dry season precipitation is low. The higher values (16.2 kg N ha^{-1} $year^{-1}$) obtained by Jordan et al. (1982) might be attributable to the use of the different methods.

Heath forest. The only estimate (35 kg N ha^{-1} $year^{-1}$), available for N_2-fixation in a heath forest formation is by Jordan et al. (1982), although high, when compared with that found by the same authors for *terra firme* forest, seems to agree with the fact that heath forests tend to be more N-limited than *terra firme* forests with N_2-fixation being lower in the latter one.

The use of the natural ^{15}N abundance method is dependent on local conditions: ^{15}N values of non-nodulating species should be higher than those of nodulating species and sufficiently different from atmospheric N_2 (Pons et al. 2007). These assumptions are not met for heath forests where soil ^{15}N values are close to unity and there is an absence of a relationship of foliar ^{15}N of nodulating species with nodulation (Martinelli et al. 1999; Roggy et al. 1999; Pons et al. 2007). Nodulating legumes in heath forest on heavily eluted white sand soils are different from nodulating legumes on other soils because they do not show a higher leaf N concentration compared with non-nodulating trees (Pons et al. 2007). Low foliar

^{15}N in plants on heath forest formations with low concentration of extractable P are possibly attributed to a low availability of N in these soils that, as well as flooded forests, are N-limited due to low rates of mineralisation and nitrification (Vitousek and Matson 1988; Luizão et al. 2004; Nardoto et al. 2008; Mardegan et al. 2009). While it is likely that species that nodulate abundantly on white sand in heath forests, such as *Ormosia coutinhoi*, fix atmospheric N_2, this could be limited at low availability of P due to the high P requirement of the process (Pons et al. 2007). Therefore, while individual trees may or may not fix atmospheric N_2 depending on N availability, P may ultimately constrain biological N_2-fixation in tropical soils (Binkley et al. 2003; van Groenigen et al. 2006).

Flooded forest. Isotopic evidence suggests that N_2-fixing may contribute significantly to N input in floodplain forests (Martinelli et al. 1992; Kreibich et al. 2006). Kreibich et al. (2006) estimated that most nodulated legume species of floodplain forests obtained more than 30 % of their N from the atmosphere with no seasonal variation in ^{15}N values. The authors estimated also that N input derived from atmosphere for *várzea* forests was c. 4–5 %. Considering that net biomass production in *várzea* requires, on average, 323 kg N ha^{-1} year^{-1} (Furch 1999), Kern et al. (2011) converted the values in a range of N gain via N_2-fixation between 12.9 and 16.1 kg N ha^{-1}year^{-1}.

Montane forest. Data on biological N_2-fixation in montane forests of the Amazon basin are not available. However, considering that leguminous species are less common in these ecosystems than in lowland vegetation (Crews 1999; ter Steege et al. 2000, 2006, 2013) their overall potential for N_2-fixation is likely to be low.

While the ^{15}N natural abundance method is the only method available for quantitative estimations of symbiotic N_2-fixation in natural forest ecosystems (Pons et al. 2007), it is, at present, insufficient for clearly resolving the rate of N_2-fixation at the individual plant level (Hedin et al. 2009). Its validity is questioned on the grounds that it is a two-source mixing model and that it is not possible to exclude that variation in ^{15}N is caused by reasons other than N_2-fixation only (Roggy et al. 1999; Pons et al. 2007). The considerable complexity observed in ^{15}N across different soil N pools and fractionating paths of the plant-soil N cycle is therefore oversimplified by the method (Hedin et al. 2009).

11.4.1.2 Atmospheric Deposition and Canopy Leaching

Rainfall and fog can contain airborne nutrients that can reach the vegetation during precipitation/fog events. Studies on aerosol and rainfall chemistry in the central part of the Brazilian Amazon basin showed that N-containing aerosols in the atmosphere are primarily derived from oceanic and biogenic sources that are the main determinant of rainfall solute composition in the wet season (Williams et al. 1997; Pauliquevis et al. 2012). In contrast, in the dry season biomass burning emissions have a high impact on rainfall solute composition.

In general, rainfall in Amazonia is characterised by low concentrations of major ions but it contributes important quantities of weak organic acids (Andreae et al. 1990; Williams et al. 1997; Pauliquevis et al. 2012). While Lesack and Melack (1991) reported that overall deposition is seasonally similar, owing to the higher frequency of large precipitation events in the wet season and their relatively large loads, Williams et al. (1997) found seasonal differences in ionic deposition, which occurred in proportion to total rainfall amount. The concentrations of most solutes in rainfall above a *terra firme* forest in central Brazilian Amazonia declined as the wet season advanced (Forti and Moreira-Nordemann 1991). Biomass burning (Germer et al. 2007) is an important source of both NO_3^- and NH_4^+ (Andreae et al. 1988a). Concentrations of aerosol at the transition zone from open *terra firme* rainforest to Cerrado in southern Amazonia undergo pronounced seasonal changes, with peaks occurring at the end of the dry season, when biomass burning is most widespread (Artaxo et al. 2002). In *terra firme* values of NH_4^+ in rainfall average 6.6 kg ha^{-1} year^{-1}, with NO_3^- being 6.1 kg ha^{-1} year^{-1} (Table 11.3). Values of PO_4^{3-} are much lower (0.11 kg ha^{-1} year^{-1}) and are generally of biogenic origin (Table 11.4). While only single studies on nutrient input with rainfall in a tall heath forest are available (Tables 11.3 and 11.4), we expect input to be in the range of values for *terra firme* with variations as a function of the distance from oceanic sources and from anthropogenic activities, as it has also been shown for montane forests.

In montane forests, biomass burning and industrial emissions contribute to high values of NO_3^- in rainfall, while air masses from African deserts carry PO_4^{3-} (Gerold 2008). Boy et al. (2008) found a significant increase in element concentrations of rainfall for total N as well as NO_3^-, with high values of dry deposition of biomass burning-related elements in a montane forest in Ecuador. Values of NH_4^+ plus NO_3^- in rainfall reaching lower montane forests in southern Ecuador range between 9 and 12 kg ha^{-1} year^{-1}, while PO_4^{3-} ranges between 0.3 and 1.8 kg ha^{-1} year^{-1} (Tables 11.3 and 11.4).

Precipitation that reaches the forest floor in *terra firme* forest by throughfall (78–91 %) and stemflow (0.3–1.8 %) appears to be similar to values on other forest types: c. 79 % for flooded blackwater forest or *igapó* (Filoso et al. 1999) and 84–94 % for tall heath forest formations (Herrera 1979; Cornu et al. 1998) with stemflow values of 1.2 % and 1.5 %. Throughfall values in montane forests vary considerably within lower and upper montane forests ranging from 43 % to 91 % and 74–92 %; stemflow ranges between 0.08 % and 1.1 %.

Element concentrations in throughfall are greater than in rainfall as a result of leaching from the leaves, particulate dry deposition (NH_4^+), and gaseous dry deposition (NO_3^-, total N). Values of NH_4^+ and NO_3^- for throughfall in *terra firme* forests average 11.8 and 18.3 kg ha^{-1} year^{-1} (Table 11.3), while PO_4^{3-} averages 1.2 kg ha^{-1} year^{-1} (Table 11.4); in lower montane forests total N ranges from 19 and 29 kg ha^{-1} year^{-1} and PO_4^{3-} from 2.3 to 5.3 kg ha^{-1} year^{-1} (Tables 11.3 and 11.4).

However, throughfall in upper montane forests is often much reduced in NO_3^- and in some cases PO_4^{3-} compared to concentrations in incident rainfall and cloud water (Giambelluca and Gerold 2011). Negative values were reported for difference in annual flux in throughfall and rainfall for total N, NO_3^-, and PO_4^{3-} in upper

Table 11.3 Nitrogen cycling in different forest formations in the Amazon basin

		Terra firme	Open *Terra firme*	Seasonal forest
Pools (kg ha^{-1})				
AGB[a]		1063 ± 278 (2)	1476 ± 124 (2)	–
WD[e]		85 ± 10 (10)	–	–
Fine roots[d]		41 ± 3 (6) (0–10 cm)	–	–
Coarse roots		540 (1)	–	–
Soil[d]				
TOTAL[d]				
Inputs (kg ha^{-1} $year^{-1}$)				
NH_4^+ in precipitation		6.6 ± 2.6 (3)	4.5 (1)	–
NO_3^- in precipitation		6.1 ± 5.9 (2)	0.8 (1)	–
N_2-fixation		4–7[b]	3[b]	–
Outputs (kg ha^{-1} $year^{-1}$)				
Surface flow		–	–	–
Leaching		14 (1)	–	–
Base flow		–	–	–
Deep groundwater flow		–	–	–
Stream export		–	–	–
Denitrification		–	–	–
	N_2O	3.7 ± 1.8 (5)	2.5 ± 0.2 (9)	–
	NO	2 ± 0.5 (4)	1.9 ± 0.5 (2)	–
Balance (INPUTS–OUTPUTS)[c]				
Internal cycling (kg ha^{-1} $year^{-1}$)				
Litterfall		83 ± 4 (18)	5.7 (1)	–
Fine root turnover[d]		28 ± 3 (4) (0–10 cm)	2.1 (1)	–
NH_4^+ in throughfall		11.8 ± 2.8 (6)	–	–
NO_3^- in throughfall		18.3 ± 6.8 (4)	–	–
TOTAL[d]				

THF tall heath forest; *LSHF* low-stature heath forest; *TN* total N

[a]Trunks and branches

[b]Estimated value

[c]NH_4^+ plus NO_3^-

[d]All underground components have been reported in an individualistic manner based on sampling to different depths. For this reason, no total estimates are given. We call for standardised protocols to allow comparative biogeochemical/earth system science studies

[e]The values include the undetermined proportion relative to annual decomposed wood, which should be considered as part of flux. The rate of decomposition, however, is highly variable (see values and references in text Section 'Carbon Fluxes')

THF	LSHF	*Várzea*	*Igapó*	Pre-montane forest	Lower montane	Upper montane
–	–	–	–	–	–	–
–	–	–	46.6 (1)	34 (1)	–	–
–	–	–	–	72 ± 19 (2) (0–30 cm)	54 ± 1 (2) (0–30 cm)	59 (1) (0–30 cm)
–	–	–	–	–	–	–
21 (1)[TN]	–	2.3 ± 0.8 (2)	0.8 (1)	–	8.7–11.9 (3)[TN]	–
	–	6.3 ± 0.9 (2)	3.5 (1)	–		–
35 (1)	–	12.9–16.1[b]	–	–	–	–
–	–	4.3 (1)	–	–	4 (1)[TN]	–
9 (1)	–	2.5 (1)[c]	–	–	–	–
–	–	16.7 (1)	–	–	–	–
–	–	–	–	–	–	–
–	–	12 (1)[TN]	–	–	3.8 (1)[TN]	–
–	–	–	–	0.32 ± 0 (2)	–	–
–	–	–	–	–	–	–
–	–	–	–	–	0.07(1)	0.01 (1)
44 ± 16 (2)	12 (1)	115 ± 18 (2)	–	117 (1)	129 ± 44 (3)	45 (1)
–	–	–	–	48 ± 3 (2) (0–30 cm)	29 ± 4 (2) (0–30 cm)	51 (1) (0–30 cm)
9 (1)[c]	–	13 (1)	0.7 (1)	–	19.2–28.7 (3)[TN]	–
–	–	21 (1)	2.7 (1)	–		–

Table 11.4 Phosphorus cycling in different forest formations in the Amazon basin

	Terra firme	Open *Terra firme*	Seasonal forest
Pools (kg ha^{-1})			
AGB[a]	27 ± 1 (2)	39 ± 2 (2)	–
WD[c]	0.9± 0.1 (10)	–	–
Fine roots[b]	1 ± 0.6 (2) (0–10 cm)	–	–
Coarse roots	17 (1)	–	–
Soil[b]			
TOTAL[b]			
Inputs (kg ha^{-1} $year^{-1}$)			
PO_4^{3-} in precipitation	0.11 ± 0.06 (3)	–	–
Outputs (kg ha^{-1} $year^{-1}$)			
Surface flow	–	–	–
Leaching	–	–	–
Base flow	–	–	–
Deep groundwater flow	–	–	–
Stream export	0.008 (1)	–	–
Balance (INPUTS–OUTPUTS)[c]			
Internal cycling (kg ha^{-1} $year^{-1}$)			
Litterfall	2 ± 0.2 (14)	–	–
Fine root turnover[b]	–	–	–
PO_4^{3-} in throughfall	1.2 ± 0.4 (5)	–	–
TOTAL[b]			

THF tall heath forest; *LSHF* low-stature heath forest

Source references for Tables 11.2, 11.3 and 11.4: Adis et al. (1979); Araujo-Murakami et al. (2014); Baker et al. (2007) (raw data from Chao et al. 2009); Barbosa and Fearnside (1996); Barbosa and Ferreira (2004); Barlow et al. (2007); Bongers et al. (1985); Boy et al. (2008); Brown et al. (1995); Chambers et al. (2000); Chao et al. (2008, 2009, unpublished); Chave et al. (2010); Cleveland et al. (2010); Coolman (1994); Cuevas and Medina (1986); Cummings et al. (2002); da Costa et al. (2014); Davidson et al. (2008); del Aguila-Pasquel et al. (2014); Ferraz et al. (1997) with biomass data from (Chambers et al. 2001a); Filoso et al. (1999); Garcia-Montiel et al. (2001, 2004); Germer et al. (2007); Gibbon et al. (2010); Girardin et al. (2010, 2014); Gouveia Neto (2006); Gurdak et al. (2014); Huasco et al. (2014); Jiménez et al. (2009); Johnson et al. (2006); Jordan et al. (1982); Kauffman et al. (1988, 1995); Keller et al. (2004); Kern et al. (2011); Kesselmeier et al. (2002a); Klinge and Herrera (1983); Klinge and Rodrigues (1968); Kreibich et al. (2003); Leopoldo et al. (1987); Lesack (1993); Lesack and Melack (1996); Lips and Duivenvoorden (1996); Lopez-Gonzalez et al. (2014); Lugli (2013); Lugli (personal communication); Luizão (1989, 1996); Luizão and Schubart (1987); Luizão et al. (2004); Malhi et al. (2009, 2014); Martins (personal communication); Martins et al. (2014); Martinson et al. (2013); Martius (1997); Martius and Bandeira (1998); Melillo et al. (2001); Metcalfe et al. (2007, 2008); Mitchard et al. (2014); Monteiro et al. (2014); Moser et al. (2011); Nardoto et al. (2014); Nascimento and Laurance (2002); Nebel et al. (2001); Neu (2009); Neu et al. (2011); Palace et al. (2007, 2008); Pauliquevis et al. (2012); Pavlis and Jeník (2000); Pyle et al. (2008); Rice et al. (2004); Röderstein et al. (2005); Roman-Cuesta et al. (2011); Saldarriaga et al. (1988); Schöngart and Wittmann (2011); Scott et al. (1992); Selva et al. (2007); Silver et al. (2000, 2005); Smith et al. (1998); Steudler et al. (2002); Tobón et al. (2004); Uhl and Jordan (1984); Unger et al. (2012); Vasconcelos and Luizão (2004); Villela and Proctor (1999); Waterloo et al. (2006); Wilcke et al. (2005, 2008, 2013); Williams and Melack (1997); Williams et al. (1997); Wolf et al. (2011); Wullaert et al. (2010); Yasin (2001)

[a]Trunks and branches

[b]All underground components have been reported in an individualistic manner based on sampling to different depths. For this reason, no total estimates are given. We call for standardised protocols to allow comparative biogeochemical/earth system science studies

[c]The values include the undetermined proportion relative to annual decomposed wood, which should be considered as part of flux. The rate of decomposition however is highly variable (see values and references in text Section 'Carbon Fluxes')

THF	LSHF	*Várzea*	*Igapó*	Pre-montane forest	Lower montane	Upper montane
–	–	–	–	–	–	–
–	–	–	0.9 (1)	0.9 (1)	–	–
–	–	–	–	–	–	–
–	–	–	–	–	–	–
–	–	0.14(1)	0.13 (1)	–	0.31–1.77 (3)	–
–	–	0.08 (1)[TP]	–	–	–	0.55 (1)
–	–	–	–	–	–	–
–	–	0.19 (1)[TP]	–	–	–	–
–	–	–	–	–	–	–
–	–	0.06 (1)[TP]	–	–	–	–
1.4 ± 0.6 (2)	0.4 (1)	2.3 ± 0.6 (2)	–	5.7 (1)	10.5 ± 1 (2)	2 (1)
–	–	–	–	–	–	
–	–	2.9 (1)	0.9 (1)	–	2.3–5.3 (3)	

montane forests in Bolivia (outside the Amazon basin), suggesting that these nutrients are being absorbed by the canopy through foliar uptake by trees and uptake/storage by epiphytes (Gerold et al. 2008). The authors showed that main macronutrients decreased in throughfall from lower to upper montane despite higher rainfall with elevation.

A decrease in total N from rainfall to throughfall (Table 11.3) was also reported in a heath forest formation by Jordan et al. (1982). However, more data would be required to be able to confirm the possible absorption of N by the canopy.

11.4.2 Nitrogen and Phosphorus Stocks and Fluxes

11.4.2.1 Stocks

Above-ground biomass (AGB) Data on N and P stocks are available exclusively for dense (1063 kg N ha^{-1} and 27 kg P ha^{-1}) and open *terra firme* forests (1476 kg N ha^{-1} and 39 kg P ha^{-1}) (Tables 11.3 and 11.4).
Wood debris (WD)-Data on N and P stocks in wood debris are available from two studies, for *terra firme* and flooded forests in central Brazilian Amazonia and for a lower montane forest site in southern Ecuador (Tables 11.3 and 11.4). N stocks average 85 kg N ha^{-1} for *terra firme*, while single values are available for flooded forests (47 kg N ha^{-1}) and lower montane forest (34 kg N ha^{-1}). P stocks for *terra firme* average 0.9 kg P ha^{-1}. The same value was found for a flooded forest and a lower montane forest (Table 11.4).
Coarse roots A single study conducted in forest on *terra firme* on ferralsols in Venezuela showed stocks of 540 kg ha^{-1} for N and 17 kg ha^{-1} for P in total roots (coarse and fine).
Fine roots Few data on N stocks in fine roots are available, exclusively for *terra firme* forests and along an elevation gradient from pre- to upper montane forests. Values average 41 kg N ha^{-1} for *terra firme* (0–10 cm), 72 kg N ha^{-1} in pre-montane forests, and 54 and 59 kg N ha^{-1} in lower and upper montane forests (0–30 cm; Table 11.3). Data on P stocks in fine roots were reported exclusively for *terra firme* (0–10 cm) and average 1 kg P ha^{-1} (Table 11.4).

11.4.2.2 Fluxes

Litterfall Litterfall nutrient concentrations have been shown to be related to soil fertility, with forests on the ferralsol/acrisol groups returning smaller amounts of P and Ca than moderately fertile sites, although N concentrations were similar (Vitousek and Sanford 1986). Nitrogen concentrations in leaf litterfall vary substantially within *terra firme* forests, both at small and across large scales. Forests on relatively fertile soils and on clayey ferralsols tend to have relatively high litterfall N concentrations and recycle large amounts of N annually, while forests on sandy ferralsols are characterised by low litterfall N concentrations and they annually recycle less N (Vitousek and Sanford 1986). Within ferralsols, leaf litterfall from

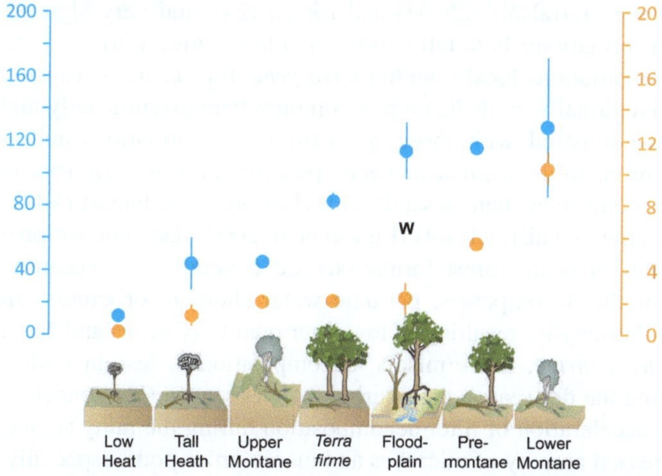

Fig. 11.3 Nitrogen (*blue circles*) and phosphorus (*orange circles*) contents (expressed in kg ha^{-1} year^{-1}) in litterfall of different forest type formations within the Amazon basin. W, white water (locally known as *várzea*)

sandy soils are characterised by lower N content than clayey soils (69 vs. 100 kg ha^{-1} year^{-1}) and values for acrisols are on average 70 kg ha^{-1} year^{-1}. Within ferralsols the same trend was observed for living mature leaves in clayey and sandy soils and the differences were not related to possible differences in the amount of leguminous trees, which was similar in valley (sand) and plateau (clay) (Luizão et al. 2004). Lower values were found for heath forests, both in tall (44 kg ha^{-1} year^{-1}) and low-stature (12 kg ha^{-1} year^{-1}) formations, while higher N litterfall contents were observed in flooded *várzea* (115 kg N ha^{-1} year^{-1}; Table 11.3, Fig. 11.3).

The available data for montane forests in Ecuador show values of 129 kg ha^{-1} year^{-1} in lower montane and 45 kg N ha^{-1} year^{-1} in upper montane forests (Table 11.3, Fig. 11.3).

Phosphorus content in leaf litterfall of *terra firme* forests is on average 2.8 kg ha^{-1} year^{-1} on acrisols, 1.7 kg ha^{-1} year^{-1} on clayey ferralsols, and 1.6 kg ha^{-1} year^{-1} on sandy ferralsols. Higher P content is found in pre-montane (5.7 kg ha^{-1} year^{-1}) and lower montane forest (10.5 kg ha^{-1} year^{-1}), while values for upper montane forest (2 kg ha^{-1} year^{-1}) are comparable to the values for *terra firme* forests (Table 11.4). Litterfall P content in *várzea* is on average 2.3 kg ha^{-1} year^{-1}, while it is lower in low-stature (0.4 kg ha^{-1} year^{-1}) and tall (1.4 kg ha^{-1} year^{-1}) heath forests (Table 11.4).

Decomposition of litter on the forest floor Decomposition rates are highly variable which can be linked to the quality of organic material—high when the C:N ratio is low and decomposers are not N limited and low when C:N is high (Vitousek 1982). Values of the C:N ratio for leaf litter vary across forest types, being intermediate in *terra firme* forests (25–44)—on acrisols (28–44) and on sandy ferralsols (33–43), lower on clayey ferralsols (25–33) and *várzea* (24), and very high on podzols in heath forest formations, both tall (41–64) and low stature (78).

There are variations locally within *terra firme* forests and seasonally inundated forests, and regionally, with decomposition rates being significantly higher in *terra firme* when compared with those of heath forest formations and flooded and montane forests. At the local scale decomposition rates in *terra firme* on ferralsols can be higher on clayey than on sandy soils (Luizão and Schubart 1987), or they can be similar in the two distinct soils (Luizão et al. 2004). Decomposition and nutrient turnover rates in heath forest formations are generally low because of nutrient limitation of the decomposers, periodic water shortage or excess, and the high degree of sclerophylly, resulting in low litter quality (Cuevas and Medina 1986).

Within *terra firme*, on ferralsols, decomposition is less in both plateau and valley, during the dry season than during the wet season (Vasconcelos and Luizão 2004). The acceleration of litter decomposition during the rainy season is attributable to increased activity of microbes and macro-arthropods, especially termites of the genus *Syntermes* (Luizão and Schubart 1987). Humidity stimulates the growth of fine roots (Luizão and Schubart 1987) which directly, or through mycorrhizae, can penetrate the decomposing material and remove essential nutrients (Smith and Read 2008). In contrast, decomposition processes in floodplain forests during the flooded period are slowed down (Adis et al. 1979). There is evidence, at the local scale, that leaf litter in seasonally inundated forest decomposes slower than that in adjacent forest on *terra firme* (Rueda-Delgado et al. 2006; Capps et al. 2011; dos Santos and Nelson 2013). The reduced litter breakdown and decomposition in inundated forests could be attributable to the scarcity of invertebrate leaf shredders and to their probable minor role in organic matter processing (Wantzen et al. 2008; Capps et al. 2011). While the presence of more sclerophyllous leaves in *várzea* and *igapó* forests (Prance 1979; Waldhoff and Furch 2002) could partly explain their lower litter decomposition rates than those in adjacent *terra firme* forests, studies that compare leaf chemistry in these environments are still scant (Klinge et al. 1983) to allow any sound conclusion.

Tropical montane forest soils are typically covered by thick layers of organic matter, in which a considerable amount of nutrients is locked up (Grubb 1977). Lower litter decomposition and soil organic turnover rates are observed with increasing elevation (Vitousek et al. 1994; Tanner et al. 1998; Leuschner et al. 2007), and these have been linked to a decrease in temperature, increase in water logging, increases in the concentrations of hydrogen ions, free aluminium or phenols in the soil solution, and low N and high lignin concentrations of the litter (Berg and McClaugherty 2008). Montane forests in Ecuador are characterised by an increasing C:N ratio of the uppermost organic horizons (L/Of1) along an elevation gradient—22 at 1050 m a.s.l. to 63 at 3060 m a.s.l. (Leuschner et al. 2007)—and an increasing total litter C:N ratio from lower (31) to upper montane forests (59) (Wolf et al. 2011) that indicates possible N limitation of decomposer activity at high elevations and thus low mineralisation rates (Leuschner et al. 2007).

Nitrogen mineralisation, nitrification, and immobilisation N mineralisation involves the breaking down of dissolved organic N by microbes and release into the soil of NH_4^+. When dissolved organic N is insufficient to support microbial requirements, additional inorganic N (NH_4^+ or NO_3^-) is absorbed by microbes from the soil causing 'immobilisation'.

Forest soils dominated by NO_3^- over NH_4^+ have been described as having a relative excess of N with N losses through NO_3^- leaching and NO and N_2O emissions, while forests soils with high concentrations of NH_4^+ over NO_3^- are more N-limited ecosystems (Davidson et al. 2000; Amundson et al. 2003).

Within *terra firme* forests total soil N concentration is significantly higher in clayey than sandy ferralsols and it generally decreases with soil depth (Chauvel et al. 1987; Luizão et al. 2004; Nardoto et al. 2008). Higher concentrations of total mineral N in the upper soil layers are generally associated with greater biological activity and consequently a higher intensity of the mineralisation processes (Luizão et al. 2004). Concentrations of NO_3^- in *terra firme* on ferralsols ranged from ca. 4 to 14 μg N g^{-1} dw while those of NH_4^+ ranged from 3 to 40 μg N g^{-1} dw (Silver et al. 2000; Luizão et al. 1992). These values are comparable in magnitude with those of NO_3^- (2–17 μg N g^{-1} dw) and NH_4^+ (1–14 μg N g^{-1} dw) in soils of open *terra firme* on ferralsols in the state of Rondônia, Brazil (Piccolo et al. 1994; Neill et al. 1997; Hughes et al. 2002). Clayey ferralsols have generally higher NO_3^- and lower NH_4^+ concentrations than those on sandy soils (Livingston et al. 1988; Silver et al. 2000; Luizão et al. 2004). It has been shown that inorganic N in soil of seasonally flooded forests is generally dominated by NH_4^+ indicating the prevalence of reducing conditions throughout the year (Koschorreck and Darwich 2003); NH_4^+ tends to increase during the terrestrial phase, reaching the highest values (25 μg N g^{-1} dw) shortly after flooding and then decreasing until the end of the aquatic phase. Different trends were found by Kern et al. (2011) with highest values of NH_4^+ occurring during the aquatic phase and NO_3^- values declining during the terrestrial phase and increasing during the inundation period. Conversely, NO_3^- accumulated during the terrestrial phase in a seasonally flooded forest in Venezuela, probably owing to the fine texture of the soils that prevented both desiccation and oxygenation (Barrios and Herrera 1994). NH_4^+ is also the dominant form of mineral N in pre-montane (4 μg N g^{-1} dw), lower montane (10 μg N g^{-1} dw), and upper montane forests (13 μg Ng^{-1} dw) in southern Ecuador, with NO_3^- values of 1, 3, and 0.3 μg N g^{-1} dw at 1000 m, 2000 m, and 3000 m, NO_3 being below detection limit in 60 % of all extracts and in 90 % of the extracts at 3000 m (Wolf et al. 2011).

Rates of net N mineralisation and nitrification in soils from open *terra firme* on ferralsols range from 0.3 to 6 μg g^{-1} d^{-1} and from 0.2 to 5 μg g^{-1} d^{-1} (Piccolo et al. 1994; Neill et al. 1997; Hughes et al. 2002). These values are comparable in magnitude to those found in *terra firme* on ferralsols that tend to have higher rates of net N mineralisation and nitrification on clayey plateaux than in sandy valleys, with ranges of 0.5–2 μg g^{-1} d^{-1} and 0.7–2 μg g^{-1} d^{-1} (Silver et al. 2000). Low rates of net N mineralisation and net nitrification (as well as low N concentrations in tree leaves) were reported from sandy valley soil in *terra firme* (Montagnini and Buschbacher 1989; Silver et al. 2000; Luizão et al. 2004). Sotta et al. (2008) found in *terra firme* forest high gross N mineralisation rates and large microbial biomass in plateau clay soils, while sandy valley soils were characterised by low gross N mineralisation rates and lower microbial biomass. Clay soils also had higher gross nitrification rates than sandy soil, suggesting that substrate availability rather than insufficient aeration limited nitrification in sandy soils. Conversely, Vitousek and Matson (1988) showed that N immobilisation, which could result from a proliferation of soil microorganisms, was greater in sandy than in clayey soils in *terra firme* forests. Fluctuations in the availability of soil N can also be driven by seasonal changes in soil moisture, which appears to be related to the delicate balance between the processes of immobilisation and mineralisation. Luizão et al. (1992) observed, for example, that rewetting of soil samples collected after a dry period for the determination of microbial biomass appeared to induce net immobilisation, whereas mineral-N accumulated in situ during dry periods due to net mineralisation. Slow N cycling was found in heath forest formations, where values for net N mineralisation were c. 0.3 μg g^{-1} d^{-1} for low-stature and 0.7 μg g^{-1} d^{-1} for tall heath forest, with net nitrification close to 0 and 0.1 μg g^{-1} d^{-1} (Luizão et al. 2007). Slow mineralisation of organic matter has been proposed as a primary mechanism that limits nutrient supply to vegetation in tropical montane forests (Grubb 1977; Vitousek and Sanford 1986). Net mineralisation and nitrification rates ranged, respectively, between 0.1–0.7 μg g^{-1} d^{-1} and 0–0.4 μg g^{-1} d^{-1} in pre-montane forests, 0.6–1.6 μg g^{-1} d^{-1} and 2 μg g^{-1} d^{-1} in lower montane forests, and 0.1–1.1 μg g^{-1} d^{-1} and 0.1 μg g^{-1} d^{-1} in upper montane forests in north-eastern Ecuador (Wolf et al. 2011; Martinson et al. 2013). With mean daily net N mineralisation rates in the range of 1.5–1.6 μg g^{-1} d^{-1}, these montane forests (1500–2000 m) were supplied with similar amounts of mineral N to that found in moderately fertile tropical lowland forest soils. While soil mass-related N net mineralisation and nitrification rates remained invariant between 500 m and 2000 m, net N mineralisation and nitrification per stand area in montane forests (2000 m a.s.l.) was about 40 % lower than in *terra firme* (500 m a.s.l.).

Fine root turnover Some data on nutrient values for fine root production are available for *terra firme* and along an elevation gradient. In *terra firme* forests N averages 28 kg N $ha^{-1}year^{-1}$ (Table 11.3). Nitrogen values in pre-, lower, and upper montane forests are on average 48 kg N $ha^{-1}year^{-1}$, 29 kg N $ha^{-1}year^{-1}$, and 51 kg N $ha^{-1}year^{-1}$; no data are available for heath and flooded forests.

11.4.3 Losses

11.4.3.1 Leaching, Surface and Subterranean Flows, and Stream Export of N, P, and C

Theoretically, the low cation exchange capacity of the most common Amazonian soils (the dominance of iron and aluminium oxides and kaolinite clay minerals as well as low soil organic matter contents and low pH values) makes these soils prone to become 'leaky', retaining low amounts of mineral nutrients at the exchange sites. On weathered and heavily leached tropical soils, throughfall and surface flows are generally richer in solutes than either incident rainfall or base flow stream water. Available data for *terra firme* on ferralsols showed rates of N leaching of 8 kg ha^{-1} $year^{-1}$ NH_4^+ plus 6 kg ha^{-1} $NO_3^-$$year^{-1}$ (Table 11.3) in the Rio Negro-Casiquiare region of Venezuela (Jordan et al. 1982). Data on N losses (NH_4^+ plus NO_3^-) in a tall heath forest and in *várzea* are available exclusively for leaching with values of 9 kg N ha^{-1} $year^{-1}$ (Jordan et al. 1982) and 3 kg N ha^{-1} $year^{-1}$ (Kreibich et al. 2003). Studies conducted in lower montane forests show values of total N in surface flow (Wilcke et al. 2008) and stream export (Wilcke et al. 2013) of 4 kg N ha^{-1} $year^{-1}$ (Table 11.3). Total surface and base flow N accounted in a central Amazonian flooded forest for 4 kg N ha^{-1} $year^{-1}$ (Williams and Melack 1997) and 17 kg N ha^{-1} $year^{-1}$ (Lesack 1993), while stream water N export was estimated to be of 12 kg N ha^{-1} $year^{-1}$ (Lesack and Melack 1996).

Studies on P export, available exclusively for *terra firme*, flooded forests, and lower montane forests show that P concentrations in soil solution and in forest streams are typically very low. Values of total P in surface and base flow in a flooded forest had values of 0.08 and 0.19 kg P ha^{-1} $year^{-1}$ (Lesack 1993; Williams and Melack 1997) with stream export of 0.06 kg P ha^{-1} $year^{-1}$(Lesack and Melack 1996) (Table 11.4). Total P export in surface flow was found to be 0.6 kg ha^{-1} $year^{-1}$ in lower montane forest in Ecuador (Wilcke et al. 2008), while stream export in *terra firme* showed a value of 0.01 kg ha^{-1} $year^{-1}$ (Leopoldo et al. 1987).

Terra firme forests showed average values of stream DOC export of 1.5 kg C ha^{-1} $year^{-1}$ (Table 11.2). Seasonal forests at the transition with Cerrado showed values of surface flows of 0.5 kg C ha^{-1} $year^{-1}$ for dissolved inorganic C (DIC) and 6 kg C ha^{-1} $year^{-1}$ for dissolved organic C (DOC) (Neu 2009). Values of DOC in deep groundwater flows were comparable to those in surface flows, while DIC values increased to 113 kg C ha^{-1} $year^{-1}$ (Neu et al. 2011). Stream C export reported for seasonal forests showed values of 8 kg DIC ha^{-1} $year^{-1}$ and 13 kg DOC ha^{-1} $year^{-1}$.

11.4.3.2 N Trace Gas Emissions

Ammonia volatilisation, nitrification, and denitrification are related to the rates of soil microbial processes and to soil and environmental characteristics, such as soil N availability, water content, soil texture and structure, moisture-holding capacity, pH, and organic C content. Soil water content, N availability, and labile organic C control microbial nitrification and denitrification and hence the relative proportion of gaseous products of these processes (Verchot et al. 1999; Chapin et al. 2002). While nitrification often produces greater quantities of NO relative to N_2O, the opposite is true for denitrification (Davidson 1993). The first process is often predominant at water content below field capacity, while denitrification increases in wet soils, under conditions of high nitrate and low oxygen.

N_2O. Garcia-Montiel et al. (2004) developed an empirical relationship between N_2O and CO_2 emissions from tropical forest soils based on 7 years of field measurements in Rondônia, Brazil, and combined this relationship with estimates of forest soil respiration across the Amazon basin from a process-based biogeochemistry model to estimate annual regional emissions and spatial patterns of N_2O. The authors concluded that N_2O emissions were spatially variable, with average basin-wide estimates from 7.4×10^5 Mg year^{-1} to 8.3×10^5 Mg year^{-1}, depending on the fraction attributed to root respiration (0.3 and 0.4 respectively). The highest N_2O emissions were estimated in the wettest north-west of the basin and decreased with drier conditions towards the east and south.

Estimates for annual emissions of N_2O from soils of *terra firme* and open transitional forest average 3.7 kg N ha^{-1} year^{-1} and 2.5 kg N ha^{-1} year^{-1}, respectively (Table 11.3). N_2O fluxes show seasonal variation that corresponds to strong seasonal variation in soil water content and are higher in the wet season than in the dry season (Garcia-Montiel et al. 2001; Davidson et al. 2008). A small number of studies have shown that N_2O fluxes were up to 10 times higher in clayey than sandy ferralsols in *terra firme* forests (Luizão et al. 1989; Matson et al. 1990; Silver et al. 2005). These findings, together with the low rates of mineralisation and nitrification that characterise white sand soils, suggest that annual emissions of N_2O from podzols underlying heath forests are likely to be lower than those found for forests on ferralsols and acrisols.

Soil N_2O emissions decreased along an elevation gradient (1000–3000 m a.s.l.) in southern Ecuador and averaged 0.36 kg N ha^{-1} year^{-1} in lower montane and 0.06 kg N ha^{-1} year^{-1} in upper montane forests (Table 11.3). Gas fluxes decreased towards ridge positions and showed negative annual N_2O fluxes at the 3000 m site and N_2O fluxes exceeded NO fluxes at all elevations (Wolf et al. 2011).

NO. Estimates of annual emissions of NO from *terra firme* and open transitional forest averaged 2 kg ha^{-1} year^{-1} (Table 11.3). NO emissions were lower than N_2O emissions from most forest formations and soils, especially during the wet season, because moist conditions favour reduction of NO to N_2O (Garcia-Montiel et al. 2001; Davidson et al. 2008).

The only available estimates of annual emissions of NO from montane forests in the Amazon basin ranged from 0.01 N ha^{-1} $year^{-1}$ in upper montane forests to 0.08 kg N ha^{-1} $year^{-1}$ in pre-montane forests (Wolf et al. 2011) (Table 11.3). NO emissions decrease with elevation with increasing thickness of organic layers, and from lower slopes to ridge positions.

11.4.3.3 Biogenic Volatile Organic Compounds

Biogenic volatile organic compounds (BVOCs) include organic atmospheric trace gases (e.g. isoprenoids, alkanes, alkenes, carbonyls, alcohols, esters, ethers, and acids) (Kesselmeier and Staudt 1999; Jardine and Jardine 2016). Their emissions are affected by a variety of factors, such as light intensity, temperature, and biotic and abiotic stress. Their composition and concentration are determined by seasonal changes as a consequence of distinct seasonal vegetation emissions, climate factors, and anthropogenic influences, such as fires (Kesselmeier et al. 2009). Emission inventories in *terra firme* forests in central and south-western Amazonia showed isoprene, formic acid, and formaldehyde as the most prominent BVOC species present in air at the canopy top, as well as above forests during wet season and wet-to-dry season transition, while a significant increase in formic and acetic acids and formaldehyde and acetaldehyde was observed at the transition phase from dry to wet season (Kesselmeier et al. 2000, 2002b).

Bracho-Nunez et al. (2012) showed that under flooding stress condition, seedlings of tree species from Amazonian floodplain forests produced ethanol and acetaldehyde at the beginning of root anoxia after short-term inundation, while emission of isoprenoids, acetone, and methanol exhibited distinct behaviour related to the origin of the tree species (i.e. *igapó* or *várzea*). The length of the waterlogging period affected the emission of BVOCs, which decreased considerably together with photosynthesis, transpiration, and stomatal conductance.

Biogenic VOCs constitute an integral component of the global carbon cycle (Guenther 2002). Kesselmeier et al. (2002a) estimated an annual loss of C via BVOCs for tropical forests of 0.54 % of gross primary production (GPP). Kesselmeier et al. (2002a), using GPP values of 36 Mg C ha^{-1} $year^{-1}$ for eastern Amazonian forests on *terra firme* (Carswell et al. 2002), calculated that 0.19 Mg C ha^{-1} $year^{-1}$ released as BVOCs are emitted into the atmosphere, representing 3.4 % of net ecosystem production estimated for central Amazonian *terra firme* forests at 5.6 Mg C ha^{-1} $year^{-1}$ by Malhi et al. (1998).

While it is known that BVOCs can be internally recycled with or without intermediate chemical transformation into other organic compounds (e.g. Jardine and Jardine 2016), they can be transformed in the atmosphere to CO_2 by chemical reactions or be deposited to the oceans following atmospheric transformation to water-soluble species; estimates of their fate and the consequences for the carbon cycle remain uncertain (Kesselmeier et al. 2002a). It is known that some potential oxidation products, such as organic acids and aldehydes, contribute significantly to the acidity of the precipitation in central Amazonia (Andreae et al. 1988b). However, while it can be assumed that large amounts of these carbon emissions are recycled within the biosphere, a substantial part is probably transformed into longer-lived oxidation products that are lost from the terrestrial biosphere by transport (Kesselmeier et al. 2009).

11.5 Synthesis and Conclusions

11.5.1 Emerging Patterns for Elemental Cycles Across Forest Formations

11.5.1.1 Carbon

Above-ground biomass (AGB)—There is a clear decreasing pattern for AGB C, from seasonally flooded, *terra firme*, and pre-montane forests, followed by tall heath forests, seasonal and open *terra firme* forests, montane forests, and low-stature heath forest formations. **Wood debris (WD)**—Values for WD C tend to be higher in *terra firme*, tall heath, and upper montane forests and decrease in seasonally flooded, lower montane, and low-stature heath forests. **Fine roots/fine root turnover**—There is an insufficient number of studies for comparison owing to the lack of standard protocols. **Coarse roots**—Values are calculated as a proportion of AGB; there are wide-ranging values from destructive sampling/allometric equations for *terra firme* in the Amazonas State, Brazil (Higuchi et al. 2016). **Input from rainfall/throughfall**—They seem higher in lowland than in montane forest; however, the number of samples is very limited to allow any sound conclusion. **Losses through BCOCs—**Estimates are available exclusively for *terra firme*. **Losses through leaching/streamflow**—Data are available exclusively for dense and open *terra firme* forests and for seasonal forests and are insufficient. **Fine litterfall**—There is a decreasing pattern for fine litterfall, from seasonally flooded, dense and open *terra firme*, and lower montane forests, followed by pre- and upper montane forests and tall and low-stature heath forest formations.

11.5.1.2 Nitrogen

AGB—Scant information is available for dense and open *terra firme* forests only. **WD**—N in WD is tentatively higher in *terra firme* forests, but for other forest types (i.e. seasonally flooded and lower montane forests) only one study is available. **Fine roots/fine root turnover**—Available data are insufficient to allow to draw sound conclusions. **Coarse roots**—The available data are limited to a single study in a forest on *terra firme*. **Input from rainfall**—This does not appear to vary with forest type. **N_2-fixation**—The available data are insufficient to allow to draw sound conclusions. **Losses through leaching/streamflow**—Data from very few studies are available and this makes comparisons difficult once more. **Litterfall**—There is a decreasing pattern from *várzea* and lower montane forests followed *terra firme* and pre-montane forests with the lowest values for tall heath forest formations and upper montane forests. **Input from throughfall**—Values for *terra firme*, *várzea*, and lower montane forests appear similar and higher than those for the other forest types.

11.5.1.3 Phosphorus

AGB—Scant information is available exclusively for dense and open *terra firme* forests. **WD**—Information is available for *terra firme* and appears similar to values for lower montane forests and seasonally flooded forests where only one study is available. **Fine roots/fine root turnover**—Few data are available exclusively for forests on *terra firme*. **Coarse roots**—Only a single study is available for *terra firme*. **Input from rainfall**—Geographic location seems to play an important role in determining the concentration of P in rainfall; however, the lack of data does not allow the identification of potential patterns. **Losses through leaching/streamflow**—There is a general lack of data. **Litterfall**—There is a decreasing pattern from lower to pre-montane forests, *várzea, terra firme*, and upper montane forests followed by tall and low-stature forest formations. **Input from throughfall**—There is a general lack of data.

11.5.2 Hiatuses and the Way Forward

The above synthesis of available information on our understanding of biogeochemistry in Amazonia shows that despite a large body of literature, additional research effort is required for reliable estimates of stocks and fluxes of elements in distinct forest formations. Once such data are available, it will allow us to acquire spatially explicit estimates of the biogeochemistry of the principal vegetation formations and thus refine/extend a range of models (e.g. dynamic vegetation models, or land surface models), currently parameterised using global Amazon basin-wide estimates.

The existing body of work on biogeochemistry is biased towards certain vegetation types and locales. Some early studies are available from lowland forests in Venezuela and Colombia; montane forests studies are few and are largely confined to Ecuador and Peru. For floodplain forests (*várzea* and *igapó*), only few data are available for biomass stocks and estimates of the different compounds of the above-ground and below-ground stocks and production. However, for some other wetland types in Amazonia almost no data exist (e.g. palm swamps (*buritizal*), floodplains along the clear-water rivers, and the extensive net of small streams—c. 1 million km^2).

The great majority of lowland studies on N and P have been conducted in dense forest on *terra firme* in a handful of locations, primarily around Manaus and Santarem in central Brazilian Amazonia, with some additional studies in Rondônia (transition between dense, open, and seasonal forest), sites that formed part of the site network of the Large-scale Biosphere-Atmosphere experiment between 1998 and 2007. Given that dense humid lowland evergreen forests account for c. 40 % of the vegetation of Brazilian Amazonia, this focus appears to have resulted in under-studying other important forest vegetation types, especially the various open low-land forest formations (26 % cover) and alluvial forests that cover an estimated 12 % of the basin in Brazil (IBGE 2012). This is by no means to imply that the dense lowland forest has been over-studied. Far from it, for example, long-term integrated studies of linked biogeochemical cycles are acutely missing. Carbon stock studies within the RAINFOR network have encompassed 130 1-km^2 cells across the basin, with some concentration near research bases, more balanced in terms of coverage of vegetation type than dedicated nutrient cycling studies, however, still lacking proportional coverage in relation to the cover of each vegetation formation. The regional studies of C cycling in eight regions of the Amazon basin by the GEM network (http://gem.tropicalforests.ox.ac.uk/) along with permanent LBA sampling sites broadly represent changing temperature, water, and nutrient availability. However, their coverage is biased towards dense lowland evergreen rainforest on *terra firme* and is not representative of open *terra firme*, floodplain, and heath forest types. The inclusion of additional sites for representativeness of all major Amazon basin vegetation types and studying N and P cycles in tandem with the detailed studies of C cycle would be desirable since the cycles of C, N, and P are interlinked.

There exists an imbalance in the number of studies whose focus has been C, N, and P. There has been relatively more focus on C and N, than on P, and there is an apparent lack of linked studies of C, N, and P cycles. Townsend et al. (2011) pointed out some considerations for the importance of linked biogeochemical cycles, such as (1) understanding nutrient 'limitation' is key to predicting how the C cycle will respond to environmental change; (2) multiple nutrients may limit aspects of the tropical C cycle, yet most models generally consider only single-element limitation; and (3) nutrient interactions are largely absent from widely used projections of future atmospheric CO_2 levels, representing a critical gap in the ability to predict climate change.

The Amazon basin is important in the earth system. Model projections that link vegetation to atmosphere cannot reliably be based on spot studies that are biased towards one or other vegetation type. Equally, most spatially explicit models would require validation in the field, the number of minimum ground validation points depending partly on the spatial coverage of biogeochemically distinct vegetation units and partly on the variance within each unit. Such a spatially representative coverage would allow obtaining more realistic estimates of basin-wide ecosystem structure and function, including vegetation dynamics, biomass, and carbon balance in function of N and P (and other nutrient cycles), and would implicitly include a measure of biodiversity in a broad sense, often undeservedly neglected in biogeochemical studies. Such an information base is sorely missing for coupled vegetation–biogeochemistry–atmosphere models (e.g. Kruijt et al. 2016).

Acknowledgements EB held a post-doctoral grant (SFRH/BPD/77795/2011) from Fundação para a Ciência e Tecnologia, Portugal, and carried out this review while a member of the Centro de Ecologia Funcional, Universidade de Coimbra, based at the Instituto Nacional de Pesquisas da Amazônia, INPA, Brasil. We thank Oliver Phillips for supplying the biomass data referenced in Mitchard et al. (2014) and Lopez-Gonzalez et al. (2014).

References

Adis J, Furch K, Irmler U (1979) Litter production of a central Amazonian black water inundation forest. Trop Ecol 20:236–245

Amundson R, Austin AT, Schuur EAG, Yoo K, Matzek V, Kendall C, Uebersax A, Brenner D, Baisden WT (2003) Global patterns of the isotopic composition of soil and plant nitrogen. Global Biogeochem Cycles 17:1031

Andreae MO, Browell EV, Garstang M, Gregory GL, Harriss RC, Hill GF, Jacob DJ, Pereira MC, Sachse GW, Setzer AW, Dias PLS, Talbot RW, Torres AL, Wofsy SC (1988a) Biomass-burning emissions and associated haze layers over Amazonia. J Geophys Res Atmos 93:1509–1527

Andreae MO, Talbot RW, Andreae TW, Harriss RC (1988b) Formic and acetic acid over the central Amazon region, Brazil: 1. Dry season. J Geophys Res Atmos 93:1616–1624

Andreae MO, Talbot RW, Berresheim H, Beecher KM (1990) Precipitation chemistry in central Amazonia. J Geophys Res Atmos 95:16987–16999

Aragão LEOC et al (2009) Above- and below-ground net primary productivity across ten Amazonian forests on contrasting soils. Biogeosciences 6:2759–2778

Araújo AC, von Randow C, Restrepo-Coupe N (2016) Ecosystem-atmosphere exchanges of CO_2 in dense and open 'terra firme' rain forests in Amazonia. In: Nagy L, Forsberg BR, Artaxo P (eds) Interactions between biosphere, atmosphere and human land use in the Amazon basin. Springer, Heidelberg, pp 147–167

Araujo-Murakami A, Doughty CE, Metcalfe DB, Silva-Espejo JE, Arroyo L, Heredia JP, Flores M, Sibler R, Mendizabal LM, Pardo-Toledo E, Vega M, Moreno L, Rojas-Landivar VD, Halladay K, Girardin CAJ, Killeen TJ, Malhi Y (2014) The productivity, allocation and cycling of carbon in forests at the dry margin of the Amazon forest in Bolivia. Plant Ecol Divers 7:55–69

Artaxo P, Martins JV, Yamasoe MA, Procópio AS, Pauliquevis TM, Andreae MO, Guyon P, Gatti LV, Leal AMC (2002) Physical and chemical properties of aerosols in the wet and dry seasons in Rondônia, Amazonia. J Geophys Res Atmos 107:8081

Baker TR, Coronado ENH, Phillips OL, Martin J, van der Heijden GMF, Garcia M, Espejo JS (2007) Low stocks of coarse woody debris in a southwest Amazonian forest. Oecologia 152:495–504

Barbosa RI, Fearnside PM (1996) Carbon and nutrient flows in an Amazonian forest: fine litter production and composition at Apiaú, Roraima, Brazil. Trop Ecol 37:115–125

Barbosa RI, Ferreira CAC (2004) Biomassa acima do solo de um ecossistema de "campina" em Roraima, norte da Amazônia brasileira. Acta Amazonica 34:577–586

Barlow J, Gardner TA, Ferreira LV, Peres CA (2007) Litter fall and decomposition in primary, secondary and plantation forests in the Brazilian Amazon. For Ecol Manage 247:91–97

Barrios E, Herrera R (1994) Nitrogen cycling in a Venezuelan tropical seasonally flooded forest: soil nitrogen mineralization and nitrification. J Trop Ecol 10:399–416

Barron A, Purves D, Hedin L (2011) Facultative nitrogen fixation by canopy legumes in a lowland tropical forest. Oecologia 165:511–520

Berg B, McClaugherty C (2008) Plant litter: decomposition, humus formation, carbon sequestration, 2nd edn. Springer, Berlin

Binkley D, Senock R, Cromack K (2003) Phosphorus limitation on nitrogen fixation by Facaltaria seedlings. For Ecol Manage 186:171–176

Bongers F, Engelen D, Klinge H (1985) Phytomass structure of natural plant communities on spodosols in southern Venezuela: the Bana woodland. Vegetatio 63:13–34

Boy J, Rollenbeck R, Valarezo C, Wilcke W (2008) Amazonian biomass burning-derived acid and nutrient deposition in the north Andean montane forest of Ecuador. Global Biogeochem Cycles 22, GB4011

Bracho-Nunez A, Knothe N, Costa W, Maria Astrid L, Kleiss B, Rottenberger S, Piedade MT, Kesselmeier J (2012) Root anoxia effects on physiology and emissions of volatile organic compounds (VOC) under short- and long-term inundation of trees from Amazonian floodplains. Springerplus 1:9

Brown IF, Martinelli LA, Thomas WW, Moreira MZ, Ferreira CAC, Victoria RA (1995) Uncertainty in the biomass of Amazonian forests: an example from Rondônia, Brazil. For Ecol Manage 75:175–189

Cairns MA, Brown S, Helmer EH, Baumgardner GA (1997) Root biomass allocation in the world's upland forests. Oecologia 111:1–11

Capps KA, Graça MAS, Encalada AC, Flecker AS (2011) Leaf-litter decomposition across three flooding regimes in a seasonally flooded Amazonian watershed. J Trop Ecol 27:205–210

Carswell FE, Costa AL, Palheta M, Malhi Y, Meir P, Costa JDPR, Ruivo MDL, Leal LDSM, Costa JMN, Clement RJ, Grace J (2002) Seasonality in CO_2 and H_2O flux at an eastern Amazonian rain forest. J Geophys Res Atmos 107:8076

Chambers JQ, Higuchi N, Schimel JP, Ferreira LV, Melack JM (2000) Decomposition and carbon cycling of dead trees in tropical forests of the central Amazon. Oecologia 122:380–388

Chambers JQ, Santos JD, Ribeiro RJ, Higuchi N (2001a) Tree damage, allometric relationships, and above-ground net primary production in central Amazon forest. For Ecol Manage 152:73–84

Chambers JQ, Schimel JP, Nobre AD (2001b) Respiration from coarse wood litter in central Amazon forests. Biogeochemistry 52:115–131

Chao KJ, Phillips OL, Baker TR (2008) Wood density and stocks of coarse woody debris in a northwestern Amazonian landscape. Can J For Res (Revue Canadienne De Recherche Forestiere) 38:795–805

Chao KJ, Phillips OL, Baker TR, Peacock J, Lopez-Gonzalez G, Martinez RV, Monteagudo A, Torres-Lezama A (2009) After trees die: quantities and determinants of necromass across Amazonia. Biogeosciences 6:1615–1626

Chapin FS III, Matson P, Mooney HA (2002) Principles of terrestrial ecosystem ecology. Springer, New York, NY

Chauvel A, Lucas Y, Boulet R (1987) On the genesis of the soil mantle of the region of Manaus, Central Amazonia, Brazil. Experientia 43:234–241

Chave J, Navarrete D, Almeida S, Alvarez E, Aragao L, Bonal D, Chatelet P, Silva-Espejo JE, Goret JY, von Hildebrand P, Jimenez E, Patiño S, Penuela MC, Phillips OL, Stevenson P, Malhi Y (2010) Regional and seasonal patterns of litterfall in tropical South America. Biogeosciences 7:43–55

Cleveland C, Houlton B, Neill C, Reed S, Townsend A, Wang Y (2010) Using indirect methods to constrain symbiotic nitrogen fixation rates: a case study from an Amazonian rain forest. Biogeochemistry 99:1–13

Coolman RM (1994) Nitrous oxide emissions from Amazonian ecosystems. North Carolina State University, Raleigh, NC

Coomes DA, Grubb PJ (1996) Amazonian caatinga and related communities at La Esmeralda, Venezuela: Forest structure, physiognomy and floristics, and control by soil factors. Vegetatio 122:167–191

Cornu S, Lucas Y, Ambrosi JP, Desjardins T (1998) Transfer of dissolved Al, Fe and Si in two Amazonian forest environments in Brazil. Eur J Soil Sci 49:377–384

Crews TE (1999) The presence of nitrogen fixing legumes in terrestrial communities: evolutionary vs ecological considerations. Biogeochemistry 46:233–246

Cuevas E, Medina E (1986) Nutrient dynamics within Amazonian forest ecosystems. I. Nutrient flux in fine litter fall and efficiency of nutrient utilization. Oecologia 68:466–472

Cuevas E, Medina E (1988) Nutrient dynamics within amazonian forests. II. Fine root growth, nutrient availability and leaf litter decomposition. Oecologia 76:222–235

Cummings DL, Kauffman JB, Perry DA, Hughes RF (2002) Aboveground biomass and structure of rainforests in the southwestern Brazilian Amazon. For Ecol Manage 163:293–307

da Costa ACL, Metcalfe DB, Doughty CE, de Oliveira AAR, Neto GFC, da Costa MC, Silva Junior JDA, Aragão LEOC, Almeida S, Galbraith DR, Rowland LM, Meir P, Malhi Y (2014) Ecosystem respiration and net primary productivity after 8–10 years of experimental through-fall reduction in an eastern Amazon forest. Plant Ecol Divers 7:7–24

Davidson EA (1993) Soil-water content and the ratio of nitrous-oxide to nitric-oxide emitted from soil. In: Oremland RS (ed) Biogeochemistry of global change: radiatively active trace gases. Chapman and Hall, London, pp 369–386

Davidson EA, Keller M, Erickson HE, Verchot LV, Veldkamp E (2000) Testing a conceptual model of soil emissions of nitrous and nitric oxides. Bioscience 50:667–680

Davidson EA, de Carvalho CJ, Figueira AM, Ishida FY, Ometto JP, Nardoto GB, Saba RT, Hayashi SN, Leal EC, Vieira IC, Martinelli LA (2007) Recuperation of nitrogen cycling in Amazonian forests following agricultural abandonment. Nature 447:995–998

Davidson EA, Nepstad DC, Ishida FY, Brando PM (2008) Effects of an experimental drought and recovery on soil emissions of carbon dioxide, methane, nitrous oxide, and nitric oxide in a moist tropical forest. Glob Chang Biol 14:2582–2590

Davidson EA, de Araujo AC, Artaxo P, Balch JK, Brown IF, Bustamante MMC, Coe MT, DeFries RS, Keller M, Longo M, Munger JW, Schroeder W, Soares BS, Souza CM, Wofsy SC (2012) The Amazon basin in transition. Nature 481:321–328

de Faria SM, Diedhiou AG, de Lima HC, Ribeiro RD, Galiana A, Castilho AF, Henriques JC (2010) Evaluating the nodulation status of leguminous species from the Amazonian forest of Brazil. J Exp Bot 61:3119–3127

del Aguila-Pasquel J, Doughty CE, Metcalfe DB, Silva-Espejo JE, Girardin CAJ, Chung Gutierrez JA, Navarro-Aguilar GE, Quesada CA, Hidalgo CG, Reyna Huaymacari JM, Halladay K, del Castillo Torres D, Phillips O, Malhi Y (2014) The seasonal cycle of productivity, metabolism and carbon dynamics in a wet aseasonal forest in north-west Amazonia (Iquitos, Peru). Plant Ecol Divers 7:71–83

dos Santos AR, Nelson BW (2013) Leaf decomposition and fine fuels in floodplain forests of the Rio Negro in the Brazilian Amazon. J Trop Ecol 29:455–458

Doughty CE, Metcalfe DB, da Costa MC, de Oliveira AAR, Neto GFC, Silva JA, Aragão LEOC, Almeida SS, Quesada CA, Girardin CAJ, Halladay K, da Costa ACL, Malhi Y (2014) The production, allocation and cycling of carbon in a forest on fertile terra preta soil in eastern Amazonia compared with a forest on adjacent infertile soil. Plant Ecol Divers 7:41–53

Fearnside P (2016) The impact of land use on carbon stocks and fluxes in Brazilian Amazonia: Implications for policy. In: Nagy L, Forsberg BR, Artaxo P (eds) Interactions between biosphere, atmosphere and human land use in the Amazon basin. Springer, Heidelberg, pp 385–405

Ferraz J, Higuchi N, dos Santos J, Biot Y, Marques F, Baker K, Baker R, Hunter I, Proctor J (1997) Distribuição de nutrientes nas árvores e exportação de nutrientes pela exploração seletiva de madeira. In: Higuchi N, Ferraz JBS, Antony L, Luizão F, Luizão R, Biot Y, Hunter I, Proctor J,

Ross S (eds) Bionte: Biomassa e Nutrientes Florestais. Relatório Final. Instituto Nacional de Pesquisas da Amazônia (INPA), Manaus, Brazil, pp 133–149

Filoso S, Williams M, Melack J (1999) Composition and deposition of throughfall in a flooded forest archipelago. Biogeochemistry 45:169–195

Forti MC, Moreira-Nordemann LM (1991) Rainwater and throughfall chemistry in a "terra firme" rain forest: central Amazonia. J Geophys Res Atmos 96:7415–7421

Furch K (1999) Zur Biogeochemie Überschwemmungsgebietes Zentralamazoniens, der *Várzea* auf der Ilha de Marchantaria nahe Manaus, Brasilien. Universität Hamburg, Hamburg

Fyllas NM, Patiño S, Baker TR, Bielefeld Nardoto G, Martinelli LA, Quesada CA, Paiva R, Schwarz M, Horna V, Mercado LM, Santos A, Arroyo L, Jiménez EM, Luizão FJ, Neill DA, Silva N, Prieto A, Rudas A, Silviera M, Vieira ICG, Lopez-Gonzalez G, Malhi Y, Phillips OL, Lloyd J (2009) Basin-wide variations in foliar properties of Amazonian forest: phylogeny, soils and climate. Biogeosci Discuss 6:3707–3769

Garcia-Montiel DC, Steudler PA, Piccolo MC, Melillo JM, Neill C, Cerri CC (2001) Controls on soil nitrogen oxide emissions from forest and pastures in the Brazilian Amazon. Global Biogeochem Cycles 15:1021–1030

Garcia-Montiel DC, Melillo JM, Steudler PA, Tian H, Neill C, Kicklighter DW, Feigl B, Piccolo M, Cerri CC (2004) Emissions of N_2O and CO_2 from terra firme forests in Rondonia, Brazil. Ecol Appl 14:S214–S220

Gehring C, Vlek PLG, de Souza LAG, Denich M (2005) Biological nitrogen fixation in secondary regrowth and mature rainforest of central Amazonia. Agric Ecosyst Environ 111:237–252

Germer S, Neill C, Krusche A, Neto SG, Elsenbeer H (2007) Seasonal and within-event dynamics of rainfall and throughfall chemistry in an open tropical rainforest in Rondônia, Brazil. Biogeochemistry 86:155–174

Gerold G (2008) Soil, climate and vegetation of tropical montane forests—a case study from the Yungas, Bolivia. In: Gradstein SR, Homeier J, Gansert D (eds) The tropical mountain forest—patterns and processes in a biodiversity hotspot, vol 2, Biodiversity and Ecology Series. Göttingen Centre for Biodiversity and Ecology, Göttingen, pp 137–162

Gerold G, Schawe M, Bach K (2008) Hydrometeorologic, pedologic and vegetation patterns along an elevational transect in the montane forest of the Bolivian Yungas. Die Erde 139:141–168

Giambelluca TW, Gerold G (2011) Hydrology and biogeochemistry of tropical montane cloud forests. In: Levia DF, Carlyle-Moses D, Tanaka T (eds) Forest hydrology and biogeochemistry. Springer, Netherlands, pp 221–259

Gibbon A, Silman MR, Malhi Y, Fisher JB, Meir P, Zimmermann M, Dargie GC, Farfan WR, Garcia KC (2010) Ecosystem carbon storage across the grassland-forest transition in the High Andes of Manu National Park, Peru. Ecosystems 13:1097–1111

Girardin CAJ, Malhi Y, Aragao LEOC, Mamani M, Huasco WH, Durand L, Feeley KJ, Rapp J, Silva-Espejo JE, Silman M, Salinas N, Whittaker RJ (2010) Net primary productivity allocation and cycling of carbon along a tropical forest elevational transect in the Peruvian Andes. Glob Chang Biol 16:3176–3192

Girardin CAJ, Aragao LEOC, Malhi Y, Huaraca Huasco W, Metcalfe DB, Durand L, Mamani M, Silva-Espejo JE, Whittaker RJ (2013) Fine root dynamics along an elevational gradient in tropical Amazonian and Andean forests. Global Biogeochem Cycles 27:252–264

Girardin CAJ, Espejob JES, Doughty CE, Huasco WH, Metcalfe DB, Durand-Baca L, Marthews TR, Aragao LEOC, Farfán-Rios W, García-Cabrera K, Halladay K, Fisher JB, Galiano-Cabrera DF, Huaraca-Quispe LP, Alzamora-Taype I, Eguiluz-Mora L, Revilla NS, Silman MR, Meir P, Malhi Y (2014) Productivity and carbon allocation in a tropical montane cloud forest in the Peruvian Andes. Plant Ecol Divers 7:107–123

Gloor E (2016) Climate and the Amazonian carbon balance. In: Nagy L, Forsberg BR, Artaxo P (eds) Interactions between biosphere, atmosphere and human land use in the Amazon basin. Springer, Heidelberg, pp 99–115

Gouveia Neto SC (2006) Concentrações e balanços de carbono orgânico dissolvido em duas bacias do estado de Rondônia: uma comparação entre floresta e pastagem. Universidade de São Paulo, Piracicaba

Grace J (2016) The Amazon carbon balance: an evaluation of methods and results. In: Nagy L, Forsberg BR, Artaxo P (eds) Interactions between biosphere, atmosphere and human land use in the Amazon basin. Springer, Heidelberg, pp 77–98

Grubb PJ (1977) Control of forest growth and distribution on wet tropical mountains with special reference to mineral nutrition. Annu Rev Ecol Syst 8:83–107

Guenther A (2002) The contribution of reactive carbon emissions from vegetation to the carbon balance of terrestrial ecosystems. Chemosphere 49:837–844

Gurdak DJ, Aragão L, Rozas-Davila A, Huasco WH, Cabrera KG, Doughty CE, Farfan-Rios W, Silva-Espejo JE, Metcalfe DB, Silman MR, Malhi Y (2014) Assessing above-ground woody debris dynamics along a gradient of elevation in Amazonian cloud forests in Peru: balancing above-ground inputs and respiration outputs. Plant Ecol Divers 7:143–160

Hedin LO, Brookshire ENJ, Menge DNL, Barron AR (2009) The nitrogen paradox in tropical forest ecosystems. Annu Rev Ecol Evol Syst 40:613–635

Herrera R (1979) Nutrient distribution and cycling in an Amazon caatinga forest on spodosols in southern Venezuela. University of Reading, England

Herrera R, Jordan C, Klinge H, Medina E (1978) Amazon ecosystems: their structure and functioning with particular emphasis on nutrients. Interciencia 3:223–232

Higuchi N, Suwa R, Higuchi FG, Lima AJN, dos Santos J, Noguchi H, Kajimoto T, Ishizuka M (2016) Overview of forest carbon stocks study in Amazonas State, Brazil. In: Nagy L, Forsberg BR, Artaxo P (eds) Interactions between biosphere, atmosphere and human land use in the Amazon basin. Springer, Heidelberg, pp 169–185

Hijmans RJ, Cameron SE, Parra JL, Jones PG, Jarvis A (2005) Very high resolution interpolated climate surfaces for global land areas. Int J Climatol 25:1965–1978

Huasco WH, Girardin CAJ, Doughty CE, Metcalfe DB, Baca LD, Silva-Espejo JE, Cabrera DG, Aragão LEOC, Davila AR, Marthews TR, Huaraca-Quispe LP, Alzamora-Taype I, Mora LE, Farfán-Rios W, Cabrera KG, Halladay K, Salinas-Revilla N, Silman MR, Meir P, Malhi Y (2014) Seasonal production, allocation and cycling of carbon in two mid-elevation tropical montane forest plots in the Peruvian Andes. Plant Ecol Divers 7:125–142

Hughes RF, Kauffman JB, Cummings DL (2002) Dynamics of aboveground and soil carbon and nitrogen stocks and cycling of available nitrogen along a land-use gradient in Rondônia, Brazil. Ecosystems 5:244–259

IBGE (2012) Manual Técnico da Vegetação Brasileira. Sistema fitogeográficoInventário das formações florestais e campestres. Técnicas e manejo de coleções botânicas. Procedimentos para mapeamentos. Instituto Brasileiro de Geografia e Estatística—IBGE Diretoria de Geociências Coordenação de Recursos Naturais e Estudos Ambientais, Rio de Janeiro

Jackson RB, Canadell J, Ehleringer JR, Mooney HA, Sala OE, Schulze ED (1996) A global analysis of root distributions for terrestrial biomes. Oecologia 108:389–411

James EK, Loureiro MF, Pott A, Pott VJ, Martins CM, Franco AA, Sprent JI (2001) Flooding-tolerant legume symbioses from the Brazilian Pantanal. New Phytol 150:723–738

Jardine K, Jardine A (2016) Biogenic volatile organic compounds in Amazonian forest ecosystems. In: Nagy L, Forsberg B, Artaxo P (eds) Interactions between biosphere, atmosphere and human land use in the Amazon basin. Springer, Heidelberg, pp 17–31

Jiménez EM, Moreno FH, Peñuela MC, Patiño S, Lloyd J (2009) Fine root dynamics for forests on contrasting soils in the Colombian Amazon. Biogeosciences 6:2809–2827

Johnson MS, Lehmann J, Selva EC, Abdo M, Riha S, Couto EG (2006) Organic carbon fluxes within and streamwater exports from headwater catchments in the southern Amazon. Hydrol Process 20:2599–2614

Jordan C, Caskey W, Escalante G, Herrera R, Montagnini F, Todd R, Uhl C (1982) The nitrogen cycle in a terra firme rainforest on oxisol in the Amazon territory of Venezuela. Plant Soil 67:325–332

Kauffman JB, Uhl C, Cummings DL (1988) Fire in the Venezuelan Amazon. 1. Fuel biomass and fire chemistry in the evergreen rainforest of Venezuela. Oikos 53:167–175

Kauffman JB, Cummings DL, Ward DE, Babbitt R (1995) Fire in the Brazilian Amazon. 1. Biomass, nutrient pools, and losses in slashed primary forests. Oecologia 104:397–408

Keller M, Palace M, Asner GP, Pereira R, Silva JNM (2004) Coarse woody debris in undisturbed and logged forests in the eastern Brazilian Amazon. Glob Chang Biol 10:784–795

Kern J, Kreibich H, Koschorreck M, Darwich A (2011) Nitrogen balance of a floodplain forest of the Amazon River: the role of nitrogen fixation. In: Junk WJ, Piedade MTF, Wittmann F, Schöngart J, Parolin P (eds) Amazonian floodplain forests. Springer, Netherlands, pp 281–299

Kesselmeier J, Staudt M (1999) Biogenic Volatile Organic Compounds (VOC): an overview on emission, physiology and ecology. J Atmos Chem 33:23–88

Kesselmeier J, Kuhn U, Wolf A, Andreae MO, Ciccioli P, Brancaleoni E, Frattoni M, Guenther A, Greenberg J, De Castro Vasconcellos P, de Oliva T, Tavares T, Artaxo P (2000) Atmospheric Volatile Organic Compounds (VOC) at a remote tropical forest site in central Amazonia. Atmos Environ 34:4063–4072

Kesselmeier J, Ciccioli P, Kuhn U, Stefani P, Biesenthal T, Rottenberger S, Wolf A, Vitullo M, Valentini R, Nobre A, Kabat P, Andreae MO (2002a) Volatile organic compound emissions in relation to plant carbon fixation and the terrestrial carbon budget. Global Biogeochem Cycles 16:1126

Kesselmeier J, Kuhn U, Rottenberger S, Biesenthal T, Wolf A, Schebeske G, Andreae MO, Ciccioli P, Brancaleoni E, Frattoni M, Oliva ST, Botelho ML, Silva CMA, Tavares TM (2002b) Concentrations and species composition of atmospheric Volatile Organic Compounds (VOCs) as observed during the wet and dry season in Rondônia (Amazonia). J Geophys Res Atmos 107:8053

Kesselmeier J, Guenther A, Hoffmann T, Piedade MT, Warnke J (2009) Natural volatile organic compound emissions from plants and their roles in oxidant balance and particle formation. In: Keller M, Bustamante M, Gash J, Dias PS (eds) Amazonia and global change. American Geophysical Union, Washington, DC, pp 183–206

Klinge H, Herrera R (1983) Phytomass structure of natural plant communities on spodosols in southern Venezuela: the tall Amazon caatinga forest. Vegetatio 53:65–84

Klinge H, Rodrigues MA (1968) Litter production in an area of Amazonian terra firme forest part 1. Litterfall, organic carbon and total nitrogen contents of litter. Amazoniana 1:287–302

Klinge H, Furch K, Harms E, Revilla J (1983) Foliar nutrient levels of native tree species from Central Amazonia. I. Inundation forests (varzea, igapo). Amazoniana 8:19–45

Koschorreck M, Darwich A (2003) Nitrogen dynamics in seasonally flooded soils in the Amazon floodplain. Wetl Ecol Manage 11:317–330

Kreibich H, Lehmann J, Scheufele G, Kern J (2003) Nitrogen availability and leaching during the terrestrial phase in a várzea forest of the Central Amazon floodplain. Biol Fertil Soils 39:62–64

Kreibich H, Kern J, Camargo P, Moreira M, Victória R, Werner D (2006) Estimation of symbiotic N_2 fixation in an Amazon floodplain forest. Oecologia 147:359–368

Kriticos DJ, Webber BL, Leriche A, Ota N, Macadam I, Bathols J, Scott JK (2012) CliMond: global high-resolution historical and future scenario climate surfaces for bioclimatic modelling. Methods Ecol Evol 3:53–64

Kruijt B, Meir P, Johnson M, Rammig A, Fauset S, Baker T, Galbrait D, von Randow C, Verbeeck H (2016) Modelling Amazonian carbon budgets and vegetation dynamics in a changing climate. In: Nagy L, Forsberg BR, Artaxo P (eds) Interactions between biosphere, atmosphere and human land use in the Amazon basin. Springer, Heidelberg, pp 331–366

Leopoldo PR, Franken W, Salati E, Ribeiro MN (1987) Towards a water-balance in the central Amazonian region. Experientia 43:222–233

Lesack LFW, Melack JM (1991) The deposition, composition, and potential sources of major ionic solutes in rain of the Central Amazon Basin. Water Resour Res 27:2953–2977

Lesack LFW (1993) Export of nutrients and major ionic solutes from a rain forest catchment in the Central Amazon Basin. Water Resour Res 29:743–758

Lesack LFW, Melack JM (1996) Mass balance of major solutes in a rainforest catchment in the Central Amazon: implications for nutrient budgets in tropical rainforests. Biogeochemistry 32:115–142

Leuschner C, Moser G, Bertsch C, Röderstein M, Hertel D (2007) Large altitudinal increase in tree root/shoot ratio in tropical mountain forests of Ecuador. Basic Appl Ecol 8:219–230

Lima AJN, Suwa R, de Mello Ribeiro GHP, Kajimoto T, dos Santos J, da Silva RP, de Souza CAS, de Barros PC, Noguchi H, Ishizuka M, Higuchi N (2012) Allometric models for estimating above- and below-ground biomass in Amazonian forests at São Gabriel da Cachoeira in the upper Rio Negro, Brazil. For Ecol Manage 277:163–172

Lips JM, Duivenvoorden JF (1996) Fine litter input to terrestrial humus forms in Colombian Amazonia. Oecologia 108:138–150

Livingston GP, Vitousek PM, Matson PA (1988) Nitrous-oxide flux and nitrogen transformations across a landscape gradient in Amazonia. J Geophys Res Atmos 93:1593–1599

Lopez-Gonzalez G, Mitchard ETA, Feldpausch TR, Brienen RJW, Monteagudo A, Baker TR, Lewis SL, Lloyd J, Quesada CA, Gloor E, ter Steege H, Meir P, Alvarez E, Araujo-Murakami-A, Aragão LEOC, Arroyo L, Aymard G, Banki O, Bonal D, Brown S, Brown FI, Ceron CE, Chama Moscoso V, Chave J, Comiskey JM, Cornejo F, Corrales Medina M, Da Costa L, Costa FRC, Di Fiore A, Domingues T, Erwin TL, Fredericksen T, Higuchi N, Honorio Coronado EN, Killeen TJ, Laurance WF, Levis C, Magnusson WE, Marimon BS, Marimon-Junior BH, Mendoza Polo I, Mishra P, Nascimento M, Neill D, Nunez Vargas MP, Palacios WA, Parada-Gutierrez A, Pardo Molina G, Pena-Claros M, Pitman N, Peres CA, Poorter L, Prieto A, Ramirez-Angulo H, Restrepo Correa Z, Roopsind A, Roucoux KH, Rudas A, Salomao RP, Schietti J, Silveira M, De Souza PF, Steiniger M, Stropp J, Terborgh J, Thomas RP, Toledo M, Torres-Lezama A, Van Andel TR, van der Heijden GMF, Vieira ICG, Vieira S, Vilanova-Torre E, Vos VA, Wang O, Zartman CE, de Oliveira EA, Morandi PS, Malhi Y, Phillips OL (2014) Amazon forest biomass measured in inventory plots. Plot Data from "Markedly divergent estimates of Amazon forest carbon density from ground plots and satellites." doi:10.5521/FORESTPLOTS.NET/2014_1

Lugli LF (2013) Estoque de nutrientes na serapilheira fina e grossa em função de fatores edáficos em florstas do Amazonas. Instituto Nacional de Pesquisas da Amazônia, Brazil

Luizão FJ (1989) Litter production and mineral element input to the forest floor in a Central Amazonian forest. GeoJournal 19:407–417

Luizão FJ (1996) Ecological studies in three contrasting forest types in central Amazonia. University of Stirling, Stirling

Luizão FJ, Schubart HOR (1987) Litter production and decomposition in a terra-firme forest of central Amazonia. Experientia 43:259–265

Luizão F, Matson P, Livingston G, Luizao R, Vitousek P (1989) Nitrous oxide flux following tropical land clearing. Global Biogeochem Cycles 3:281–285

Luizão RCC, Bonde TA, Rosswall T (1992) Seasonal variation of soil microbial biomass—the effects of clearfelling a tropical rainforest and establishment of pasture in the central Amazon. Soil Biol Biochem 24:805–813

Luizão RCC, Luizão FJ, Paiva RQ, Monteiro TF, Sousa LS, Kruijt B (2004) Variation of carbon and nitrogen cycling processes along a topographic gradient in a central Amazonian forest. Glob Chang Biol 10:592–600

Luizão FJ, Luizão RCC, Proctor J (2007) Soil acidity and nutrient deficiency in central Amazonian heath forest soils. Plant Ecol 192:209–224

Malhi Y, Nobre AD, Grace J, Kruijt B, Pereira MGP, Culf A, Scott S (1998) Carbon dioxide transfer over a Central Amazonian rain forest. J Geophys Res Atmos 103:31593–31612

Malhi Y, Aragão LEOC, Metcalfe DB, Paiva R, Quesada CA, Almeida S, Anderson L, Brando P, Chambers JQ, Da Costa ACL, Hutyra LR, Oliveira P, Patiño S, Pyle EH, Robertson AL, Teixeira LM (2009) Comprehensive assessment of carbon productivity, allocation and storage in three Amazonian forests. Glob Chang Biol 15:1255–1274

Malhi Y, Farfán Amézquita F, Doughty CE, Silva-Espejo JE, Girardin CAJ, Metcalfe DB, Aragão LEOC, Huaraca-Quispe LP, Alzamora-Taype I, Eguiluz-Mora L, Marthews TR, Halladay K, Quesada CA, Robertson AL, Fisher JB, Zaragoza-Castells J, Rojas-Villagra CM, Pelaez-Tapia Y, Salinas N, Meir P, Phillips OL (2014) The productivity, metabolism and carbon cycle of two lowland tropical forest plots in south-western Amazonia, Peru. Plant Ecol Divers 7:85–105

Mardegan SF, Nardoto GB, Higuchi N, Moreira MZ, Martinelli LA (2009) Nitrogen availability patterns in white-sand vegetations of Central Brazilian Amazon. Trees 23:479–488

Marengo JA, Alves L, Soares W, Rodriguez DA (2016) Extreme seasonal climate variations in the Amazon basin: droughts and floods. In: Nagy L, Forsberg BR, Artaxo P (eds) Interactions between biosphere, atmosphere and human land use in the Amazon basin. Springer, Heidelberg, pp 53–74

Martinelli LA, Victoria RL, Trivelin PCO, Devol AH, Richey JE (1992) 15N natural abundance in plants of the Amazon River floodplain and potential atmospheric N2 fixation. Oecologia 90:591–596

Martinelli LA, Piccolo MC, Townsend AR, Vitousek PM, Cuevas E, McDowell W, Robertson GP, Santos OC, Treseder K (1999) Nitrogen stable isotopic composition of leaves and soil: tropical versus temperate forests. Biogeochemistry 46:45–65

Martins DL, Schietti J, Feldpausch TR, Luizão FJ, Phillips OL, Andrade A, Castilho CV, Laurance SG, Oliveira Á, Amaral IL, Toledo JJ, Lugli LF, Purri Veiga Pinto JL, Mendoza EMO, Quesada CA (2014) Soil-induced impacts on forest structure drive coarse woody debris stocks across central Amazonia. Plant Ecol Divers 1–13

Martinson GO, Corre MD, Veldkamp E (2013) Responses of nitrous oxide fluxes and soil nitrogen cycling to nutrient additions in montane forests along an elevation gradient in southern Ecuador. Biogeochemistry 112:625–636

Martius C (1997) Decomposition of wood. In: Junk W (ed) The central Amazon floodplain. Springer, Berlin, pp 267–276

Martius C, Bandeira AG (1998) Wood litter stocks in tropical moist forest in central Amazonia. Ecotropica 4:115–118

Matson PA, Vitousek PM, Livingston GP, Swanberg NA (1990) Sources of variation in nitrous-oxide flux from Amazonian ecosystems. J Geophys Res Atmos 95:16789–16798

Melack J (2016) Aquatic ecosystems. In: Nagy L, Forsberg BR, Artaxo P (eds) Interactions between biosphere, atmosphere and human land use in the Amazon basin. Springer, Heidelberg, pp 117–146

Melillo JM, Steudler PA, Feigl BJ, Neill C, Garcia D, Piccolo MC, Cerri CC, Tian H (2001) Nitrous oxide emissions from forests and pastures of various ages in the Brazilian Amazon. J Geophys Res Atmos 106:34179–34188

Metcalfe DB, Meir P, Williams M (2007) A comparison of methods for converting rhizotron root length measurements into estimates of root mass production per unit ground area. Plant Soil 301:279–288

Metcalfe D, Meir P, Aragão LC, Costa AL, Braga A, Gonçalves PL, Athaydes Silva Junior J, Almeida S, Dawson L, Malhi Y, Williams M (2008) The effects of water availability on root growth and morphology in an Amazon rainforest. Plant Soil 311:189–199

Mitchard ETA, Feldpausch TR, Brienen RJW, Lopez-Gonzalez G, Monteagudo A, Baker TR, Lewis SL, Lloyd J, Quesada CA, Gloor M, ter Steege H, Meir P, Alvarez E, Araujo-Murakami-A, Aragão LEOC, Arroyo L, Aymard G, Banki O, Bonal D, Brown S, Brown FI, Cerón CE, Chama Moscoso V, Chave J, Comiskey JA, Cornejo F, Corrales Medina M, Da Costa L, Costa FRC, Di Fiore A, Domingues TF, Erwin TL, Frederickson T, Higuchi N, Honorio Coronado EN, Killeen TJ, Laurance WF, Levis C, Magnusson WE, Marimon BS, Marimon Junior BH, Mendoza Polo I, Mishra P, Nascimento MT, Neill D, Núñez Vargas MP, Palacios WA, Parada A, Pardo Molina G, Peña-Claros M, Pitman N, Peres CA, Poorter L, Prieto A, Ramirez-Angulo H, Restrepo Correa Z, Roopsind A, Roucoux KH, Rudas A, Salomão RP, Schietti J, Silveira M, de Souza PF, Steininger MK, Stropp J, Terborgh J, Thomas R, Toledo M,

Torres-Lezama A, van Andel TR, van der Heijden GMF, Vieira ICG, Vieira S, Vilanova-Torre E, Vos VA, Wang O, Zartman CE, Malhi Y, Phillips OL (2014) Markedly divergent estimates of Amazon forest carbon density from ground plots and satellites. Glob Ecol Biogeogr 23:935–946

Mooney HA (2001) 1.22 – Biological diversity, evolution, and biogeochemistry. In: Schulze ED, Heimann M, Harrison SP, Holland EA, Lloyd J, Prentice IC, Schimel DSS (eds) Global biogeochemical cycles in the climate system. Academic Press, San Diego, pp 279–284

Montagnini F, Buschbacher R (1989) Nitrification rates in 2 undisturbed tropical rain forests and 3 slash-and-burn sites of the Venezuelan Amazon. Biotropica 21:9–14

Monteiro MTF, Oliveira SM, Luizão FJ, Cândido LA, Ishida FY, Tomasella J (2014) Dissolved organic carbon concentration and its relationship to electrical conductivity in the waters of a stream in a forested Amazonian blackwater catchment. Plant Ecol Divers 7:205–213

Moreira FMS, da Silva MF, de Faria SM (1992) Occurrence of nodulation in legume species in the Amazon region of Brazil. New Phytol 121:563–570

Moser G, Leuschner C, Hertel D, Graefe S, Soethe N, Iost S (2011) Elevation effects on the carbon budget of tropical mountain forests (S Ecuador): the role of the belowground compartment. Glob Chang Biol 17:2211–2226

Nardoto GB, Ometto JPHB, Ehleringer JR, Higuchi N, Bustamante MMC, Martinelli LA (2008) Understanding the influences of spatial patterns on N availability within the Brazilian Amazon forest. Ecosystems 11:1234–1246

Nardoto GB, Quesada CA, Patiño S, Saiz G, Baker TR, Schwarz M, Schrodt F, Feldpausch TR, Domingues TF, Marimon BS, Marimon Junior B-H, Vieira ICG, Silveira M, Bird MI, Phillips OL, Lloyd J, Martinelli LA (2014) Basin-wide variations in Amazon forest nitrogen-cycling characteristics as inferred from plant and soil 15N/14N measurements. Plant Ecol Divers 7:173–187

Nascimento HEM, Laurance WF (2002) Total aboveground biomass in central Amazonian rainforests: a landscape-scale study. For Ecol Manage 168:311–321

Nebel G, Dragsted J, Vega AS (2001) Litter fall, biomass and net primary production in flood plain forests in the Peruvian Amazon. For Ecol Manage 150:93–102

Neill C, Piccolo MC, Cerri CC, Steudler PA, Melillo JM, Brito M (1997) Net nitrogen mineralization and net nitrification rates in soils following deforestation for pasture across the southwestern Brazilian Amazon Basin landscape. Oecologia 110:243–252

Neu V (2009) O ciclo do carbono na bacia do Alto Xingu: interações entre ambientes terrestre, aquático e atmosférico. Universidade de São Paulo, Piracicaba

Neu V, Neill C, Krusche AV (2011) Gaseous and fluvial carbon export from an Amazon forest watershed. Biogeochemistry 105:133–147

Ometto JHB, Ehleringer J, Domingues T, Berry J, Ishida F, Mazzi E, Higuchi N, Flanagan L, Nardoto G, Martinelli L (2006) The stable carbon and nitrogen isotopic composition of vegetation in tropical forests of the Amazon Basin, Brazil. In: Martinelli L, Howarth R (eds) Nitrogen cycling in the Americas: natural and anthropogenic influences and controls. Springer, Netherlands, pp 251–274

Palace M, Keller M, Asner GP, Silva JNM, Passos C (2007) Necromass in undisturbed and logged forests in the Brazilian Amazon. For Ecol Manage 238:309–318

Palace M, Keller M, Silva H (2008) Necromass production: studies in undisturbed and logged Amazon forests. Ecol Appl 18:873–884

Pauliquevis T, Lara LL, Antunes ML, Artaxo P (2012) Aerosol and precipitation chemistry measurements in a remote site in Central Amazonia: the role of biogenic contribution. Atmos Chem Phys 12:4987–5015

Pavlis J, Jeník J (2000) Roots of pioneer trees in the Amazonian rain forest. Trees 14:442–455

Phillips OL, Lewis SL, Higuchi N, Baker T (2016) Recent changes in Amazon forest biomass and dynamics. In: Nagy L, Forsberg BR, Artaxo P (eds) Interactions between biosphere, atmosphere and human land use in the Amazon basin. Springer, Heidelberg, pp 189–222

Piccolo MC, Neill C, Cerri CC (1994) Net nitrogen mineralization and net nitrification along a tropical forest-to-pasture chronosequence. Plant Soil 162:61–70

Pons TL, Perreijn K, Van Kessel C, Werger MJA (2007) Symbiotic nitrogen fixation in a tropical rainforest: 15N natural abundance measurements supported by experimental isotopic enrichment. New Phytol 173:154–167

Prance GT (1979) Notes on the vegetaion of Amazonia. 3. Terminology of Amazonian forest types subject to inundation. Brittonia 31:26–38

Pyle EH, Santoni GW, Nascimento HEM, Hutyra LR, Vieira S, Curran DJ, van Haren J, Saleska SR, Chow VY, Carmago PB, Laurance WF, Wofsy SC (2008) Dynamics of carbon, biomass, and structure in two Amazonian forests. J Geophys Res Biogeosci 113, G00B08

Quesada CAN, Lloyd J (2016) Soil vegetation interactions in Amazonia. In: Nagy L, Forsberg BR, Artaxo P (eds) Interactions between biosphere, atmosphere and human land use in the Amazon basin. Springer, Heidelberg, pp 267–299

Rice AH, Pyle EH, Saleska SR, Hutyra L, Palace M, Keller M, de Camargo PB, Portilho K, Marques DF, Wofsy SC (2004) Carbon balance and vegetation dynamics in an old-growth Amazonian forest. Ecol Appl 14:55–71

Rocha W, Metcalfe DB, Doughty CE, Brando P, Silvério D, Halladay K, Nepstad DC, Balch JK, Malhi Y (2014) Ecosystem productivity and carbon cycling in intact and annually burnt forest at the dry southern limit of the Amazon rainforest (Mato Grosso, Brazil). Plant Ecol Divers 7:25–40

Röderstein M, Hertel D, Leuschner C (2005) Above- and below-ground litter production in three tropical montane forests in southern Ecuador. J Trop Ecol 21:483–492

Roggy JC, Prévost MF, Garbaye J, Domenach AM (1999) Nitrogen cycling in the tropical rain forest of French Guiana: comparison of two sites with contrasting soil types using δ15N. J Trop Ecol 15:1–22

Roman-Cuesta RM, Salinas N, Asbjornsen H, Oliveras I, Huaman V, Gutierrez Y, Puelles L, Kala J, Yabar D, Rojas M, Astete R, Jordan DY, Silman M, Mosandl R, Weber M, Stimm B, Gunter S, Knoke T, Malhi Y (2011) Implications of fires on carbon budgets in Andean cloud montane forest: the importance of peat soils and tree resprouting. For Ecol Manage 261:1987–1997

Rueda-Delgado G, Wantzen KM, Tolosa MB (2006) Leaf-litter decomposition in an Amazonian floodplain stream: effects of seasonal hydrological changes. J N Am Benthol Soc 25:233–249

Saldarriaga JG, West DC, Tharp ML, Uhl C (1988) Long-term chronosequence of forest succession in the upper rio Negro of Colombia and Venezuela. J Ecol 76:938–958

Schöngart J, Wittmann F (2010) Biomass and net primary production of central Amazonian floodplain forests. In: Junk WJ, Piedade MTF, Wittmann F, Schöngart J, Parolin P (eds) Amazonian floodplain forests. Springer, Netherlands, pp 347–388

Scott DA, Proctor J, Thompson J (1992) Ecological studies on a lowland evergreen rain forest on Maracá Island, Roraima, Brazil. II. Litter and nutrient cycling. J Ecol 80(4):705–717

Selva EC, Couto EG, Johnson MS, Lehmann J (2007) Litterfall production and fluvial export in headwater catchments of the southern Amazon. J Trop Ecol 23:329–335

Silver WL, Neff J, McGroddy M, Veldkamp E, Keller M, Cosme R (2000) Effects of soil texture on belowground carbon and nutrient storage in a lowland Amazonian forest ecosystem. Ecosystems 3:193–209

Silver WL, Thompson AW, McGroddy ME, Varner RK, Dias JD, Silva H, Crill PM, Keller M (2005) Fine root dynamics and trace gas fluxes in two lowland tropical forest soils. Glob Chang Biol 11:290–306

Smith SE, Read DJ (2008) Mycorrhizal symbiosis, 3rd edn. Academic, New York, NY

Smith K, Gholz HL, Oliveira F d A (1998) Litterfall and nitrogen-use efficiency of plantations and primary forest in the eastern Brazilian Amazon. For Ecol Manage 109:209–220

Sotta ED, Corre MD, Veldkamp E (2008) Differing N status and N retention processes of soils under old-growth lowland forest in Eastern Amazonia, Caxiuanã, Brazil. Soil Biol Biochem 40:740–750

Sprent JI (1999) Nitrogen fixation and growth of non-crop legume species in diverse environments. Perspect Plant Ecol Evol Syst 2:149–162

Sprent JI (2009) Legume nodulation. Wiley-Blackwell, United Kingdom

Steudler PA, Garcia-Montiel DC, Piccolo MC, Neill C, Melillo JM, Feigl BJ, Cerri CC (2002) Trace gas responses of tropical forest and pasture soils to N and P fertilization. Global Biogeochem Cycles 16:7-1–7-12.

Sylvester-Bradley R, De Oliveira LA, De Podestá Filho JA, John TVS (1980) Nodulation of legumes, nitrogenase activity of roots and occurrence of nitrogen-fixing Azospirillum ssp. in representative soils of central Amazonia. Agro-Ecosystems 6:249–266

Tanner EVJ, Vitousek PM, Cuevas E (1998) Experimental investigation of nutrient limitation of forest growth on wet tropical mountains. Ecology 79:10–22

ter Steege H, Sabatier D, Castellanos H, Van Aandel T, Duivenvoorden J, Adalardo de Oliveira A, Ek R, Lilvah R, Maas P, Mori S (2000) An analysis of the floristic composition and diversity of Amazonian forests including those of the Guiana Shield. J Trop Ecol 16:801–828

ter Steege H, Pitman NCA, Phillips OL, Chave J, Sabatier D, Duque A, Molino J-F, Prevost M-F, Spichiger R, Castellanos H, von Hildebrand P, Vasquez R (2006) Continental-scale patterns of canopy tree composition and function across Amazonia. Nature 443:444–447

ter Steege H, Pitman NCA, Sabatier D, Baraloto C, Salomão RP, Guevara JE, Phillips OL, Castilho CV, Magnusson WE, Molino J-F, Monteagudo A, Núñez Vargas P, Montero JC, Feldpausch TR, Coronado ENH, Killeen TJ, Mostacedo B, Vasquez R, Assis RL, Terborgh J, Wittmann F, Andrade A, Laurance WF, Laurance SGW, Marimon BS, Marimon B-H, Guimarães Vieira IC, Amaral IL, Brienen R, Castellanos H, Cárdenas López D, Duivenvoorden JF, Mogollón HF, Matos FDDA, Dávila N, García-Villacorta R, Stevenson Diaz PR, Costa F, Emilio T, Levis C, Schietti J, Souza P, Alonso A, Dallmeier F, Montoya AJD, Fernandez Piedade MT, Araujo-Murakami A, Arroyo L, Gribel R, Fine PVA, Peres CA, Toledo M, Aymard CGA, Baker TR, Cerón C, Engel J, Henkel TW, Maas P, Petronelli P, Stropp J, Zartman CE, Daly D, Neill D, Silveira M, Paredes MR, Chave J, Lima Filho DDA, Jørgensen PM, Fuentes A, Schöngart J, Cornejo Valverde F, Di Fiore A, Jimenez EM, Peñuela Mora MC, Phillips JF, Rivas G, van Andel TR, von Hildebrand P, Hoffman B, Zent EL, Malhi Y, Prieto A, Rudas A, Ruschell AR, Silva N, Vos V, Zent S, Oliveira AA, Schutz AC, Gonzales T, Trindade Nascimento M, Ramirez-Angulo H, Sierra R, Tirado M, Umaña Medina MN, van der Heijden G, Vela CIA, Vilanova Torre E, Vriesendorp C, Wang O, Young KR, Baider C, Balslev H, Ferreira C, Mesones I, Torres-Lezama A, Urrego Giraldo LE, Zagt R, Alexiades MN, Hernandez L, Huamantupa-Chuquimaco I, Milliken W, Palacios Cuenca W, Pauletto D, Valderrama Sandoval E, Valenzuela Gamarra L, Dexter KG, Feeley K, Lopez-Gonzalez G, Silman MR (2013) Hyperdominance in the Amazonian Tree Flora. Science 342, 1243092

Thompson J, Proctor J, Viana V, Milliken W, Ratter JA, Scott DA (1992) Ecological studies on a lowland evergreen rain forest on Maraca Island, Roraima, Brazil. I. Physical environment, forest structure and leaf chemistry. J Ecol 80(4):689–703

Tobón C, Sevink J, Verstraten JM (2004) Solute fluxes in throughfall and stemflow in four forest ecosystems in northwest Amazonia. Biogeochemistry 70:1–25

Townsend AR, Cleveland CC, Houlton BZ, Alden CB, White JWC (2011) Multi-element regulation of the tropical forest carbon cycle. Front Ecol Environ 9:9–17

Uhl C, Jordan CF (1984) Succession and nutrient dynamics following forest cutting and burning in Amazonia. Ecology 65:1476–1490

Unger M, Homeier J, Leuschner C (2012) Effects of soil chemistry on tropical forest biomass and productivity at different elevations in the equatorial Andes. Oecologia 170:263–274

van Groenigen KJ, Six J, Hungate BA, de Graaff MA, van Breemen N, van Kessel C (2006) Element interactions limit soil carbon storage. Proc Natl Acad Sci U S A 103:6571–6574

Vasconcelos HL, Luizão FJ (2004) Litter production and litter nutrient concentrations in a fragmented Amazonian landscape. Ecol Appl 14:884–892

Verchot LV, Davidson EA, Cattanio JH, Ackerman IL, Erickson HE, Keller M (1999) Land use change and biogeochemical controls of nitrogen oxide emissions from soils in eastern Amazonia. Global Biogeochem Cycles 13:31–46
Villela DM, Proctor J (1999) Litterfall mass, chemistry, and nutrient retranslocation in a monodominant forest on Maracá island, Roraima, Brazil. Biotropica 31:198–211
Vitousek P (1982) Nutrient cycling and nutrient use efficiency. Am Nat 119:553–572
Vitousek PM, Matson PA (1988) Nitrogen transformations in a range of tropical forest soils. Soil Biol Biochem 20:361–367
Vitousek PM, Sanford RL Jr (1986) Nutrient cycling in moist tropical forest. Annu Rev Ecol Syst 17:137–167
Vitousek PM, Turner DR, Parton WJ, Sanford RL (1994) Litter decomposition on the Mauna Loa environmental matrix, Hawaii—patterns, mechanisms, and models. Ecology 75:418–429
Waldhoff D, Furch B (2002) Leaf morphology and anatomy in eleven tree species from Central Amazonian floodplains (Brazil). Amazoniana-Limnologia Et Oecologia Regionalis Systemae Fluminis Amazonas 17:79–94
Wantzen KM, Yule CM, Mathooko JM, Pringle CM (2008) 3: Organic matter processing in tropical streams. In: Dudgeon D (ed) Tropical stream ecology. Academic, London, pp 43–64
Waterloo MJ, Oliveira SM, Drucker DP, Nobre AD, Cuartas LA, Hodnett MG, Langedijk I, Jans WWP, Tomasella J, de Araujo AC, Pimentel TP, Munera Estrada JC (2006) Export of organic carbon in run-off from an Amazonian rainforest blackwater catchment. Hydrol Process 20:2581–2597
Wilcke W, Hess T, Bengel C, Homeier E, Valarezo C, Zech W (2005) Coarse woody debris in a montane forest in Ecuador: mass, C and nutrient stock, and turnover. For Ecol Manage 205:139–147
Wilcke W, Yasin S, Fleischbein K, Goller R, Boy J, Knuth J, Valarezo C, Zech W (2008) Nutrient status and fluxes at the field and catchment scale. In: Beck E, Bendix J, Kottke I, Makeschin F, Mosandl R (eds) Gradients in a tropical mountain ecosystem of Ecuador. Springer, Berlin, pp 203–215
Wilcke W, Leimer S, Peters T, Emck P, Rollenbeck R, Trachte K, Valarezo C, Bendix J (2013) The nitrogen cycle of tropical montane forest in Ecuador turns inorganic under environmental change. Global Biogeochem Cycles 27:1194–1204
Williams MR, Melack JM (1997) Solute export from forested and partially deforested catchments in the central Amazon. Biogeochemistry 38:67–102
Williams MR, Fisher TR, Melack JM (1997) Chemical composition and deposition of rain in the central Amazon, Brazil. Atmos Environ 31:207–217
Wolf K, Veldkamp E, Homeier J, Martinson GO (2011) Nitrogen availability links forest productivity, soil nitrous oxide and nitric oxide fluxes of a tropical montane forest in southern Ecuador. Global Biogeochem Cycles 25, GB4009
Wullaert H, Homeier J, Valarezo C, Wilcke W (2010) Response of the N and P cycles of an old-growth montane forest in Ecuador to experimental low-level N and P amendments. For Ecol Manage 260:1434–1445
Yasin S (2001) Water and nutrient dynamics in microcatchments under montane forest in the south Ecuadorian Andes. University of Bayreuth, Bayreuth

Chapter 12
Soil–Vegetation Interactions in Amazonia

Carlos A. Quesada and Jon Lloyd

12.1 Historical Perspective

The nature and properties of the soils associated with the apparently luxuriant rainforest vegetation of the Amazon basin have been of interest since the earliest visits of European naturalists to the region. Explorers such as Coudreau, Bates, Katzer, Wallace, and LeCointe were amongst the first to describe Amazonian vegetation and soils, with some of these early observations giving rise to the now discredited idea that such dense and prolific vegetation cover must be associated with a high level of soil fertility (Sombroek 1966; Richter and Babbar 1991). For example, Wallace (1853) claimed that Amazonia's plant production and the universal fertility of the associated soils were 'unequalled anywhere on Earth'. Nevertheless, once agricultural endeavours associated with the colonisation of the interior regions of Brazil began to fail in Amazonia, this view became challenged, and the role of a tight nutrient cycling between vegetation and soils was perceived (Sombroek 1966).

The first real interest in describing the various soils of the Amazon region followed the aftermath of the rubber boom (1879–1912) with the Crude Rubber Survey (Schurz and Hargis 1923) and Marbut and Manifold (1926) first describing the agricultural potentials of Amazonia in relation to its soils. After the collapse of the rubber boom in Amazonia in the early 1900s, little new data were gathered until

C.A. Quesada (✉)
Instituto Nacional de Pesquisas da Amazônia, Manaus, AM, Brazil
e-mail: carlos.quesada@inpa.gov.br

J. Lloyd
School of Marine and Tropical Biology, James Cook University, QLD, Australia

Department of Life Sciences, Imperial College London, London, UK

© Springer-Verlag Berlin Heidelberg 2016
L. Nagy et al. (eds.), *Interactions Between Biosphere, Atmosphere and Human Land Use in the Amazon Basin*, Ecological Studies 227,
DOI 10.1007/978-3-662-49902-3_12

the end of the Second World War when Brazilian Federal initiatives were set in motion to actively colonise the Amazon basin. It is thus the soil survey work of Day and Bennema (1958), Day and Santos (1958), Day (1959, 1961), Day et al. (1964), and Sombroek (1966) that provides the foundations for our current knowledge of Amazonian soils. This knowledge was further expanded by the large-scale *RADAMBRASIL* soil survey of the Brazilian Amazon (1978) and the survey studies from Aubert and Tavernier (1972). The discovery in the 1960s of oil and gas reserves east of the Andes in Peru and Bolivia brought an increased interest in the soils of this 'Low Jungle' region of these countries with Sanchez and Buol (1975) and Cochrane et al. (1985) first reporting on the soils found there. These two studies modified to some extent our knowledge of basin-wide soil patterns, giving rise to a new appreciation of a wider diversity of Amazonian soils than had been previously the case.

Most early motivation for soil science was directed by a perceived need to develop agriculture, and with the exception of Sombroek (1966), there was little interest in the relationships between the natural vegetation and soils. Sombroek (1966) discussed in detail the relationships between soils and the different timber volumes of their forests. He also showed preliminary information about the distribution of some timber species in relation to soils and tentatively related some vegetation formation types to edaphic factors. Later, Sombroek (1984) reviewed and expanded information on Amazonian soils in the light of new soil surveys becoming available in Brazil and Colombia as well as reconnaissance surveys in Peru, Venezuela, and Guiana. In a concluding synthesis study, he suggested that 13 different landforms could be found in Amazonia, each having its unique soil and vegetation associations (Sombroek 2000).

Associated with the International Biological Programme in the 1960s, ecologists began to investigate the functioning of tropical forest ecosystems, e.g. Went and Stark (1968), Stark (1971a, b), Stark and Jordan (1978), Jordan and Stark (1978), Herrera et al. (1978a, b), Medina et al. (1978), Jordan and Herrera (1981), Cuevas and Medina (1986, 1989), Jordan (1989) and Golley (1983, 1986) and Golley et al. (1978). This work, described in more detail in Sect. 1.3, gave the first information about nutrient budgets, direct nutrient cycling, and mechanisms of nutrient conservation on oligotrophic soils and resulted in the first comparative studies in different locations in Amazonia. More recently, the Large-scale Biosphere-Atmosphere Experiment/Programme in Amazonia (LBA) has brought attention once more to Amazonian soils and their influence of tropical forest ecosystem structure and functioning. In this chapter, we review literature with relevance for soil–vegetation interactions, with our main focus being on soils under primary forest vegetation; see Davidson and Martinelli (2009) concerning the biogeochemistry of secondary forests, Luizão et al. (2009) for managed systems, and Trumbore and Camargo (2009) for a discussion on soil carbon stocks and flows in natural and anthropogenic systems in Amazonia.

12.2 The Geology and Geomorphology of the Amazon Basin

Detailed information on the geological history of the Amazon basin can be found in Hoorn and Wesselingh (2010), and a review of interactions between geology and soils is given in Quesada et al. (2011). Here we summarise the three major geological phases that have influenced the Amazonian landscape and shaped the modern physiography of Amazonia and the properties of its soils.

The formation of the Amazon Craton occurred during the Proterozoic (3 to 1 Ga BP), thus being the oldest nucleus of the South American continent. The tectonic processes associated with its formation gave rise to the modern configuration of the basin (de Brito Neves 2002) as well as being an important source of sediments in the intra- and pericratonic sedimentary basins (Kroonenberg and Roever 2010). The Craton itself consists of a stable basement, with the main units of the Guiana Shield stretching below the Amazon basin into the Brazilian Shield, but also continuing westwards under the cover of the sub-Andean basins (Kroonenberg and Roever 2010). Throughout geological history, reactivation processes of the Craton formed arches and belts that eventually acted as drainage divides, thus forming much of the current drainage system of the basin (Hoorn and Wesselingh 2010). The processes of uplift and denudation since the break-up of Gondwana (ca. 200 Ma BP) have resulted in an increased sediment flux towards the basin from the Mesozoic onwards (Kroonenberg and Roever 2010). Pre-drift reconstructions of the Amazon Craton suggest that it was once the western part of Gondwana, thus being connected at that time to what is nowadays western Africa.

Secondly, during the late Ordovician (ca. 440 million years ago) an east-west rift formed a lower zone in the Amazon Craton, separating the Guiana Shield in the north from the Brazilian Shield in the south, with this rift giving rise to the Solimões/Amazonas sedimentary basins (Brito Neves 2002; Hoorn et al. 2010). The process of separation of continents was resumed during the Jurassic (ca. 195 Ma BP), with the sedimentation of the intra-cratonic basin completed at around 120 Ma BP (region IV in Fig. 12.1).

The third important phase in the geological history of the Amazon basin involved the break-up and rifting of the Pangaea supercontinent, with the associated opening of the Atlantic and considerable tectonic movement which eventually led to the Andean uplift and to the formation of the Amazon basin itself (Hoorn et al. 2010). This eventual uplift of the Andes is likely to be the most important geological event in Amazonia, ultimately shaping its geography and having had an influence on the evolution of its living organisms (Hoorn et al. 2010).

The uplift of the Andes led to a reorganisation of drainage divides through time with this having profound influences on the physical environment (Hoorn et al 2010; Vonhof and Kaandorp 2010). Before the rise of the Andes, the drainage divide was located in eastern Amazonia, but during the Palaeogene (ca. 65 Ma BP) it migrated westwards, and by the end of the Palaeogene, the drainage divide separating east and west flowing rivers was located in central Amazonia. At

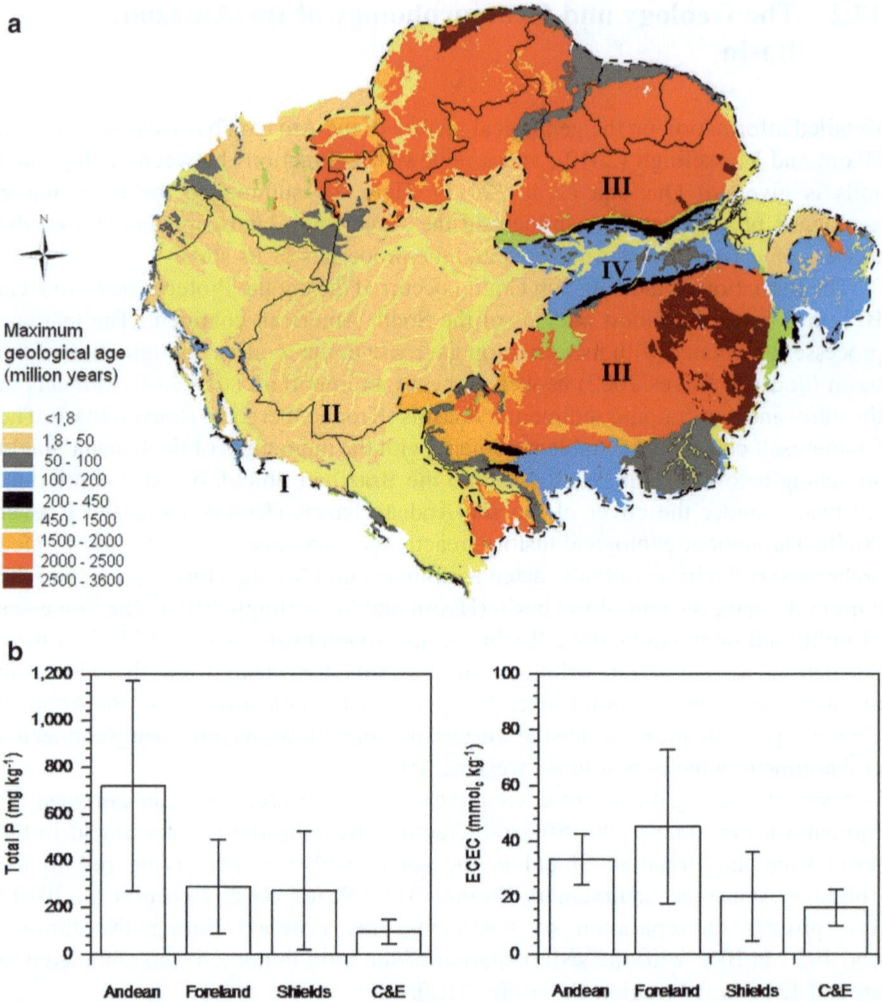

Fig. 12.1 Interactions between soil nutrient concentrations and Amazonian geological history. (**a**) Map of maximum geological age in Amazonia (modified from Quesada et al. 2011), showing the distinct geological provinces. (**b**) Total soil phosphorus concentrations and effective cation exchange capacity (ECEC) of soils occurring on the four different geological zones. Andean (zone I) includes forests occurring between 500 and 3000 m asl in the Andes eastern flank ($n = 17$). Foreland (zone II) consists of forests placed on the sub-Andean basins ($n = 63$). Shields (zone III) include forests occurring on both the Brazilian and Guiana Shields and C&E (zone IV) includes forests occurring on the Central–Eastern intra-cratonic basin ($n = 27$)

about 30 Ma BP, tectonic adjustment and plate subduction along the Pacific margins then caused the uplift of the central Andes. The northern Andes were uplifted later (ca. 23 Ma BP) due to plate break-ups and collision with the South American and Caribbean plates. During that period, a large wetland (of swamps and shallow lakes) was formed in western Amazonia, the so-called Pebas system.

Further uplift of the Andes at about 12 Ma BP has led to a high accumulation of sediment load eroded from the Andes in the western sub-Andean forelands (region II in Fig. 12.1a), generating alluvial megafans. About 10 Ma BP, the water divide did change to the west, pushing sediments from the megafans eastward. The disappearance of the megawetlands in western Amazonia occurred around 7 Ma BP, with the system changing from swamp/lacustrine environment to a fluvial system (the Acre system). This way, the landscape in western Amazonia changed from a negative relief which was accumulating sediments to a positive, river incised relief that has been transporting high sediment load in its waters.

The processes described above resulted in large differences in the age of parent material across the basin which are reflected in today's variations in soil nutrient status (Fig. 12.1b). For example, western Amazonian soils are generally of a higher nutrient status than those of the east because they were formed on more recent sediments that had eroded from the Andes. Indeed, the recently formed alluvial soils of western Amazonia are still receiving Andean sediments (Quesada et al. 2011). Much of the sediment being deposited in western Amazonia over recent millennia had previously been protected from weathering during submersion in water during the Pebas phase. Thus, processes of soil formation in the region have only become significant from the Pliocene onwards and with much of the region having soils that are less than 2 million years old (Quesada et al. 2011).

Conversely, parent material on the Guiana and Brazilian shields (region III in Fig. 12.1a) is very old (Proterozoic). Nevertheless, the fact that many of these soils have developed over crystalline rocks seems to counterbalance the effect of old-age parent material. This results in many soils being of a somewhat higher nutrient status when compared to the comparatively younger intra-cratonic basin. Especially as this is a relatively intensely inhabited and widely studied area of the basin, it is important to note that the sediments which fill the intra-cratonic basin (region IV in Fig. 12.1a) have originated from ancient weathering of the Brazilian and Guiana Shields. Thus, the soils which eventually evolved in the intra-cratonic basin have been based on already pre-weathered material that has then suffered subsequent sedimentation and that subsequently underwent another cycle of weathering. This has resulted in very low soil fertilities in that area (Quesada et al. 2011).

The above regional variations in geology, lithology, and geomorphology have resulted in a high diversity of soils with varied physical and chemical properties. For example, Quesada et al. (2011) have shown that Amazonian soils are highly diverse, with 19 of the 32 World Reference Base (WRB) soil groups being found within the basin. These authors also questioned the general notion that the vast majority of tropical soils are intensively weathered and nutrient poor and showed that Amazonian soils spanned almost all soil taxa, from younger to the most-developed soil groups, with many of them having a relatively high nutrient content by global standards. For instance, total soil P, which is the most likely element to limit tropical forest productivity (Quesada et al. 2012), ranges from as little as 10 mg kg^{-1} in some white sands of Guiana, to about 850 mg kg^{-1} in Peruvian lowland alluvial soils, and as much as 1630 mg kg^{-1} in the soils of Andean cloud

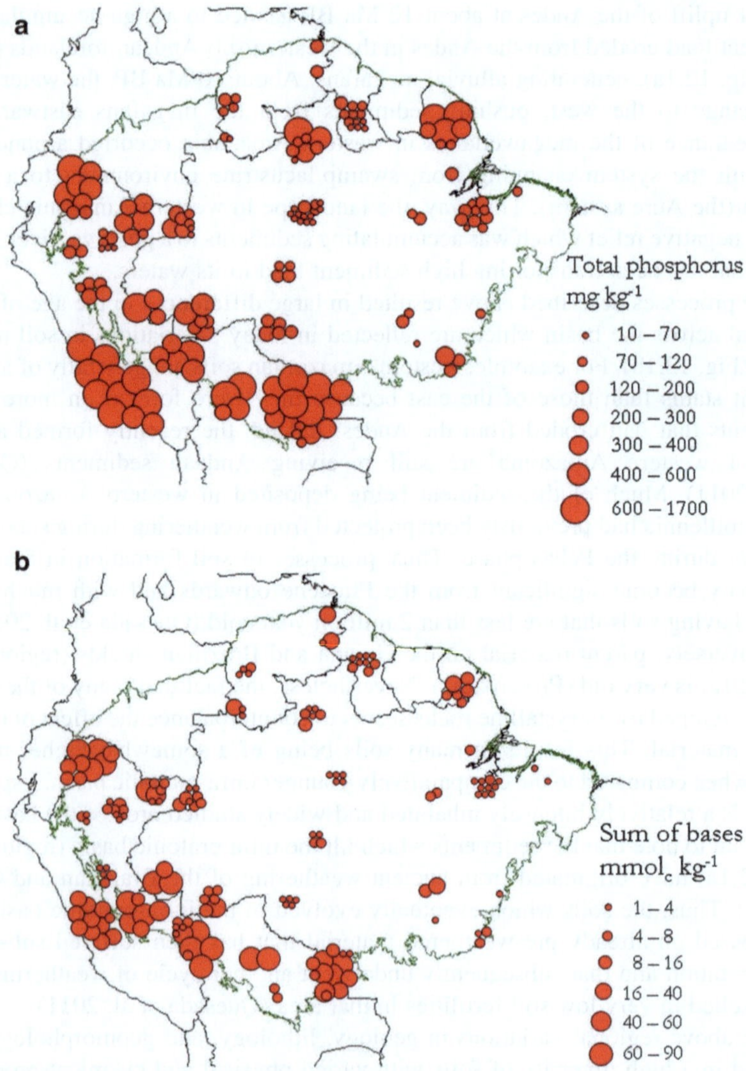

Fig. 12.2 Geographical distribution of (**a**) total soil phosphorus and (**b**) sum of bases (exchangeable) in Amazonia. Each point is a permanent plot from the RAINFOR network, where the size of circles indicates variations in concentrations

forests (Fig. 12.2a). A similar geographic pattern is discernible for soil exchangeable cations (Fig. 12.2b).

This wide fertility spectrum across Amazonia occurs in parallel with large variation in soil physical properties. Younger soil types almost invariably show a lower degree of profile development, often being shallow and with hard subsurface horizons that restrict root growth. Soil types that have resulted from weathering for many millions of years usually have favourable physical properties, such as good soil structure, good drainage, and, due to their depth, a high water storage capacity. Across the Amazon basin, soil chemical and physical properties are thus correlated as both are controlled by common pedogenic processes (Quesada et al. 2010).

The broad variations in chemical and physical properties mentioned above are also reflected by differences in soil carbon content and in the mechanisms by which soil carbon is stabilised. For instance, soil carbon concentration tends to show a strong positive correlation with soil clay content for highly weathered soils, such as Ferralsols and Acrisols (Telles et al. 2003; Dick et al. 2005). This relationship for soils dominated by a 1:1 clay mineralogy is thought to exist because more finely textured soils also have higher specific surface area (SSA), with a higher abundance of exchange sites on the mineral surface; this then results in more soil carbon stabilisation through mineral–organic matter associations (Saidy et al. 2012). Nevertheless, this relationship only occurs because Ferralsols and Acrisols have their mineralogy dominated by only one mineral, kaolinite. Thus, their specific surface area varies with changes in soil clay content in a simple way (Fig. 12.3a). On the other hand, soil carbon content in young and intermediate age soil groups with a more complex mineralogy is not simply dependent on the amount of clay found in the soil (Fig. 12.3b). Rather, it is the nature of the minerals present in the soil that underlies differences in SSA and hence the carbon-holding capacity of these soils (Denef et al. 2009). This difference may have important implications for our

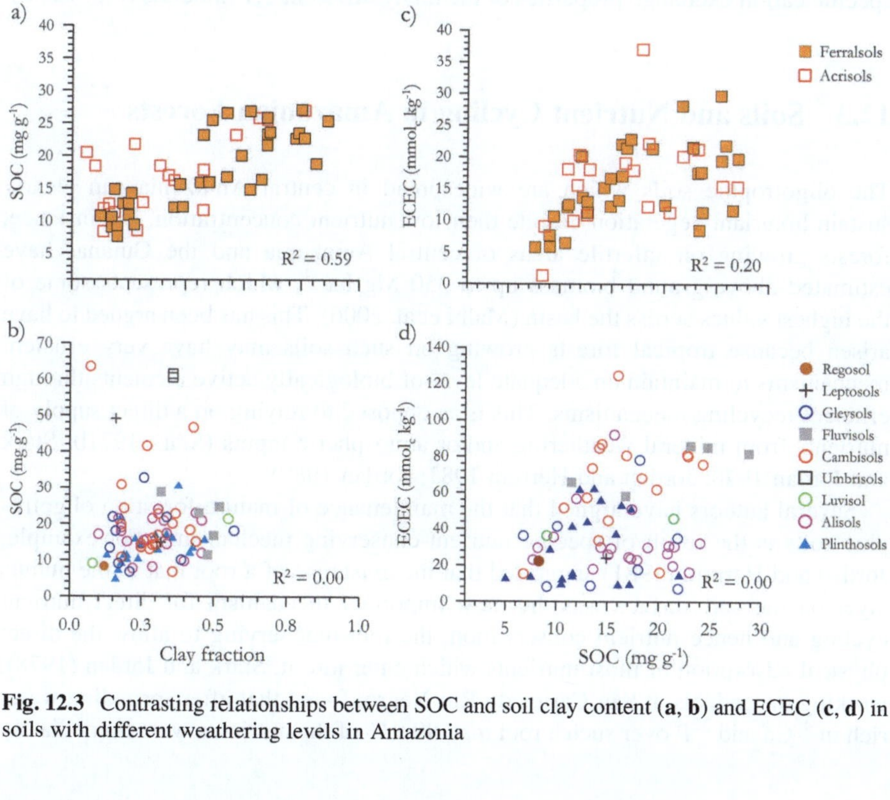

Fig. 12.3 Contrasting relationships between SOC and soil clay content (**a**, **b**) and ECEC (**c**, **d**) in soils with different weathering levels in Amazonia

understanding of soil carbon turnover since determinations of soil carbon residence time in Amazonia are still rare and when available have been restricted to highly weathered soils, such as Ferralsols and Acrisols (Telles et al. 2003). What little information is available indicates a low residence time of bulk soil carbon of ca. 10 years as indicated by ^{14}C dating studies (Trumbore and Camargo 2009). This indicates that most of the carbon in these soils is readily mineralised and returned to the atmosphere. This is, however, likely to be different for young and intermediate age soil groups since mechanisms for soil carbon stabilisations are different, with these soils likely to be able to hold carbon in larger quantities and/or at greater strength, such as it has been observed in smectite-rich soils, where considerably longer mean residence times have been found (Wattel-Koekkoek et al. 2003).

In addition to its biogeochemical importance, soil organic matter is also responsible for the maintenance of soil fertility in the tropics, with a considerable fraction of soil cation exchange capacity (a surrogate for soil fertility) of Ferralsols being associated with organic matter exchange sites (Tiessen et al. 1994). Although this is observed for both Ferralsols and Acrisols in Amazonia (Fig. 12.3c), younger and intermediate soil types generally have a much higher cation exchange capacity, which, rather than being dependent on soil organic matter, is dependent on their highly variable mineralogical composition. This occurs with a much greater range of variation than the more highly weathered soils as a consequence of the different specific cation exchange properties of the many different 2:1 minerals (Fig. 12.3d).

12.3 Soils and Nutrient Cycling in Amazonian Forests

The oligotrophic soils which are widespread in central Amazonia can clearly sustain luxuriant vegetation despite their low nutrient concentration. For instance, forests growing on infertile areas of central Amazonia and the Guianas have estimated above-ground biomass up to 350 Mg ha^{-1}, which represents some of the highest values across the basin (Malhi et al. 2006). This has been argued to have arisen because tropical forests growing on such soils may have very efficient mechanisms to maintain an adequate level of biologically active elements through efficient recycling mechanisms. This is as opposed to relying on a direct supply of nutrients from mineral weathering and/or atmospheric inputs (Stark 1971b; Stark and Jordan 1978; Jordan and Herrera 1981; Jordan 1989).

Several authors have argued that the maintenance of mature forest on oligotrophic soils is the result of specific nutrient-conserving mechanisms. For example, Jordan and Herrera (1981) suggested that the existence of a root mat in the humus layer on the soil surface was the most important mechanism for direct nutrient cycling and hence nutrient conservation, the root mat serving to allow the direct physical adsorption of most nutrients which enter into it. Stark and Jordan (1978), working on podzols at San Carlos do Rio Negro, found that after sprinkling water rich in ^{45}Ca and ^{32}P over such a root mat, 99.9 % of the nutrients were immediately absorbed and only 0.1 % leached into the soil beneath. When these isotopes were applied to the experimental plot in the form of radioactive labelled leaves, no detectable activity was found in the leachate (Jordan and Stark 1978). Stark and Jordan (1978) also showed that detectable levels of radioactivity occurred in samples outside as well as inside the experimental plots, showing that the isotopes were actually taken up and translocated elsewhere by the root systems. Roots directly outside the plots had isotope concentrations in the same range as the ones inside the experimental plot, the authors suggesting that this reflected a large investment in lateral roots for nutrient uptake.

Herrera et al. (1978b) further provided evidence for mycorrhizal fungi providing an important mechanism by which nutrients are transferred from organic matter to living roots. In that study, ^{32}P was incorporated into a leaf which was later placed in isolation with a living root in a Petri dish on the forest floor. Radioactivity was later found both in the root and in the mycorrhizal connection between leaf and root. In a set of time series photographs, Herrera et al. (1978a) showed how a fresh leaf can become enmeshed by roots with, after several weeks of decomposition, only a skeleton remaining. The authors' inference was that nutrient transfer could occur directly from decomposing leaf tissue to the roots of growing plants through mycorrhizal associations. This is in accordance with the hypothesis of direct cycling by Went and Stark (1968), which suggested that nutrients are obtained from litter on the forest floor by mycorrhizal fungi which then pass the nutrients onto living roots.

Other nutrient-conserving mechanisms are thought to be associated with the root mat itself. These include a rapid growth of feeder roots (Jordan and Escalante 1980), the existence of algae on the root mat which absorb nutrients from throughfall and subsequently releasing it as they decompose (Jordan and Herrera 1981), nitrogen fixation by microorganisms (Jordan and Herrera 1981), and an absence of denitrifying bacteria (the latter thought to be a consequence of large amounts of tannins in organic matter and a continuous low pH—(Jordan 1979). Herrera et al. (1978b) also reported that the direct element adsorption on surfaces of organic material probably prevents losses by leaching until nutrients can be taken up by roots.

Other nutrient-conserving mechanisms have been postulated to be associated with oligotrophic environments. These include long-lived evergreen scleromorphic leaves (Medina et al. 1978), nutrient resorption, notably P, N, and K before leaves are shed (Small 1972), nitrogen fixation, and scavenging of rainfall nutrients by algae and lichens in canopy leaves (Herrera et al. 1978a). Also important may be the production of phenols and other toxic secondary compounds. These are thought to reduce nutrient losses by herbivory and thus extend leaf lifespan leading to a more efficient use of nutrients (Jordan and Herrera 1981; Coley et al. 1985) and a fast recovery by secondary vegetation following forest gap formation (Jordan and Herrera 1981)—all of which can be seen to contribute to the avoidance of nutrient losses by leaching.

Following the observation of such nutrient-conserving mechanisms, a perception that tropical forests growing on poor soils are decoupled from their soil nutritional environment ensued (Golley 1983, 1986; Golley et al. 1978). Jordan (1982) presented data from nutrient input through rainfall and output by leaching for a forest growing on infertile soil in Venezuela. He found that rates of nutrient leaching were equal to or less than the inputs from atmosphere, thus concluding that once the forest was not successional and aggrading, this meant that the weathering of parent material was not contributing to the forest nutrient economy. It is thus generally considered that most tropical forests can maintain themselves solely on inputs from the atmosphere and nutrient cycling (Vitousek and Sanford 1986). However, many of the studies reported above, and especially those at San Carlos in Venezuela (Medina et al. 1977), focused their attention on sandy podzols (Vitousek and Sanford 1986). The podzols around San Carlos de Rio Negro represent an extreme situation where plants grow on soils of unusually low nutrient concentrations (mass basis) even by Amazon basin standards (Fyllas et al. 2009; Quesada et al. 2010).

The nutrient cycles of tropical forests may then be best considered as only sometimes reflecting an efficient, closed or semi-closed system where a tight internal nutrient cycling serves to hold nutrients within the biomass and with minimal inputs or outputs via weathering and leaching, respectively. In terms of quantifying efficiency, Vitousek (1982, 1984) suggested that the nutrient use of a tropical tree or stand of trees should be considered efficient if a relatively large amount of organic matter is fixed per unit of nutrient taken up by the plant. One process determining this efficiency would be the capacity of a given plant to reabsorb nutrients from structures that undergo subsequent senescence. At the ecosystem level, the nutrients released from organic matter decomposition should be rapidly absorbed by roots, mycorrhizae, and soil decomposer organisms or adsorbed by surface charges in organic matter and thus retained by the system. This would minimise nutrient losses despite the potentially large amount of nutrients circulating in the forest system.

Vitousek (1984) further suggested that an efficient within-stand nutrient economy could reflect a limitation on net primary production (NPP) while an inefficient within-stand nutrient economy would be suggestive of an adequate or better nutrient supply. Based on such assumptions, he determined that most lowland forests in Amazonia have a relaxed nitrogen economy as they are inefficient regarding nitrogen use, cycling large amounts of it in litterfall (dry mass:nitrogen ratio of *ca.* 80). On the other hand, montane forests have high efficiency for nitrogen (dry mass:nitrogen ratio of *ca.* 160). Regarding phosphorus and calcium, he found signs that those nutrients could limit forest development as some of the lowland forests have higher efficiency for both P (dry mass:phosphorus ratio ranges from 1000 to 8000) and Ca (dry mass:calcium ratio ranges from 100 to 800). Vitousek's results were subsequently confirmed by Silver (1994) who tested Vitousek's indices of nutrient use efficiency (litterfall mass:nutrient ratio) against soil nutrient indices. She found that extractable P was strongly and significantly correlated to the litterfall mass:P ratio. A weaker but significant relationship was found with Ca, while no relationship was found for N. Soil P concentration was also strongly correlated with production of litterfall mass in lowland tropical forests.

Grubb (1989) was strongly critical of the nutrient use efficiency (NUE) approach as in his view, at least as applied by its proponents, it failed to take into account differing allocations of forest production to leaves or stem wood. Furthermore, NUE omits the uptake of nutrients followed by leaching, as well as ignoring variations in NUE due to the plant age which may be very important in mixed-age communities. Nutrient use efficiency thus has been a debated concept, and a number of different definitions and indices have emerged (Chapin 1980; Vitousek 1982; Berendse and Aerts 1987; Bridgham et al. 1995; Aerts and Chapin 2000; Hiremath and Ewel 2001). At the forest stand level, the most appropriate approaches may include the production of biomass per unit of nutrient taken up (Berendse and Aerts 1987) and production per unit of nutrient available in the soil (Bridgham et al. 1995). Berendse and Aerts (1987) suggested two mechanisms by which a tree could be more efficient than another: (1) intrinsic nutrient productivity (the rate of production per unit of nutrient) and (2) the mean residence time of a unit of nutrient within plants.

Herbert et al. (1999) studied nutrient use efficiencies across a soil age gradient in Hawaii and found that N and P productivity increased about twofold across the age gradient. Greater nutrient productivity contributed substantially to overall NUE in the oldest substrate when compared to the youngest one. In contrast, a longer nutrient residence time within the canopy contributed to NUE in both old and younger substrates. Despite Grubb's criticisms, a considerable part of recent literature about forest nutrient economy is still based on the NUE approach (see Cuevas and Medina 1986, 1989; Tanner et al. 1998; Smith et al. 1998; Sollins 1998; Roggy et al. 1999; McGrath et al. 2001).

Although no nutrient cycling system can be assumed to be leak proof, it does seem logical that nutrient cycles should become tighter as nutrient availability through weathering declines, with this at least partially decoupling the nutrient economy of a given forest from its underlying soil fertility. Other factors may also be relevant, such as the decline of plant-accessible soil phosphorus forms during soil formation and ecosystem evolution. These occur not only through long-term losses in total P amount through leaching and loss of soil mass but also by the transformations of P pools to unavailable, recalcitrant forms (Walker and Syers 1976; Quesada et al. 2010). In most situations, the phosphorus economy is still, however, to a large degree, dependent on the P retention and exchange characteristics of the soil. For example, even when organic forms are the dominant source of phosphorus this still requires that transformations from organic to inorganic P species occur within the mineral soil matrix. This is because mineralisation of organic phosphorus is thought to occur mostly through interactions with extracellular phosphohydrolases excreted into the soil by roots and litter-decomposing and soil microorganisms, with abiotic mineralisation of organic phosphorus compounds being far less efficient (Quiquampoix and Mousain 2005).

As outlined in Sect. 1.2, some soils in the Amazon basin do indeed have high nutrient concentrations or have the products of parent material weathering close to the plant rooting zone. Forests on such soils might not, therefore, be expected to exhibit so strong internal nutrient cycling characteristics as the forests on dystrophic soils. That is to say, on relatively fertile soils, forest nutrient cycles may have important weathering inputs from the soil, this potentially being the case for all elements other than N. With respect to the processes discussed above, Baillie (1989) reported that ideally, there were two types of nutrient cycle. The first is an 'open' system where there are losses of nutrients to the atmosphere, surface water, and groundwater. These losses are balanced by gains from rainfall and aerosol depositions, but mainly from mineral weathering in the soil and/or upper part of the saprolite (Jordan 1985; Kellman and Carty 1986). The second, a 'closed' system, occurs where there are virtually no nutrient inputs from weathering. This system is more efficient and is almost leak proof for much of the time. Any losses experienced are replenished by atmospheric inputs (Jordan 1982). In this type of system, nutrients absorbed by roots are almost totally derived from organic matter decomposition. The two types of cycling correspond to the coupled and uncoupled models of Burnham (1989), but, rather than being clearly defined groups, they represent the extremes of a range in nutrient cycles. Indeed, as has been shown by Quesada et al. (2010, 2011) many tropical forests are actually likely to be in intermediate situations of partial weathering inputs to the system, and thus, a wide range of nutrient-conserving mechanisms probably exist with an associated continuum in extent of nutrient cycle openness/closure.

Baillie (1989) reported that although the factors affecting the nature and scale of weathering inputs include forest successional status, site stability, and climate, it is the soil's physical and chemical characteristics which are the more important. Soil morphological features, such as poor subsoil drainage, or a considerable soil depth—the latter being typical of highly weathered tropical soils (Burnham 1989)—or mechanical impedance to root penetration may exclude roots from zones where weathering is occurring—thus leading to closed systems. Likewise, soils at early stages of weathering, soils with volcanic ash deposition, some alluvial soils, soils on erosional slopes, and soils exposed to instabilities, such as earthquakes or landslides, are expected to be characterised by a regular (though perhaps erratic) contribution of parent material weathering to the nutrient cycle and should therefore approach open systems (Baillie 1989; Burnham 1989). In such systems with weathering inputs, the type of parent material, its rate of weathering, the type and rate of nutrient release, and its depth of placement in the soil are all important in determining the rate of the release of plant-available nutrients. There are also soil effects in the working of closed cycles. For example, texture and mineralogy can influence the ability of an ecosystem to retain and release nutrients.

Full measures of ecosystem nutrient cycle 'openness' or 'closure' effectively require measures of both the rates of nutrient uptake and the release of it by the vegetation as well as input/outputs, such as those associated with wet and dry deposition, weathering, and leaching losses. Nevertheless, an examination of the relationship between concentrations of nutrients in the leaves versus the soil can be informative, especially when—rather than plotting the same element on different bases—e.g. phosphorus per unit leaf dry weight versus phosphorus per unit soil dry weight as in Cleveland et al. (2011)—both plant and soil values are converted to the common units per ground area. For the plant canopy, this can be simply estimated as

$$[\Theta]_{\mathrm{C,a}} = L\left\langle M_{\mathrm{a}}[\Theta]_{\mathrm{L,m}}\right\rangle \tag{12.1}$$

where $[\Theta]_{C,a}$ is the (total) canopy nutrient concentration (expressed as g m^{-2} or mol m^{-2}), where L is the leaf area index (m^{-2} m^{-2}), M_a is the leaf mass per unit area (typically g m^{-2}), and $[\Theta]_{L,m}$ with the triangular brackets indicating mean values. For the soil nutrients again only a simple transformation is required, this being

$$[\Theta]_{\mathrm{S,a}} = \int_{z=0}^{z=d} \rho_{\mathrm{b}}[\Theta]_{\mathrm{ex,m}}dz \tag{12.2}$$

where $[\Theta]_{S,a}$ is the soil nutrient concentration (in g m^{-2} or mol m^{-2}), ρ_b is the soil bulk density (typically in kg m^{-3}), $[\Theta]_{ex,m}$ is the plant available soil nutrient on mass basis (typically g g^{-1} or mmol g^{-1}), z is the soil depth (below the soil surface), and d is the depth of soil nutrient availability considered. Using values of $[\Theta]_{L,m}$ and M_a published by Fyllas et al. (2009), along with associated leaf area index measurements (Fyllas et al. 2012), Fig. 12.4 shows estimates of $[\Theta]_{C,a}$ for many of the Amazonian sites of Fig. 12.2 as a function of $[\Theta]_{S,a}$, the latter estimated to 0.3 m depth and calculated using the data of Quesada et al. (2010) in Fig. 12.4.

Here, in all cases we have attempted to fit a Michaelis–Menten type relationship viz.

$$[\Theta]_{\mathrm{C,a}} = [\Theta]_{\mathrm{max}}/\left(1 + K_{\mathrm{m}}/[\Theta]_{\mathrm{S,a}}\right) \tag{12.3}$$

where $[\Theta]_{max}$ is the maximum canopy nutrient (Θ) concentration observed—any value close to or beyond which presumably reflects some sort of $[\Theta]_{S,a}$ availability excess for that nutrient—and with K_m a constant reflecting the $[\Theta]_{S,a}$ at which $[\Theta]_{C,a} = 0.5[\Theta]_{max}$.

For phosphorus, a clear positive but saturating relationship is evident with the estimates of both $[\Theta]_{max}$ and K_m both significant at $p < 0.005$ and with $[\Theta]_{max}$ estimated at 0.026 ± 0.003 mol m^{-2} and $K_m = 0.45 \pm 0.12$ mol m^{-2} Less than 0.1 of the sites sampled are characterised as having $[\Theta]_{C,a} < 0.5[\Theta]_{max}$ (Fig. 12.3a).

For potassium, whether the relationship is saturating or not is less clear, though with significant terms again obtained ($p < 0.05$) and with $[\Theta]_{max}$ estimated at 0.167 ± 0.034 mol m^{-2} and $K_m = 0.291 \pm 0.127$ mol m^{-2} (Fig. 12.4b). Here there are relatively more sites for which $[\Theta]_{C,a} = 0.5[\Theta]_{max}$ (in all *ca.* 0.4 of the dataset).

In the case of calcium, a Michaelis–Menten type relationship clearly applicable ($[\Theta]_{max} = 0.143 \pm 0.011$ mol m^{-2} and $K_m = 0.242 \pm 0.069$ mol m^{-2}) there are—as for phosphorus—again a few sites for which $[\Theta]_{C,a} < 0.5[\Theta]_{max}$. Most markedly,

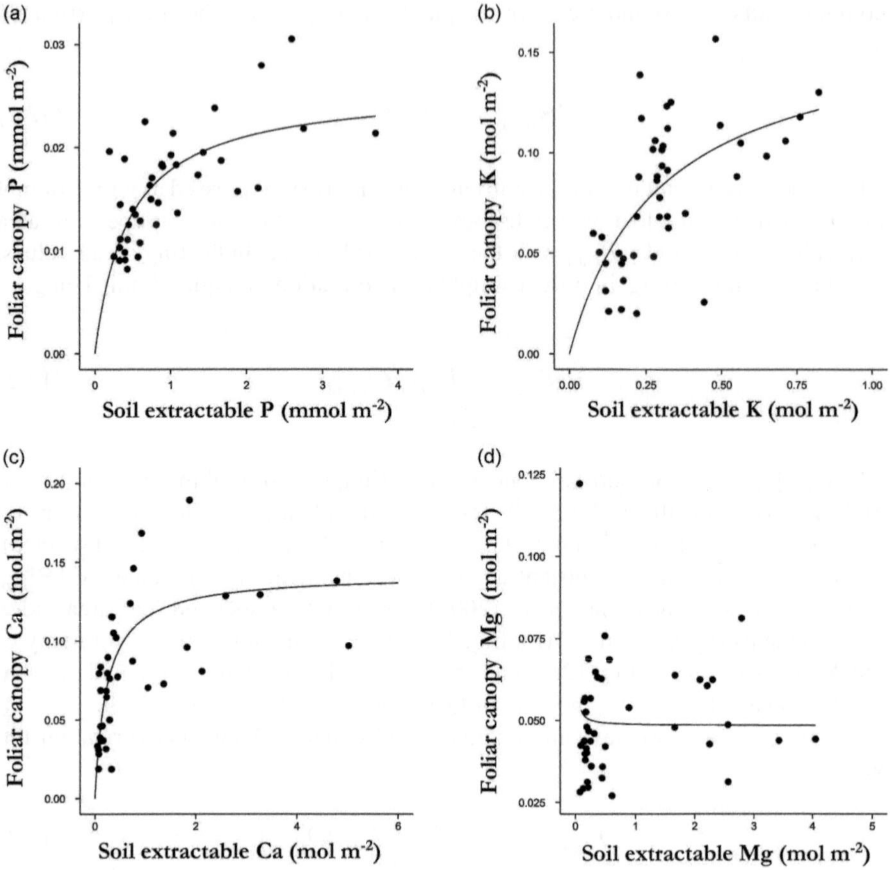

Fig. 12.4 Relationships between soil and leaf nutrient concentrations with both expressed on a ground area basis. (**a**) phosphorus; (**b**) potassium; (**c**) calcium; (**d**) magnesium. Nitrogen has not been included in this analysis as no appropriate plant availability measures were available

it is also clear that, unlike phosphorus or potassium, there are at least 10 sites in our dataset for which $[\Theta]_{\mathrm{C,a}} < [\Theta]_{\mathrm{max}}$ and with calcium then presumably in excess.

There is no relationship between the availability of Mg in the soil and canopy Mg content (Fig 12.4d). This suggests that in all cases there may be more than sufficient for the vegetation requirements for this element.

Also of interest is a comparison of the absolute values of the soil versus canopy nutrient concentrations when both are expressed in the same units. Taking the ratios at $[\Theta]_{\mathrm{S,a}} = K_{\mathrm{m}} \Leftrightarrow [\Theta]_{\mathrm{C,a}} = 0.5[\Theta]_{\mathrm{max}}$ for this cross-elemental comparison, we find $[\Theta]_{\mathrm{C,a}}/[\Theta]_{\mathrm{S,a}} = 0.05, 0.28$, and 0.29 for phosphorus, potassium, and calcium, respectively. Thus, it is clear that the soil:plant relationship for phosphorus—with there usually being much less phosphorus partitioned to the canopy foliage than is present in the soil—is very different to that for the cations examined where the soil typically contains only about five times as much available nutrient that is present in the leaves.

Also interesting to consider in this context the amount of nutrient stored in other woody plant components, especially the tree boles; for which—although overall concentrations are lower typically, contain around 10 times the total foliar nutrient stock as a consequence of their higher biomass (Bond 2010). For example, from data cited by Vitousek and Sanford (1986) total above-ground calcium stocks of tropical forests can be expected to range from *ca.* 0.7 to 10 mol m^{-2} whilst for magnesium and potassium the equivalent values are 0.3–2 mol m^{-2} and 0.8--8 mol m^{-2}, respectively. A consideration of these values in terms of soil availability (Fig. 12.4) then suggests that for Ca and Mg the magnitude of nutrient stocks in the soil and above-ground biomass is more or less similar. But that for potassium only a small fraction of the biogeochemically active pool is retained in the soil. With little additional available potassium available below 0.3 m depth in the less fertile soils of the Amazon basin (Quesada et al. 2011), this raises the interesting possibility that, for some tropical forests at least, total above-ground carbon stocks might be limited by potassium availability (but see also Bond 2010).

In this respect, the situation for phosphorus seems somewhat different with typical total above-ground values of 0.1–1.0 mol m^{-2} nearly an order of magnitude less than those of the soil. This suggests that any limitation of P on tropical forest productivity is likely to be via expensive carbon costs of acquisition as opposed to a simple lack of abundance. Indeed, it is worth noting that our estimates of plant available P here have included not only that already available in the soil solution but also both the inorganic and organic pools extracted with bicarbonate and then NaOH using the Hedley procedure (Quesada et al. 2010). As is discussed in Quesada et al. (2012), there is good evidence that all these pools contain P that is accessible to plants, should it be required. But the carbon costs for the acquisition of at least some of these plant-accessible pools may be considerable in terms of fine root production, mycorrhizal associations, organic acid exudation, and phosphatase production (see also Sect. 1.4.2).

Figure 12.4 shows that when nutrients become scarce, leaf-level nutrient concentrations start to decrease, and nutrient-conserving mechanisms presumably begin to assume a greater importance, but never to the extent that stand-level nutrient content is no longer independent of soil nutrient availability. Thus, as soil nutrient availability declines it is not surprising that one observes lower levels of wood productivity (Quesada et al. 2012). This coupling of soils and vegetation nutrient economy with growth is paralleled—and from an anthropogenic carbon viewpoint to some extent even offset—by another important characteristic of the Amazon basin rainforests: a transition from fast-growing, nutrient-demanding forest ecosystem on relatively young soils to a slow-growing ecosystem, with low nutrient demands on older soils. This is because, associated with this transition, there is also a significant increase in woody biomass residence time (longer-living trees) as soils become more weathered. The net result of this is that biomass stocks are, on average, greater on low productivity and low nutrient soils. The growth rates of such forests can then be viewed as adjusted to the rate in which nutrients are made available through decomposition and nutrient mineralisation, thus suggesting a view that the coupling of soils and tree nutrient economy becomes partial when slow growth strategies are adopted by the forest. These differences in stand-level growth and dynamics seem strongly mediated though trait-mediated species–soil associations (Fyllas et al. 2012; Quesada et al. 2012), and the great intrinsic variability in potential soil fertility niche across the Amazon forest biodiversity spectrum may be one reason why nowhere in the basin do we see the sort of 'ecosystem degradation' associated with intensively weathered soils as reported by many chronosequence studies (e.g. Peltzer et al. 2010).

12.4 Influence of Soils on Forest Species Composition, Structure, and Dynamics

12.4.1 Species Composition

Historically, the literature has been controversial with regard to the importance of soil nutrient status in relation to tree species richness in Amazonia. Some studies have reported that species richness was generally negatively correlated with soil nutrient status (Huston 1980; Faber-Langendoen and Gentry 1991); others have reported a positive correlation between species richness and soil nutrient status (Stark 1970; Gentry 1982, 1988; Phillips et al. 2003). For example, Ruokolainen et al. (2007) reported that beta diversity of three different plant groups in the Ecuadorian Amazon were highly correlated and that much of this association could be explained by variations in edaphic conditions. Yet others emphasised relationships with rainfall and the length of dry season (Clinebell et al. 1995; ter Steege et al. 2003).

Species distributions are often reported to be associated with soil properties. John et al. (2007) showed significant relationships between tree distribution and soil nutrient concentrations for at least a third of the tree species in lowland forests of Colombia, Ecuador, and Panama. More recently, Higgins et al. (2011) have suggested that floristic patterns in Amazonian forests might be associated with edaphic variation across geological formations. The authors used remote sensing techniques and associated soil and plant inventories to document geological and floristic discontinuities occurring in Amazonia. The geological discontinuity identified by Landsat and SRTM data corresponded to a 15-fold change in soil cation concentrations and an almost total change in plant species composition. The authors concluded that floristic patterns of Amazonian forests were partitioned into large-area units on the basis of geological formations and their edaphic properties. Other studies of Amazonian forests have found strong relationships between plant species composition and edaphic properties. This suggests that, to a large degree, floristic patterns may be related to underlying geological patterns (Tuomisto et al. 1995, 2003; Phillips et al. 2003; ter Steege et al. 2006; Pitman et al. 2008).

Irrespective of the underlying mechanism(s), it is clear that basin-wide variations in plant community composition occur, with the central and eastern Amazon areas typically dominated by plant families (Caesalpiniaceae, Lecythidaceae, Chrysobalanaceae, and Sapotaceae) which have 'stress-tolerant' growth strategies, such as high wood density, large seeds, and seedling banks. Trees in these families usually have intrinsically slow growth rates and with characteristically low foliar nutrient concentrations (dry weight basis) and a high leaf mass per unit area (Fyllas et al. 2012), which may be taken as indication that they are adapted to nutrient-poor environments. Conversely, the dominant families of the western Amazon, such as Cecropiaceae, Mimosaceae, Malvaceae, Arecaceae, Moraceae, and Myristicaceae, are generally characterised by a low wood density and high growth/mortality rates (Phillips et al. 2004; Baker et al. 2004b) and higher foliar nutrient concentrations (Fyllas et al. 2009, 2012). Phillips et al. (2004) suggested that the floristic gradient is likely to be an effect of the edaphic resource gradient and that the effects of soil quality on tree turnover rates may be favouring some phylogenetically conserved growth and regeneration strategies over others. In agreement with the above, ter Steege et al. (2006) suggested that there existed two independent gradients of species composition across Amazonia. The first gradient stretches from the Guiana Shield to south-west Amazonia and is congruent with parallel gradients of soil fertility, seed mass, and wood density. The second gradient of species composition stretches from southern Colombia to south-eastern Amazonia (the southern fringe of the basin) and appears to be related to dry season length. Interestingly, recent work from ter Steege et al. (2013) estimated that only relatively few species (227) accounted for about half of all trees >10 cm DBH occurring in Amazonia. Most of these 'hyperdominant' trees are, however, habitat specialists, having wide ranges but being typically associated with one or two forest types (e.g. white sand forests, seasonally flooded forests) and limited to specific geographical regions of Amazonia.

12.4.2 Forest Dynamics

Stem turnover can be defined as the average rate at which trees die and are recruited in a forest population. The rate of stem turnover increases across the Amazon basin from the east (average turnover rate of 1.35 % $year^{-1}$) to 2.6 % $year^{-1}$ in the more fertile west and south-west (Phillips et al. 2004). This coincides with the gradients of soil fertility and geology described above in Sect. 1.2, and, indeed lower rates of stem turnover, net biomass gain, and productivity are associated with the nutrient-poor eastern and central areas with higher rates found in the more fertile west and south-west portions of Amazonia. This has resulted in the hypothesis that soil fertility may play a defining role in determining the threefold difference in productivity, turnover, and net biomass gain between the western and central-east areas of Amazonia (Phillips et al. 1998; Malhi et al. 2004; Baker et al. 2004a, b; Phillips et al. 2004; Stephenson and Van Mantgen 2005).

However, a recent interpretation of soil–plant relationships in Amazonia has argued that soil fertility itself may not be the main driver of stem turnover rate variations; instead, those authors provided evidence that soil physical quality is most likely the principal driver of stem turnover rates (Quesada et al. 2012). Soil physical properties, such as shallow soil depth, poor drainage, and physical impediments in the subsoil, can be an important source of limitation to forest growth, directly or indirectly influencing tree mortality and turnover rates. Such physical conditions are strongly correlated to soil fertility (Fig. 12.5) since both nutrient losses and profile development occur simultaneously (Quesada et al. 2010). Highly weathered soils with low nutrient concentrations are typically of several metres depth above the parent material and usually have good physical structure (e.g. for water relations, root penetration) as a result of millennia of soil development (Sanchez 1987). On the other hand, richer soils in Amazonia are closely related to lower levels of pedogenesis, consequently having poorly developed horizons, and are often shallow (Quesada et al. 2010, 2011).

Limiting soil physical conditions often result in harsh root environments and can exert a major influence on plant growth and survival (Arshad et al. 1996; Schoenholtz et al. 2000). Shallow soil depth and steep topography often result in limited root space and short and stunted root systems, with these characteristics highly increasing the possibility of tree death by wind blow (Dietrich et al. 1996).

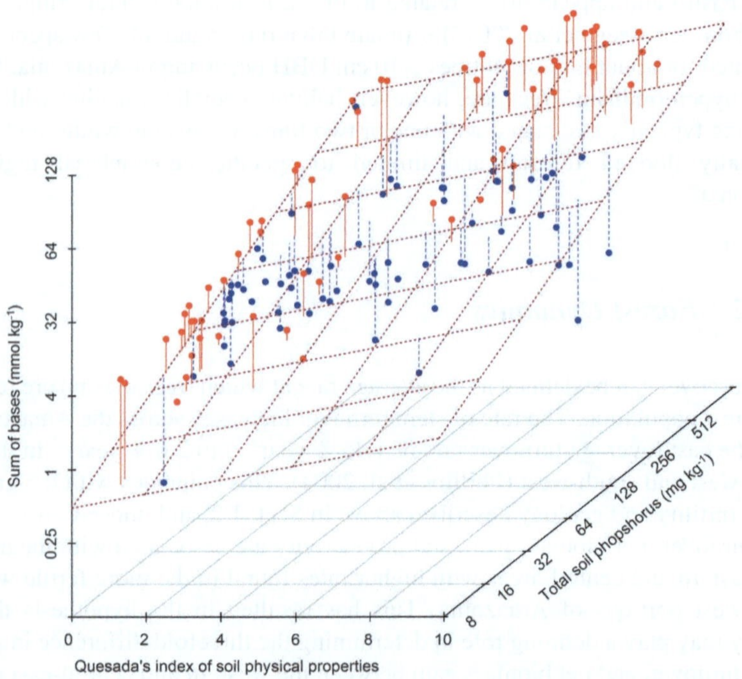

Fig. 12.5 Convergent variation of soil physical properties (represented here by the index Π_1) and the concentration of soil P and exchangeable bases in Amazonia

For example, Lieberman and Lieberman (1987) reported that steep slopes in Costa Rica set a limit in tree size, above which trees are more likely to collapse. Soil structure is another physical characteristic of prime importance, which has been associated with restrictions for root development (Van Wambeke 1992; Arshad et al. 1996) and reductions in productivity, as was shown in tree plantations in the tropics (Dedecek et al. 2001; Hirai et al. 2003). Any of these adverse properties may, alone, be positively associated with high tree turnover rates in Amazonia (Quesada et al. 2012); accounting for the likely interactions amongst them increases their importance in accounting for the observed tree turnover in Amazonia.

The dominant mode of tree death in a tropical forest and the level of disturbance associated with it can also be strongly related to the soil physical environment. For instance, trees in tropical forests of Asia and South America were found to die standing on plateaux with deep, well-drained soils, while trees in areas with steep topography and soil limitations for root anchorage, such as flooding and shallow soil depth tended, to die uprooted (Gale and Barford 1999; Gale 2000; Gale and Hall 2001). The resulting gaps created by tree death are usually different too, with those resulting from gradual fragmentation of standing dead trees being smaller and formed over an extended period of time, which results in lower levels of disturbance after tree death as compared to gaps formed by uprooted and snapped trees. The latter, mostly found in association with poor soil physical conditions, are large in size and produce a dense layer of debris, resulting in much higher disturbance level than gradual decay of standing dead trees (Gale and Barford 1999; Carey et al. 1994; Sugden et al. 1985).

Soil physical properties change the structural assemblage of the forest, and how the total biomass is stored in the trees (Martins et al. 2014). Physically constrained soils with high rates of stem turnover tend to be dominated by smaller trees, often having more stems per unit area, while forests growing in favourable physical and low-disturbance soil conditions allow trees to live longer and thus accumulate more biomass per individual. Soil physical properties have also been reported to be related to the abundance of palms in Amazonia (Emilio et al. 2014), pan-tropical tree height:diameter relationships (Feldpausch et al. 2011), forest demographic structure Cintra et al. (2013), and necromass stocks (Martins et al. 2014).

Wood production, as opposed to turnover, is thought to be directly influenced by soil nutrient availability, and both experimental and correlational studies support the existence of nutrient limitation. Considering montane forests first, the particular nutrient or nutrients that limit production seem to vary with sites (Tanner et al. 1998). For example, nutrient addition experiments in montane forests reported that stem growth was 77 % higher after nitrogen addition and increased by 32 % after phosphorus addition in Jamaica and 133 % higher after simultaneous nitrogen and phosphorus addition in Venezuela (Tanner et al. 1990, 1992). Other experiments in montane forests have demonstrated that forest growth and litterfall production may be limited by nitrogen or by both nitrogen and phosphorus (Herbert and Fownes 1995; Raich et al. 1996; Vitousek and Farrington 1997; Vitousek 2004). For lowland tropical forests, nutrient addition with phosphorus and other nutrients have been shown to increase litterfall, sapling growth, and the growth of successional vegetation (Walker et al. 1996; Mirmanto et al. 1999; Kitayama et al. 2004; Davidson et al. 2004; Wright et al. 2011).

Vitousek (1984) reported that the production of fine litterfall in 62 areas of tropical forests were significantly correlated with litter phosphorus concentration, suggesting that phosphorus, but not nitrogen, limited litterfall production in most areas in his study. He also showed that Amazonian forests generally had lower P circulation and higher dry mass:litter P ratios than other lowland tropical forests. Paoli and Curran (2007) reported that litterfall in Borneo was positively related to soil extractable phosphorus (Olsen P) and the sum of bases. The authors further reported that basal area and biomass growth and forest NPP (litterfall + biomass growth in their study) showed a strong positive relationship with soil nutrients, particularly with phosphorus. They concluded that the spatial variation of soil phosphorus supply was the primary determinant of NPP in their study area. Kitayama et al. (2000) using soil phosphorus fractionation techniques and associated litterfall measurements reported that P use efficiency (PUE) increased with elevation on Mount Kinabalu.

In a first study of Amazon basin-wide patterns of wood production, Malhi et al. (2004) reported that above-ground coarse wood productivity (W_P) varied between 1.5 and 5.5 Mg C ha^{-1} year^{-1}, with the pattern of productivity varying regionally. The authors demonstrated that central and eastern Amazonian forests had relatively low rates of production, and an intermediate group of moderate W_P was observed occurring in northern Amazonia as well as in the southern fringe of Amazonia (Brazilian and Guiana Shields). Highest W_P occurred in western Amazonia. Overall, the lowest W_P was found in the caatinga forest growing on sandy soils in Venezuela (stunted forest physiognomy; see Anderson (1981) for detailed description). Those authors then proposed a division of the study sites in broad soil categories which were associated with variations in W_P: the lowest W_P occurred on white sands, while the Ferralsols of central and eastern Amazonia had 24 % higher rates of W_P. Forests on late Pliocene and Holocene sediments showed no differences between their W_P values, which were both 50 % higher than those on Ferralsols. Cambisols from submontane regions had the highest rates of W_P and were 75 % higher than the Ferralsols while alluvial and seasonally flooded soils varied in their values of W_P depending on the kind of sediment deposited by the floods. Malhi et al. (2004) concluded that soil factors may be very important in determining the W_P at a basin-wide scale. More direct evidence of nutrient limitation on forest productivity has been reported by Quesada et al. (2012) who demonstrated that rates of basal area and biomass growth were best accounted for by variations in total soil phosphorus concentrations across 59 permanent plots across Amazonia (with additional effects of exchangeable bases and dry season precipitation also evident).

There is also evidence that low soil P concentrations can limit microbial processes in moist tropical forests. Cleveland et al. (2002) showed that microbial C decomposition, especially from the least labile C fractions, was limited by low soil P concentration in forests over Ferralsols in Costa Rica, but with a similar interaction not observed in soils with high P. Similarly, microbial mineralisation of dissolved organic matter in soils in response to P addition has been shown to be related to higher P immobilisation in decomposing material and subsequent higher P concentration in dissolved organic matter (Cleveland et al. 2006). By contrast, McGroddy et al. (2004) found no difference in litter decomposition rates in substrates of contrasting soil phosphorus status in eastern Amazonia.

In terms of nitrogen availability, Tanner et al. (1998) suggested, based on patterns of foliar and litterfall concentrations, that nitrogen was in excess supply in most lowland tropical forests as a product of high rates of N transformations, indicated by the relatively high rates of N mineralisation and N-containing trace gas emissions from soils. Using plant and soil $^{15}N:^{14}N$ ratios from a range of tropical and temperate forests, Martinelli et al. (1999) also showed that tropical forests tended to be N-rich environments with an open N budget and with N inputs and outputs larger than the internal N cycle within the systems. Although N is often considered to be an excess nutrient in lowland tropical forests, it is likely to limit growth in montane forests and forests on white sand (Jordan and Herrera 1981; Vitousek and Sanford 1986; Martinelli et al. 1999). Contrary to previous suggestions (i.e. Cleveland et al. 1999), there is, however, little evidence that N fixation is a common process in Amazonia (Nardoto et al. 2014), with the exception of some particular conditions that may lead to eventual N shortage or when fixing N gives a clear competitive advantage, such as legume trees in gaps (Vitousek et al. 2002; Hedin et al. 2009) or during secondary forest regeneration (Batterman et al. 2013).

Overall, phosphorus has increasingly emerged as the most likely nutrient to constrain tropical lowland forest productivity (Vitousek 1982, 1984, 2004; Cuevas and Medina 1986; Vitousek and Sanford 1986; Silver 1994; Reich et al. 1995; McGrath et al. 2001; Paoli and Curran 2007; Quesada et al. 2012). Such a limitation by P is conceptually attractive as P is only supplied by parent material. Thus, the evolution of soils during weathering tends to reduce plant available P pools either by parent material weathering and leaching and/or modification to the chemical state of both organic and inorganic P towards non-available forms such as the P occluded by Fe and Al oxides (Walker and Syers 1976). In addition, because P is essential for high-energy P bounds and triose phosphates, deficiency in this nutrient could potentially limit carboxylation in photosynthesis and thus community-level primary production (Raaimakers et al. 1995; Crews et al. 1995; Herbert and Fownes 1995; Raich et al. 1996; Kitayama et al. 2004). However, generalisations deserve caution as not all forests may be effectively limited by P (Lloyd et al. 2001). For example, Mirmanto et al. (1999) reported that N and P fertilisation during 4 years had no effect on overall stem growth (but did stimulate growth in the 'meranti' group of fast-growing Dipterocarpaceae) but increased litterfall (i.e. likely, leaf production) in an evergreen lowland tropical rainforest in Indonesia—a forest where a high amount of nutrients is in circulation despite the low soil nutrient concentration. Clark et al. (1998) found no difference in adult tree growth rate between two *terra firme* sites where soil P concentration varied by a factor of two. However, the role of P may have been obscured in many studies that only took into account the amount readily available P in the soil solution (see discussion in Quesada et al. 2012). This generally represents the smallest fraction of P in soils and is unlikely to reflect in any meaningful way the total phosphorus pool available to trees. This is because solution P is in dynamic equilibrium with labile P pools (Cross and Schlesinger 1995; Johnson et al. 2003). This means that once P is depleted from soil solution, adsorbed P is released from the labile pools replacing the P in solution (Lloyd et al. 2001; Hinsinger 2001). Therefore, any one point measure of readily available P in time potentially has little meaning on its own.

To access these other forms of P, mycorrhizal associations may allow forest trees to explore more soil area as well as to overcome the P depletion zones that are formed around roots (Alexander 1989). Mycorrhizae raise the affinity of infected roots for soil solution P and lower its concentration limits for absorption (Mosse et al. 1973). This potentially leads to more desorption of P from the labile pools. Mycorrhizal association can also increase P availability by making more phosphorus available of that adsorbed by iron oxides and otherwise unavailable to plant roots (Alexander 1989). Moreover, although it is not completely clear whether or not arbuscular mycorrhizae (AM) are able to use organic P directly, it has been shown that ectomycorrhizae (ECM) can. Thus, organic P, the largest P pool in most strongly weathered tropical soils (up to 80 %, Sanchez 1976), may also be available to mycorrizal plants (Alexander 1989, but also see Lloyd et al. 2001). Mycorrhizal hyphae (including AM and ECM) can make connections between individual trees of the same or different species and these individuals thus can exchange C and P through these hyphal connections (Whittingham and Read 1982; Brownlee et al. 1983).

Low levels of exchangeable bases may constrain tree growth in tropical forests. For instance, calcium, one of the most mobile elements, can be depleted in soils during pedogenesis (Thomas 1974), provoking occasional suggestions of Ca (Vitousek and Sanford 1986; Cuevas and Medina 1986, 1989; Marrs et al. 1991; McGrath et al. 2001) and K limitation (Wright et al. 2011; Quesada et al. 2012). Direct evidence of this has, however, to date only come from managed forest plantations for which there are significant losses of Ca from the ecosystem as a result the biomass removal by harvesting (Hase and Foelster 1983; Bruijnzeel 1984; Spangenberg et al. 1996; Nykvist 1998). This calcium ‘deficiency’ effect is likely mediated through a role for this cation in wood formation (Fromm 2010) which raises the important point that nutrient limitation to forest growth need not necessarily be mediated through an effect on photosynthetic productivity itself. Indeed, as is discussed below, a high soil potassium status might even serve to reduce stand-level carbon acquisition rates through a favouring of low wood density species (Quesada et al. 2012). This is because such species do not necessarily have a faster growth rate than species with high wood density when considered on a carbon (as opposed to height/volume) basis (Keeling et al. 2008).

12.4.3 Forest Biomass

Studies on the effect of soil nutrients on above-ground biomass (AGB) of lowland tropical forests have yielded contrasting results. In a landscape-scale study near Manaus, Brazil, Laurance et al. (1999) found that AGB was positively correlated with total N, total exchangeable bases, K, Mg, and clay content while it was negatively correlated with Zn, Al saturation, and sand content. Roggy et al. (1999) reported a positive relationship between soil nitrogen and AGB in lowland tropical forests in French Guiana. On the other hand, several studies have found no relationship between soils and AGB. Clark and Clark (2000) found no difference in AGB amongst forest plots on three apparently different soils in La Selva, Costa Rica, and Proctor et al. (1983) found no relationship between soil nutrient concentration and AGB in four lowland tropical forests in Sarawak. Similarly, Chave et al. (2001) reported no relationship between soil type and AGB in French Guiana, and DeWalt and Chave (2004) also found no relationship between different soil types and AGB in permanent plots within and amongst sites at Manaus, Barro Colorado, and La Selva, but they did find a positive correlation in Cocha Cachu, Peru, where differences in soil fertility were greater; however, their observed differences in AGB there could also be related to soil physical limitation at the low fertility site where there was a hardpan at 0.7 m. Paoli et al. (2007) found that the AGB of lowland forests in Borneo was positively correlated with exchangeable phosphorus (Olsen P) and negatively with sand fraction. Similar results were reported by Kitayama et al. (2004) for two lowland forests in Borneo in soils of contrasting age, where differences in AGB were positively associated with differences in P and N.

In the first Amazon basin-wide study relating soils to AGB, Quesada et al. (2012) found that soil fertility was generally negatively correlated with AGB. This means that the old and nutrient-poor soils with low tree growth rates had the highest biomass, while the nutrient-rich and most productive forests had much less AGB. The authors attributed this observation to the fact that the most productive forests were also those with usually having high stem turnover rates, with the balance between growth rates and tree residence times then being the main factor defining variations in AGB. This means that, in a less than straightforward way, AGB is influenced by edaphic properties related to both stem turnover rate and wood production (physical properties and phosphorus availability, respectively). Through the correlations of Fig. 12.5, they then suggested that a series of feedback mechanisms initiated by soil properties may exist contributing to the maintenance of different of biomass densities in slow versus fast growth systems (Fig. 12.6). Soil fertility and soil physical properties are thus considered to both play an important role either by controlling resource availability and/or forest disturbance rates, respectively (Quesada et al. 2012).

Large variations in above-ground biomass have been reported in forests with similar basal area (and by implication wood volume) across the Amazon basin, with variations in AGB mostly explained by regional variations in wood density (Baker et al. 2004a, b), and it is for this reason that Quesada et al. (2012) speculated that a

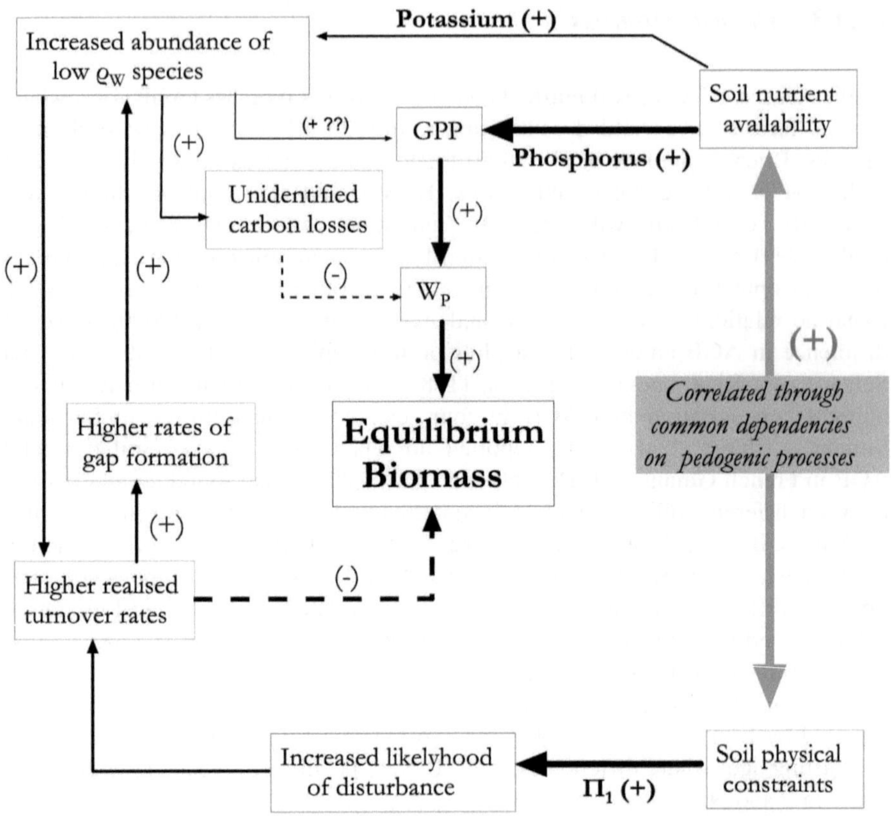

Fig. 12.6 Differential effects of soil nutrients and adverse physical conditions on above-ground coarse wood productivity (W_P), stem turnover rates, and wood density (ρ_W). *GPP* Gross Primary Productivity, the annual rate of carbon gain by the stand through photosynthesis (from Quesada et al. 2012)

negative relationship between soil potassium availability and wood density might be a contributing factor (other things being equal) the biomass of forests on soil of high cation status actually having a lower biomass than those on more dystrophic soils. This effect is probably mediated directly—through high K status soils favouring trees of a low wood density—and indirectly—through low wood density trees ending to have intrinsically shorter lifetimes. Overall, Figure 12.6 suggests that soil cation availability, soil phosphorus status, and soil physical conditions all interact together to influence stand biomass with quite complex interactions possible. Thus, understanding the variation seen in the relationships between these three key tropical soil properties as shown in Fig. 12.5—themselves likely to depend on factors such as soil parent material as well as the nature of long-term atmospheric inputs and weathering trajectories—may turn out to be critical to our understanding of edaphic controls on tropical forest carbon stocks at the local, regional, and global scale.

References

Aerts R, Chapin FS III (2000) The mineral nutrition of wild plants revisited: a re-evaluation of processes and patterns. Adv Ecol Res 30:1–67

Alexander I (1989) Mycorrhizas in tropical forests. In: Proctor J (ed) Mineral nutrients in tropical forests and Savanna ecosystems. Blackwell, Oxford, pp 169–188

Anderson AB (1981) White-sand vegetation of Brazilian Amazonia. Biotropica 13(3):199–210

Arshad MA, Lowery B, Grossman B (1996) Physical tests for monitoring soil quality. In: Doran JW, Jones AJ (eds) Methods for assessing soil quality. Soil Science Society of America, Madison, WI, pp 123–141, SSSA Special Publication, 49

Aubert G, Tavernier R (1972) The soils of Africa. In: Soils of the humid tropics. National Academy of Science, Washington, DC, pp 17–44

Baillie IC (1989) Soil characteristic and classification in relation to the mineral nutrition of tropical wooded ecosystems. In: Proctor J (ed) Mineral nutrients in tropical forests and Savanna ecosystems. Blackwell, Oxford, pp 15–26

Baker TR, Phillips OL, Malhi Y, Almeida S, Di Fiori A, Erwin T, Higuchi N, Killeen TJ, Laurance SG, Laurance WF, Lewis SL, Monteagudo A, Neil DA, Vargas PN, Pitman NCA, Silva JNM, Martinez RV (2004a) Increasing biomass in Amazonian forest plots. Philos Trans R Soc London B 359:353–365

Baker TR, Phillips OL, Malhi Y, Almeida S, Arroyo L, Di Fiore A, Erwin T, Killeen TJ, Laurance SG, Laurance WG, Lewis S, Lloyd J, Monteagudo A, Neill DA, Patiño S, Pitman NCA, Silva JMN, Martines RV (2004b) Variation in wood density determines spatial patterns in Amazonian forest biomass. Glob Chang Biol 10:545–562

Batterman SA, Hedin LO, van Breugel M, Ransijn J, Craven DJ, Hall JS (2013) Key role of symbiotic dinitrogen fixation in tropical forest secondary succession. Nature 502:224–227

Berendse F, Aerts R (1987) Nitrogen use efficiency: a biologically meaningful definition? Funct Ecol 1:293–296

Bond WJ (2010) Do nutrient-poor soils inhabit development of forests? A nutrient stock analysis. Plant Soil 334:47–60

Bridgham SD, Pastor J, McClaugherty CA, Richardson CJ (1995) Nutrient-use efficiency: a litterfall index, a model, and a test along a nutrient-availability gradient in North Carolina Peatlands. Am Nat 145:1–21

Bruijnzeel LA (1984) Immobilization of nutrients in plantation forest of Pinus merkusii and Agathis dammara growing on volcanic soils in central Java, Indonesia. In: Batchik AT, Pushparahja E (eds) Soils and nutrition of perennial crops. Malaysian Soil Science Society, Kuala Lumpur, pp 13–15

Brownlee C, Duddridge JA, Malibari A, Read DJ (1983) The structure and function of micelial systems of ectomicorrhizal roots with special reference to their role in forming interplant connections and providing pathways for assimilate and water transport. Plant Soil 71:433–443

Burnham CP (1989) Pedological processes and nutrient supply from parent material in tropical soils. In: Proctor J (ed) Mineral nutrients in tropical forests and Savanna ecosystems. Blackwell, Oxford, pp 27–42

Carey EV, Brown S, Gillespie AJR, Lugo AE (1994) Tree mortality in mature lowland tropical moist and tropical lower montane moist forests of Venezuela. Biotropica 26:255–265

Chapin FS (1980) The mineral nutrition of wild plants. Annu Rev Ecol System 11:233–260

Chave J, Riera B, Bubois M (2001) Estimation of biomass in a Neotropical forest of French Guiana: spatial and temporal variability. J Trop Ecol 17:79–96

Cintra BBL, Schietti J, Emillio T, Martins D, Moulatlet G, Souza P, Levis C, Quesada CA, Schöngart J (2013) Soil physical restrictions and hydrology regulate stand age and wood biomass turnover rates of Purus–Madeira interfluvial wetlands in Amazonia. Biogeosciences 10(11):7759–7774

Clark DA, Clark DB (2000) Landscape scale variation in forest structure and biomass in a tropical forest. For Ecol Manage 14:185–198

Clark DB, Clark DA, Read JM (1998) Edaphic variations and the mesoscale distribution of tree species in a Neotropical rainforest. J Ecol 86:101–112

Cleveland CC, Townsend AR, Schimel DS, Fisher H, Howarth RW, Hedin LO, Perakis S, Latty E, Fisher JV, Elseroad A, Wasson MF (1999) Global patterns of terrestrial biological nitrogen (N_2) fixation in natural ecosystems. Global Biogeochem Cycles 13(2):623–645

Cleveland CC, Townsend AR, Schmidt SK (2002) Phosphorus limitation of microbial processes in moist tropical forests: evidence from short-term laboratory incubations and field studies. Ecosystems 5:680–691

Cleveland CC, Reed SC, Townsend AR (2006) Nutrient regulation of organic matter decomposition in a tropical rain forest. Ecology 87:492–503

Cleveland CC, Townsend AR, Taylor P, Alvarez Clare S, Bustamante M, Chuyong G, Dobrowski S, Grierson P, Harms KE, Houlton BZ, Marklein A, Parton W, Porder S, Reed S, Sierra CA, Silver W, Tanner EVJ, Wieder WR (2011) Relationships among net primary productivity, nutrients and climate in tropical rain forest: a pan-tropical analysis. Ecol Lett 14(9):939–947

Clinebell RR, Phillips OL, Gentry AH, Stark N, Zuuring H (1995) Prediction of Neotropical tree and liana species richness from soil and climatic data. Biodivers Conserv 4:56–90

Cochrane TT, Sanchez LG, Azevedo LG, de Porras JA and Garver CL (1985) Land in tropical America: guide to climate, landscapes, and soils for agronomists in Amazonia, the Andean Peidmont, Central Brazil in Orinoco, 3 vols. CIAT/EMBRAPA-CPAC

Coley PD, Bryant JP, Chapin FS III (1985) Resource availability and plant antiherbivore defense. Science 230:895–899

Crews TE, Kitayama K, Fownes JH, Riley RH, Herbert DA, Mueller-Dombois D, Vitousek PM (1995) Changes in soil phosphorus fractions and ecosystem dynamics across a long chronosequence in Hawaii. Ecology 76:1407–1424

Cross AF, Schlesinger WH (1995) A literature review and evaluation of the Hedley fractionation: applications to the biogeochemical cycle of soil phosphorus in natural ecosystems. Geoderma 64:197–214

Cuevas E, Medina E (1986) Nutrient dynamics within Amazonian forest ecosystems. Oecologia 68:466–472

Cuevas E, Medina E (1989) Patterns of nutrient accumulation and release in Amazonian forests of the upper Rio Negro basin. In: Proctor J (ed) Mineral nutrients in tropical forests and Savanna ecosystems. Blackwell, Oxford, pp 217–240

Davidson EA, Martinelli LA (2009) Nutrient limitations to secondary forest regrowth. In: Keller M, Bustamante M, Gash J, Dias P (eds) Amazonia and global change, vol 186, Geophysical Monographs. American Geographical Union, Washington, DC, pp 299–310

Davidson EA, Ishida FY, Nepstad DC (2004) Effects of an experimental drought on soil emissions of carbon dioxide, methane, nitrous oxide, and nitric oxide in a moist tropical forest. Glob Chang Biol 10:718–730

Day TH (1959) Report for the reconnaissance soil survey of the Caete-Maracassume area. Stencilled report FAO/SPVEA Mission Belém (Brasil)

Day TH (1961) Soil investigations conducted in the Lower Amazon valley. FAO, EPTA report 1395, Rome

Day TH, Bennema J (1958) Report on an excursion to the Rio Gurupi. Typewritten FAO files, Rome

Day TH, Santos WH (1958) Levantamento detalhado dos solos da Estação Experimental de São Salvador, Marajó. Stencilled report. Inst. Agron. do Norte and FAO/SPVEA Mission Belém (Brasil)

Day TH, Bennema J, Santos WH (1964) British Guyana Soil Survey. Report UNSF/FAO Rome

de Brito Neves BB (2002) Main stages of the development of the sedimentary basins of South America and their relationships with the tectonics of supercontinents. Gondwana Res 5:175–196

Dedecek RA, Bellote AFJ, Gava JL, Menegol O (2001) Site characterisation and the effects of harvesting on soil tillage on the productivity of *Eucalyptus grandis* plantations in Brazil. In: Kobayashi S, Turnbull JW, Toma T, Mori T, Majid NMNA (eds) Rehabilitation of degraded tropical forest ecosystems: workshop proceedings, 2–4 November 1999, CIFOR, Bogor, Indonesia, pp 157–164

Denef K, Plante AF, Six J (2009) Characterization of soil organic matter. In: Kutsch WL, Bahn M, Heinemeyer A (eds) Soil carbon dynamics: an integrated methodology. Cambridge University Press, London, pp 91–126

DeWalt SJ, Chave J (2004) Structure and biomass of four lowland Neotropical forests. Biotropica 36:7–19

Dick DP, Gonçalves CN, Dalmolin RS, Knicker H, Klamt E, Kögel-Knabner I, Simões ML, Martin-Neto L (2005) Characteristics of soil organic matter of different Brazilian Ferralsols under native vegetation as a function of soil depth. Geoderma 124(3):319–333

Dietrich WE, Windsor DM, Dunne T (1996) Geology, climate and hydrology of Barro Colorado Island. In: Leigh EG, Rand AS, Windsor DM (eds) The ecology of a tropical rain forest: seasonal rhythms and long term changes, 2nd edn. Smithsonian Institution Press, Washington, DC, pp 101–108

Emilio T, Quesada CA, Costa FRC, Magnusson WE, Schietti J, Feldpauschd TR, Brienen RJW, Baker TR, Chave J, Álvarez E, Araújo A, Bánki O, Castilho CV, Honorio E, Killeen T, Malhi Y, Oblitas Mendoza EM, Monteagudo A, Neill D, Parada GA, Peña-Cruz A, Ramirez-Angulo H, Schwarz M, Silveira M, ter Steege H, Terborgh JW, Thomas R, Torres-Lezama A, Vilanova E, Phillips OL (2014) Soil physical conditions limit palm and tree basal area in Amazonian forests. Plant Ecol Divers 1–2(7):215–229

Faber-Langendoen D, Gentry AH (1991) The structure and diversity of rain forests at Bajo Calima, Chocó Region, Western Colombia. Biotropica 23:2–11

Feldpausch TR, Banin L, Phillips OL, Baker TR, Lewis SL, Quesada CA, Affum-Baffoe K, Arets EJMM, Berry NJ, Bird M, Brondizio ES, de Camargo P, Chave J, Djagbletey G, Domingues TF, Drescher M, Fearnside PM, França MB, Fyllas NM, Lopez-Gonzalez G, Hladik A, Higuchi N, Hunter MO, Iida Y, Salim KA, Kassim AR, Keller M, Kemp J, King DA, Lovett JC, Marimon BS, Marimon-Junior BH, Lenza E, Marshall AR, Metcalfe DJ, Mitchard ETA, Moran EF, Nelson BW, Nilus R, Nogueira EM (2011) Height-diameter allometry of tropical forest trees. Biogeosciences 8:1081–1106

Fromm J (2010) Wood formation of trees in relation to potassium and calcium nutrition. Tree Phys 30(9):1140–1147

Fyllas NM, Patino S, Baker TR, Nardoto GB, Martinelli L, Quesada CA, Paiva R, Schwartz M, Horna V, Mercado LM, Santos AJB, Arroyo L, Jimenez EM, Luizao FJ, Neil DA, Silva N, Prieto A, Rudas A, Silveira M, Vieira I, Lopez-Gonzalez G, Malhi Y, Phillips OL, Lloyd J (2009) Basin-wide variations in foliar properties of Amazonian forest: phylogeny, soils and climate. Biogeosciences 6:2677–2708

Fyllas NM, Quesada CA, Lloyd J (2012) Deriving plant functional types for Amazonian forests for use in vegetation dynamics models. Perspect Plant Ecol Evol Syst 14:97–110

Gale N (2000) The relationship between canopy gaps and topography in a western Ecuadorian rain forest. Biotropica 32:653–661

Gale N, Barford AS (1999) Canopy tree mode of death in western Ecuadorian forests. J Trop Ecol 15:415–436

Gale N, Hall P (2001) Factors determining the mode of tree death in three Bornean rain forests. J Veg Sci 12(3):337–346

Gentry AH (1982) Patterns of tropical plant diversity. Evol Biol 15:1–84

Gentry AH (1988) Tree species richness of upper Amazonian forests. Proc Nat Acad Sci U S A 85:156–159

Golley FB (1983) Nutrient cycling and nutrient conservation. In: Golley FB (ed) Tropical rain forest ecosystems. Elsevier, Amsterdam

Golley FB (1986) Chemical plant-soil relationships in tropical forests. J Trop Ecol 2:219–229

Golley FB, Richardson T, Clements R (1978) Elemental concentrations in tropical forests and soils of north-western Colombia. Biotropica 10:144–151

Grubb PJ (1989) Mineral nutrients: a plant ecologist view. In: Proctor J (ed) Mineral nutrients in tropical forests and Savanna ecosystems. Blackwell, Oxford, pp 417–440

Hase H, Foelster H (1983) Impact of plantation forestry with teak (*Tectona grandis*) on the nutrient status of young alluvial soils in West Venezuela. For Ecol Manage 6(1):33–57

Hedin LO, Brookshire EJ, Menge DN, Barron AR (2009) The nitrogen paradox in tropical forest ecosystems. Annu Rev Ecol Evol Syst 40:613–635

Herbert DA, Fownes JH (1995) P limitation on forest leaf area and net primary productivity on highly weathered tropical montane soils in Hawaii. Biogeochemistry 29:23–235

Herbert DA, Fownes JH, Vitousek PM (1999) Hurricane damage to Hawaiian forest: nutrient supply rate affects resistance and resilience. Ecology 80:908–920

Herrera R, Jordan CF, Klinge H, Medina E (1978a) Amazon ecosystems: their structure and functioning with particular emphasis on nutrients. Interciencia 3:223–232

Herrera R, Merida T, Stark N, Jordan CF (1978b) Direct P transfer from leaf litter to roots. Naturwissenschaften 65:208–209

Higgins MA, Ruokolainen K, Tuomisto H, Llerena N, Cardenas G, Phillips OL, Vasquez R, Rasanen M (2011) Geological control of floristic composition in Amazonian forests. J Biogeogr 38:2136–2149

Hinsinger P (2001) Bioavailability of soil inorganic P in the rhizosphere as affected by root-induced chemical changes: a review. Plant Soil 237:173–195

Hirai, K, Ferraz JBS, Kobayashi M, Ferreira SJF, Sales PC, Lopes MC, Hotta I (2003) Physical properties of the soils under degraded areas in the Central Amazon. In: Projeto Jacaranda Fase II: Pesquisas Florestais na Amazonia Central. CPST-INPA, Manaus, Brazil, pp 153–167

Hiremath AJ, Ewel JJ (2001) Ecosystem nutrient use efficiency, productivity, and nutrient accrual in model tropical communities. Ecosystems 4:669–682

Hoorn C, Wesselingh FP (2010) Amazonia: landscape and species evolution. Wiley-Blackwell, Chichester

Hoorn C, Wesselingh FP, ter Steege H, Bermudez MA, Mora A, Sevink J, Sanmartín I, Sanchez-Meseguer A, Anderson CL, Figueiredo JP, Jaramillo C, Riff D, Negri FR, Hooghiemstra H, Lundberg J, Stadler T, Särkinen T, Antonelli A (2010) Amazonia through time: Andean uplift, climate change, landscape evolution, and biodiversity. Science 330(6006):927–931

Huston M (1980) Soil nutrients and tree species richness in Costa Rican forests. J Biogeogr 7:147–157

John R, Dalling JW, Harms KE, Yavitt JB, Stallard RF, Mirabello M, Hubbell SP, Valencia R, Navarrete H, Vallejo M, Foster RB (2007) Soil nutrients influence spatial distributions of tropical tree species. Proc Natl Acad Sci U S A 104:864–869

Johnson AH, Frizano J, Vann DR (2003) Biogeochemical implications of labile phosphorus in forest soils determined by the Hedley fractionation procedure. Oecologia 135:487–499

Jordan CF (1979) Stem flow and nutrient transfer in a tropical rain forest. Oikos 31:255–268

Jordan CF (1982) The nutrient balance of an Amazonian rain forest. Ecology 63:647–654

Jordan CF (1985) Nutrient cycling in tropical forest ecosystems. Wiley, Chichester

Jordan CF (1989) Are process rates higher in tropical forest ecosystems? In: Proctor J (ed) Mineral nutrients in tropical forests and Savanna ecosystems. Blackwell, Oxford, pp 205–216

Jordan CF, Escalante G (1980) Root productivity in an Amazonian rain forest. Ecology 61:14–18

Jordan CF, Herrera R (1981) Tropical rain forests: are nutrients really critical? Am Nat 117:167–180

Jordan CF, Stark N (1978) Retencion de nutrients en la estera de raices de un bosque pluvial Amazonico. Acta Cient Venezoelana 29:263–267

Keeling HC, Baker TR, Martinez RV, Monteagudo A, Phillips OL (2008) Contrasting patterns of diameter and biomass increment across tree functional groups in Amazonian forests. Oecologia 158(3):521–534

Kellman M, Carty A (1986) Magnitude of nutrient influxes from atmospheric sources to a Central America *Pinnus caribea* woodland. J Appl Ecol 23:211–226

Kitayama K, Majalap-Lee N, Aiba SI (2000) Soil phosphorus fractionation and phosphorus-use efficiencies of tropical rain forests along altitudinal gradients of Mount Kinabalu, Borneo. Oecologia 123:342–349

Kitayama K, Aiba SI, Takyu M, Majalap N, Wagai R (2004) Soil phosphorus fractionation and phosphorus-use efficiency of a Bornean tropical montane rain forest during soil aging with podozolizatio. Ecosystems 7:259–274

Kroonenberg SB, Roever EWF (2010) Geological evolution of the Amazonian Craton. In: Hoorn C, Wesselingh FP (eds) Amazonia: landscape and species evolution. Wiley-Blackwell, Chichester, pp 9–28

Laurance WF, Fearnside PM, Laurance SG, Delamonica P, Lovejoy T, Merona JR, Chambers JQ, Gascon C (1999) Relationship between soils and Amazon forest biomass: a landscape study. For EcolManage 118:127–138

Lieberman D, Lieberman M (1987) Forest tree growth and dynamics at La Selva, Costa Rica (1969-1982). J Trop Ecol 3:347–369

Lloyd J, Bird MI, Veenendaal EM, Kruijt B (2001) Should phosphorus availability be constraining moist tropical forest responses to increasing CO_2 concentrations? In: Schulze ED, Harrison SP, Heimann M, Holland EA, Lloyd J, Prentice IC, Schimel D (eds) Global biogeochemical cycles in the climate system. Academic, San Diego, CA, pp 96–114

Luizão FJ, Fernside P, Cerri CEP, Lehmann J (2009) The maintenance of soil fertility in Amazonian managed systems. In: Keller M, Bustamante M, Gash J, Dias P (eds) Amazonia and global change, vol 186, Geophysical Monographs. American Geographical Union, Washington, DC, pp 311–336

Malhi Y, Baker T, Phillips OL, Almeida S, Alvares E, Arroyo L, Chave J, Czimczik C, Di Fiore A, Higuchi N, Killeen T, Laurance SG, Laurance WF, Lewis S, Montoya LMM, Monteagudo A, Neill D, Nunes Vargas P, Panfil SN, Patino S, Pitman N, Quesada CA, Salomao R, Silva N, Lezama AT, Vasquez Martinez R, Terborgh J, Vinceti B, Lloyd J (2004) The above-ground coarse wood productivity of 104 Neotropical forest plots. Glob Change Biol 10:1–29

Malhi J, Wood D, Baker TR, Wright J, Phillips OL, Cochrane T, Meir P, Chave J, Almeida S, Arroyo L, Higuchi N, Killeen TJ, Laurance SG, Laurance WF, Lewis SL, Monteagudo A, Neill DA, Núñez Vargas P, Pitman NCA, Quesada CA, Salomão R, Silva JNM, Torres Lezama A, Terborgh J, Vásquez Martínez R, Vinceti B (2006) The regional variation of aboveground live biomass in old-growth Amazonian forests. Glob Chang Biol 12:1107–1138

Marbut CF, Manifold CB (1926) The soils of the Amazon basin in relation to their agricultural possibilities. Geogr Rev 15:617–643

Marrs RH, Thompson J, Scott D, Proctor J (1991) Nitrogen mineralization and nitrification in terra firme forests and savanna soils in ilha de Maraca, Roraima. Braz J Trop Ecol 7:123–137

Martinelli LA, Piccolo MC, Townsend AR, Vitousek PM, Cuevas E, McDowell W, Robertson GP, Santos OC, Treseder K (1999) Nitrogen stable isotopic composition of leaves and soil: tropical versus temperate forests. Biogeochemistry 46:45–65

Martins DL, Schietti J, Feldpausch T, Luizão FJ, Phillips OL, Andrade A, Castilho C, Laurance SG, Oliveira A, Amaral IL, Toledo JJ, Lugli LF, Pinto JLPV, Oblitas Mendoza EM, Quesada CA (2014) Soil-induced impacts on forest structure drive coarse woody debris stocks across central Amazonia. Plant Ecol Divers, accepted manuscript. doi:10.1080/17550874.2013.879942#_blank

McGrath DA, Smith CK, Gholz HL, Oliveira FA (2001) Effects of land-use change on soil nutrient dynamics in Amazonia. Ecosystems 4:625–645

McGroddy ME, Silver WL, Cosme de Oliveira R Jr (2004) The effect of phosphorus availability on decomposition dynamics in a seasonal lowland Amazonian forest. Ecosystems 7:172–179

Medina E, Delgado M, Troughton JH, Medina JD (1977) Physiological ecology of CO_2 fixation in Bromeliaceae. Flora 166:137–152

Medina E, Sobrado M, Herrera R (1978) Significance of leaf orientation for leaf temperature in an Amazonian sclerophyll vegetation. Radiat Environ Biophys 15:131–140

Mirmanto E, Proctor J, Green J, Nagy L, Suriantata (1999) Effects of nitrogen and phosphorus fertilization in a lowland evergreen rainforest. Philos Trans R Soc London B Biol Sci 354:1825–1829

Mosse B, Hayman DS, Arnold DJ (1973) Plant growth responses to VA micorrhizas. New Phytol 72:809–815

Nardoto GB, Quesada CA, Patiño S, Saiz G, Baker TR, Schwarz M, Schrodt F, Feldpausch TR, Domingues TF, Marimon BS, Marimon Junior BH, Vieira I, Silveira M, Bird MI, Phillips OL, Lloyd J, Martinelli LA (2014) Basin-wide variations in Amazon forest nitrogen-cycling characteristics as inferred from plant and soil 15N/14N measurements. Plant Ecol Divers. doi: 10.1080/17550874.2013.807524

Nykvist N (1998) Logging can cause a serious lack of calcium in tropical rainforest ecosystems: an example from Sabah, Malaysia. In: Schulte A, Ruhiyat D (eds) Soils of tropical forest ecosystems. Springer, Berlin, pp 87–91

Paoli GD, Curran LM (2007) Soil nutrients limit fine litter production and tree growth in mature lowland forest of southwestern Borneo. Ecosystems 10:503–518

Paoli GD, Curran LM, Slik JWF (2007) Soil nutrients affect spatial patterns of aboveground biomass and emergent tree density in southwestern Borneo. Oecologia 155:287–299

Peltzer DA, Wardle DA, Allison VJ, Baisden T, Bardgett RD, Chadwick OA, Condron L, Parfitt RL, Porder S, Richardson SJ (2010) Understanding ecosystem retrogression. Ecol Monogr 80 (4):509–529

Phillips OL, Malhi Y, Higuchi N, Laurance WF, Nuñez VP, Vásquez MR, Laurance SG, Ferriera LV, Stern M, Brown S, Grace J (1998) Changes in the carbon balance of tropical forests: evidence from long-term plots. Science 282:439–442

Phillips OL, Vargas PN, Monteagudo AL, Cruz AP, Zans MC, Sanchez WG, Yli-Halla M, Rose S (2003) Habitat association among Amazonian tree species: a landscape-scale approach. J Ecol 91:757–775

Phillips OL, Baker T, Arroyo L, Higuchi N, Killeen T, Laurance WF, Lewis SL, Lloyd J, Malhi Y, Monteagudo A, Neill D, Núñez Vargas P, Silva N, Terborgh J, Vásquez Martínez R, Alexiades M, Almeida S, Brown S, Chave J, Comiskey JA, Czimczik CI, Di Fiore A, Erwin T, Kuebler C, Laurance SG, Nascimento HEM, Olivier J, Palacios W, Patiño S, Pitman N, Quesada CA, Saldias M, Torres Lezama A, Vinceti B (2004) Patterns and process in Amazon tree turnover, 1976-2001. Philos Trans R Soc Lond B 359:437–462

Pitman NCA, Mogollón H, Dávila N, Ríos M, García-Villacorta R, Guevara J, Baker TR, Monteagudo A, Phillips OL, Vásquez-Martinez R, Ahuite M, Aulestia M, Cardenas D, Ceron CE, Loizeau PA, Neill DA, Núñez VP, Palacios WA, Spichiger R, Valderrama E (2008) Tree community change across 700 km of lowland Amazonian forest from the Andean foothills to Brazil. Biotropica 40:525–535

Proctor J, Anderson JM, Chai P, Vallack HW (1983) Ecological studies in four contrasting lowland rainforests in Gunung Mulu National Park Sarawak. J Ecol 71:237–260

Quesada CA, Lloyd J, Schwarz M, Patino S, Baker TR, Czimczik CI, Fyllas NM, Martinelli L, Nardoto GB, Schmerler J, Santos AJB, Hodnett MG, Herrera R, Luizao FJ, Arneth A, Lloyd G, Dezzeo N, Hilke I, Kuhlmann I, Raessler M, Brand WA, Geilmann H, Moraes Filho JO, Carvalho FP, Araujo Filho RN, Chaves JE, Cruz Junior OF, Pimentel TP, Paiva R (2010) Variations in chemical and physical properties of Amazon forest soils in relation to their genesis. Biogeosciences 7:1515–1541

Quesada CA, Lloyd J, Anderson LO, Fyllas NM, Schwarz M, Czimczik CI (2011) Soils of Amazonia with particular reference to the RAINFOR sites. Biogeosciences 8:1415–1440

Quesada CA, Phillips OL, Schwarz M, Czimczik CI, Baker TR, Patiño S, Fyllas NM, Hodnett MG, Herrera R, Almeida S, Alvarez Dávila E, Arneth A, Arroyo L, Chao KJ, Dezzeo N, Erwin T, Di Fiore A, Higuchi N, Honorio Coronado E, Jimenez EM, Killeen T, Lezama AT, Lloyd G, López-González G, Luizão FJ, Malhi Y, Monteagudo A, Neill DA, Núñez Vargas P, Paiva R,

Peacock J, Peñuela MC, Peña Cruz A, Pitman N, Priante Filho N, Prieto A, Ramírez H, Rudas A, Salomão R, Santos AJB, Schmerler J, Silva N, Silveira M, Vasquez-Martínez R, Vieira I, Terborgh J, Lloyd J (2012) Basin-wide variations in Amazon forest structure and function are mediated by both soils and climate. Biogeosciences 9:2203–2246

Quiquampoix H, Mousain D (2005) Enzymatic hydrolysis of organic phosphorus. In: Turner BL, Frossard E, Baldwin DS (eds) Organic phosphorus in the environment. CAB International, Wallingford, pp 89–112

Raaimakers D, Boot RGA, Dijkstra P, Pot S (1995) Photosynthetic rates in relation to leaf phosphorus content in pioneer versus climax tropical rainforest trees. Oecologia 102:120–125

RADAMBRASIL (1978) Levantamento dos Recursos Naturais. Ministério de Minas e Energia, Departamento Nacional de Produção Mineral, Rio de Janeiro

Raich JW, Russel AE, Crews TE, Farrington H, Vitousek P (1996) Both N and P limit plant production on young Hawaiian lava flows. Biogeochemistry 32:1–14

Reich PB, Ellsworth DS, Uhl C (1995) Leaf carbon and nutrient assimilation and conservation in species of different successional status in an oligotrophic Amazonian forest. Funct Ecol 9:65–76

Richter DD, Babbar LI (1991) Soil diversity in the tropics. Adv Ecol Res 21:315–389

Roggy JC, Prevost MF, Garbaye J, Domenach AM (1999) Nitrogen cycling in the tropical rainforest of French Guiana: comparison of two sites with contrasting soil types ^{15}N. J Trop Ecol 15:1–22

Ruokolainen K, Tuomisto H, Macia MJ, Higgins MA, Yli-Halla M (2007) Are floristic and edaphic patterns in Amazonian rain forests congruent for trees, pteridophytes and Melastomataceae? J Trop Ecol 23:13–25

Saidy AR, Smernik RJ, Baldock JA, Kaiser K, Sanderman J, Macdonald LM (2012) Effects of clay mineralogy and hydrous iron oxides on labile organic carbon stabilisation. Geoderma 173:104–110

Sanchez PA (1976) Properties and management of soils in the tropics. John Wiley, New York, NY, 618pp

Sanchez PA (1987) Soil productivity and sustainability in Agroforestry systems. In: Steppler HA, Nair PKR (eds) Agroforestry: a decade of development. International Council for Research in Agroforestry, Nairobi, pp 205–223

Sanchez PA, Buol SW (1975) Properties of some soils of the upper Amazon Basin of Peru. Soil Sci Soc Am J 38:117–121

Schoenholtz SH, Van Miegroet H, Burger JA (2000) A review of chemical and physical properties as indicators of forest soil quality: challenges and opportunities. For Ecol Manage 138:335–356

Schurz WL, Hargis OD (1923) Rubber production in the Amazon Valley. The Crude Rubber Survey, US Department of Commerce

Silver WL (1994) Is nutrient availability related to plant nutrient use in humid tropical forests? Oecologia 98:336–343

Small E (1972) Photosynthetic rates in relation to N recycling as adaptations to nutrient deficiency in bog peat plants. Can J Bot 50:2227–2233

Smith K, Gholz HL, de Assis Oliveira F (1998) Litterfall and nitrogen-use efficiency of plantations and primary forest in the eastern Brazilian Amazon. For Ecol Manag 109:209–220

Sollins P (1998) Factors influencing species composition in tropical lowland rainforest: does soil matter? Ecology 79:23–30

Sombroek WG (1966) A reconnaissance of the soils of the Brazilian Amazon region. Centre for Agricultural Publications and Documentation, Wageningen

Sombroek WG (1984) Soils of the Amazon Region. In: Sioli H (ed) The Amazon: limnology and landscape ecology of a mighty tropical river and its basin. Junk, Dordrecht, pp 521–535

Sombroek WG (2000) Amazon landforms and soils in relation to biological diversity. Acta Amazonica 30:81–100

Spangenberg A, Grimm U, Sepeda da Silva J, Fölster H (1996) Nutrient store and export rates of *Eucalyptus urograndis* plantations in eastern Amazonia (Jari). For Ecol Manage 80 (1):225–234
Stark N (1970) The nutrient content of plants and soils from Brazil and Suriname. Biotropica 2:51–60
Stark N (1971a) Nutrient cycling I nutrient distribution in some Amazonian soils. Trop Ecol 12:24–50
Stark N (1971b) Nutrient cycling II nutrient distribution in Amazonian vegetation. Trop Ecol 12:177–201
Stark N, Jordan CF (1978) Nutrient retention by the root mat of an Amazonian forest. Ecology 59:434–437
Stephenson N, Van Mantgen PJ (2005) Forest turnover rates follow global and regional patterns of productivity. Ecol Lett 8:524–531
Sugden AM, Tanner EVJ, Kapos V (1985) Regeneration following clearing in a Jamaican montane forest: results of a ten year study. J Trop Ecol 1:329–351
Tanner EVJ, Kapos V, Freskos S, Healey J, Theobold AM (1990) N and P fertilization of Jamaican montane forest. J Trop Ecol 6:231–238
Tanner EVJ, Kapos V, Franco W (1992) N and P fertilization on Venezuelan montane forest trunk growth and litterfall. Ecology 73:78–86
Tanner EVJ, Vitousek PM, Cuevas E (1998) Experimental investigation of nutrient limitation of growth on wet tropical mountains. Ecology 79:10–22
Telles ECC, Camargo PB, Martinelli LA, Trumbore SE, Costa ES, Santos J, Higuchi N, Oliveira RC Jr (2003) Influence of soil texture on carbon dynamics and storage potential in tropical forest soils of Amazonia. Global Biogeochem Cycles 17(2):1040. doi:10.1029/2002GB001953
ter Steege H, Pitman N, Sabatier D, Castellanos H (2003) A spatial model of tree α-diversity and tree density for the Amazon. Biodivers Conserv 12:2255–2277
ter Steege H, Pitman N, Phillips OL, Chave J, Sabatier D, Duque A, Molino JF, Prévost MF, Spichiger R, Castellanos H, von Hildebrand P, Vásquez R (2006) Continental-scale patterns of canopy tree composition and function across Amazonia. Nature 443:444–447
ter Steege H, Pitman NC, Sabatier D, Baraloto C, Salomão RP, Guevara JE et al (2013) Hyperdominance in the Amazonian tree flora. Science 342(6156):1243092
Thomas MF (1974) Tropical geomorphology. McMillan, London, 331pp
Tiessen H, Cuevas E, Chacon P (1994) The role of soil organic matter in sustaining soil fertility. Nature 371:783–785. doi:10.1038/371783a0
Trumbore S, Camargo PB (2009) Soil carbon dynamics. In: Keller M, Bustamante M, Gash J, Dias P (eds) Amazonia and global change, vol 186, Geophysical Monographs. American Geographical Union, Washington, DC, pp 451–465
Tuomisto H, Ruokolainen K, Kalliola R, Linna A, Danjoy W, Rodriguez Z (1995) Dissecting Amazonian biodiversity. Science 269:63–66
Tuomisto H, Poulsen AD, Ruokolainen K, Moran RC, Quintana C, Celi J, Canas G (2003) Linking floristic patterns with soil heterogeneity and satellite imagery in Ecuadorian Amazonia. Ecol Appl 13:352–371
Van Wambeke A (1992) Soils of the tropics—properties and appraisal. McGraw-Hill, New York, NY
Vitousek PM (1982) Nutrient cycling and nutrient use efficiency. Am Nat 119:553–572
Vitousek PM (1984) Litterfall, nutrient cycling, and nutrient limitation in tropical forests. Ecology 65:285–298
Vitousek PM (2004) Nutrient cycling and limitation: Hawai'i as a model system. Princeton University Press, Princeton, NJ, 223pp
Vitousek PM, Farrington H (1997) Nutrient limitation and soil development: experimental test of a biogeochemical theory. Biogeochemistry 37:63–75
Vitousek PM, Sanford RL (1986) Nutrient cycling in moist tropical forest. Annu Rev Ecol Syst 17:137–167

Vitousek PM, Cassman K, Cleveland C, Crews T, Field CB, Grimm NB, Howarth RW, Marino R, Martinelli L, Rastetter EB, Sprent JI (2002) Towards an ecological understanding of biological nitrogen fixation. Biogeochemistry 57–58:1–45

Vonhof HB, Kaandorp RJG (2010) Climate variations in Amazonia during the neogene and the quaternary. In: Hoorn C, Wesselingh FP (eds) Amazonia: landscape and species evolution. Wiley-Blackwell, Chichester, pp 201–210

Walker TW, Syers JK (1976) The fate of phosphorus during pedogenesis. Geoderma 15:1–19

Walker L, Zimmerman JK, Lodge DJ, Grajales SG (1996) An altitudinal comparison of growth and species composition in hurricane-damaged forests in Puerto Rico. J Ecol 84:77–89

Wallace (1853) A narrative of travels on the Amazon and Rio Negro, with an account of the native tribes, and observations on the climate, geology and natural history of the Amazon Valley, 2nd edn. Haskell House, London

Wattel-Koekkoek EJW, Buurman P, Van Der Plicht J, Wattel E, Van Breemen N (2003) Mean residence time of soil organic matter associated with kaolinite and smectite. Eur J Soil Sci 54 (2):269–278

Went FW, Stark N (1968) Micorrhiza. Bioscience 18:1035–1039

Whittingham J, Read DJ (1982) Vesicular-arbuscular micorrhiza in natural vegetation systems. III. Nutrient transfer between plants with mycorrhizal interconnections. New Phytol 90:277–284

Wright SJ, Yavitt JB, Wurzburger N, Turner BL, Tanner EV, Sayer EJ, Corre MD (2011) Potassium, phosphorus, or nitrogen limit root allocation, tree growth, or litter production in a lowland tropical forest. Ecology 92(8):1616–1625

Chapter 13
Fires in Amazonia

Luiz E.O.C. Aragão, Liana O. Anderson, André Lima, and Egidio Arai

13.1 Introduction

Tropical forest fires are an emerging important environmental issue of the twenty-first century. In Amazonia, this recent preoccupation is, in part, related to the fact that some global climate models predict an increase in the frequency and intensity of droughts (see e.g. Chap. 4), due to changes in atmospheric circulation induced by planetary warming (Li et al. 2006), which may turn the world's largest tropical forest into a more fire-prone system (Malhi et al. 2008). A reduction in rainfall is expected to exacerbate the synergism between climate, deforestation, and fires (Cochrane and Laurance 2002; Hutyra et al. 2005). This drought–deforestation–fire interaction may increase the likelihood of fires to leak into surrounding undisturbed forests, consequently magnifying the contribution of Amazonian fires to global carbon emissions from land use. Fire frequency in Amazonia induced by ongoing human activities has already been observed to have increased during

L.E.O.C. Aragão (✉)
Remote Sensing Division, National Institute for Space Research—INPE, 12227-010 São José dos Campos, SP, Brazil

College of Life and Environmental Sciences, University of Exeter, Exeter, EX4 4RJ, UK
e-mail: laragao@dsr.inpe.br

L.O. Anderson
Remote Sensing Division, National Institute for Space Research—INPE, 12227-010 São José dos Campos, SP, Brazil

Environmental Change Institute, University of Oxford, Oxford, UK

A. Lima • E. Arai
Remote Sensing Division, National Institute for Space Research—INPE, 12227-010 São José dos Campos, SP, Brazil

© Springer-Verlag Berlin Heidelberg 2016

L. Nagy et al. (eds.), *Interactions Between Biosphere, Atmosphere and Human Land Use in the Amazon Basin*, Ecological Studies 227,
DOI 10.1007/978-3-662-49902-3_13

periods of prolonged dry seasons during the 1998, 2005, and 2010 droughts (Aragão and Shimabukuro 2010a).

The Large Scale Biosphere-Atmosphere Programme (LBA) has studied the multiple facets of fire since the mid-1990s. These and other studies have focused on understanding the frequency of fire occurrence, their impacts on vegetation and consequent C emissions (Chaps. 5 and 6), as well as the effects of environmental changes on fire events and potential feedbacks between climate, deforestation, and fire (Uhl and Kauffman 1990; Cochrane and Schulze 1999; Rosenfeld 1999; Cochrane et al. 1999; Ackerman et al. 2000; Laurance et al. 2001a, b; Laurance and Williamson 2001; Cochrane and Laurance 2002; Barlow and Peres 2004a; Nepstad et al. 2004; Artaxo et al. 2005; Alencar et al. 2006; Aragão et al. 2007a, 2008; Bowman et al. 2009).

With increasing international demand for C emission reductions to avoid passing dangerous climate change thresholds, controlling the indiscriminate use of fires in the Amazon region can be an efficient strategy to reduce carbon emissions. The key challenge is to improve our understanding of fire regimes and on how they may change with future changes in land use and climate. This will improve our ability to forecast fire incidence at spatial and temporal scales that permit operational interventions for minimising the impacts of fires on carbon emissions, ecosystem services, and human health.

In this chapter, we start by providing an overview of the state of our knowledge on the spatial and temporal patterns of fires, focusing on fire incidence. We briefly introduce the history of the use of fire in Amazonia, including pre-Columbian fires, and depict its configuration in space and time with a particular consideration about its relationship with land use and land cover and the influence of climate seasonality and recent droughts on these patterns. We subsequently focus on the impacts of fire, examining the extent of burned forests during major droughts and describing the main impacts of fire on forest carbon stocks, forest structure, and composition as well as Amazonian people. We then review the main modelling approaches for quantifying and predicting fire occurrence. We conclude by providing a comprehensive view of the processes that influence fire occurrence, potential feedbacks, and impacts in Amazonia, centred on human actions, fire, deforestation, and climate and feedbacks among them.

Most of the analyses presented in this chapter refer to the Brazilian Legal Amazon (BLA), which is the administrative boundary defined by law by the Brazilian government, including not only the Amazon forest 'biome', but also part of the Cerrado (savanna) and Pantanal (hyperseasonal flooded cerrado) 'biomes' within the national frontiers of Brazil. However, some analyses and discussion are referred to Amazonia. In this case, we analysed the data in the context of the whole Amazon forest 'biome', which includes closed and open evergreen broadleaf lowland rainforests across the Amazon basin.

13.1.1 Fire Incidence

Deforestation has been for years the major green house gas (GHG) emission source in Brazil, contributing to c. 77 % of all GHG emissions of the country (MCT 2010). The deforestation process in BLA, which relies on clear cut of the native vegetation and the subsequent use of fire to remove the slashed material, was a key contributor for the 1,614,970 fire occurrences detected by the Terra/MODIS sensor between 2001 and 2010 in Brazil (Fig. 13.1a and b).

Despite a reduction in deforestation rates in 2010 by 64 % below its 5-year average from 2005 to 2009 (PRODES 2013), fire incidence has increased in 59 % of BLA with decreasing trends in deforestation rates (Aragão and Shimabukuro 2010a). Possibly this pattern relates to the high probability that deliberate fires, used for managing pastures and suppressing regrowth in deforested areas, were leaking into surrounding intact forests, helped by recent droughts, and by ongoing increase of forest edge area, number of fragments, and secondary forest area (Aragão and Shimabukuro 2010b).

To disentangle the potential interacting factors and processes, we first depict the historical usage of fire in Amazonia, from pre-Columbian time to recent days. We then examine the recent spatial and temporal footprint of fire in Amazonia, giving particular attention to the effects of land use and land cover, climate seasonality, and droughts on fire patterns.

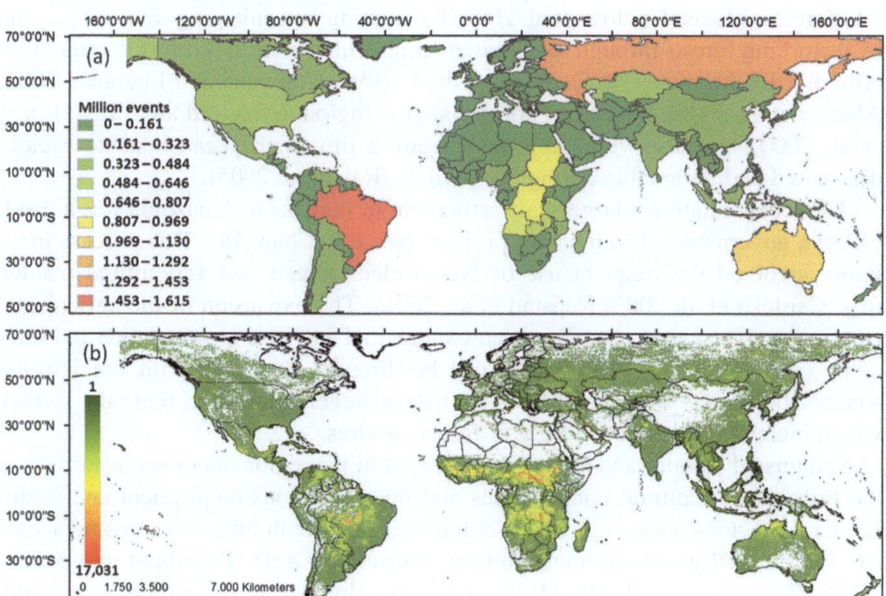

Fig. 13.1 Global view of (**a**) total number of active fire detections per country and (**b**) the spatial configuration of total number of active fires per 0.25° × 0.25° grid cells from Terra/MODIS sensor between 2001 and 2010

13.1.2 Fire Usage in Amazonia

Most of the historical evidence suggests that wildfires in tropical forests were rare, with return time intervals typically ranging from hundreds to thousands of years (Sanford et al. 1985; Meggers 1994; Bush et al. 2007). Despite the possibility of natural fire occurrence in pre-Columbian times, it is well accepted that the presence of charcoal is an indicative of the use of fire by humans (Bush et al. 2008).

Establishing the causes and consequences of paleo-wildfires in Amazonia is compromised by the small number of sites investigated and their spatial configuration. Radiocarbon dating of charcoal collected from soil samples around San Carlos de Rio Negro, Venezuela, indicates the occurrence of wildfire events at 6000, 3000, 1500, 650, 400, and 250 years ago (Sanford et al. 1985; Saldarriaga and West 1986). These dates match with dry climatic phases in the late Holocene as confirmed by pollen dating (Sanford et al. 1985).

In the past 50 years or so, fires have become more frequent (Bowman et al. 2009), with the vast majority of burning events resulting from human-lit fires (Cochrane and Schulze 1999; Cochrane et al. 1999; Uhl and Kauffman 1990). In the Amazon basin, fire is widely used for the initial conversion of extensive areas of natural vegetation into agricultural fields and pasture areas and for the subsequent suppression of secondary succession (Cochrane and Schulze 1999; Kodandapani et al. 2004; Giglio et al. 2006; Bowman et al. 2008; Sorrensen 2008).

Anthropogenic activities can facilitate and directly increase the spread of fire into forest systems (Cardoso et al. 2003) by creating and enlarging forest edges and by disturbing forests through selective logging, which increases forest flammability (Uhl and Bushbacher (1985); Nepstad et al. 1999; Cochrane and Laurance 2002; Alencar et al. 2006). Secondary forests, regrowing on deforested areas (e.g. Lucas et al. 2000), are also vulnerable to spreading fire as they can become rapidly desiccated and flammable during dry periods (Ray et al. 2005).

The recent relatively large-scale settlement of humans in Amazonia was a result of large government development projects (see e.g. Chap. 18). This process inevitably involved the frequent use of fire to clear forests and agricultural residue (e.g. Cardoso et al. 2003; Nepstad et al. 2004). The expansion of the agricultural frontier, therefore, led not only to an exponential increase in population in Amazonia (Aragão et al. 2014), which can be directly associated with fire ignition sources, but also to fragmentation of the natural vegetation matrix (cerrado, forest), which increased the susceptibility of forests to fires.

Land use dynamics and related fire patterns in the region may vary according to the price of agricultural commodities and due to various biophysical and socioeconomic factors, such as planned settlement, changes in infrastructure and accessibility, as well as government policies (Sorrensen 2008; Brondizio and Moran 2008; Carmenta et al. 2011). Clearing for intensive agriculture is usually characterised by repeated burning. This process can take up to 3 years for achieving the complete removal of the slashed vegetation (Morton et al. 2006).

Several studies have demonstrated a temporal association between fire and deforestation, in the Brazilian Amazon and elsewhere (Sorrensen 2000, 2004; Bowman et al. 2008; Sorrensen 2008; Aragão et al. 2008; Morton et al. 2013). This relationship is consistent with the fact that burning events in Amazonia are usually man-made. Spatial patterns of fire occurrence are also expected to follow the patterns of forest conversion and subsequent land use (Lima et al. 2012). Despite the extensive evaluation of the temporal links between fire and deforestation, there is still a need for better understanding the spatial structure of the association between fire and deforestation and evaluating how past and present land use and land cover change (LULCC) influence these spatial patterns.

13.1.3 Fire, Land Use, and Land Cover

Natural fires are rare in the Amazon forest 'biome' (Cochrane 2001; Bush et al. 2007), but common in the Cerrado 'biome' (Ramos-Neto and Pivello 2000), in the south and east of BLA (Fig. 13.2a). In the Cerrado, the long dry season period (5–7 months) with rainfall lower than 100 mm month^{-1} (Sombroek 2001) makes climatic conditions suitable for natural fires. This 'natural' pattern of low fire incidence in the rainforest of Amazonia and high incidence in the Cerrado appears to have changed in recent years.

Deforestation in BLA was responsible for the transformation of around 760,000 km^2 of pristine seasonal, open, and closed canopy forests into pastures for cattle ranching and agricultural lands by 2012 (PRODES 2013; Chap. 15). This value corresponds to c. 15 % of the original forested area of BLA. Forest conversion in BLA has mostly affected the contact zones between the Cerrado and Amazon forest 'biomes', expanding deep into the forest where access routes were available (e.g. Fig. 13.2b). The spatial configuration of fires detected by satellites is highly linked to the distribution of deforestation (Fig. 13.2c). This is expected because of two widespread land use practices, acting as ignition sources for fires in the region: (1) land clearing by slash and burn and (2) management of pastures using fire.

Fire recurrence time, as the time needed for fire to strike in the same area, has been reduced to around 5–15 years (Cochrane et al. 1999; Alencar et al. 2006) because of the amplification of land use in Amazonia. Within BLA, Mato Grosso (MT) has been the state with the highest fire occurrence, detected by MODIS/TERRA satellite, with an average (mean $\pm$ standard deviation) of 1432 $\pm$ 1838 and 754 $\pm$ 1109 active fires per month detected between November 2000 and July 2011 in the biomes Amazonia and Cerrado, respectively (Fig. 13.3).

Observing fire patterns across 'biomes' within BLA for the same period, it is clear that despite the dominance of fires in the contact zone between Amazonia and Cerrado 'biomes' (1443 $\pm$ 1716 active fires), closed and open broadleaf forests are also exposed to extensive fire occurrence (1692 active fires) (Table 13.1).

In the BLA region, as a whole, fire incidence has been directly related to deforestation (Aragão et al. 2008). The annual rates of deforestation in BLA have

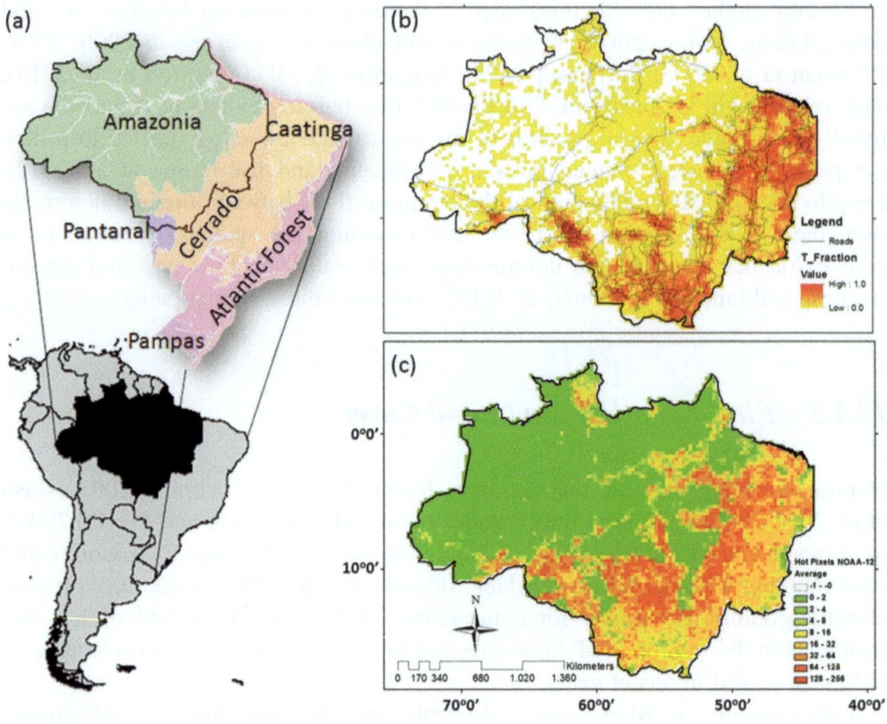

Fig. 13.2 (**a**) Map of South America, displaying the Brazilian Legal Amazon (BLA, *black* limits) and a detailed distribution of Brazilian 'biomes'. (**b**) Fraction of the total grid cell area (0.25° × 0.25°) that has been deforested by 2007, where 0 means no deforestation and 1 complete conversion of the grid cell. (**c**) Mean number of active fires detected by the NOAA-12 sensor from 1998 to 2006 in each grid cell with similar spatial resolution as (**b**)

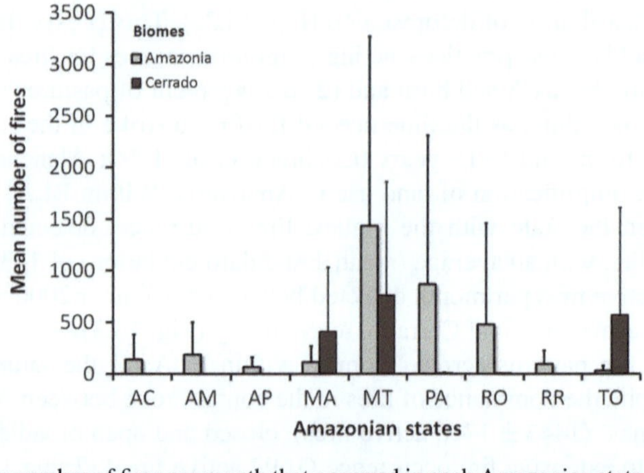

Fig. 13.3 Mean number of fires per month that occurred in each Amazonian state from November 2000 to July 2011. Error bars correspond to the standard deviation

Table 13.1 Mean (±SD) number of active fires detected by MODIS/Terra in 0.25° × 0.25° grid cells analysed in each forest type (count) in Brazilian Amazonia from November 2000 to July 2011

	Contact zone	Open broadleaf	Closed broadleaf	Seasonal semi-deciduous	Seasonal deciduous	Campinarana
Mean	1443	1055	637	112	31	4
Sd	1716	2072	870	158	52	4
Count	129	123	126	129	79	42
Percent of the total	100	95	98	100	61	33

Percent of the total, the percentage of grid cells within each forest type that have at least one fire recorded within the analysed period

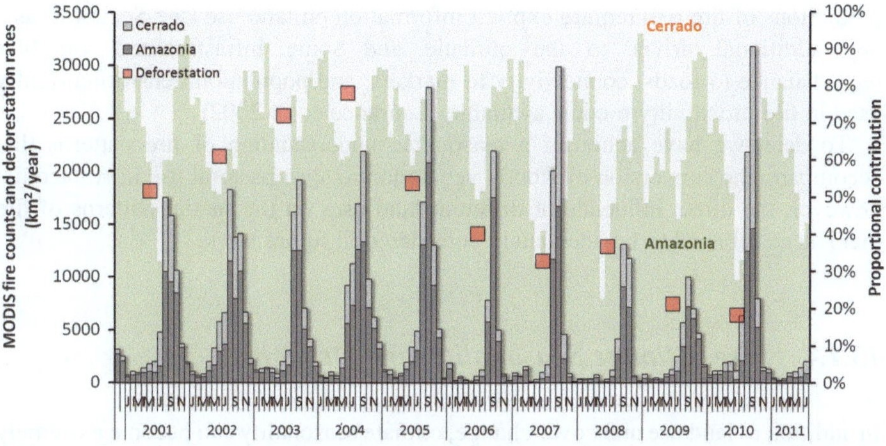

Fig. 13.4 Total number of monthly active fire occurrences within the domain of the Brazilian Legal Amazonia. *Grey bars* correspond to the sum of fire occurrence in the Cerrado (*light grey*) and Amazon forest (*dark grey*) 'biomes'. The *coloured areas* correspond to the proportional contribution of fires in the Cerrado (*light magenta*) and in the Amazon forest (*light green*) to the total number of fire events detected in each month. *Magenta squares* correspond to annual deforestation rates

decreased from 21,400 km^2 $years^{-1}$ (1980s) to 4571 km^2 recorded in 2012 (PRODES 2013). Nevertheless, this significant reduction in deforestation did not involve a proportional reduction in fire incidence (Fig. 13.4). This result corroborates the study by Vasconcelos et al. (2013), which, analysing MODIS active fire data from 2003 to 2012 for the state of Amazonas, did not observe the deforestation–fire relationship proposed by Aragão et al. (2008). Moreover, Aragão and Shimabukuro (2010a) also quantified a decoupling between fire and deforestation in BLA, analysing active fire data from AVHRR and MODIS sensors.

The decoupling between fire and deforestation in the past 10 years may be related to the fact that secondary forests are not included in forest loss monitoring programmes (PRODES 2013). Their conversion to agricultural use, as reported for

BLA (Fearnside et al. 2007), is not accounted for as new deforestation, while fires associated with this conversion are quantified. In addition, the increased number of forest fragments and area of forest edges (Broadbent et al. 2008), which are more vulnerable to escapee fires, and the increased frequency of droughts in recent decades (Marengo et al. 2011) could also have contributed to an absolute increase in fire occurrences.

In contrast to the positive trends observed in 42 % of the forested area in BLA from 1998 to 2006, in the Cerrado within the BLA, fire occurrence has decreased during the same period (Aragão et al. 2013). The Cerrado covers a climatic region with high risk of fires (Arima et al. 2007); however, once the natural vegetation has been cleared by fire, mechanised intensive agriculture tends to reduce fires (use for burning residues) in comparison with agro-pastoral land use. Therefore, robust predictions of fire risk require explicit information on land use (see Sect. 4.3) as a key additional driver to the climatic and some infrastructural variables (e.g. distance to roads, connectivity to markets, and population density) currently used in fire probability models available (Laurance et al. 2002).

To date, we have achieved a reasonable understanding of fire patterns that accompany the conversion of woody vegetation to agro-pastoral use in Amazonia; however, the direct influence of different land uses on the spatial patterns of fire incidence still need to be adequately considered in future work.

13.1.4 Fire, Climate Seasonality, and Droughts

In addition to land use and cover change, climate seasonality can become extremely relevant for determining fire occurrence in Amazonian forests if human-related ignition sources are active. Rainfall, temperature, and relative humidity (Cardoso et al. 2003; Sismanoglu and Setzer 2005), plant available water (PAW) (Nepstad et al. 2004), and vapour pressure deficit (VPD) inside the canopy (Ray et al. 2005) are some of the most important factors directly related to forest fires in Amazonia. The seasonality of these variables, which normally co-vary, defines the period of occurrence and the intensity of fires (the intensity also depends on the availability of fuel loads). Overall, fires tend to intensify during July, August, and September (Fig. 13.4). This period corresponds to the dry season in most of the Amazon area, with rainfall lower than 100 mm month^{-1}, high VPD, and low PAW, especially in the south and east of the region.

Severe droughts, moreover, can exacerbate fire incidence and severity, as observed recently in 1997/1998, 2005, and 2010. The majority of droughts in the region are associated with extreme El Niño events, which is characterised by the anomalous warming of the equatorial Pacific Ocean near the coast of Peru (Marengo 1992; Uvo et al. 1998; Ronchail et al. 2002; Marengo 2004). Recent El Niño events occurred in 1982/1983, 1986/1987, and 1997/1998 (Fig. 13.5). During the last decade, contrarily, droughts in Amazonia have been associated with anomalously warm waters in the tropical Atlantic Ocean, following the Atlantic

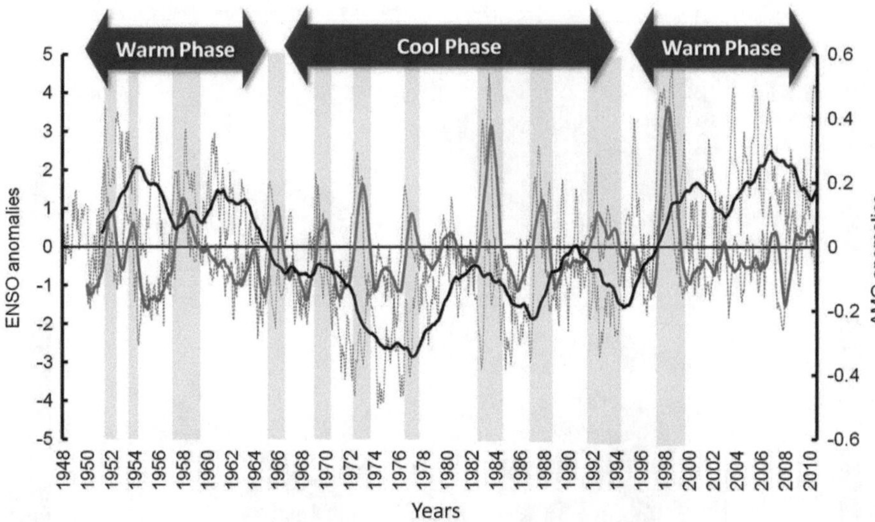

Fig. 13.5 Monthly sea surface temperature anomalies in the tropical Pacific Ocean (*dotted grey line*), related to the El Niño Southern Oscillation (ENSO) event and measured using the Multivariated El Niño Index (MEI) and in the north Atlantic Ocean (*slashed black lines*) represented by the Atlantic Multidecadal Oscillation index (AMO). Thick lines represent moving averages of 12 months for the ENSO (*grey*) and 40 months for the AMO (*black*)

Multidecadal Oscillation (AMO) cycle (Li et al. 2006; Good et al. 2008; Marengo et al. 2008). The AMO was identified as a partial driver of the 1997/98 drought and the main driver of the 2005 and 2010 droughts (Marengo et al. 2008). Amazonia constantly experiences El Niño or AMO-driven cycles that can cause droughts (Fig. 13.5). As some of these droughts can manifest as extreme events (see e.g. Chap. 4), refining our current understanding on how fire patterns respond to such extremes is critical for predicting future impacts of fire on Amazonian ecosystems and human populations.

El Niño-driven droughts normally affect northern Amazonia during the boreal winter, which corresponds to the dry season in parts of South America, north of the Equator. Moreover, eastern Amazonia is affected by these El Niño-driven droughts during the austral winter, due to the opposite timing of the dry season. AMO-driven droughts are related to droughts in the south-west of Amazonia during the Austral winter (Saatchi et al. 2013).

The 1997/1998 El Niño-driven drought created perfect conditions for the widespread occurrence of extensive wide-spreading fires. The total area of forests burned by understory fires in BLA, for instance, was 14 times higher than during an average non-El Niño year (Alencar et al. 2004, 2006). These fires followed the drought pattern affecting areas in northern (Barbosa and Fearnside 1999) and south-eastern flanks of BLA (Alencar et al. 2004).

The drought in 2005, conversely, was driven by the warming up of the tropical Atlantic. This drought also led to a large reduction in rainfall during the dry season

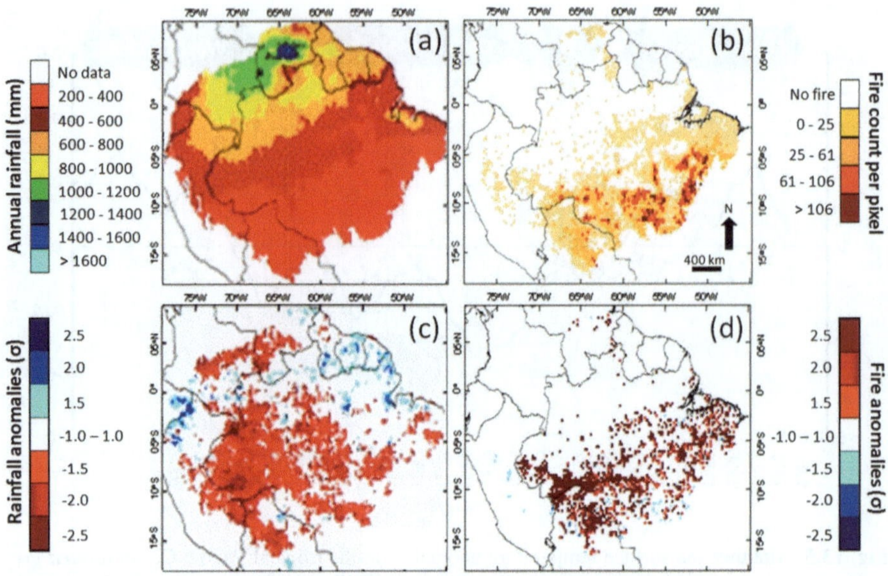

Fig. 13.6 *Top left panel* (**a**) shows the long-term (1997–2006) mean of accumulated rainfall for the months of July, August, and September (dry season in most of the Amazon region). *Top right panel* (**b**) shows the same as top *left panel* but for active fire dataset. *Bottom left panel* (**c**) displays rainfall anomalies (units in standard deviation of the long-term mean) for the trimester of July, August, and September of 2005 and the *bottom right panel* (**d**) shows the same as *bottom left* using the active fire dataset. Adapted from Aragão et al. (2007a)

(July, August, September) (Fig 13.6a and c). This anomalous water shortage created ideal conditions for the widespread occurrence of fires in south-western Amazonia (Aragão et al. 2007a; Cardoso and Oliveira 2007). Fire occurrence increased by 33 % in relation to the long-term average with anomalies reaching values larger than two standard deviations (σ) of the mean (Fig. 13.6b and d).

In 2010, despite the drought being more severe than in 2005, fire incidence was 26 % lower. A total of 34,484 active fires were recorded for the Amazon forest 'biome' in 2005 in comparison with 25,612 in 2010. Cerrado areas within BLA were more affected by fire than the Amazon forest 'biome' (Fig. 13.4). The lower number of fires in 2010 in relation to 2005 was probably a reflexion of the 66 % reduction in deforestation rates in comparison to 2005, limiting ignition sources. Nonetheless, the number of fires normalised by the area of deforested land in BLA increased from 1.83 fires km^{-2} of deforested land in 2005 to 3.97 fires km^{-2} of deforested land in 2010. Morton et al. (2013) quantified a 22 % increase in the area affected by understory forest fires from 2005 to 2010. These results clearly indicate that fire counts detected from satellites are not restricted to forested areas quantified by the INPE/PRODES deforestation programme and other land cover types may be increasingly exposed to fire impacts (Aragão and Shimabukuro 2010b; Lima et al. 2012).

The key to understanding fire in Amazonia is that although drought exacerbates fire occurrence, the ignition of fires is man-made; even during extreme droughts Amazonian forests would not be affected by natural wildfires. This is evident by the lack of significant positive fire anomalies in the Peruvian Amazon (low human occupation/activity) during the 2005 drought (Fig. 13.6d), as opposed to fire anomalies observed in eastern and southern BLA (high levels of human activity)

Where ignition sources are present, two climatic variables are critical in shaping monthly and annual fire incidence behaviour: monthly rainfall and the length of the dry season. Monthly rainfall explains around 60 % of the variance in fire incidence between 1997 and 2006 (Fig. 13.7a). Rainfall lower than evapotranspirative loss (mean c. 103.4 ± 9.1 mm month^{-1} (Shuttleworth et al. 1989; Malhi et al. 2002; Cox et al. 2004; Rocha et al. 2004; Hutyra et al. 2005) causes water deficit, which is correlated with a high fire incidence. Over 50,000 active fires per month were detected in one single month for the whole BLA during the 2005 drought event. The relationship between fire and monthly rainfall follows an exponential decay function, decreasing fire incidence with increase in rainfall (Aragão et al. 2008), as for example observed between 2003 and 2012 in Amazonas State (Vasconcelos et al. 2013).

At the annual scale, the length of the dry season explains c. 70 % of the variance in the maximum monthly number of active fires (Fig. 13.7b). As the length of dry season length increases, and consequently enhancing the water deficit of many forests in Amazonia, which are common features of droughts in the region, leaf shedding is exacerbated (Alencar et al. 2004; Phillips et al. 2009). This boost of organic matter on the forest floor, with associated increase in canopy gaps (Ray et al. 2005), favours the rise of temperature inside the forest and the reduction in soil and litter moisture. With the increase in drying the accumulated combustible fuel on the forest floor during extended dry periods conditions becomes ideal for the spread

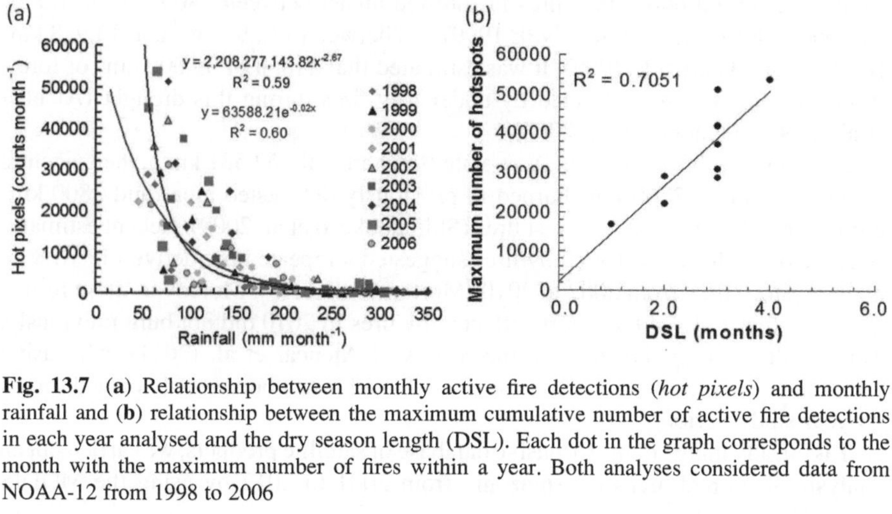

Fig. 13.7 (**a**) Relationship between monthly active fire detections (*hot pixels*) and monthly rainfall and (**b**) relationship between the maximum cumulative number of active fire detections in each year analysed and the dry season length (DSL). Each dot in the graph corresponds to the month with the maximum number of fires within a year. Both analyses considered data from NOAA-12 from 1998 to 2006

of fires into forest (Uhl and Kauffman 1990; Cochrane and Schulze 1999; Cochrane et al. 1999; Barlow and Peres 2004b; Nepstad et al. 2004; Ray et al. 2005).

Understanding relationships between fire incidence, human activities, and climate in Amazonia may offer an approximation of the expected changes in fire activity and an indication of the likelihood of forests to be impacted by fires during these extreme events under future climate conditions.

13.2 Fire Impacts

13.2.1 Extent of Burned Areas

Fire in the late Holocene has been associated with the increased adoption of agriculture (c. AD 200 and AD 800), with El Niño-related droughts (c. AD 800 and AD 1000–1100), and with insolation minima (Bush et al. 2008; Mauas et al. 2008). Fires that occurred in this period and in non-drought years were likely to be small fires that almost always extinguished themselves, at most, within 100 m inside the forest (Uhl and Kauffman 1990). However, in the early 1970s, colonisation and settlement projects in BLA changed this pattern.

The beginning of the large-scale burned area estimates is associated with the use of satellite imageries for detecting anomalous high temperature and fire plumes. One of the first studies for BLA (c. 5×10^6 km^2) suggested that in the dry season of 1987, 350,000 independent fires were detected, possibly corresponding to about 200,000 km^2 of area burned (Setzer and Pereira 1991).

During the El Niño event in 1997/1998, it was estimated that in Roraima state (total area of 224,299 km^2) fires burned over an area between 33,000 km^2 and 38,144 km^2 (UNDAC 1998; Barbosa and Fearnside 1999). From this total, three studies quantified understory fires in forested areas: between 7800 and 9200 km^2 (Barbosa 1998), 11,730 km^2 (INPE 1999) and between 11,394 km^2 and 13,928 km^2 (Barbosa and Fearnside 1999). It was estimated that a total of 39,000 km^2 of forest in the whole BLA was affected by understory fires during this drought (Alencar et al. 2006; Mendonça et al. 2004)

During the 2005 drought, in Acre state (total area of 152,581 km^2), the epicentre of the drought, c. 3700 km^2 burned in previously deforested areas and 2800 km^2 corresponded to understory forest fires (Shimabukuro et al. 2009). Recent estimates of Amazonian forests affected by fires suggested a repeated fire activity in 16 % of all understory fires from 2002 to 2010 (Morton et al. 2013). Moreover, these results indicated that 73 % of the forests affected by fires in 2010 did not burn previously. This result is in agreement with the results of Alencar et al. (2011), who using Landsat data showed that 72 % of the fire-affected forest burned only once during a 23-year study period.

Taking advantage of the extensive database of satellite products, we carried out an analysis of burned areas in Amazonia from 2001 to 2012 by using the MODIS

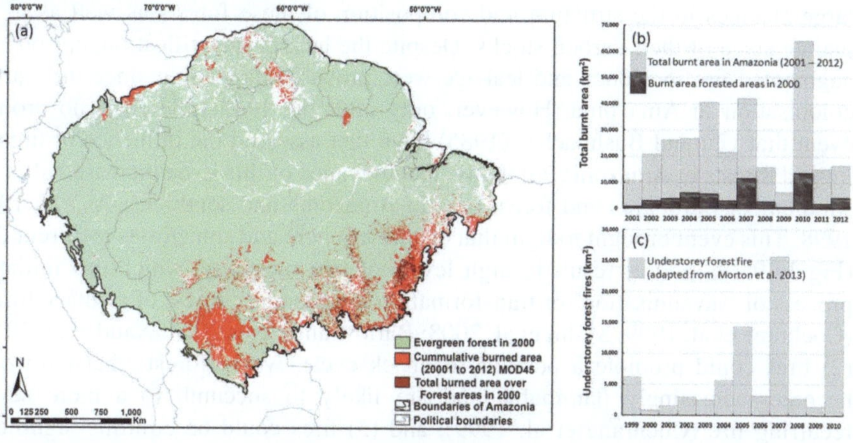

Fig. 13.8 (**a**) Spatial distribution of the cumulative burned area from 2001 to 2012 based on the MODIS MOD45 c5.1 product. The boundaries of Amazonia cover an area of c. 6.76 million km^2. (**b**) Extent of annual burned areas for Amazonia and over primary forests areas in 2000 based on MOD45 collection 5.1. (**c**) Extent of understory forest fires, adapted from Morton et al. (2013)

product MOD45 collection 5.1, for the geographical boundaries defined by Achard et al. 2005 (Fig. 13.8a). Our results showed peaks in burned areas in 2005 (40,500 km^2), 2007 (42,000 km^2), and 2010 (64,000 km^2) (Fig. 13.8b). Considering only the areas that were primary forest or recently deforested, by masking the data outside forest boundaries in 2000, we quantified that burned areas associated with these two land covers peaked in 2004, 2007, and 2010 with an area of c. 6000 km^2, 11,500 km^2, and 13,600 km^2, respectively (Fig. 13.8b). Based on this, c. 4.5 % of Amazonia has burned at least once in the last 12 years. Our results also indicate that c. 60,000 km^2 of the burned area recorded during the studied period was related to forest conversion, land maintenance, and fire leakage to forests. It is interesting to note that, although monitoring an area extent less than half that considered in our analysis, Morton et al. (2013) detected peaks in burned forests of 14,300 km^2, 25,600 km^2, and 18,500 km^2 in 2005, 2007, and 2010, respectively (Fig 13.8c). The differences between the estimates provided by these two independent studies indicate that there are still high uncertainties related to the detection of burned areas, particularly over primary forests. Cloud coverage, data availability for only part of the area, the time window selected, and methods used for detecting burn scar, especially in forest areas, are a major source of uncertainty for quantifying the area burnt (Box 1).

13.2.2 Impact Fires on the Structure, Composition, and Carbon Stocks of Forests

The amplified incidence of large forest areas affected by fire in recent years, because of the leakage of agricultural fires into surrounding forests, has caused large changes in the structure and composition of these forests as well as in the maintenance of their carbon stocks. Despite the lack of quantification, episodes of augmented fire incidence and leakage were probably happening since the start of colonisation in Amazonia. However, only after the 1982/1983 El Niño drought event that Uhl and Bushbacher (1985) have first assessed the influence of fires on logged forests in Amazonia. Globally, the relevance of this issue increased after the simultaneous droughts and forest fires in Amazonia and south-east Asia in 1997/1998. This event brought to light that (1) these severe and sometimes recurrent fires (Fig 13.9a, b, and c) result in high levels of tree mortality, which can initiate a process of 'savannisation' or transformation to secondary forest of primary forests (Cochrane et al. 1999; Malhi et al. 2008; Barlow and Peres 2008; Xaud et al. 2013), (2) fires could promote a positive-feedback cycle, where forests that burn once become increasingly flammable and are likely to succumb to a more severe recurring fire (Cochrane et al. 1999), and (3) fires could be emitting significant levels of CO_2 globally (Nepstad et al. 1999).

In transitional semi-deciduous Amazonian forests between evergreen rainforest and cerrado small trees are highly vulnerable to low-intensity understory fires. About 50 % of stems with a diameter at breast height (DBH) < 10 cm can die within 1 year after fire (Balch et al. 2011). Larger trees can also suffer high levels of fire-induced mortality and biomass loss, which tends to increase with fire intensity

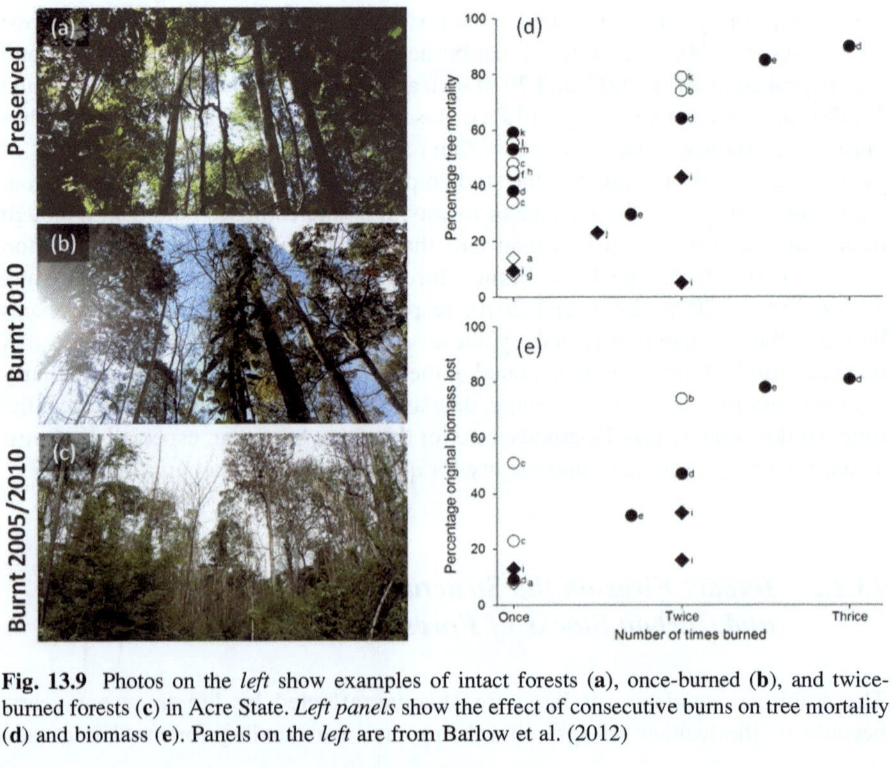

Fig. 13.9 Photos on the *left* show examples of intact forests (**a**), once-burned (**b**), and twice-burned forests (**c**) in Acre State. *Left panels* show the effect of consecutive burns on tree mortality (**d**) and biomass (**e**). Panels on the *left* are from Barlow et al. (2012)

and decrease in return time interval (Brando et al. 2012; Barlow et al. 2012, Fig. 13.9d and e). Substantial variation in the vulnerability of trees to fire has been observed in Amazonia (Balch et al. 2011). This variation is dependent on species-specific traits that can protect trees against fire damage. For instance, Brando et al. (2012) quantified that < 20 % of individuals with bark thicker than 18 mm died from fire damage. Moreover, mortality decreases as tree diameter and height increase and species with dense wood survive better than species with light wood. The large variation in fire-induced tree mortality, determined by the canopy and fire characteristics, is also reflected in the results of biomass loss in Amazonian forests (Fig. 13.9d and e).

Primary forests affected by successive fire events tend to undergo a complete turnover in species composition (Barlow and Peres 2008, Fig. 13.10). Fire-induced mortality and consequent gap formation favour the establishment of fast-growing pioneer species (Barlow and Peres 2008). These species usually have lower wood density than slow-growing, late-successional species (Baker et al. 2004). This characteristic cannot only directly increase the vulnerability of these forests to recurrent fire but also indirectly create a feedback, as trees with low wood density are more susceptible to mortality during droughts (Phillips et al. 2009) and as a result the increase in organic debris can facilitate fire spread into these forests during subsequent drought events.

Annually recurring experimental fires over 5 years have reduced the number and diversity of regenerating stems, and the species pool tended to change towards cerrado-like vegetation in MT (Massad et al. 2013). The reduction in species

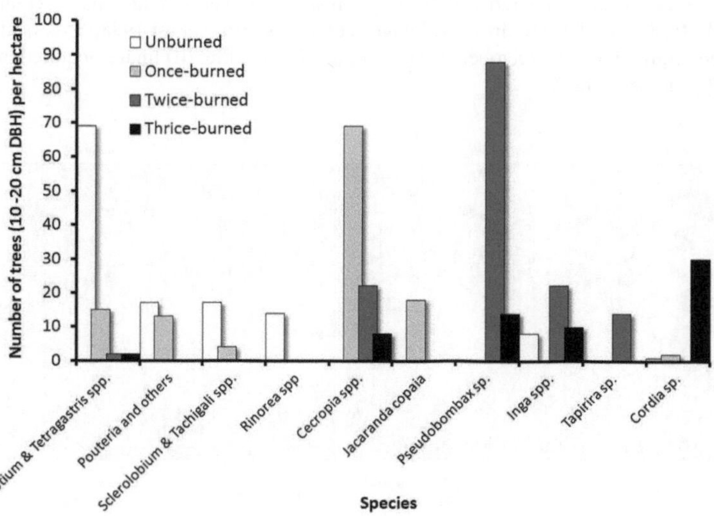

Fig. 13.10 Tree species and genera from the 10–20 cm DBH size class (and shrubs and samplings <10 cm DBH) that were most abundant in each burn treatment, showing a high degree of turnover in community composition with each additional burn. All species (or genera) with a density greater than 10 trees per hectare are shown for trees ≥10 cm DBH, in once-, twice-, and thrice-burned forest plots. Data in the figure are from (Barlow and Peres 2008)

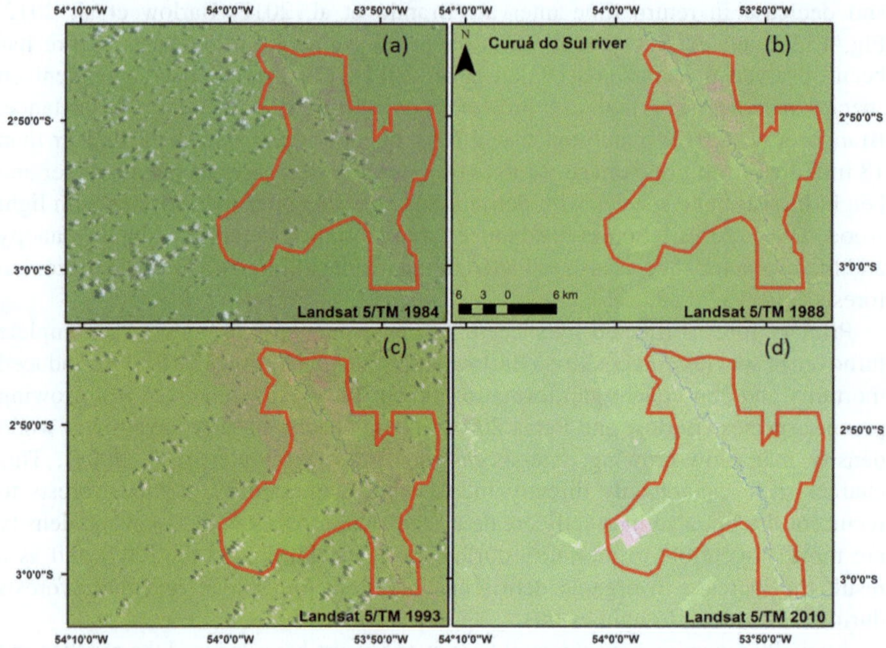

Fig. 13.11 Landsat 5/TM images, path/row 227/62, composition RGB 543, showing the change in the reflectance of a forest area affected by fires during the El Niño in 1983 (Nelson 1994), highlighted by the *red polygon*. Burned forests exhibit *higher shades* of *red colour* or *lighter shades* of *green* when compared with healthy forests (*darker green*). (**a**) Image acquired on 24th August 1984, 1 year after the fires. (**b**) Image acquired on 3rd August 1988, 5 years after the fire. (**c**) Image acquired on 20th October 1993, 10 years after the fire. (**d**) Image acquired on 29th June 2010, 27 years after the fire

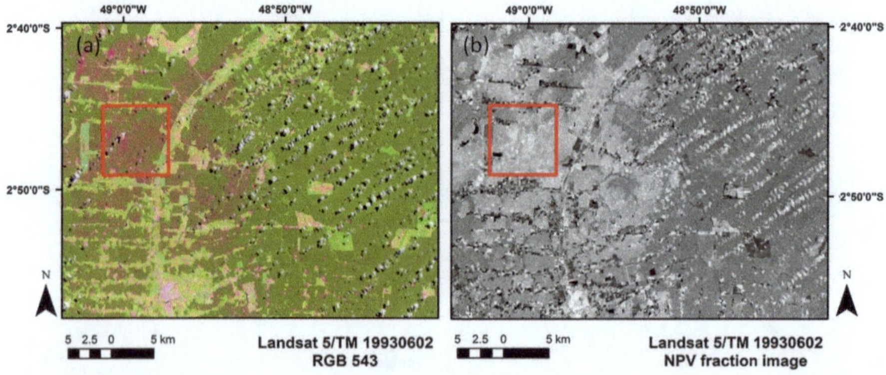

Fig. 13.12 Example of a forest burned area (*left* image) and the non-photosynthetic vegetation fraction image (*right* image) for the same area and date evaluated in the pioneer work of Cochrane and Souza (1998). Burned forests are highlighted by the *red rectangle* and appear brightened in the NPV fraction image. Landsat 5, path/row: 223/62, 2nd of June 1993

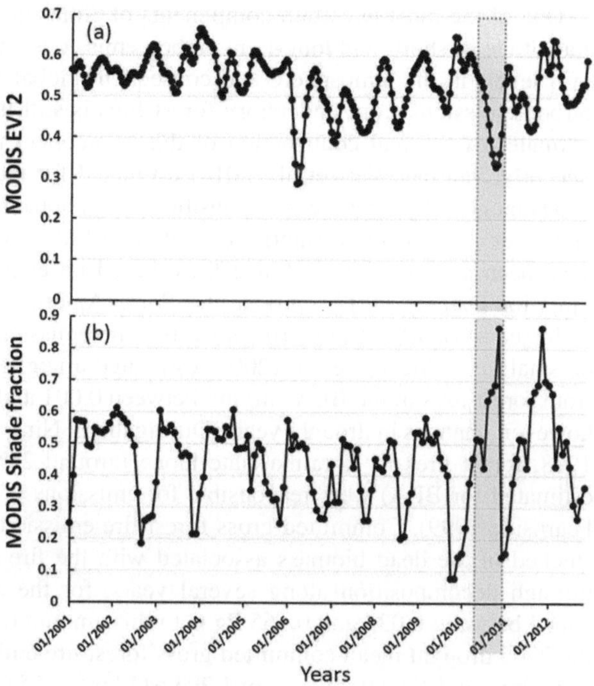

Fig. 13.13 Example of a forest affected by fires in 2010, detected by two-image processing techniques applied to MODIS data in Southern Para State. (**a**) The Enhanced Vegetation Index (EVI) 2, derived from MODIS trajectory in time acquired through the website: http://www.dsr.inpe.br/laf/series/ exhibiting the decrease in the index after 2008, due to the forest degradation probably by logging, and lowest values at the end of the dry season in 2010, after the forest fires. (**b**) Same area as "a" showing the increase in the shade fraction, derived from the unmixing model, due to the burn scar in the forest

diversity in burned plots can enhance the susceptibility of remaining species to herbivory (Massad et al. 2013), which can slow down or impede the long-term recovery of fire-affected forests. So, in regions along the Amazon forest—Cerrado ecotone, fire may be an important factor shaping the boundary between the two 'biomes' (Staver et al. 2011; Silvério et al. 2013).

Fire-induced modifications in the canopy structure and floristic characteristics are precursors of changes in ecosystem processes. In forests submitted to experimental annual fires total net primary productivity (NPP) was reduced by 15 % in years following the fire and autotrophic respiration was reduced by 4 % in comparison with the adjacent intact forest (Rocha et al. 2013). Litter production in these fire-affected forests can be reduced by 50 % (4.3 Mg ha^{-1} $year^{-1}$) in comparison with that in intact forests across Amazonia (Balch et al. 2008). Decomposition rates in areas affected by successive annual fires have tended to be slower (Silveira et al. 2009), resulting, potentially, from a drier microclimate and lower litter moisture (Balch et al. 2008). In contrast, a single low-intensity fire had no such effect (Silveira et al. 2009).

The decline in litter production after successive burns seems to suppress the spread of fires, even when microclimatic conditions are favourable (Balch et al. 2008). However, the inhibition of fire by shortage of fine fuel loads may not be sustained in the long term, as delayed tree mortality (Barlow et al. 2003) can increase fuel availability in subsequent years.

One of the most uncertain components of Amazonian forest fire impacts is the magnitude of short- and long-term carbon emissions and potential implications for CO_2 levels in the atmosphere and consequent global warming. Quantification of carbon emissions from understory forest fires is still lacking, preventing accurate estimates of the real contribution of this component to the global carbon cycle. Recently, van der Werf et al. (2010) estimated for the period between 1997 and 2009 that globally, fires were responsible for an annual mean carbon emission of 2.0 Pg C year^{-1}, with South America contributing 14.5 %. Of this, about 8 % appears to have been associated with forest fires, based on estimates from the Global Fire Emission Dataset (GFED) product for South America.

In years not affected by droughts forest fire emissions in Amazonia are likely to be small, e.g. Alencar et al. (2006) estimated a negligible amount of C emission from forest fires in the BLA varying between 0.001 and 0.011 Pg C for 1995. This, however, changes in drought years. During the El Niño event that occurred in 1997/1998, forest fires in Roraima state alone (around 25 % of the total burned area estimated for BLA) were responsible for emissions of c. 0.03 Pg C (Barbosa and Fearnside 1999). Committed gross forest fire emissions, which include all carbon stocked in the dead biomass associated with the fire event that will be released through decomposition along several years, for the southern part of Amazonia added between 0.024 and 0.165 Pg C to this amount (Alencar et al. 2006). During the 2005 drought mean committed gross forest fire emissions for the states of MT, Rondônia, and Acre (total area of 1,293,515 km^2 or 25 % of BLA) were estimated to be 0.21 (0.04–0.34) Pg C (Aragão et al. 2007b).

Studies on the long-term effect of forest fires in Amazonia are few. Barlow et al. (2003) suggested that mortality, especially from large trees, could increase after 3 years of a fire event, possibly doubling the amount of biomass loss and consequent carbon emissions. To address the many hiatuses in our knowledge, we suggest that future work should concentrate on (1) quantifying short- and long-term carbon dynamics in burned forests; (2) developing systematic mapping of the whole spatial extent of burned forests; (3) estimating the recovery rates of carbon stocks and species composition in forests affected by fire; and (4) better quantifying the burning efficiency for live and dead components of the biomass including charcoal formation rates. Integration of field-based surveys, remote sensing information, and ecological modelling is critically important for progressing towards a more accurate estimate of the contribution of forest fires to the global carbon cycle.

13.3 Modelling Fire Occurrence in Amazonia

Current rates of human-induced environmental changes and climate variability, in addition to predictions regarding the future climate in Amazonia, indicate that the conditions for increased forest fire frequency and propagation have already been established. With this in mind, models for forecasting fire occurrence are critical for quantitatively estimating the alteration in the magnitude and spatio-temporal configuration of fires. In addition, the forecast at the correct scale would allow operational actions for potentially avoiding fire episodes and their consequent impacts on carbon emissions, ecosystem services, and human health.

Models have been developed to predict fire risk for Amazonia, aiming to provide information for preemptive actions and for evaluating potential changes in fire pattern as a response to environmental changes (Cardoso et al. 2003; Nepstad et al. 2004; Alencar et al. 2004; Sismanoglu and Setzer 2005; Silvestrini et al. 2011; Chen et al. 2011). These studies have advanced the understanding of spatial and temporal dynamics of fires.

The Brazilian Centre for Weather Forecasts and Climate Studies (CPTEC/INPE) operates a system to predict daily fire risk in South America (Justino et al. 2002; Sismanoglu and Setzer 2002). Their approach is based on meteorological information about cumulative precipitation, minimum relative humidity, and maximum temperature of the preceding 120 days of the prediction date. In addition, active fire data detected by using AVHRR and MODIS sensors are also used (CPTEC 2013). This is a unique systematic product and probably the most comprehensive to date. Forecasts of fire risk are produced daily and can be consulted at http://www.inpe.br/queimadas/risco.php.

Other approaches exist too. Cardoso et al. (2003) developed a model for evaluating the impact of forest conversion on fire occurrence. Along with climate, variables related to human activities are included to predict contemporary patterns of fire incidence. The authors used active fire data from the non-interpolated Automated Biomass Burning Algorithm (ABBA), based on GOES-8 (Prins et al. (1998), to calibrate the fire risk model as a function of total and minimum precipitation, distance from paved roads, forest cover, and deforestation. The model was applied under two scenarios to estimate fire occurrence within a $2.5° \times 2.5°$ grid cell for the dry seasons of 1995 and 1997: one scenario with moderate deforestation rates, following the deforestation model developed by Laurance et al. (2001a), and another, extreme scenario, where forest conversion to degraded pastures and paved roads are present in each grid cell analysed. The analysis of both cases indicated that the frequency and spatial configuration of fires in Amazonia are susceptible to extensive changes related to agricultural development.

Nepstad et al. (2004) developed the fire susceptibility system RisQue to map the vulnerability of Amazonian forests to fire in response to the 1997/1998 El Niño drought. RisQue derived information about PAW, which the authors considered as the main driver of fire risk, based on information about soil properties, climate, evapotranspiration, and land use (especially selective logging). The resulting monthly maps of fire risk with a spatial resolution of 8 km showed that small declines in rainfall and increases in evapotranspiration could significantly augment fire risk during drought periods.

More recently, Silvestrini et al. (2011) integrated climate (VPD) with land use variables to model mid-twenty-first century fire responses to climate change and land use. Maps based on VPD to forecast fire risk were integrated with an annual probability of anthropogenic fires. Distance to deforested areas, distance to forest, distance to urban areas, distance to roads, elevation, and protected areas were the key variables used to calculate an annual anthropogenic fire probability. Calibration and validation of this model was based on the night-time AVHRR/NOAA-12 active fire data. The model indicated that extremely wet areas in the north-western Amazonia might become vulnerable to the spread of fires under future climate change. Furthermore, fire occurrence might double in Amazonia by 2050 if trends in climate change and deforestation rates were sustained at the levels that prevailed in the early 2000s.

Chen et al. (2011) have recently proposed a model based on the fact that fire intensity is strongly correlated to changes in sea surface temperatures of the Pacific and Atlantic Oceans. However, the use of it as a fire-risk warning system is limited as the best spatial resolution achievable is 5° by 5°, and the information does not allow accurate spatial planning for effective actions to prevent and curtail fire. Importantly, the model does not account for the influence of human-ignition sources, which have been consistently associated with fire occurrence in the region.

All of these studies have produced significant advances in modelling methods and understanding of spatial and temporal dynamics of fires. It is necessary that future models combine relevant climatic, anthropogenic, and biophysical variables that best forecast fires in the Amazon. Large amounts of freely available satellite-derived and geospatial information allow refining operational systems for analysing fire risk in Amazonia. For effective application of these models for supporting activities to restrain fire in Amazonia, the choice of the spatial resolution is critical and must be tailored to Amazonian geographic and political conditions.

13.4 Conclusions

Historically, fire occurrence was rare in Amazonia. However, with the increasing rates of settlement of humans in the region fire became a common feature of the system. Humans provide ignition sources for fire, mainly associated with large-scale deforestation, slash-and-burn, and management of pastures. The combination of the presence of ignition sources with extreme droughts has enhanced the flammability of natural ecosystems in Amazonia in recent years. During the droughts in 1998, 2005, and 2010, vast areas of forest were affected by fires.

The impact of fires on Amazonian ecosystems is large, with changes in forest structure and species composition, carbon stocks, and human health. Fires can also be part of a complex feedback loop that can increase the effects of climate and human-induced environmental changes (Fig. 13.14). The quantification of long-term impacts is still understudied and requires the implementation of permanent field plots associated with the knowledge of the age and intensity of fire-affected areas.

Remote sensing has been a critical tool in accessing the frequency of fire events and the extent of burned areas. Future work must combine field and remote sensing information to produce a more synoptic quantification of the extent of fire impacts. Moreover, operational programmes must be put in place to monitor the long-term

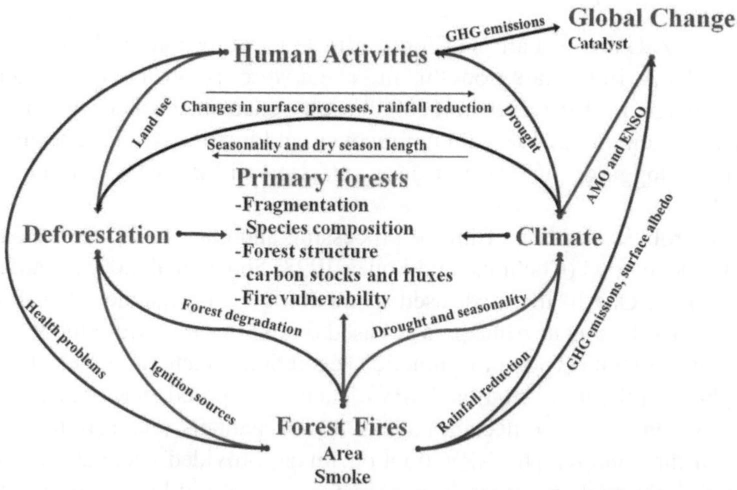

Fig. 13.14 Diagram depicting plausible feedbacks between land use, climate, fire, forests, and humans

impacts of fires in Amazonia. This would be a crucial step towards a better understanding of the resilience of ecosystems to fires in terms of biodiversity, carbon stocks, and ecosystems services. This is essential as climate change may exacerbate drought incidence and intensity in the region. So, understanding the functioning of disturbed systems may provide a clue to the consequences of environmental change on Amazonian forests, critically important for the management of these ecosystems in the future.

Box 1: Challenges for Estimating Burnt Forest Areas in the Satellite Era

Advances in mapping. The detection of burnt areas in Amazonia was first carried out by testing the relationships of fire pixels, detected in the 1-km^2 infrared channels of the Advanced Very High Resolution Radiometer (AVHRR) with the area burnt (Myers 1989; Setzer and Pereira 1991). However, high uncertainty and problems related to the method were detected (Fearnside 1990). Reflectance data from optical sensors (e.g. AVHRR) began to be used later for estimating large-scale burnt areas (e.g. Setzer et al. 1994; Razafimpanilo et al. 1995) and to describe the temporal patterns and trends of fire occurrence in Amazonia (e.g. Prins and Menzel 1994; Holben et al. 1996). Matson and Holben (1987) evaluated the possibility of using the 1-km^2 spatial resolution Normalised Difference Vegetation Index (NDVI) for detecting burnt areas at the subpixel scale. Their study found that burned forests had consistent lower values of NDVI than undisturbed forests.

(continued)

Nelson (1994) mapped areas of forest affected by fires during the El Niño in 1983 in BLA. Burn scars from this fire event were still visible in the Landsat 5/TM image in 1984 and in 1988. Forest reflectance values in the area affected by the fire in 1983 did not recover, and it is likely that reoccurrence of fires, logging, or other causes affected these areas subsequently (Fig. 13.11).

More robust methods of image processing and classification were subsequently developed (Cochrane and Souza 1998; Souza et al. 2003; Matricardi et al. 2010). One of the most used methodologies to separate intact forests from burned forests in Amazonia is based on the linear mixture model using three end members: photosynthetic vegetation (such as green leaves), non-photosynthetic vegetation (NPV), such as exposed tree branches, and shade, given by low reflectance areas in all channels representing shaded areas in the canopy. The NPV fraction image provided adequate means to separate unburned and recently burned forests, but old burned forests were not completely differentiated from intact and recently burned forests (Fig. 13.12). The information provided by the linear mixture model was then combined in the Normalised Difference Fraction Index (NDFI) that allows the detection and mapping of burned forests using Landsat 5 and 7 images (Souza et al. 2005).

The use of multiple dates, taking advantage of the high temporal resolution of MODIS images, brought considerable advances in detecting burned forests. Anderson et al. (2005) and Shimabukuro et al. (2009) have accurately identified burned forests in Amazonia (Fig. 13.13) by analysing the MODIS intra-annual variability of shade fraction images derived from a linear mixture model. Burned pixels exhibit lower reflectance (darker surfaces) and higher proportion of shade which make it possible to separate them from pixels that represent unburnt forest. Recently, Morton et al. (2011) used both intra- and inter-annual mean NDVI, derived from MODIS data to create a burn damage and recovery algorithm (BDR) to separate burned forests from selective logging and deforestation.

Uncertainties. Early estimates of burned area derived from active fire information and retrieved by thermal sensors have several limitations. They overestimate the burned area for the following reasons (Setzer and Pereira 1991): (1) small fires (less than 50 m^2) with high flame temperatures can be detected by the thermal sensor and by converting the fire pixel size (c. 1 km^2) directly to burned area would lead to an overestimation; (2) false fire pixel detection, in sandy areas, rocks, and bare soils that can reach high temperatures; (3) false detection in contrasting surface temperatures (e.g. boundaries between forest and bare ground); and (4) thermal sensor saturation generating false fire detection in regions close to the original fire. Conversely, omissions in fire pixels have also been observed due to (1) cloud coverage,

(continued)

(2) atmospheric attenuation due to smoke derived from fires obscuring detection, (3) low-intensity fire lines occurring over grasslands or pastures areas, and (4) timing of acquisition in relation to the start–end time of the fire (Schroeder et al. 2005, 2008a, b).

Although the methods have improved progressively through time by using reflectance data and series of intra- and inter-annual images, there are still many sources of uncertainties. The main factor affecting optical data is the presence of cloud shades, which produces a spectral signature similar to burned areas (low reflectance, low vegetation index, and high shade values). Although cloud and cloud shades filtering methods are available, if the algorithm is too restrictive, areas can remain with no data over long periods (weeks and months). On the other hand, if the algorithm is too permissive, commission error will occur. Another source of uncertainty is related to the seasonal dynamics of the forest. There is a lack of field data for tracking natural seasonal changes in the canopy structure. These changes, particularly in years of prolonged droughts, may be detected by the sensor of a satellite and may be erroneously interpreted due to the lack of knowledge and field data. For example, MODIS sensor data, which allows multi-temporal procedures for monitoring the forest canopy with medium spatial resolution, have been available only since 2000. Not enough time has passed since to have allowed an adequate evaluation of its use for detecting natural phenomena, such as extreme droughts.

Many global initiatives for mapping burned area started in the early 2000s (e.g. GLOBSCAR project, Global Burned Area GBA-2000, ATSR-2 World Fire Atlas, Global VGT burnt area product—L3JRC, MODIS burned area product—MCD45). However, differences among these products are evident (Simon et al. 2004; Giglio et al. 2005; Jain 2007; Chuvieco et al. 2008; Chang and Song 2009). For example, according to GLOBSCAR, 4333 km^2 of forests were burned in Brazil in the year 2000; GBA-2000 detected 846 km^2 (Tansey et al. 2004). It is expected that in the next years new methods will emerge, as there is a considerable increase in the freely available long-term time series of remotely sensed data.

References

Achard F, Balslev H, Beck S et al (2005) A proposal for defining the geographical boundaries of Amazonia (EUR 21808-EN), 39

Ackerman A, Toon O, Taylor J et al (2000) Effects of aerosols on cloud albedo: evaluation of Twomey's parameterization of cloud susceptibility using measurements of ship tracks. J Atmos Sci 57:2684–2695

Alencar A, Solorzano LA, Nepstad DC, Solórzano LA (2004) Modeling forest understory fires in a eastern Amazonian landscape. Ecol Appl 14:139–149. doi:10.1890/01-6029

Alencar A, Vera Diaz MDC, Nepstad D, Diaz MDV (2006) Forest understory fire in the Brazilian Amazon in ENSO and non-ENSO years: area burned and committed carbon emissions. Earth Interact 10:1–17

Alencar A, Asner GP, Knapp D, Zarin D (2011) Temporal variability of forest fires in eastern Amazonia. Ecol Appl 21(7):2397–2412

Anderson LO, Aragão L, Lima A, Shimabukuro YE (2005) Detecção de cicatrizes de áreas queimadas baseada no modelo linear de mistura espectral e imagens índice de vegetação utilizando dados multitemporais do sensor MODIS. Acta Amaz 35:445–456

Aragão L, Malhi Y, Roman-Cuesta RM et al (2007a) Spatial patterns and fire response of recent Amazonian droughts. Geophys Res Lett. doi:10.1029/2006gl028946

Aragão LEO, Shimabukuro YE, Lima A, Malhi Y, Anderson LO (2007b) Burn scars in Amazonian forests under extreme drought conditions. In: Anais do XIII Simposio Brasileiro de Sensoriamento Remoto. INPE, Florianópolis, Brazil, pp. 6605–6608

Aragão L, Malhi Y, Barbier N et al (2008) Interactions between rainfall, deforestation and fires during recent years in the Brazilian Amazonia. Philos Trans R Soc Lond B Biol Sci 363:1779–1785. doi:10.1098/rstb.2007.0026

Aragão L, Shimabukuro YE (2010a) The incidence of fire in Amazonian forests with implications for REDD. Science 328:1275–1278. doi:10.1126/science.1186925

Aragão L, Shimabukuro YE (2010b) Response to comment on "The incidence of fire in Amazonian Forests with implications for REDD.". Science 330:1627. doi:10.1126/science.1195063

Aragão LEOC, Shimabukuro YE, Cardoso M, Anderson LO, Lima A, Poulter B (2013) Frequência de queimadas durante as secas recentes. Secas na Amazônia: causas e consequências (ed. Borma LDS, Nobre CA). Oficina de Textos (ISBN:9788579750786), 367 p.

Aragão LEOC, Poulter B, Barlow J, Anderson LO, Malhi Y, Saatchi S, Phillips O, Gloor E (2014) Environmental change and the carbon balance of Amazonian forests. Biol Rev. doi:10.1111/brv.12088

Arima EY, Simmons CS, Walker RT, Cochrane MA (2007) Fire in the Brazilian Amazon: a spatially explicit model for policy impact analysis. J Reg Sci 47:541–567. doi:10.1111/j.1467-9787.2007.00519.x

Artaxo P, Gatti L, Leal A et al (2005) Química atmosférica na Amazônia: A floresta e as emissões de queimadas controlando a composição da atmosfera amazônica. Acta Amaz 35:185–196

Baker TR, Phillips OL, Malhi Y et al (2004) Increasing biomass in Amazonian forest plots. Philos Trans R Soc Lond B Biol Sci 359:353–365. doi:10.1098/rstb.2003.1422

Balch JK, Nepstad DC, Brando PM et al (2008) Negative fire feedback in a transitional forest of southeastern Amazonia. Glob Chang Biol 14:2276–2287. doi:10.1111/j.1365-2486.2008.01655.x

Balch JK, Nepstad DC, Curran LM et al (2011) Size, species, and fire behavior predict tree and liana mortality from experimental burns in the Brazilian Amazon. For Ecol Manage 261:68–77. doi:10.1016/j.foreco.2010.09.029

Barbosa RI (1998) Avaliação da área dos sistemas naturais e agroecossistemas atingida pelo fogo no Estado de Roraima (01.12.97 a 30.04.98). http://agroeco.inpa.gov.br/reinaldo/RIBarbosa_ProdCient_Usu_Visitantes/1998RelatFog_2.pdf

Barbosa I, Fearnside P (1999) Fires in the Brazilian Amazon: greenhouse gas emissions from burning of different ecosystems in Roraima during the 1997-98 El Ninõ event. Acta Amaz 29:513–534

Barlow J, Lagan BO, Peres CA (2003) Morphological correlates of fire-induced tree mortality in a central Amazonian forest. J Trop Ecol 19:291–299. doi:10.1017/s0266467403003328

Barlow J, Peres CA (2004a) Avifaunal responses to single and recurrent wildfires in Amazonian forests. Ecol Appl 14:1358–1373

Barlow J, Peres CA (2004b) Ecological responses to El Nino – induced surface fires in central Brazilian Amazonia: management implications for flammable tropical forests. Philos Trans R Soc B Biol Sci 359:367–380. doi:10.1098/rstb.2003.1423

Barlow J, Peres CA (2008) Fire-mediated dieback and compositional cascade in an Amazonian forest. Philos Trans R Soc Lond B 363(1498):1787–1794
Barlow J, Parry L, Gardner TA et al (2012) The critical importance of considering fire in REDD+ programs. Biol Conserv. doi:10.1016/j.biocon.2012.03.034
Bowman DMJS, Amacher GS, Merry FD (2008) Fire use and prevention by traditional households in the Brazilian Amazon. Ecol Econ 67:117–130. doi:10.1016/j.ecolecon.2007.12.003
Bowman DMJS, Balch JK, Artaxo P et al (2009) Fire in the Earth system. Science 324:481–484. doi:10.1126/science.1163886
Brando PM, Nepstad DC, Balch JK et al (2012) Fire-induced tree mortality in a neotropical forest: the roles of bark traits, tree size, wood density and fire behavior. Glob Chang Biol 18:630–641. doi:10.1111/j.1365-2486.2011.02533.x
Broadbent EN, Asner GP, Keller M et al (2008) Forest fragmentation and edge effects from deforestation and selective logging in the Brazilian Amazon. Biol Conserv 141:1745–1757. doi:10.1016/j.biocon.2008.04.024
Brondizio ES, Moran EF (2008) Human dimensions of climate change: the vulnerability of small farmers in the Amazon. Philos Trans R Soc Lond B Biol Sci 363:1803–1809. doi:10.1098/rstb.2007.0025
Bush MB, Silman MR, de Toledo MB et al (2007) Holocene fire and occupation in Amazonia: records from two lake districts. Philos Trans R Soc B Biol Sci 362:209–218. doi:10.1098/rstb.2006.1980
Bush MB, Silman MR, McMichael C et al (2008) Fire, climate change and biodiversity in Amazonia: a Late-Holocene perspective. Philos Trans R Soc Lond B Biol Sci 363:1795–1802. doi:10.1098/rstb.2007.0014
Cardoso MF, Hurtt GC, Moore BM et al (2003) Projecting future fire activity in Amazonia. Glob Chang Biol 9:656–669
Cardoso M, Oliveira G (2007) Comparação entre a ocorrência de fogo durante a seca de 2005 eo El Niño em 1998 na Amazônia. 13 SBSR, pp 4425–4429
Carmenta R, Parry L, Blackburn A et al (2011) Understanding human-fire interactions in tropical forest regions: a case for interdisciplinary research across the natural and social sciences. Ecol Soc 16:1–22
Chang D, Song Y (2009) Comparison of L3JRC and MODIS global burned area products from 2000 to 2007. J Geophys Res 114:1–20. doi:10.1029/2008JD011361
Chen Y, Randerson JT, Morton DC et al (2011) Forecasting fire season severity in South America using sea surface temperature anomalies. Science 334:787–791. doi:10.1126/science.1209472
Chuvieco E, Opazo S, Sione W et al (2008) Global burned-land estimation in Latin America using MODIS composite data. Ecol Appl 18:64–79
Cochrane M, Souza C Jr (1998) Linear mixture model classification of burned forests in the Eastern Amazon. Int J Remote Sens 17:3433–3440
Cochrane MA, Alencar A, Schulze M et al (1999) Positive feedbacks in the fire dynamic of closed canopy tropical forest. Science 284:1832–1835. doi:10.1126/science.284.5421.1832
Cochrane MA, Schulze MD (1999) Fire as a recurrent event in tropical forests of the Eastern Amazon: effects on forest structure, biomass, and species composition. Biotropica 31:2–16. doi:10.1111/j.1744-7429.1999.tb00112.x
Cochrane MA (2001) Synergistic interactions between habitat fragmentation and fire in evergreen tropical forests. Conserv Biol 15:48823
Cochrane MA, Laurance WFF (2002) Fire as a large-scale edge effect in Amazonian forests. J Trop Ecol 18:311–325. doi:10.1017/S0266467402002237
Cox PM, Betts RA, Collins M et al (2004) Amazonian forest dieback under climate-carbon cycle projections for the 21st century. Theor Appl Climatol 78:137–156. doi:10.1007/s00704-004-0049-4
CPTEC (2013) Monitoramento de queimadas e incêndios. http://www.inpe.br/queimadas/
Fearnside PM (1990) Fire in the tropical rain forest of the Amazon basin. In: Goldammer JG (ed) Fire in the tropical biota. Springer, Berlin, pp 106–116

Fearnside PM, Barbosa RI, Graça PMLA (2007) Burning of secondary forest in Amazonia: Biomass, burning efficiency and charcoal formation during land preparation for agriculture in Apiaú, Roraima, Brazil. For Ecol Manage 242(2):678–687

Giglio L, van der Werf G, Randerson J et al (2005) Global estimation of burned area using MODIS active fire observations. Atmos Chem Phys Discuss 5:11091–11141

Giglio L, Csiszar IA, Justice CO (2006) Global distribution and seasonality of active fires as observed with the Terra and Aqua Moderate Resolution Imaging Spectroradiometer (MODIS) sensors. J Geophys Res 111:12. doi:10.1029/2005jg000142

Good P, Lowe JA, Collins M, Moufouma-Okia W (2008) An objective tropical Atlantic sea surface temperature gradient index for studies of south Amazon dry-season climate variability and change. Philos Trans R Soc B Biol Sci 363:1761–1766. doi:10.1098/rstb.2007.0024

Holben BN, Setzer A, Eck TF et al (1996) Effect of dry-season biomass burning on Amazon basin aerosol concentrations and optical properties, 1992–1994. J Geophys Res Atmos 101:19465–19481. doi:10.1029/96JD01114

Hutyra L, Munger J, Nobre C et al (2005) Climatic variability and vegetation vulnerability in Amazônia. Geophys Res Lett 32, L24712. doi:10.1029/2005GL024981

INPE (1999) Estimativa da área de cobertura florestal afetada pelo incêndio em Roraima a partir de dados de satélite; janeiro de 1999, 1–8

Jain A (2007) Global estimation of CO emissions using three sets of satellite data for burned area. Atmos Environ 41:6931–6940

Justino F, Souza S, Setzer A (2002) Relação entre focos de calor e condições meteorológicas no Brasil. XII Congr. Bras. Meteorol, pp 2086–2093

Kodandapani N, Cochrane MA, Sukumar R (2004) Conservation threat of increasing fire frequencies in the Western Ghats, India. Conserv Biol 18:1553–1561

Laurance WFL, Williamson GB (2001) Positive feedbacks among forest fragmentation, drought, and climate change in the Amazon. Conserv Biol 15:1529–1535. doi:10.1046/j.1523-1739.2001.01093.x

Laurance W, Cochrane MA, Bergen S (2001a) The future of the Brazilian Amazon. Science 291:438–439

Laurance WF, Williamson GB, Delamônica P et al (2001b) Effects of a strong drought on Amazonian forest fragments and edges. J Trop Ecol 17:771–785

Laurance WF, Albernaz AK, Schroth G, Fearnside PM, Bergen S, Venticinque EM, Da Costa C (2002) Predictors of deforestation in the Brazilian Amazon. J Biogeogr 29(5–6):737–748

Li W, Fu R, Dickinson RE (2006) Rainfall and its seasonality over the Amazon in the 21st century as assessed by the coupled models for the IPCC AR4. J Geophys Res 111:1–14. doi:10.1029/2005JD006355

Lima A, Silva TSF, Aragão L et al (2012) Land use and land cover changes determine the spatial relationship between fire and deforestation in the Brazilian Amazon. Appl Geogr 34:239–246. doi:10.1016/j.apgeog.2011.10.013

Lucas RM, Honzak M, Curran PJ et al (2000) Mapping the regional extent of tropical forest regeneration stages in the Brazilian Legal Amazon using NOAA AVHRR data. Int J Remote Sens 21:2855–2881

Malhi Y, Pegoraro E, Nobre AD et al (2002) Energy and water dynamics of a central Amazonian rain forest. J Geophys Res Atmos 107:LBA 45–1–LBA 45–17. doi: 10.1029/2001JD000623

Malhi Y, Roberts JT, Betts RA et al (2008) Climate change, deforestation, and the fate of the Amazon. Science 319:169–172. doi:10.1126/science.1146961

Marengo JA (1992) Interannual variability of surface climate in the Amazon Basin. Int J Climatol 12:853–863. doi:10.1002/joc.3370120808

Marengo JA (2004) Interdecadal variability and trends of rainfall across the Amazon basin. Theor Appl Climatol 78:79–96. doi:10.1007/s00704-004-0045-8

Marengo JA, Nobre CA, Tomasella J et al (2008) The drought of Amazonia in 2005. J Clim 21:495–516. doi:10.1175/2007jcli1600.1

Marengo JA, Tomasella J, Alves LM et al (2011) The drought of 2010 in the context of historical droughts in the Amazon region. Geophys Res Lett 38:1–5. doi:10.1029/2011GL047436

Massad T, Balch J, Davidson E et al (2013) Interactions between repeated fire, nutrients, and insect herbivores affect the recovery of diversity in the southern Amazon. Oecologia 172:219–229. doi:10.1007/s00442-012-2482-x

Matricardi EA, Skole DL, Pedlowski MA et al (2010) Assessment of tropical forest degradation by selective logging and fire using Landsat imagery. Remote Sens Environ 114:1117–1129. doi:10.1016/j.rse.2010.01.001

Matson M, Holben B (1987) Satellite detection of tropical burning in Brazil. Int J Remote Sens 8:509–516

Mauas PJD, Flamenco E, Buccino AP (2008) Solar forcing of the stream flow of a continental scale South American River. Phys Rev Lett. doi:10.1103/PhysRevLett.101.168501

MCT (Ministério da Ciência e Tecnologia) (2010) Segunda Comunicação Nacional do Brasil à Convenção-Quadro das Nações Unidas sobre Mudança do Clima, vol. 1. MCT, Coordenação Geral de Mudanças Globais de Clima, Brasília, DF, Brazil, 280 pp

Meggers B (1994) Archeological evidence for the impact of mega-Nino events on Amazonia during the past two millennia. Clim Change 28:321–338

Mendonça M, Vera Diaz M, Nepstad D et al (2004) The economic cost of the use of fire in the Amazon. Ecol Econ 49:89–105

Morton DC, DeFries RS, Shimabukuro YE et al (2006) Cropland expansion changes deforestation dynamics in the southern Brazilian Amazon. Proc Natl Acad Sci U S A 103:14637–14641. doi:10.1073/pnas.0606377103

Morton D, DeFries R, Jyoteshwar N et al (2011) Mapping canopy damage from understory fires in Amazon forests using annual time series of Landsat and MODIS data. Remote Sens Environ 115:1706–1720

Morton DC, Le Page Y, DeFries R et al (2013) Understorey fire frequency and the fate of burned forests in southern Amazonia. Philos Trans R Soc B 368:1471–2970. doi:10.1098/rstb.2012.0163

Myers N (1989) Deforestation rates in tropical forests and their climatic implications: a friends of the Earth report, 116 p.

Nelson BW (1994) Natural forest disturbance and change in the Brazilian Amazon. Remote Sens Rev 10:105–125

Nepstad DC, Moreira AG, Alencar AA (1999) Flames in the rain forest: origins, impacts and alternatives to Amazonian fires. Pilot Program to Conserve the Brazilian Rain Forest. Conservation and development of Brazil's tropical forest regions, Brasilia (Brasil), 161 p.

Nepstad D, Lefebvre P, Lopes da Silva U et al (2004) Amazon drought and its implications for forest flammability and tree growth: a basin-wide analysis. Glob Chang Biol 10:704–717. doi:10.1111/j.1529-8817.2003.00772.x

Phillips OL, Aragão L, Lewis SL et al (2009) Drought sensitivity of the Amazon rainforest. Science 323:1344–1347. doi:10.1126/science.1164033

Prins E, Menzel W (1994) Trends in South American biomass burning detected with the GOES visible infrared spin scan radiometer atmospheric sounder from 1983 to 1991. J Geophys Res 99:716–719

Prins EM, Feltz JM, Menzel WP, Ward DE (1998) An overview of GOES-8 diurnal fire and smoke results for SCAR-B and 1995 fire season in South America. J Geophys Res Atmos 103 (D24):31821–31835

PRODES (2013) Monitoring of the Brazilian Amazon Forest by satellite. http://www.obt.inpe.br/prodes/. Accessed 15 Aug 2013

Ramos-Neto MB, Pivello VR (2000) Lightning fires in a Brazilian savanna National Park: rethinking management strategies. Environ Manage 26(6):675–684

Ray D, Nepstad D, Moutinho P (2005) Micrometeorological and canopy controls of fire susceptibility in a forested Amazon landscape. Ecol Appl 15:1664–1678

Razafimpanilo H, Frouin R, Iacobellis SF, Somerville CJ (1995) Methodology for estimating burned area from AVHRR reflectance data. Remote Sens Environ 54:273–289
Rocha HR, Goulden ML, Miller SD et al (2004) Seasonality of water and heat fluxes over a tropical forest in eastern Amazonia. Ecol Appl 14:22–32. doi:10.1890/02-6001
Rocha W, Metcalfe DB, Doughty CE et al (2013) Ecosystem productivity and carbon cycling in intact and annually burnt forest at the dry southern limit of the Amazon rainforest (Mato Grosso, Brazil). Plant Ecol Divers 7:1–16. doi:10.1080/17550874.2013.798368
Ronchail J, Cochonneau G, Molinier M et al (2002) Interannual rainfall variability in the Amazon basin and sea-surface temperatures in the equatorial Pacific and the tropical Atlantic Oceans. Int J Climatol 22:1663–1686. doi:10.1002/joc.815
Rosenfeld D (1999) TRMM observed first direct evidence of smoke from forest fires inhibiting rainfall. Geophys Res Lett 26:3105–3108
Saatchi S, Asefi-Najafabady S, Malhi Y et al (2013) Persistent effects of a severe drought on Amazonian forest canopy. Proc Natl Acad Sci U S A 110:565–570. doi:10.1073/pnas.1204651110
Saldarriaga JG, West DC (1986) Holocene fires in the northern Amazon basin. Quat Res 26:358–366
Sanford RL, Saldarriaga J, Clark KE et al (1985) Amazon Rainforest fires. Science 227:53–55
Schroeder W, Morisette J, Csiszar I et al (2005) Characterizing vegetation fire dynamics in Brazil through multisatellite data: common trends and practical issues. Earth Interact 9:1–26
Schroeder W, Csiszar I, Morisette J (2008a) Quantifying the impact of cloud obscuration on remote sensing of active fires in the Brazilian Amazon. Remote Sens Environ 112:456–470
Schroeder W, Prins E, Giglio L et al (2008b) Validation of GOES and MODIS active fire detection products using ASTER and ETM+ data. Remote Sens Environ 112:2711–2726. doi:10.1016/j.rse.2008.01.005
Setzer A, Pereira M (1991) Amazonia Biomass burning in 1987 and an estimate of their tropospheric emissions. Ambio 20:19–22
Setzer A, Pereira M, Pereira AC Jr (1994) Satellite studies of biomass burning in Amazonia—some practical aspects. Remote Sens Rev 10:91–103
Shimabukuro YE, Duarte V, Arai E, Freitas RM, Lima A, Valeriano DM, Brown IF, Maldonado MLR (2009) Fraction images derived from Terra Modis data for mapping burnt areas in Brazilian Amazonia. Int J Remote Sens 30:1537–1546. doi:10.1080/01431160802509058
Shuttleworth WJ, Leuning R, Black TA et al (1989) Micrometeorology of temperate and tropical forest (and discussion). Philos Trans R Soc B Biol Sci 324:299–334. doi:10.1098/rstb.1989.0050
Silveira JM, Barlow J, Krusche AV et al (2009) Effects of experimental fires on litter decomposition in a seasonally dry Amazonian forest. J Trop Ecol 25:657. doi:10.1017/S0266467409990150
Silvério DV, Brando PM, Balch JK et al (2013) Testing the Amazon savannization hypothesis: fire effects on invasion of a neotropical forest by native cerrado and exotic pasture grasses. Philos Trans R Soc B Biol Sci 368:1–9
Silvestrini RA, Soares-Filho BS, Nepstad D et al (2011) Simulating fire regimes in the Amazon in response to climate change and deforestation. Ecol Appl 21:1573–1590
Simon M, Plummer S, Fierens F et al (2004) Burnt area detection at global scale using ATSR-2: the GLOBSCAR products and their qualification. J Geophys Res 109:1–16. doi:10.1029/2003jd003622
Sismanoglu R, Setzer A (2002) Avaliação inicial do desempenho do risco de fogo gerado no CPTEC. XII Congr Bras Meteorol 1991–1999
Sismanoglu R, Setzer A (2005) Risco de fogo da vegetação na América do Sul: comparação de três versões na estiagem de 2004. XII Simpósio Bras. Sensoriamento Remoto, Goiânia, pp 3349–3355

Sombroek W (2001) Spatial and temporal patterns of Amazon rainfall: consequences for the planning of agricultural occupation and the protection of primary forests. AMBIO 30 (7):388–396
Sorrensen CL (2008) Potential hazards of land policy: conservation, rural development and fire use in the Brazilian Amazon. Land Use Policy. doi:10.1016/j.landusepol.2008.10.007
Sorrensen CL (2000) Linking smallholder land use and fire activity: examining biomass burning in the Brazilian Lower Amazon. For Ecol Manage 128:11–25
Sorrensen CL (2004) Contributions of fire use study to land use/cover change frameworks: understanding landscape change in agricultural frontiers. Hum Ecol 32:395–420
Souza C Jr, Firestone L, Silva M, Roberts D (2003) Mapping forest degradation in the Eastern Amazon from SPOT 4 through spectral mixture models. Remote Sens Environ 87:494–506. doi:10.1016/j.rse.2002.08.002
Souza C Jr, Roberts D, Cochrane M (2005) Combining spectral and spatial information to map canopy damage from selective logging and forest fires. Remote Sens Environ 98:329–343
Staver AC, Archibald S, Levin SA (2011) The global extent and determinants of Savanna and forest as alternative biome states. Science 334:230–232. doi:10.1126/science.1210465
Tansey K, Gregoire JM, Stroppiana D et al (2004) Vegetation burning in the year 2000: global burned area estimates from SPOT VEGETATION data. J Geophys Res. doi:10.1029/2003jd003598
Uhl C, Buschbacher R (1985) A disturbing synergism between cattle ranch burning practices and selective tree harvesting in the eastern Amazon. Biotropica 17:265–268
Uhl C, Kauffman JB (1990) Deforestation, fire susceptibility, and potential tree responses to fire in the eastern Amazon. Ecology 71:437–449
UNDAC (1998) Brazil fires in the state of Roraima (August 1997–April 1998). Final report, Geneva, Switzerland
Uvo C, Repelli CA, Zebiak SE, Kushnir Y (1998) The relationships between tropical Pacific and Atlantic SST and northeast Brazil monthly precipitation. J Clim 11:551–562
Vasconcelos SS, Fearnside PM, Graça PMLA et al (2013) Variability of vegetation fires with rain and deforestation in Brazil's state of Amazonas. Remote Sens Environ 136:199–209. doi:10.1016/j.rse.2013.05.005
Van der Werf GR, Randerson JT, Giglio L et al (2010) Global fire emissions and the contribution of deforestation, Savanna, forest, agricultural, and peat fires (1997–2009). Atmos Chem Phys 10:11707–11735. doi:10.5194/acp-10-11707-2010
Xaud HAM, Martins FDSRV, Santos JRD (2013) Tropical forest degradation by mega-fires in the northern Brazilian Amazon. For Ecol Manage 294:97–106

Chapter 14
Modelling Amazonian Carbon Budgets and Vegetation Dynamics in a Changing Climate

Bart Kruijt, Patrick Meir, Michelle Johnson, Anja Rammig, Sophie Fauset, Tim Baker, David Galbraith, Celso von Randow, and Hans Verbeeck

14.1 Introduction

The forests of the Amazon region are under threat of both climate change and land use change, with risks of accelerated degradation involving positive feedbacks through moisture, CO_2, and temperature (Davidson et al. 2012). The potential of accelerated degradation under twenty-first century climate change was flagged first by White et al. (1999) and Cox et al. (2000), followed by a series of modelling studies showing large inter-model variability in climate sensitivity of the Amazon forest biome as well as in the climate drivers themselves (e.g. Friedlingstein et al. 2006; Nobre and Borma 2009; Sampaio et al. 2007). Modelled changes in biomass depend on a number of factors. First, the assumed future greenhouse gas

B. Kruijt (✉)
ALTERRA, Wageningen University and Research centre, Wageningen, Netherlands

Brazilian National Space Research Institute (INPE), Cachoeira Paulista, SP, Brazil
e-mail: Bart.Kruijt@wur.nl

P. Meir
University of Edinburgh, Edinburgh, Scotland, UK

Australian National University, Canberra, Australia

M. Johnson • S. Fauset • T. Baker • D. Galbraith
University of Leeds, Leeds, UK

A. Rammig
TUM School of Life Sciences Weihenstephan, Technical University Munich, Freising, Germany

C. von Randow
Brazilian National Space Research Institute (INPE), Cachoeira Paulista, SP, Brazil

H. Verbeeck
University of Gent, Ghent, Belgium

© Springer-Verlag Berlin Heidelberg 2016
L. Nagy et al. (eds.), *Interactions Between Biosphere, Atmosphere and Human Land Use in the Amazon Basin*, Ecological Studies 227,
DOI 10.1007/978-3-662-49902-3_14

emissions scenario is important, second the climate model, third the vegetation model used, and, finally, the inclusion of land-surface feedbacks. The early studies relied on a single climate model (HADCM3), which forecast extreme changes in moisture circulation and temperature, and a surface model (MOSES-TRIFFID), in which carbon loss was very sensitive to temperature (Huntingford et al. 2008; Galbraith et al. 2010). Recent studies appear to indicate a more modest range of climate predictions, and recently acknowledged uncertainty in the sensitivity of forest biomass and productivity to climate, partly based on incomplete model development (Good et al. 2013; Cox et al. 2013; Huntingford et al. 2013).

In this chapter, we focus on model representation of the sensitivity of vegetation to climate, reflecting the literature until the time of writing (2012–2013, with a few additions from 2014). Some recent studies have highlighted the strong sensitivity to CO_2 and temperature in vegetation models used in simulations of climate sensitivity in tropical forests (e.g. Rammig et al. 2010; Lapola et al. 2009; Galbraith et al. 2010; Huntingford et al. 2013). The degree of forest sensitivity to climate, especially temperature, differs greatly among models. Moreover, although a subject of discussion (Poulter et al. 2010), recent work has suggested that the uncertainty associated with the physiologically driven ecosystem-scale responses in the models is higher than the uncertainty associated with future climate projections (Huntingford et al. 2013), although this is partly based on unbalanced information resources relating to each component. An increased atmospheric CO_2 concentration potentially reduces water stress, but also may lead to changes in vegetation structure by changing competition among individuals and species. The effects of increased CO_2 on tropical vegetation, however, remain largely unmeasured, especially because limitations by nutrients and temperature are poorly understood and also because little is known about how enhanced productivity might affect allocation patterns and demographic processes (recruitment, ageing, and mortality) (Galbraith et al. 2013). The effects of changing temperature and CO_2 on the balance of primary productivity (photosynthesis), respiration, and decomposition are poorly understood for the tropics, with most available information originating from temperate vegetation and temperate agricultural crops.

Dufresne et al. (2002) noted that the predictions of an Amazon dieback by Cox et al. (2000) depended strongly on allocation of the extra carbon gained by the higher availability of CO_2 to vegetation versus soil pools, which influenced net emissions of CO_2 via differential effects on supplying substrate to heterotrophic (microbial) or autotrophic respiration. This explained some of the differences, for example, noted between coupled runs forced using the HADCM3 climate simulations (Cox et al. 2000) and IPSL climate simulations (Friedlingstein et al. 2006). In general, DGVMs to date have not been very successful in simulating allocation processes and in particular tree mortality, and as a consequence they have demonstrated rather limited predictive skill reproducing the observed spatial variability of biomass over the Amazon basin (Delbart et al. 2010; Castanho et al. 2013; Mitchard et al. 2013). Although Castanho et al. (2013) introduce a map-based forcing of variability in productivity capacity, a more realistic representation of vegetation dynamics is necessary to move forward. If done successfully and with minimum increases in model complexity, this could lead to better simulations of spatial variability and climate-driven changes in biomass.

Several efforts have been made to develop and compare appropriate models to estimate the carbon budget, seasonal variability, and climate sensitivity of the region. A recent effort, within the scope of the Large-Scale Biosphere-Atmosphere experiment in Amazonia (LBA), is the LBA-MIP (Model Inter-comparison Project, Gonçalves et al. 2013; Von Randow et al. 2013). Other studies include an initiative funded by the World Bank (Vergara and Sholz 2010); there is ongoing work on the topic by Amazon-Andes Initiative (http://www.oeb.harvard.edu/faculty/moorcroft/andes-amazon/about.html), and AMAZALERT (www.eu-amazalert.org). A collection of studies on the climate sensitivity of the Amazon was published in a special issue of New Phytologist (Meir and Woodward 2010), where responses to drought, temperature, and atmospheric CO_2 were considered (Rammig et al. 2010; Jupp et al. 2010; Galbraith et al. 2010, also see Poulter et al. 2010). Recent reviews covering the importance of a better understanding of dynamics dependence of ecosystem productivity on environmental factors and climate change include Booth et al. (2012) and Smith and Dukes (2013) and a special issue of Plant Ecology and Diversity (Galbraith et al. 2014).

This chapter aims at giving an overview of the most important issues concerning the modelling of carbon budgets and vegetation dynamics of the Amazon forests, and tropical forests in general. In the following, we cover water relations, temperature dependence, CO_2 and nutrient dependence, and patterns in growth and mortality. Finally, it aims to contribute to setting the agenda for model improvement and data needs, to adequately equip global and regional DVGMs in assessing climate sensitivity of the vegetation of the region. A graphical overview of the most important issues in DGVMs is presented in Fig. 14.1.

14.2 DGVMs for the Amazon

14.2.1 Soil-Plant Water Relations

In the Amazon, flux tower data, data from manipulative experiments, and remote sensing indices show unexpected responses of vegetation properties and carbon fluxes to dry periods. Where soils are deep and the dry season is moderate, productivity appears to be little affected by seasonal drought, and it may even peak during the dry season (Saleska et al. 2003; Fisher et al. 2006). Originally considered largely aseasonal, net ecosystem carbon exchange (NEE) was shown to be seasonal at a site in central Amazonia, near Manaus (Malhi et al. 1998). Variability in apparent carbon uptake was correlated with soil moisture, especially where during the dry season water uptake was reduced; the moisture constraint at this site imposed by the specific soil characteristics was considered further by Fisher et al. (2008) and shown to be related to soil properties as well as climate. However, Araújo et al. (2002) found little seasonal variation in CO_2 fluxes in the

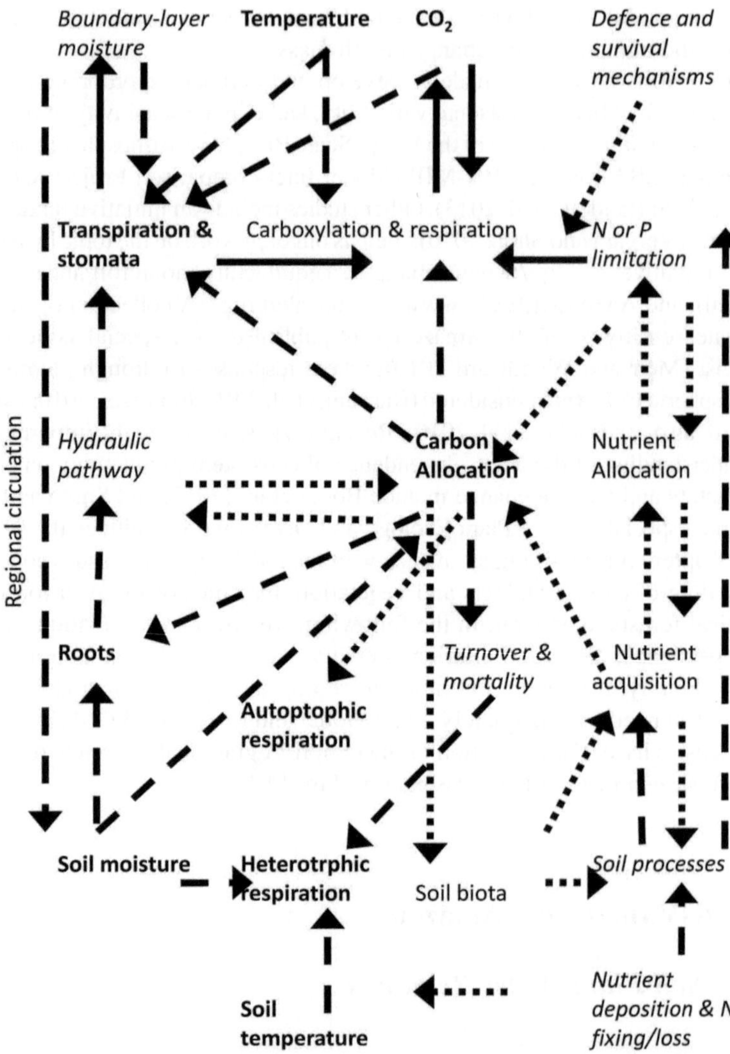

Fig. 14.1 Overview of major processes in DGVMs, their linkages, and the uncertainties therein. Processes printed in *bold* and *solid lines* represent processes that are present in all models with little uncertainty. Processes in normal type and *dashed lines* represent those present in most models but with substantial uncertainty. Processes in *italics* and *dotted lines* are not or hardly present in models, with high uncertainty

same area. After the early measurements near Manaus, it was suggested that the Central-Eastern Amazon rainforest NEE tends to be higher rather than reduced during the dry season and vegetation has been reported to 'green-up' in the late dry season. This was supported by eddy correlation flux measurements (higher NEE) and by the analysis of satellite reflectance data (MODIS EVI) (Saleska et al. 2003, 2007; Huete et al. 2006). The latter large-scale analysis, however, was later criticised because EVI values can be affected by changes in (dry season) canopy structure (Anderson et al. 2010) and by imperfect correction for clouds and aerosols (Samanta et al. 2010; Morton et al. 2014). The phenomenon of high productivity during the dry season, observed at individual flux sites, has been explained by vegetation productivity being relatively tolerant to normal dry season conditions because of deep root soil water access and soil moisture storage combined with higher insolation, and respiration processes declining as litter decomposition and surface soil microbial activity decline during the dry season (Fisher et al. 2008; Saleska et al. 2003; Bonal et al. 2008; Meir et al. 2008). These studies demonstrate some of the potential variability in the responses by GPP and respiration processes to seasonal rainfall, and this combination of responses is not represented well in large-scale and global vegetation models. In recent years, several of these models have been modified to improve the simulated seasonality of carbon fluxes for flux tower sites by introducing improved equations (e.g. Baker et al. 2008) or by optimising model parameters (e.g. Verbeeck et al. 2011; Potter et al. 2012).

The response to severe or extended drought has been studied further, using two rainfall exclusion experiments in Amazonia (Nepstad et al. 2007; Brando et al. 2008; Meir et al. 2008; Da Costa et al. 2010), and also observations of natural forest growth and mortality following the extreme natural drought of 2005 in the region (Fig. 14.2,

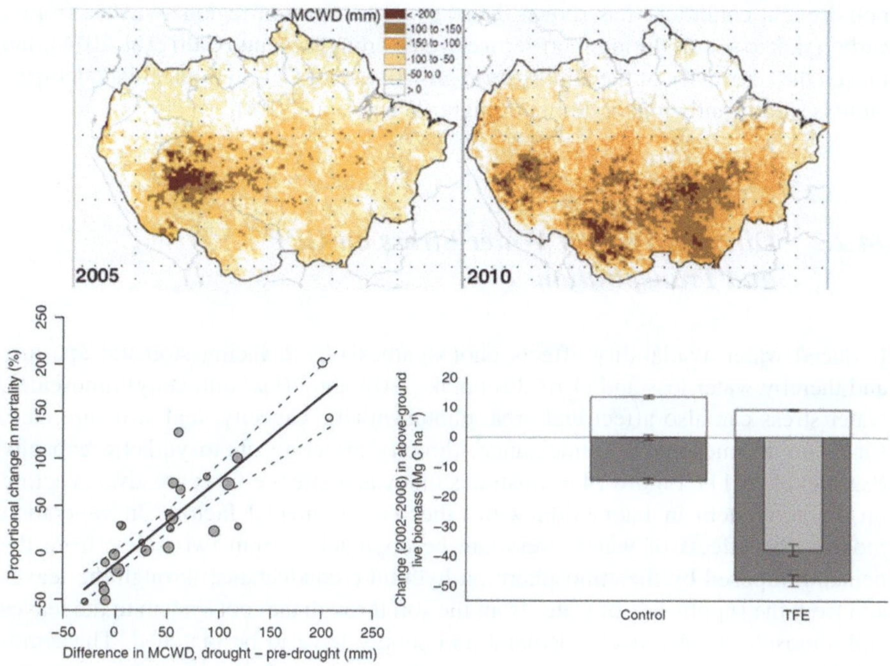

Fig. 14.2 Maps of drought index MCDW (Monthly Cumulative Water deficit) in Amazonia for two exceptionally dry years (*Top*, Lewis et al. 2011) and resulting mortality (*bottom left*, Phillips et al. 2010). The graph in the *bottom right* shows the effects after 6 years of artificial drought on mortality (da Costa et al. 2010)

Marengo et al. 2008; Phillips et al. 2009). Following these multi-year, large-scale experimental soil drought treatments at two sites in eastern Amazonia, biomass increment responded within the first year to the onset of artificial soil moisture reduction, and tree mortality increased substantially only after about 3 years of the treatment. Whilst photosynthesis and transpiration were only reported in depth from one of these years, the physiological response was consistent with the biomass response at tree and ecosystem scales (Fisher et al. 2006, 2007; Brando et al. 2008; da Costa et al. 2010). The severe natural drought of 2005 across Amazonia included an atmospheric as well as a soil drought effect; it also impacted mortality substantially but over a shorter timescale of 1 year and caused the previous regional carbon sink to turn into a source in that year, as calculated from a region-wide inventory plot network dataset (RAINFOR, Phillips et al. 2009). In both the experimental studies and the permanent plot surveys, larger trees were affected more than small ones initially (da Costa et al. 2010; Phillips et al. 2010). Hence, current understanding suggests that the response by rainforest ecosystem productivity to seasonal, chronic, or acute drought can be very substantial, varies over time and space, and is dependent on pedo-hydrological conditions, timescale, and drought severity, as well as on differences among species in vulnerability to drought (Van der Molen et al. 2011; Fisher et al. 2006; da Costa et al. 2010). Consistent with the results of these forest-based datasets, recent analysis of atmospheric measurements during drought and non-drought conditions has shown that the whole Amazon region switches from a carbon sink to source during short-term extreme drought events (Gatti et al. 2014), and indeed the sensitivity of the atmospheric growth rate of CO_2 to tropical land temperature is significantly higher during drought (Wang et al. 2014).

14.2.2 Direct Effects of Water Stress on GPP and Transpiration

Reduced water availability affects photosynthesis by reducing stomatal aperture and thereby water loss and also CO_2 uptake. At longer (leaf ontogeny) timescales, water stress can also affect leaf area, photosynthetic capacity, leaf structure, and, for example, mesophyll conductance, directly affecting photosynthetic capacity (Egea et al. 2011). Figure 14.3 illustrates that water stress effects are always acting on the ecosystem in interaction with other environmental factors. In vegetation models, the effects of water stress can be approached from two sides: from the demand imposed by the atmosphere on hydraulic conductance through the leaves and from the supply side of water from the soil through the root system to the leaves and atmosphere. In models, demand and supply have to be matched. The atmospheric demand can be represented as a radiation-dependent potential evapotranspiration with an implicit boundary layer (Priestly–Taylor or equivalent, Monteith 1995) or by only considering exchange with the lowermost layer of the atmosphere (through a Penman–Monteith equation or explicit vapour gradient diffusion).

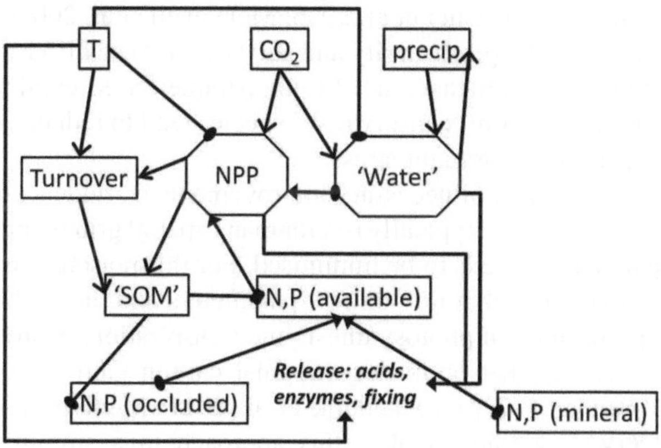

Fig. 14.3 Conceptual diagram of ecosystem functioning summarising the main pathways along which a changing CO_2, temperature, or water signal propagates through the system. It can be seen that the effects of CO_2 are largely indirect, through effects on NPP and on the plant hydraulics

In both cases, the demand is modulated by the surface (or stomatal) conductance, through which (water and CO_2) demand is made to match the supply. Most modelling effort in the past has attempted finding efficient representations of stomatal response to water stress and CO_2 demand. The main approaches have been to (1) consider stomata to respond in a linear multiplicative model to a range of independent environmental factors, where each response is determined by a set of parameters (e.g. Jarvis 1976; Stewart 1988); (2) to maintain a fixed ratio of conductance to photosynthesis and humidity (or internal to atmospheric [CO_2], Ball et al. 1987; Leuning 1995; Jacobs et al. 1996); (3) to maintain a fixed or optimal ratio of CO_2 uptake to water loss (water use efficiency, Cowan and Farquhar 1977; Medlyn et al. 2011), or to maintain leaf water potential (the SPA model, Williams et al. 1998). CO_2 concentration also affects stomatal conductance, where stomata tend to close, increasing water use efficiency (WUE), as CO_2 concentration rises. This is implicit in model approaches (2) and (3) above, through the assumption that the ratio of within-leaf to external CO_2 concentration tends to stay constant at constant humidity. Only in the first set of approaches, this dependence needs to be modelled separately. Finally, for all interactions of stomatal conductance with atmospheric gases such as water vapour or CO_2, it is important to account for feedback from the overlaying atmospheric boundary layer: stomatal closure leads to lower transpiration, but this reduces atmospheric humidity, counteracting transpiration but amplifying stomatal closure. Kruijt et al. (2008) evaluated this effect for Dutch vegetation and crop conditions. The effect of CO_2 on WUE and water (re)cycling in the Amazon may be substantial, both counteracting (lowering water stress) and enhancing (reducing rainfall) forest dieback (Spracklen et al. 2012).

Different species or groups of species economise the rate of water loss per unit of CO_2 gained differently. A distinction between 'isohydric' and 'anisohydric' strategies has recently been made, referring to species that tend to conserve water potential above a minimum critical value versus those that do not (and instead tend to conserve the acquisition of photosynthate for as long as possible), respectively (Fisher et al. 2006; McDowell et al. 2008; Van der Molen et al. 2011). To assess the productivity and survival of these different groups, the submodels for stomatal conductance need to be parameterised explicitly. It can be expected that stomata of more 'conservative' species tend to reduce conductance more rapidly on experiencing drought stress.

Such kinds of necessary improvements in models are often parameter intensive. As DGVMs are typically run for many spatial grid points, the number of site-specific parameters needs to be minimised. For this reason, it is attractive to explore model formulations that rely on few parameters and, instead, assume interdependency or optimisation of photosynthesis and transpiration. A promising candidate is the class of models that optimises marginal carbon gain per unit water lost (Cowan and Farquhar 1977; Groenendijk et al. 2011; Medlyn et al. 2011). The LPJ class of DGVMs already applies this approach in a simplified way. In this algorithm, stomatal conductance is reduced iteratively from an unstressed maximum, until water demand matches the supply from the soil. One of the major uncertainties is, however, how either water supply or stomata respond to relative soil moisture availability or water tension. Many models impose an empirical reduction function of soil moisture on surface conductance or even on photosynthesis directly (e.g. ORCHIDEE, Verbeeck et al. 2011). Usually this function is highly non-linear, leading to reductions only at low values of soil water content. Other models explicitly model the hydraulic resistances of the plant and the soil matrix as intermediaries that affect the stomata through the leaf water potential (SPA, Williams et al. 1998; Fisher et al. 2006; Medvigy et al. 2009; Christoffersen et al. 2014). These different approaches mainly affect the rate of soil drying and the reduction of transpiration as drought progresses. With extended drought periods, soil moisture eventually becomes depleted and then the critical issue is how roots access and potentially expand the supply of available water (Vermeulen et al. 2015). Adequate understanding and representation of root growth strategies is essential, especially specifying the maximum soil volume and depth that roots can access (Nepstad et al. 1994; Li et al. 2006) and whether they can cope with low-oxygen groundwater conditions. What appears to be important here is information on soil hydraulic properties (Tomasella and Hodnett 2004); such datasets are scarce for Amazonia, and a comprehensive map was published only very recently (Marthews et al. 2013). Belk et al. (2007) and Fisher et al. (2008) illustrated the importance of how these parameters cause differences in the rate at which hydraulic resistance declines with soil moisture content, comparing sites that differed in dry season soil moisture stress, although the role of the hydraulic vulnerability of roots was not examined further in these studies. Whilst specific mechanisms such as hydraulic lift have been investigated in some modelling analyses (Baker et al. 2008), soil depth, the presence of groundwater, the potential for capillary rise, and the hydraulic resistance of root conduits are likely to be important integrating parameters for large-scale models (Christoffersen et al. 2014). These aspects are poorly represented in most larger-scale models, such as DGVMs, causing systematic biases in simulations.

In terms of experimental work, to advance the issue of drought sensitivity in the Amazon, there is a need for more comprehensive basin-wide information on soil hydraulic properties, measured for a sufficient number of soil types. This would enable modellers to consistently project simulation across the basin. The strategies that roots of different ecological groups (trees, lianas, palms, herbs) take to explore and enhance the supply of water are also important, though challenging to study. Such strategies could, however, be seen as part of the whole-plant response to water stress, leading to different rates of stomatal response to drought in different ecological groups. Thus, stomatal response curves as a function of the soil water balance and atmospheric demand (e.g. VPD) for these groups could provide the empirical parameters that in fact are needed by most models. Such responses can be studied at the leaf scale (using porometers or chambers), tree scale (using sap flow sensors), or whole-ecosystem scale (using isotopic inference or eddy covariance data). Creating and studying artificial ecosystem-scale soil drought greatly helps such analyses (Fisher et al. 2007; Meir and Woodward 2010; da Costa et al. 2010; Nepstad et al. 2002; Brando et al. 2008; Markewitz et al. 2010).

14.2.3 Temperature Dependence of GPP

Temperature affects most processes in ecosystems, but most strongly those that involve enzymatic assimilation or dissimilation processes (Fig. 14.3). Gross Primary Productivity (GPP) represents the integrated uptake and assimilation of CO_2 through photosynthesis in all parts of an ecosystem. It is mainly driven by photosynthetically active radiation, the absorption of CO_2 by leaves, and the activity and abundance of the main carbon-fixing enzyme RuBisCo. This activity, however, is also sensitive to temperature. The temperature dependence of photosynthesis in dynamic global vegetation models typically takes the shape of an optimum function, bell-shaped, or composed of two exponential functions. Sensitive parameters typically include a minimum temperature, a maximum, and an optimum temperature. Some models assume a wide, flat optimum range; others assume sharp optima. A range of typical temperature functions were examined by Galbraith et al. (2010), whose work indicated that in many cases, modelled changes in productivity with twenty-first century climate forcing depend more strongly on temperature than on moisture. Several studies have highlighted the strong sensitivity of DGVM-simulated GPP (and NEP) under climate change to differences in the assumed shapes of the temperature dependence functions used (Booth et al. 2012; Huntingford et al. 2013; Vermeulen et al. 2015), though the strong sensitivity to moisture deficit remains poorly addressed.

The temperature responses as represented in most current DGVMs are based on fixed parameter settings, determined in a few species in laboratory experiments (e.g. Von Caemmerer et al. 1994; Bernacchi et al. 2001; Medlyn et al. 2002). Alternatively, temperature optima have been tuned so that they represent average growing-season temperatures of temperate (rather than tropical) ecosystems. An important question is how plastic such optima are (Smith and Dukes 2013). In theory, as Lloyd and Farquhar (2008) argued, enzyme kinetics speed up exponentially with rising temperatures until a high-temperature cell lysis or enzyme denaturation point is reached. So, for enzyme-dominated processes, such as carboxylation by RuBisCo, it is unlikely to expect a temperature optimum at ambient conditions. Conversely, the net effect of the interaction of several complex processes, such as those involving membrane transport (photon absorption, electron transport, osmosis, and active uptake), is more likely to be optimised at ambient temperatures and to be plastic (Lloyd and Farquhar 2008), resulting in temperature-related plasticity for the overall process of photosynthesis.

The temperature response of photosynthesis depends on several parameters, all of which have separate temperature dependencies. This is represented in most vegetation models, though how appropriate the parameters are for tropical forests is not known for all but a few that can be measured. The much-used Farquhar equations depend on a maximum carboxylation rate, V_{cmax}, and a maximum electron transport rate, J_{max}, both of which have been shown to vary strongly within and between ecosystems across scales, from different leaves within a single canopy to different ecosystems. There are also more intrinsic temperature-dependent parameters, such as the CO_2-compensation point in the absence of mitochondrial respiration (Γ^*), and the Michealis–Menten affinity constants for carboxylation and oxidation, K_o and K_c. The strong temperature dependence of these intrinsic parameters is hard to quantify independently, and thus they are usually fixed with respect to variation other than temperature and similar across many models, with values taken from the literature (e.g. Woodrow and Berry 1988; Bernacchi et al. 2001; Sharkey and Schrader 2006). The formulations differ among sources, however, and all these parameters are strongly interdependent. This affects the shape of the temperature dependencies of V_{cmax} and J_{max} as well, when these are fitted to data using the Farquhar equations. Although this may seem a technical issue, the consequence of this is that it is important to distinguish in model-data evaluations whether the temperature dependence of maximum photosynthetic rate (A_{max}) is considered or that of the underlying V_{cmax} and J_{max} parameters, as their respective response curves may have very different shapes. Also, it is essential that parameter sets are consistent in models, i.e. all should refer to the same set of fixed model parameters. These issues of relatedness of parameters are often overlooked (e.g. in Smith and Dukes 2013).

Figure 14.4a shows that assumed temperature dependences for V_{cmax} do vary substantially among DGVMs (parameter values taken from model descriptions). Figure 14.4b shows that, given one temperature function (that of ORCHIDEE), the effect of the Michaelis–Menten parameters causes temperature dependence of net photosynthesis to decrease with increasing intracellular CO_2 concentration. The latter depends both on atmospheric CO_2 concentration and on stomatal conductance (affected mostly by water stress).

As a consequence of the assumed optimum functions, temperature dependence can be strong in DGVMs, and therefore, in studies of climate change sensitivity, it is important that the associated parameters of temperature response curves are realistic. Also, apart from any immediate temperature response, it is important to assess how these optima acclimate to changing ambient temperatures. At the

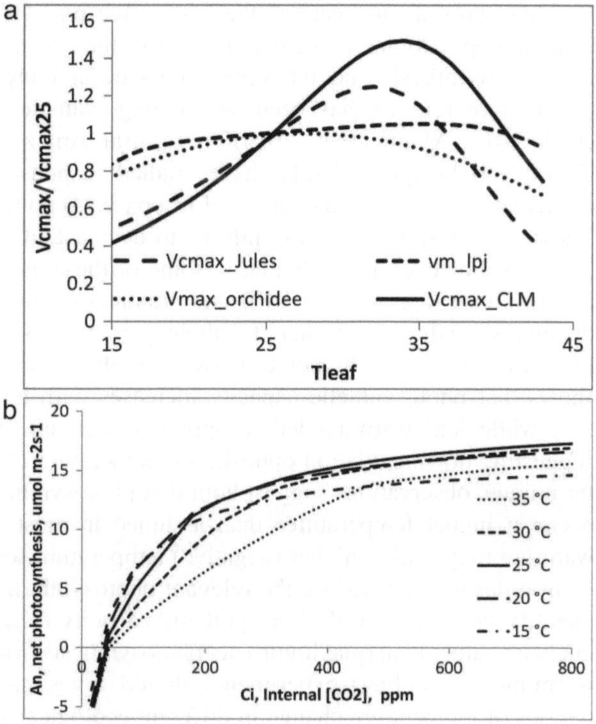

Fig. 14.4 (**a**) Temperature dependence of carboxylation capacity V_{cmax} as parameterised for Tropical forest in a range of DGVMs. (**b**) Instantaneous temperature and CO_2 dependence of net photosynthesis based upon the parameterisation of ORCHIDEE [see (**a**)] and a V_{cmax} of 40 µmol m^{-2} s^{-1} at 25 °C

timescale of interest in climate impacts on vegetation, genetic adaptation is not likely to be relevant; however, if species or ecological groups have different temperature responses and plasticity, an ecosystem may 'acclimate' through a shift in species abundance.

Such improvements in parameterisation of temperature responses require experimental data. Long-term ecosystem flux data, with naturally co-varying temperatures and GPP estimates, are of limited use because of lack of distinction in these data among species or species groups, between respiration and photosynthesis, and the inevitable covariance between temperature and other environmental variables. There is a need for photosynthetic temperature-related parameters determined at the leaf level, both for the immediate response and the long-term response. In a vast tropical rainforest domain like the Amazon, it seems almost impossible to achieve any meaningful and representative quantification of this, but with limited work at a few sites it should at least be possible to test the implicit assumption in today's models that temperature responses and optima are fairly conservative across species and regions given similar climate. If variability within ecosystems in temperature sensitivity is high and normally distributed with respect to species and abundance, then the overall temperature sensitivity of ecosystem productivity will be limited; however, in the long run, species composition may change as a consequence of different temperature dependencies.

Some work at the scale of the leaf or the plant is already under way (Kruijt et al. in prep.). Along the Peruvian slopes of the Andes into Amazonia, plasticity of leaf photosynthesis and the vegetation carbon budget under different ambient temperature regimes has been shown (e.g. van de Weg et al. 2012; Girardin et al. 2010; Malhi et al. 2014). In central Amazonia, Doughty and Goulden (2008) and Doughty (2011)) have studied response of photosynthesis to both immediate temperature increase and responses to up to 13 weeks of leaf warming. These studies showed photosynthesis to decrease at ambient temperatures above 35°. It is, however, possible that in some of these analyses temperature responses and response to correlated VPD and stomatal conductance change were hard to distinguish, while at higher T, photosynthesis also declines because its CO_2 response saturates at higher CO_2 (K_c is higher). Doughty (2011) also implicitly shows that photosynthetic capacity increases with temperature between 30° and 37°, while leaf warming led to apparent decreases in photosynthetic capacity, suggesting downregulation opposite to temperature change. In summary these, so far unique, observations suggest both that photosynthetic temperature optima may occur at higher temperatures than assumed in most models, whereas long-term warming may lead to higher (negative) temperature sensitivity.

In order to parameterise the relevant photosynthesis submodels, studies should quantify the response of photosynthetic capacity (V_{cmax}, J_{max}) and stomatal conductance rather than (maximum) net photosynthesis (Meir et al. 2002). Longer-term warming (or cooling) experiments should investigate acclimation to extended periods of temperature change in different ecological species groups.

14.2.4 Temperature Dependence of Respiration

Respiration is an integral part of ecosystem carbon budgets; however, here we offer only a brief outline of the issues concerning respiration in ecosystems. More elaborate analyses of the issues can be found elsewhere (e.g. Atkin and Tjoelker 2003, 2008; Meir et al. 2008; Chambers et al. 2004 and Smith and Dukes 2013).

By 'respiration' we refer to all biological processes in ecosystems that lead to production of CO_2, derived through the oxidation of organic material. The two distinct respiration pathways, autotrophic (plant) and heterotrophic (by consumers and decomposers of all kinds), are best modelled separately, as environmental responses can be quite distinct. Almost no process-based models, for application at large scale, have been used to comprehensively represent respiration in forest ecosystems (but see Atkin et al. 2008). Although several DGVMs distinguish autotrophic and heterotrophic respiration, in general few parameters are used to determine environmental response: a base respiration rate, approximately related to the amount of substrate or respiring organisms, whichever is limiting; and parameters describing dependence on temperature (exponential) or moisture (an optimum curve or linear function), nutrient content (linear), and photosynthetic capacity (V_{cmax}).

DVGMs often couple the (base) rate of plant respiration to productivity (photosynthesis capacity, net primary productivity, or root activity), on the grounds that the necessary enzymes in these processes turn over and need re-synthesising rapidly. This coupling of parameters in models typically operates at longer timescales (months to years), but sometimes also at daily timescales, where photosynthetic capacity also acclimates at these rates to the environment (LPJ, Sitch et al. 2003).

The temperature sensitivity of both autotrophic and heterotrophic respiration tends to be exponential in most models with little or no long-term acclimation. The fact that respiration was formulated this way did to a large extent contribute to the predicted Amazon dieback in the HADCM3 simulations of Cox et al. (2000). The physiological basis of such exponential relationships is fairly weak. In reality, there is good reason to assume that, in the absence of other limitations, such as those imposed by moisture, respiration, like most enzymatic processes, will increase with temperature monotonically up to cellular dysfunction (e.g. lysis) or potentially enzymatic denaturation. The empirical basis for specifying this kind of relationship has hitherto been limited, though this is beginning to change with new empirical datasets testing temperature responses over large temperature ranges (O'Sullivan et al. 2013). The main uncertainties here concern the exponential coefficients and how they might alter over large temperature ranges, the maximum and/or critical point beyond which respiration declines, and base respiration values defined at a common temperature. For autotrophic respiration, the base respiration mainly depends on growth, transport, and maintenance requirements and in the absence of stress responses (e.g. drought) will be generally correlated with productivity and productive capacity. Whether and why the temperature coefficient of autotrophic respiration varies much from the $Q_{10} = 2$ standard value (Atkin and Tjoelker 2003; et al. 2006) is unclear as yet. There are also indications that moisture stress (i.e. low plant water potential) can affect either base respiration or temperature coefficients (Atkin and Macherel 2009; Metcalfe et al. 2010). Finally, as with photosynthesis, it is essential to establish whether respiration rates and associated base respiration and temperature coefficients will acclimate to higher environmental temperatures. Recently, Slot et al. (2014) showed for a Panama rainforest that respiration did indeed downregulate its base rate on artificial leaf warming, with potentially important consequences for carbon uptake of ecosystems under climate change.

For soil respiration, the processes are diverse and less well understood than in plants, partly because of the complexity of the chemical and biological composition of the soil and its physical structure. First of all, while models usually split autotrophic (root-derived) respiration from heterotrophic respiration, most available data refer to bulk soil respiration, although there is an increasing amount of information on root exclusion experiments (Subke et al. 2006). While root respiration and its exponential coefficients are closely linked to productivity, heterotrophic soil respiration is related to the activity of a multitude of soil organisms and depends on accessible and decomposable soil organic matter, soil moisture, oxygen, and nutrients, as well as on temperature. Traditionally, decomposition is modelled for a few 'fractions' of soil carbon, defined by their readiness to decompose, while this description sometimes relates poorly to observable organic matter fractions (Buurman and Roscoe 2011). The dynamics of soil organic matter (SOM) breakdown are complicated at short and longer timescales by feedbacks such as occlusion of SOM in aggregates (e.g. Zimmermann et al. 2012; Stockmann et al. 2013) and this may be affected by the activity and diversity of soil organisms themselves (Whitaker et al. 2014) and at short timescales by root activity stimulating exudation and mycorrhizal activity.

Thus, respiration, in spite of the fact that it represents 30–50 % of ecosystem carbon dynamics (Loveys et al. 2003; Galbraith et al. 2014), is still poorly understood and represented in models, and there is no general model for respiration, as is available for photosynthesis (Farquhar et al. 1980). This implies that models have to rely largely on empirical data for parameterisation: respiration rates of a few main ecosystem components across a range of environmental conditions. When combined with GPP, a full suite of respiration estimates enables the calculation of forest ecosystem-scale carbon use efficiency, CUE (CUE = GPP/R), and CUE is incorporated as a fixed value in some models, usually representative for temperate regions. Empirical estimates of CUE for tropical rainforests at around 30 % are substantially lower than the 50 % originally estimated for temperate forest ecosystems, though the uncertainty on these empirically derived values is relatively high (Waring et al. 1998; Chambers et al. 2004; Malhi et al. 2009b). Notwithstanding limitations in our understanding of the multiple ecosystem components contributing to net respiratory effluxes (Meir et al. 2008), the modelling community is in urgent need of a simple well-substantiated concept, beyond the exponential response, to model the relationship between vegetation productivity and respiration, the growth rate of soil organisms in response to the environment and substrate, and the mobilisation of substrate.

14.2.5 CO_2 and Nutrient Sensitivity of Amazon Forest Productivity

Several recent studies have pointed out the strong sensitivity of DGVM predictions to the atmospheric CO_2 concentrations used in twenty-first century climate scenarios (Rammig et al. 2010; Jupp et al. 2010; Lapola et al. 2009; Booth et al. 2012; Huntingford et al. 2013), a key area of uncertainty that was identified by plant physiologists some 20 years ago (e.g. Long and Hutchin 1991). In most simulations for the twenty-first century, if atmospheric CO_2 concentration is assumed to increase, Amazonian forest biomass will increase despite likely negative effects of increased temperature and drought severity, whereas if the CO_2 concentration is kept constant, biomass is forecast to decrease (Galbraith et al. 2010; Huntingford et al. 2013). The physiology of photosynthesis implies that the effects of high CO_2 concentrations on GPP are potentially larger at tropical temperatures than for cooler climates (Hickler et al. 2008), and thus there is an urgent need to test this. Despite this theoretical sensitivity and its wide use in vegetation models, there is very little empirical information on the CO_2 response of tropical forest trees and forest ecosystems (Franks et al. 2013; Cernusak et al. 2013; Smith and Dukes 2013), mainly because of the practical difficulty of obtaining such information. As CO_2 is the primary resource for photosynthetic productivity, the immediate response to elevated CO_2 is the production of more carbohydrates in leaves, even though stomatal closure with elevated CO_2 tends to partly counteract this. The efficiency of this production depends on the usual photosynthesis parameters, including maximum carboxylation and electron transport, as well as the affinity of the photosynthetic enzyme Rubisco to CO_2 and O_2. All these parameters are temperature dependent and therefore the sensitivity of carbohydrate production to CO_2 also depends on temperature. To what extent plants and ecosystems utilise the additional carbohydrate supply, however, to increase biomass and, ultimately, to alter the total amount of carbon stored in the ecosystem depends on a suite of co-limiting factors, such as nutrient availability (to maintain the photosynthetic apparatus and other essential biomass components), ecological strategies of species present (determining their longevity, carbon turnover allocation of carbon to different plant organs, and mortality), soil biochemistry, and disturbance regime. Where carbohydrates are not used for growth or storage, respiration will potentially increase, and photosynthesis will be affected as well (Drake et al. 1997; Körner et al. 2005).

Therefore, to understand CO_2 sensitivity, the use and fate of photosynthates, including its dependence on the environment, need to be understood as well (Fig. 14.3). In the near future, a major new free-air CO_2 experiment (FACE) will be set up in the Amazon (Tollefson, 2013 and www.amazonface.org), addressing many of the issues mentioned here.

The supply and use of nutrients appears to be particularly tightly coupled to plant and ecosystem responses to elevated CO_2. Until recently most DGVMs lacked proper representation of nutrient dynamics, let alone their impact on NPP (but see below). Based on field-scale FACE experiments in even-aged temperate plantation forests, at increased CO_2 concentrations, nutrients are likely to limit the response to CO_2 in the long term, despite earlier enhancements in productivity (Norby et al. 2010). Nutrient availability may first affect photosynthetic capacity and respiration, but not necessarily to the same extent, and with differing impacts on the allocation of newly fixed carbon to above- and below-ground components. As many models rely on fixed parameters for photosynthetic capacity (V_{cmax}) and respiration, nutrient availability is only implicitly accounted for through these parameters, and not dynamically. This has as a consequence, for example, that the modelled response in productivity to elevated atmospheric CO_2 is only limited by radiation under well-watered conditions. Photosynthetic capacity of plant canopies is often simulated as an optimised response to available light and water and assuming a fixed overall nitrogen availability, with respiration changing in concert. Several studies have shown that such optimisation is likely to be an oversimplification (Meir et al. 2002; Kull and Kruijt 2002; Lloyd et al. 2010), but the concept leads to very powerful simplified model characteristics. For example, Haxeltine and

Prentice (1996) have shown that this type of model can predict canopy nitrogen quite well from absorbed radiation. In nutrient-stressed environments, or where phosphorus is limiting, such as in vegetation growing on old highly weathered soils that characterise large parts of Amazonia, this predicted capacity is likely to lead to overestimation of productivity. Also, just as with models using fixed photosynthetic capacity, the response to CO_2 is not limited and does not yet incorporate processes that might simulate alterations in the availability of the relevant limiting nutrient.

Two different approaches to modelling nutrient dependence in DGVMs can be envisaged. One approach is to link soil nutrient availability directly to concentrations in the leaves and photosynthetic capacity (Mercado et al. 2009, 2011), an approach that at the minimum requires a map of soil nutrient availability (such as in Quesada et al. 2010; Castanho et al. 2013; Yang et al. 2013), but this approach is not dynamic in their treatment of nutrient use or availability. An alternative approach is to simulate plant nutrient concentrations as the balance between demand from the plant and photosynthesis system and supply by soil processes and atmospheric deposition (Smith et al. 2014; Goll et al. 2012).

A substantial part of the demand for nutrients is determined by photosynthetic capacity, growth, and metabolism (turnover) of the leaves. Nitrogen is mainly used for the enzymes of photosynthesis and respiration, and phosphorus is required for enzyme manufacture, energy transfer, and nucleic acids. Depending on the plant's strategy for defence and drought resistance mechanisms, nutrients are also involved in the formation of additional cell wall material and to a lesser extent in secondary compounds that are not directly affecting productivity. The optimality assumption (Haxeltine and Prentice 1996), however, implies that where absorbed light is high, the demand for nutrients to build up and maintain photosynthetic capacity will also be high. In this optimisation approach for photosynthesis, this demand varies with water stress, CO_2 concentration, and temperature as well. In this model, if leaf area is high as a consequence of high productivity, then absorbed light is high (but limited by saturating absorption) and consequently the demand for nutrients is high. This approach does not fully recognise the effects of different plant strategies in the proportion of nutrients directly used for photosynthesis, however. In all DGVMs, either plant traits and nutrient allocation need to be specified more explicitly for different plant functional types, or these traits need to be modelled as parameters that respond to the climate and environment.

There are many challenges associated with modelling the transfer of nutrients between the soil and vegetation. The availability of N depends on the presence or absence of N fixation, on atmospheric deposition, and on mineralisation and denitrification. Because of the dominant role of N as a limiting nutrient in the higher latitude, temperate, and boreal regions, where data is more plentiful, a head start in incorporating N cycling feedbacks into DGVMs has been achieved (Zaehle and Friend 2010; Zaehle et al. 2010; Fisher et al. 2010b; Thornton et al 2009; Smith et al. 2014). Phosphorus on the other hand has only recently been incorporated into DGVMs (e.g. Goll et al. 2012). The numerous forms of P in the soil and the different timescales over which transformations between these pools operate provide further complications for modelling the cycling of this nutrient. The amount of plant available P in the soil depends on several factors, for example, soil properties such as pH, iron content, redox status, clay mineralogy, and the sorption capacity of the soil to name a few (Turner and Blackwell 2013; Turner and Engelbrecht 2011; Lloyd et al. 2001) and also biological processes such as organic acids, actively secreted root enzymes, and mycorrhizal associations (Hunt et al. 2007; Lloyd et al. 2001; Olander and Vitousek 2000). Therefore, not only do we need the knowledge to model the uptake of available phosphorus by vegetation and the transfer of the nutrient between organic and biomass pools, we also need knowledge of soil properties and processes. All of these processes are generally poorly represented in DGVMs, although some first attempts at including full P cycling are being tested now. These include the CASA global carbon cycle model (Wang et al. 2010), which under steady-state, present-day conditions estimates a reduction in NPP of circa 20 % in tropical evergreen broadleaf forests and tropical savannas when P is included. Phosphorus and nitrogen submodels have also been developed for use in the Community Land Model (CLM) and JSBACH (Yang et al. 2014; Goll et al. 2012).

Parameters affecting the partitioning of P among pools are assigned by soil order in these models. The majority of the Amazon Basin is represented by only two soil types in these models; thus, the spatial variation observed across the Amazon in soil P fractions (Quesada et al. 2010) will not be represented by these models. The processes and parameters in these existing nutrient cycle models need to be adapted to the Amazon basin soils and incorporated into DGVMs.

Apart from this, modelling the cycling of phosphorus between different pools is in general limited by the lack of data from tropical ecosystems to both validate and parameterise such models. The most comprehensive dataset to date, covering the whole Amazon, is from Quesada et al. (2010). Further experimental work that is needed to advance the representation of nutrient cycling in DGVMs is manifold. Apart from mapping soil N/P fractions across the Amazon basin, inventories of leaf N/P concentrations as a function of soils, absorbed light, water stress, leaf trait or strategy, and CO_2 concentration are big, but fairly straightforward prerequisites (Malhi et al. 2002). Nevertheless, such data will not provide the definitive material to dynamically model the interaction of productivity, nutrients, and CO_2. For that, nitrogen fixation, denitrification, as well as P mobilisation/immobilisation need to be quantified and parameterised. Particularly, more information is needed on the mechanisms and rates of transfer between P pools. For example, an improved quantification of the rates of biochemical mineralization and the investment by vegetation in nutrient acquisition strategies such as phosphatase enzyme production and mycorrhizal associations under different levels of soil phosphorus and moisture availability is required (Cernusak et al. 2011). This is a field of research that still has to mature and, where already existing, it needs to be linked much more closely with DGVM development.

14.2.6 Functional Types Versus Continuous Traits

Most state-of-the-art DGVMs treat vegetation at a grid cell as belonging to one of a range of plant functional types (PFTs). A PFT represents a broad group of ecological plant types by a set of model parameters determining, for example, bioclimatic limits, deciduousness, water stress sensitivity, photosynthetic capacity, and specific leaf area. Typically these PFTs have been poorly parameterised for tropical vegetation, but recent work by (Fyllas et al. 2012) has proposed a classification for Amazon forests that is much more in line with the main ecological axes of variation in plant traits along variation in relative investment into leaf and stem longevity and nutrient contents. However, simply modifying the definition of PFTs is unlikely to be sufficient, as the discussion above shows that variability in such plant traits can be large and more continuous than assumed in such simplified and discrete PFTs. Although this approach allows relatively efficient global coverage, a disadvantage may be that process representation is too coarse and modelled transitions between vegetation types in space and time (such as forecasts of 'savannisation' and multiple stable states in Amazonian vegetation) become unnecessarily sharp and unrealistic. For example, a study of the impact of a prolonged drought in Ghanaian forests (Fauset et al. 2012) demonstrated that intact forests could continue to maintain high carbon stocks and increase in biomass despite an average reduction in annual rainfall of 165 mm year^{-1} out of about 2100 mm Mean Annual Precipitation. This was associated with a shift in the composition of the forest towards more drought-tolerant species, suggesting that the resistance of the tropical forests to drought over long timescales may be increased with subtle shifts in the characteristics of the species that are present. There are two challenges to capturing more effectively these kinds of interactions within DVGMs. Firstly, changes to different vegetation types in reality occur at the level of the individual, but DGVMs currently operate at the patch scale. So if the processes represented at patch scale cannot be made to reflect individual-scale processes (and for some DVGMs this is the case), then it is inevitable that today's DGVMs cannot represent such a change appropriately. A second problem is to recognise the range of life-history strategies that exist both within individuals of a species and among species. Therefore, several attempts are being made to develop models with continuous plant traits rather than discrete ones (e.g. Scheiter et al. 2013).

Some DGVMs, especially those that aim to simulate vegetation dynamics at the patch scale, do allow several PFTs to coexist and compete for coverage within one grid cell (ED, Moorcroft et al. 2001; Medvigy et al. 2009; LPJguess, Smith et al. 2001). In such models, individual plants or 'cohorts' of individuals compete for resources within one patch. Also, some DGVMs either allow a range of plant traits within each PFT, represented by statistical distributions (Moorcroft et al. 2001), or make several traits, such as photosynthetic capacity, an internally calculated variable rather than being externally given, such as in Haxeltine and Prentice (1996).

Especially for the Amazon it may be important to study the effect of discreteness in PFT formulations and compare models with single, multiple, or plastic PFTs, as well as models that approach vegetation parameters as a continuum. It should be interesting to evaluate whether the degree of continuity in plant traits represented by each model is related to the occurrence of discrete or abrupt transitions in vegetation types, either in time or in space.

One particularly relevant option is to at least allow variation in, and competition between, plant traits that relate to drought tolerance. As described in the section on water stress, trees appear to exhibit at least two different strategies along a likely natural continuum, usually referred to as 'isohydric' and 'non-isohydric' (Tardieu and Simonneau 1998), distinguishing species that tend to avoid water stress by reducing transpiration and thus also reducing CO_2 uptake, and those that avoid carbon limitation by maintaining stomatal aperture, but risk irreversible damage from water stress. In terms of existing models, such variation between stomatal traits could be expressed by varying the sensitivity of stomata to soil/plant water potential and atmospheric VPD, as discussed above. However, variation in parameterisations of different traits needed to simulate water stress tolerance is likely to be broader than this, incorporating vulnerability to embolism in xylem, and potentially the capacity to use carbohydrate supply to ameliorate drought-related conductivity losses in xylem (Plaut et al. 2012).

The implications for experimental work of implementing more explicit plant trait diversity in models require first to quantify and better understand the key trait spectra, and to understand the mechanisms that determine these differences, so that trait variation can be modelled rather than parameterised.

14.2.7 Allocation of Primary Productivity over Vegetation Components

Above-ground biomass depends, in addition to mortality processes and net primary productivity (NPP), on the allocation of NPP into different plant tissues. Typically, models consider three vegetation carbon pools to which NPP is allocated: wood, leaves, and (fine) roots. The commonly used approaches to simulate NPP allocation in global vegetation models include (1) constant allocation patterns, where the relative fractions of NPP allocated to each plant carbon pool are determined by fixed coefficients (e.g. IBIS, Kucharik et al. 2000; and Hyland, Levy et al. 2004), (2) allocation driven by allometric constraints—the size of different plant pools are related to each other by allometric relationships that must be satisfied when NPP is allocated (e.g. JULES, Clark et al. 2011, or LPJ, Sitch et al. 2003), or (3) allocation is based on the resource limitation hypothesis where NPP allocation is governed by the relative availability of light, water, and nutrients (e.g. ORCHIDEE, Krinner et al. 2005). Castanho et al. (2013) simulated biomass for the whole Amazon, while Seiler et al. (2014) simulated the forest–dry woodland–savanna transition in

Bolivia, both showing that the settings for allocation parameters in a DGVM (INLAND and LPJGuess, respectively) can be crucial to correctly match vegetation type and biomass. Malhi et al. (2011) and Galbraith et al. (2014) reviewed allocation patterns in tropical rainforests based on field measurements and compared these to the allocation patterns simulated by a number of ecosystem models. Mean NPP allocation fractions across tropical forests were found to be $34 \pm 6\,\%$ for canopy, $39 \pm 10\,\%$ for wood, and $27 \pm 11\,\%$ for fine roots. Furthermore, canopy NPP was found to be a relatively invariant component of total NPP with the main trade-off believed to be between fine roots and wood, rather than leaves. Mean values for global vegetation models were close to the mean of the data, but varied considerably across models. For example, allocation to wood across the models reviewed ranged from 16 % to 77 % of total NPP, not inconsistent with, but varying three times more widely than the empirically derived range above of 29–49 % for tropical forests. Models whose allocation schemes were based on the resource limitation hypothesis simulated particularly high allocation to wood in Amazonian rainforests, suggesting that this type of model requires refinement for use in tropical forests.

A number of alternative, evolution- or optimality-based models of allocation have recently been developed, although these have yet to be widely applied in global vegetation models. These recent developments in allocation modelling were reviewed by Franklin et al. (2012) and include optimal response (OR) models, game-theoretic optimisation (GTO) models, and adaptive dynamics (AD) models. In OR models, such as the models of McMurtrie et al. (2008) or Franklin et al. (2009), allocation is optimised so as to maximise a fitness proxy, such as canopy photosynthesis or net growth. GTO models, such as that by King (1993), are designed around the concept of an evolutionary stable strategy which is a strategy that, when it is in equilibrium with its environment, is robust to competition from individuals following different strategies. GTO models differ fundamentally from OR models in that they explicitly include the effect of competition with other plants. Both OR and GTO models assume some a priori fitness proxy. In AD models, on the other hand, the fittest strategy is an emergent feature of the model. Furthermore, AD models allow for coexistence of different evolutionary stable strategies. Dybzinski et al. (2011) recently developed the first allocation model based on AD. One of the predictions of this model is indeed the fine root/wood trade-off observed in the review of Malhi et al. (2011) across tropical forests.

NPP allocation has to date been a remarkably understudied ecosystem process. The most comprehensive published field study on carbon allocation is that of Malhi et al. (2009b) who studied how above-ground and below-ground productivity varied across ten Amazonian lowland *terra firme* evergreen forest sites. The study identified a strong negative correlation between the fraction of NPP allocated below-ground and soil clay content. The higher below-ground allocation to NPP in sandier soils was suggested to be related to the poorer water and nutrient holding capacity of those soils as greater root productivity was necessary for access to soil moisture reserves. Our understanding of NPP allocation in tropical forests is being greatly boosted by the establishment of a network of intensive carbon cycle

Table 14.1 Percentage allocation of NPP to canopy (including leaves, fruits, flowers, and twigs), wood (including stems and coarse roots), and fine roots, from the GEM campaign. Sites are ordered by mean annual precipitation. Herbivory of leaves and branch turnover are also accounted for in canopy and wood allocation, respectively

Site	Mean annual temperature (°C)	Mean annual precipitation (mm)	Forest type	% Canopy	% Wood	% Fine root	Reference
Kenia_Wet[a]	23.4	1310	Terra firme	30.0 ± 5.0	33.9 ± 3.9	26.1 ± 3.3	Araujo-Murakami et al. (2014)
Kenia_Dry[a]	23.4	1310	Dry terra firme	41.3 ± 6.0	31.8 ± 3.5	27.0 ± 2.5	Araujo-Murakami et al. (2014)
Esperanza	13.1	1560	Montane cloud	41.6 ± 4.5	38.5 ± 3.8	19.9 ± 3.0	Girardin et al. (2014)
Wayqecha	11.8	1560	Montane cloud	49.7 ± 3.7	26.7 ± 9.3	23.7 ± 3.0	Girardin et al. 2014
Tanguro_Control	25.0	1770	Terra firme	51.5 ± 6.8	30.6 ± 4.3	18.1 ± 1.3	Rocha et al. (2014)
Tanguro_Burn[b]	25.0	1770	Terra firme	54.2 ± 8.4	35.0 ± 4.9	10.8 ± 0.6	Rocha et al. (2014)
Tambopata_05	24.4	1900	Terra firme	41.6 ± 2.7	27.7 ± 2.8	30.7 ± 4.8	Malhi et al. (2014)
Tambopata_06	24.4	1900	Terra firme	49.4 ± 4.1	32.1 ± 3.2	18.5 ± 2.7	Malhi et al. (2014)
Caxiuana_Drought[c]	25.0	1000–1250	Terra firme	29.8 ± 0.4	27.7 ± 9.3	42.5 ± 7.4	Da Costa et al. (2014)
Caxiuana_Control	25.0	2000–2500	Terra firme	30.2 ± 0.4	33.5 ± 9.5	36.3 ± 7.5	Da Costa et al. (2014)
Caxiuana_08_Terra Preta	27.3	2311	Terra preta	30.1 ± 1.2	29.0 ± 2.9	40.9 ± 6.9	Doughty et al. (2014)
Caxiuana_06	25.8	2556	Terra firme	38.8 ± 0.5	34.1 ± 3.5	27.1 ± 3.8	Doughty et al. (2014)
Allpahuayo A	25.2	2689	Terra firme	33.3 ± 6.9	45.4 ± 4.6	21.4 ± 2.1	Pasquel et al. (2014)
Allpahuayo C	25.2	2689	White sand	45.1 ± 6.4	30.3 ± 3.1	24.6 ± 2.7	Pasquel et al. (2014)
San Pedro_02	18.8	5302	Montane	50.2 ± 2.1	34.0 ± 4.0	15.8 ± 1.3	Huasco et al. (2014)
San Pedro_01	17.4	5302	Montane	54.2 ± 8.4	33.6 ± 4.0	15.7 ± 3.0	Huasco et al. (2014)

[a]Kenia_wet and Kenia_dry occur on the ecotone of humid (Kenia_wet) and dry (Kenia_dry) forest
[b]Tanguro_burn is an experimental plot burnt 6 times since 2004
[c]Caxiuana_Drought is an experimental throughfall exclusion plot receiving a 50 % reduction in rainfall

measurements across Amazonia (http://gem.tropicalforests.ox.ac.uk/), allowing the seasonal and inter-annual dynamics of carbon allocation in Amazonian rainforests to be studied for the first time. First publications from this network are summarised in Table 14.1. This data provides a much-needed testing ground for competing models of carbon allocation in tropical forests.

How increasing atmospheric CO_2 will affect allocation processes in tropical forests is important for simulation of future biomass stocks in tropical forests, but we have insufficient knowledge at present. While no FACE experiments exist for tropical regions, FACE experiments from temperate zones point to very different patterns in the response of carbon allocation to high CO_2. The comparison of loblolly pine (*Pinus taeda* in North Carolina (Duke) and sweetgum (*Liquidambar styraciflua*) in Tennessee (ORNL) in two FACE experiments has shown a stimulation of GPP and NPP of a similar magnitude (18–22 % for GPP, 18–24 % for NPP), while the impacts on carbon allocation were very different (DeLucia et al. 2005). At the pine site, >50 % of the additional NPP due to CO_2 fertilisation was allocated to long-lived woody tissues, while at the sweetgum site, the majority of the additional NPP was allocated to short-lived roots.

The consequences for long-term biomass storage at high CO_2 that result from such different patterns of carbon allocation to biomass compartments with widely different carbon residence times could be considerable, as illustrated by Körner et al. (2005).

14.2.8 Modelling Biomass: Demography of Vegetation Carbon, Mortality, Disturbance, Dispersal, and Recruitment

Changes in primary productivity and soil respiration are important for determining the ecosystem carbon content and its short-term processing, but to understand long-term dynamics it is equally important to understand the additional impacts of disturbance. Currently DGVMs do not adequately represent these types of dynamics and often exclude them or treat them in an extremely simple way (e.g. assuming a fixed proportion of disturbance per year in LPJguess, Smith et al. 2001). This can result in large bias in model estimations of biomass; conversely, including drivers of vegetation dynamics (e.g. mortality) can radically modify the response of the modelled forest to a changing climate (Fisher et al. 2010a). Inadequate representation of these processes in DGVMs is not a trivial problem for the Amazon where there are large spatial gradients in tree mortality (Phillips et al. 2004) and biomass (Baker et al. 2004; Mitchard et al. 2013). As discussed above for modelling nutrient availability, one approach might be to simply introduce observed spatial variation in biomass losses from mortality directly into the model framework (e.g. Castanho et al. 2013; Galbraith et al. 2013). Studies where this type of approach has been attempted have demonstrated the sensitivity of modelled biomass to these parameters (e.g. Delbart et al. 2010; Castanho et al. 2013). However, a more fundamental challenge is to relate forest dynamics to the underlying mechanistic controls, related to soil and climate (Quesada et al. 2012). The lack of any mechanistic relationship between environmental conditions and forest dynamics in models, in particular tree mortality, may be the reason why DGVMs have not yet been able to capture the c. 20 % and 25 % losses of biomass that occurred through the mortality of large trees in the two multi-year throughfall exclusion experiments, but instead remain either insufficiently sensitive or over-sensitive to the 50 % imposed reduction in precipitation (Galbraith et al. 2010; Powell et al. 2013). If models cannot reproduce mortality from multi-year drought, or from short-term natural drought events (Phillips et al. 2009), then we may be overestimating future biomass stocks after some of the large reductions of rainfall that have been predicted for the Amazon region (Malhi et al. 2009a).

One of the reasons that mortality is still a rudimentary process in models is because tree mortality is complex and we do not fully understand the physiological reasons of why some trees die and others survive, for example in response to drought (Mcdowell et al. 2008). Along with senescence and environmental stress-related mortality, sporadic weather events, such as wind throw, can cause death events, though it should be possible to represent these statistically, rather than by invoking physiological mechanisms. It is evident that multiple factors play a role in the Amazon so all must be considered when including dynamic processes in these models. For example, there are differences in the dominant mode and cause of death across the Amazon resulting in large differences in the scale and frequency of disturbance (Espírito-Santo et al. 2014). In western Amazonia trees die twice as fast as those in the east (Phillips et al. 2004), reflecting differences in the dynamics of these forests (Quesada et al. 2012). Observations highlight a higher frequency of small tree mortality events in the western Amazon compared to low-frequency, large tree death in the north-eastern Amazon (Chao et al. 2008). In the north-west stem breakage is the dominant cause which we may attribute to external forces such as blowdown, whereas in the north-east the majority of trees die standing perhaps a result of senescence or drought (Fig. 14.5).

Currently the inclusion of mortality in DGVMs ranges in complexity from having no explicit formulation, to having a constant rate—which can vary (0.5–5 %) across DGVMs (Galbraith et al. 2013)—and therefore there is no interaction with climate, or a carbon balance scheme where mortality results from negative NPP (Mcdowell et al. 2011; Powell et al. 2013). In the ED model, there is more complexity, as mortality also increases with decreasing wood density to reflect the hypothesised higher likelihood of mortality of low wood density PFTs (Moorcroft et al. 2001). More realistic or mechanistic ways of modelling mortality based on the physiological processes discussed in earlier sections are limited by the PFT approach. However, individual-based models will allow the exploration of some of these avenues. Alternatively, predictive methods could be used to estimate individual mortality risk based on plant characteristics (e.g. productivity, size, or plant traits), which have been developed from empirical observations (e.g. Aubry-Kientz et al. 2013; Lines et al. 2010; Chao et al. 2008).

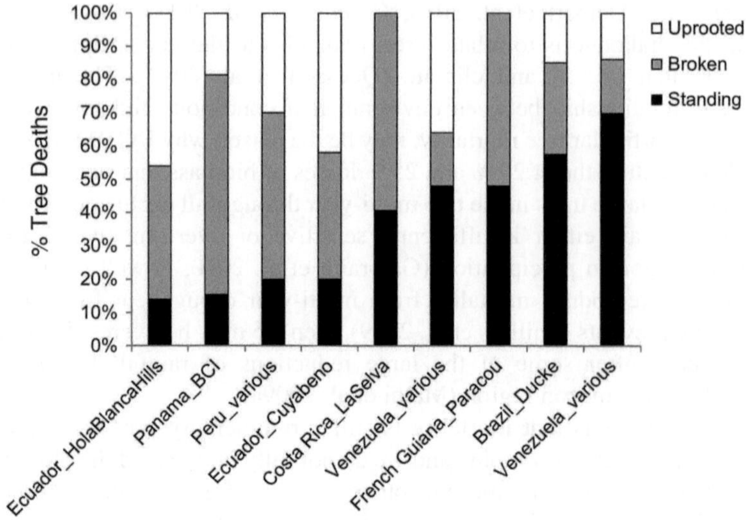

Fig. 14.5 Tree mode of death from sites in Amazonia and Central America. Trees for which mode of death was unknown were excluded. Mode of death in French Guiana and Costa Rica was not separated between uprooted or broken. Data from Putz and Milton (1982) (Panama), Lieberman et al. (1985) (Costa Rica), Carey et al. (1994) (Venezuela), Korning and Balslev (1994) (Ecuador_Cuyabeno), Gale and Barfod (1999) (Ecuador_Hola Blanca Hills), Chao et al. (2009) (Peru, Venezuela), Ferry et al. (2010) (French Guiana), and Toledo et al. (2012) (Brazil)

Recruitment is another component of forest dynamics where there is a lack of process-based understanding, particularly understanding the feedbacks with climate change. For example, Kursar et al. (2009) and Engelbrecht et al. (2007) showed that growth and survival of seedlings during drought may be a major determinant of species distribution patterns in tropical forests, and shifts in species composition following multi-decadal natural drought have been observed in African forests (Fauset et al. 2012). For modelling such dynamic vegetation processes, models would, in addition to including the above-described physiological processes, need to consider for example age structure of forest stands, competition between plants/species, establishment patterns (e.g. shade-tolerance/intolerance of seedlings), and seed dispersal, as some models have begun to do (e.g. ED: Moorcroft et al. 2001; Kim et al. 2012; LPJ-GUESS: Smith et al. 2001).

14.2.9 The Role of Fire in Natural Forests

While natural fires in Amazon rainforests play a minor role, the impacts of human-caused fires in tropical rainforests are substantial today (Uhl and Kauffman 1990). Therefore, fire is an important driver for future dynamics of remaining tropical rainforest 'fragments' and may in some circumstances act as the final trigger in vegetation change (Brando et al. 2014). The causes and impacts of fire are discussed elsewhere (Chap. 13). Biophysical and socioeconomic feedbacks from fire may also create a feedback with regional climate (Davidson et al. 2012). Shifts in temperature and precipitation patterns, altering vegetation structure, can in turn change fire probability and fire behaviour (e.g. Golding and Betts 2008). Including the role of fires in current models is crucial for understanding potential future dynamics of Amazon rainforests. Current models include the calculation of potential human ignitions, fire probability, and fire danger index (e.g. Golding and Betts 2008; Silvestrini et al. 2011). Most parameterisations of fire are yet to find their way to full coupling with global climate models. More process understanding of interaction of fire and vegetation dynamics is needed to improve simulation of fire spread and fire behaviour in different ecosystems. Only a few approaches include process-based interaction of fire and vegetation at the ecosystems scale. Parameterisations include, for example, potential fuel load (biomass or grassy biomass) and flammability (vegetation or air dryness) of the forest (Thonicke et al. 2010; Cardoso et al., pers comm). For a better understanding of the occurrence of human-caused fires, changes in current and future land use patterns need to be evaluated (e.g. Aguiar et al. 2012) and fire models should be coupled to simulations of land use change to account for edge effects and enhanced ignition risk (Cardoso et al., pers comm).

14.3 Conclusion

In conclusion, reflecting the state of the art until about the year 2014, we have identified a number of important issues that are required to be addressed in current and future research on the climate sensitivity of Amazonian vegetation. Field data collection requirements and model priorities have been indicated. These include systematic inventories of ecosystem state variables and model-driving data, experimental investigations into specific processes, as well as adapting vegetation models to better represent and accept these processes and their driver data. Focus should be centred on understanding the effects of changes in moisture, temperature, and nutrient availability, and on ecosystem demography, especially the mortality process. These are issues not only of importance to the Amazon but also relevant at the global scale and thus consist of an exciting field of study promising important improvements in the near future.

The capacity to reliably represent Amazon forest biomass and vegetation in models has improved greatly over the last decade. Also, our insight into which factors are most important to sensitively evaluate climate change impacts has improved, and it can be expected that these insights will soon lead to much more reliable DGVMs. However, data from both process-oriented experiments and ecosystem-scale observations are still sparse in comparison with the scientific challenge, so that uncertainties may remain substantial in the foreseeable future. Nevertheless, we are confident that given experimental effort and model development, within a few years we will be able to assess the effects of climate change on the Amazon for the upcoming century with greater confidence.

Acknowledgements This review has been generated with support of funding from the international EU-FP7-AMAZALERT project (no. FP7-28266). Although the authors listed have been those contributing the main text and expertise, there are several other projects that supported these authors and numerous colleagues who have indirectly contributed by generating data, model results, and evaluations. Our gratitude therefore goes to whole the Amazon climate research community.

References

Aguiar APD, Ometto JP, Nobre C, Lapola DM, Almeida C, Vieira IC, Castilla-Rubio JC (2012) Modeling the spatial and temporal heterogeneity of deforestation-driven carbon emissions: the INPE-EM framework applied to the Brazilian Amazon. Glob Chang Biol 18(11):3346–3366

Anderson LO, Malhi Y, Aragao LE, Ladle R, Arai E, Barbier N, Phillips O (2010) Remote sensing detection of droughts in Amazonian forest canopies. New Phytol 187(3):733–750

Araújo AC, Nobre AD, Culf AD, Elbers JA, Kruijt B, Von Randow C, Kabat P, Stefani P, Mendes D (2002) Comparative measurements of carbon dioxide fluxes from two nearby towers in a central Amazonian rainforest: the Manaus LBA site. J Geophys Res 107(D20):8090. doi:10.1029/2001JD000676

Araujo-Murakami A, Doughty CE, Metcalfe DB, Silva-Espejo JE, Arroyo L, Heredia JP, Flores M, Sibler R, Mendizabal LM, Pardo-Toledo E et al (2014) The productivity, allocation and cycling of carbon in forests at the dry margin of the Amazon forest in Bolivia. Plant Ecol Divers 7(1–2):55–69

Atkin OK, Macherel D (2009) The crucial role of plant mitochondria in orchestrating drought tolerance. Ann Bot 103:581–597

Atkin OK, Tjoelker MG (2003) Thermal acclimation and the dynamic response of plant respiration to temperature. Trends Plant Sci 8:343–351

Atkin OK, Atkinson LJ, Fisher RA, Campbell CD, Zaragoza-Castells J, Pitchford JW, Woodward FI, Hurry V (2008) Using temperature- dependent changes in leaf scaling relationships to quantitatively account for thermal acclimation of respiration in a coupled global climate–vegetation model. Glob Chang Biol 14:2709–2726

Aubry-Kientz M, Hérault B, Ayotte-Trépanier C, Baraloto C, Rossi V (2013) Toward trait-based mortality models for tropical forests. PLoS ONE 8(5), e63678

Baker TR, Phillips OL, Malhi Y, Almeida S, Arroyo L, Di Fiore A, Erwin T, Killeen TJ, Laurance SG, Laurance WF, Lewis SL, Lloyd J, Monteagudo A, Neill DA, Patino S, Pitman NCA, Silva JNM, Martinez RV (2004) Variation in wood density determines spatial patterns in Amazonian forest biomass. Glob Chang Biol 10:545–562

Baker IT, Prihodko L, Denning AS, Goulden ML, Miller SD, Rocha HR (2008) Seasonal drought stress in the Amazon: reconciling models and observations. J Geophys Res Biogeosci 113:1–10

Ball J, Woodrow LE, Berry JA (1987) A model predicting stomatal conductance and its contribution to the control of photosynthesis under different environmental conditions. In: Biggins J (ed) Progress in photosynthesis research. Nijhoff, Dordrecht, pp 221–224

Belk EL, Markewitz D, Rasmussen TC, Maklouf Carvalho EJ, Nepstad DC, Davidson EA (2007) Modeling the effects of throughfall reduction on soil water content in a Brazilian Oxisol under a moist tropical forest. Water Resour Res 43. doi:10.1029/2006WR005493

Bernacchi CJ, Singsaas EL, Pimentel C, Portis AR Jr, Long SP (2001) Improved temperature response functions for mod- els of Rubisco-limited photosynthesis. Plant Cell Environ 24:253–260

Bonal D, Bosc A, Ponton S et al (2008) Impact of severe dry season on net ecosystem exchange in the Neotropical rainforest of French Guiana. Glob Chang Biol 14(8):1917–1933
Booth BBB, Jones CD, Collins M et al (2012) High sensitivity of future global warming to land carbon cycle processes. Environ Res Lett 7:024002
Brando PM, Nepstad DC, Davidson EA, Trumbore SE, Ray D, Camargo P (2008) Drought effects on litterfall, wood production and belowground carbon cycling in an Amazon forest: results of a throughfall reduction experiment. Philos Trans R Soc Lond B Biol Sci 363:1839–1848
Brando PM, Balch JK, Nepstad DC, Morton DC, Putz FE, Coe MT, Alencar A (2014) Abrupt increases in Amazonian tree mortality due to drought–fire interactions. Proc Nat Acad Sci 111 (17):6347–6352
Buurman P, Roscoe R (2011) Different chemical composition of free light, occluded light and extractable SOM fractions in soils of Cerrado and tilled and untilled fields, Minas Gerais, Brazil: a pyrolysis-GC/MS study. Eur J Soil Sci 62:253–266
Carey EV, Brown S, Gillespie AJR, Lugo AE (1994) Tree mortality in mature lowland tropical moist and tropical lower montane moist forests of Venezuela. Biotropica 26:255–265
Castanho A, Coe M, Costa M, Malhi Y, Galbraith D, Quesada C (2013) Improving simulated Amazon forest biomass and productivity by including spatial variation in biophysical parameters. Biogeosciences 10:2255–2272
Cernusak LA, Winter K, Turner BL (2011) Transpiration modulates phosphorus acquisition in tropical tree seedlings. Tree Physiol 31:878–885
Cernusak LA, Winter K, Dalling JW, Holtum JA, Jaramillo C, Körner C, Wright SJ (2013) Tropical forest responses to increasing atmospheric CO_2: current knowledge and opportunities for future research. Funct Plant Biol 40(6):531–551
Chambers JQ, Tribuzy ES, Toledo LC et al (2004) Respiration from a tropical forest ecosystem: partitioning of sources and low carbon use efficiency. Ecol Appl 14:72–88
Chao KJ, Phillips OL, Gloor E, Monteagudo A, Torres-Lezama A, Vasquez Martinez R (2008) Growth and wood density predict tree mortality in Amazon forests. J Ecol 96:281–292
Chao KJ, Phillips OL, Monteagudo A, Torres-Lezama A, Martinez RV (2009) How do trees die? Mode of death in northern Amazonia. J Veg Sci 20:260–268
Christoffersen BO, Restrepo-Coupe N, Arain MA, Baker IT, Cestaro BP, Ciais P, Fisher JB, Galbraith D, Guan X, Gulden L, van den Hurk B, Ichii K, Imbuzeiro H, Jain A, Levine N, Miguez-Macho G, Poulter B, Roberti DR, Sakaguchi K, Sahoo A, Schaefer K, Shi M, Verbeeck H, Yang Z-L, Araújo AC, Kruijt B, Manzi AO, da Rocha HR, von Randow C, Muza MN, Borak J, Costa MH, Gonçalves de Gonçalves LG, Zeng X, Saleska SR (2014) Mechanisms of water supply and vegetation demand govern the seasonality and magnitude of evapotranspiration in Amazonia and Cerrado. Agric For Meteorol 191:33–50
Clark DB, Mercado LM, Sitch S, Jones CD, Gedney N, Best MJ, Pryor M, Rooney GG, Essery RLH, Blyth E, Boucher O, Harding RJ, Huntingford C, Cox PM (2011) The Joint UK Land Environment Simulator (JULES), model description—part 2: carbon fluxes and vegetation dynamics. Geosci Model Dev 4:701–722. doi:10.5194/gmd-4-701-2011
Cowan IR, Farquhar GD (1977) Stomatal function in relation to leaf metabolism and environment. In: Jennings DH (ed) Integration of activity in the higher plant. Cambridge University Press, Cambridge, pp 471–505
Cox PM, Betts RA, Jones CD, Spall SA, Totterdell IJ (2000) Acceleration of global warming due to carbon-cycle feedbacks in a coupled climate model. Nature 408:184–187
Cox PM, Pearson D, Booth BB, Friedlingstein P, Huntingford C, Jones CD, Luke CM (2013) Sensitivity of tropical carbon to climate change constrained by carbon dioxide variability. Nature 494(7437):341–344
da Costa ACL, Metcalfe DB, Doughty CE, Oliveira AR, Neto GFC, da Costa MC, Silva JA Jr, Aragão LEOC, Almeida S, Galbraith DR et al (2014) Ecosystem respiration and net primary productivity after 8–10 years of experimental throughfall reduction in an eastern Amazon forest. Plant Ecol Divers 7(1–2):7–24

Davidson EA, Janssens IA, Luo Y (2006) On the variability of respiration in terrestrial ecosystems: moving beyond Q10. Glob Chang Biol 12(2):154–164

Davidson EA, de Araujo AC, Artaxo P, Balch JK, Brown IF, Bustamante MMC, Coe MT, DeFries RS, Keller M, Longo M, Munger JW, Schroeder W, Soares-Filho BS, Souza CM Jr, Wofsy SC (2012) The Amazon basin in transition. Nature 481(7381):321–328

Delbart N, Ciais P, Chave J, Viovy N, Malhi Y, Le Toan T (2010) Mortality as a key driver of the spatial distribution of aboveground biomass in Amazonian forest: results from a dynamics vegetation model. Biogeosciences 7:3017–3039

DeLucia EH, Moore DJ, Norby RJ (2005) Contrasting responses of forest ecosystems to rising atmospheric CO_2: implications for the global C cycle. Global Biogeochem Cycles 19, GB3006. doi:10.1029/2004GB002346

Doughty CE (2011) An in situ leaf and branch warming experiment in the Amazon. Biotropica 43 (6):658–665

Doughty CE, Goulden ML (2008) Are tropical forests near a high temperature threshold? J Geophys Res Biogeosci (2005–2012) 113(G1), doi:10.1029/2007JG000632

Doughty CE, Metcalfe DB, da Costa MC, de Oliveira AAR, Neto GFC, Silva JA, Aragão LEOC, Almeida SS, Quesada CA, Girardin CAJ et al (2014) Production, allocation and cycling of carbon in a forest on fertile terra preta soil in Eastern Amazonia compared with a forest on adjacent infertile soil. Plant Ecol Divers 7(1–2):41–53

Drake BG, Gonzàlez-Meler MA, Long SP (1997) More efficient plants: a consequence of rising atmospheric CO2? Annu Rev Plant Physiol 48:609–639

Dufresne JL, Friedlingstein P, Berthelot M et al (2002) On the magnitude of positive feedback between future climate change and the carbon cycle. Geophys Res Lett 29(10):43-1–43-4

Dybzinski R, Farrior C, Wolf A, Reich PB, Pacala SW (2011) Evolutionarily stable strategy carbon allocation to foliage, wood, and fine roots in trees competing for light and nitrogen: an analytically tractable, individual-based model and quantitative comparisons to data. Am Nat 177:153–166

Egea G, Verhoef A, Vidale PL (2011) Towards an improved and more flexible representation of water stress in coupled photosynthesis-stomatal conductance models. Agric For Meteorol 151:1370–1384

Engelbrecht BMJ, Comita LS, Condit R, Kursar TA, Tyree MT, Turner BL, Hubbell SP (2007) Drought sensitivity shapes species distribution patterns in tropical forests. Nature 447:80–82

Espírito-Santo FDB, Gloor M, Keller M, Malhi Y, Saatchi S, Nelson B, Oliveira Junior RC, Pereira C, Lloyd J, Frolking S, Palace M, Shimabukuro YE, Duarte V, Monteagudo Mendoza A, López-González G, Baker TR, Feldpausch T, Brienen RJW, Asner GP, Boyd DS, Phillips OL (2014) Size and frequency of natural forest disturbances and the Amazon forest carbon balance. Nat Commun 5:3434

Farquhar GD, Caemmerer SV, Berry JA (1980) A biochemical model of photosynthetic CO_2 assimilation in leaves of C_3 species. Planta 149:78–90

Fauset S, Baker TR, Lewis SL et al (2012) Drought-induced shifts in the floristic and functional composition of tropical forests in Ghana. Ecol Lett 15:1120–1129

Ferry B, Morneau F, Bontemps JD, Blanc L, Freycon V (2010) Higher treefall rates on slopes and waterlogged soils result in lower stand biomass and productivity in a tropical rain forest. J Ecol 98:106–116

Fisher RA, Williams M, Do Vale RL, Da Costa AL, Meir P (2006) Evidence from Amazonian forests is consistent with isohydric control of leaf water potential. Plant Cell Environ 29:151–165

Fisher RA, Williams M, Da Costa AL, Malhi Y, Da Costa RF, Almeida S, Meir P (2007) The response of an Eastern Amazonian rain forest to drought stress: results and modelling analyses from a throughfall exclusion experiment. Glob Chang Biol 13:2361–2378

Fisher RA, Williams M, Ruivo MLP, Costa AL, Meir P (2008) Evaluating climatic and soil water controls on evapotranspiration at two Amazonian rainforest sites. Agric For Meteorol 148:850–861

Fisher R, McDowell N, Purves D, Moorcroft P, Sitch S, Cox P, Huntingford C, Meir P, Woodward FI (2010a) Assessing uncertainties in a second-generation dynamic vegetation model caused by ecological scale limitations. New Phytol 187:666–681

Fisher JB, Sitch S, Malhi Y, Fisher RA, Huntingford C, Tan S-Y (2010b) Carbon cost of plant nitrogen acquisition: a mechanistic, globally applicable model of plant nitrogen uptake, retranslocation, and fixation. Global Biogeochem Cycles 24, GB1014

Franklin O, McMurtrie RE, Iversen CM, Crous KY, Finzi AC, Tissue DT, Ellsworth DS, Oren R, Norby RJ (2009) Forest fine-root production and nitrogen use under elevated CO_2: contrasting responses in evergreen and deciduous trees explained by a common principle. Glob Chang Biol 15:132–144

Franklin O, Johansson J, Dewar RC, Dieckmann U, McMurtrie RE, Brannstrom A, Dybzinski R (2012) Modeling carbon allocation in trees: a search for principles. Tree Physiol 32:648–666

Franks PJ, Adams MA, Amthor JS et al (2013) Sensitivity of plants to changing atmospheric CO_2 concentration: from the geological past to the next century. New Phytol 197:1077–1094

Friedlingstein P, Cox PM, Betts RA, Bopp L, Bloh WV, Brovkin V, Cadule P, Doney SC, Eby M, Fung IY, Bala G, John J, Jones CD, Joos F, Kato T, Kawamiya M, Knorr W, Lindsay K, Matthews HD, Raddatz T, Rayner PJ, Reick C, Roeckner E, Schnitzler KG, Schnur R, Strassmann K, Weaver AJ, Yoshikawa C, Zeng N (2006) Climate—carbon cycle feedback analysis: results from the C^4 MIP model intercomparison. J Climate 19:3337–3353

Fyllas NM, Quesada CA, Phillips OL, Lloyd J (2012) Coordination of physiological and structural traits in Amazon forest trees. Biogeosciences 9:775–801

Galbraith D, Levy PE, Sitch S, Huntingford C, Cox P, Williams M, Meir P (2010) Multiple mechanisms of Amazonian forest biomass losses in three dynamic global vegetation models under climate change. New Phytol 187(3):647–665

Galbraith D, Malhi Y, Affum-Baffoe K, Castanho ADA, Doughty CE, Fisher RA, Lewis SL, Peh KSH, Phillips OL, Quesada CA, Sonke B, Lloyd J (2013) Residence times of woody biomass in tropical forests. Plant Ecol Divers 6(1):139–157

Galbraith D, Malhi Y, Aragão L, Baker T (2014) The ecosystem dynamics of Amazonia and Andean forests. Plant Ecol Divers 7:1–6

Gale N, Barfod AS (1999) Canopy tree mode of death in a western Ecuadorian rain forest. J Trop Ecol 15:415–436

Gatti LV, Gloor M, Miller JB, Doughty CE, Malhi Y, Domingues LG et al (2014) Drought sensitivity of Amazonian carbon balance revealed by atmospheric measurements. Nature 506 (7486):76–80

Girardin CJ, Malhi Y, Aragao LEOC et al (2010) Net primary productivity allocation and cycling of carbon along a tropical forest elevational transect in the Peruvian Andes. Glob Chang Biol 16:3176–3192

Girardin CAJ, Silva Espejo J, Doughty CE, Huasco WH, Metcalfe DB, Durand-Baca L, Marthews T, Aragão LEOC, Farfan W, Cabrera KG et al (2014) Productivity and carbon allocation in a tropical montane cloud forest in the Peruvian Andes. Plant Ecol Divers 7 (1–2):107–123

Golding N, Betts RA (2008) Fire risk in Amazonia due to climate change in the HadCM3 climate model: potential interactions with deforestation. Global Biogeochem Cycles 22, GB4007

Goll DS, Brovkin V, Parida BR, Reick CH, Kattge J, Reich PB, van Bodegom PM, Niinemets U (2012) Nutrient limitation reduces land carbon uptake in simulations with a model of combined carbon, nitrogen and phosphorus cycling. Biogeosciences 9:3547–3569

Gonçalves LGG, Borak JS, Costa MH, Saleska SR, Baker I, Restrepo-Coupe N, Muza MN (2013) Overview of the Large-Scale Biosphere-Atmosphere Experiment in Amazônia Data Model Intercomparison Project (LBA-DMIP). Agric For Meteorol 182–183:111–127

Good P, Jones C, Lowe J, Betts R, Gedney N (2013) Comparing tropical forest projections from two generations of hadley centre earth system models, HadGEM2-ES and HadCM3LC. J Climate 26:495–511

Groenendijk M, Dolman AJ, Ammann C, Arneth A, Cescatti A, Dragoni D, Knohl A (2011) Seasonal variation of photosynthetic model parameters and leaf area index from global Fluxnet eddy covariance data. J Geophys Res 116(G4)

Haxeltine A, Prentice IC (1996) BIOME3: an equilibrium terrestrial biosphere model based on ecophysiological constraints, resource availability, and competition among plant functional types. J Global Biogeochem Cycles 10:693–709

Hickler T, Smith B, Prentice IC, Mjofors K, Miller P, Arneth A, Sykes MT (2008) CO_2 fertilization in temperate FACE experiments not representative of boreal and tropical forests. Glob Chang Biol 14:1531–1542

Huasco WH, Girardin CAJ, Doughty CE, Metcalfe DB, Baca LD, Silva-Espejo JE, Cabrera DG, Aragão LEOC, Davila AR, Marthews TR et al (2014) Seasonal production, allocation and cycling of carbon in two mid-elevation tropical montane forest plots in the Peruvian Andes. Plant Ecol Divers 7(1–2):125–142

Huete AR, Didan K, Shimabukuro YE, Ratana P, Saleska SR, Hutyra LR, Yang W, Nemani RR, Myneni R (2006) Amazon rainforests green-up with sunlight in dry season. Geophys Res Lett 33(6), L06405

Hunt JF, Ohno T, He Z, Honeycutt CW, Dail DB (2007) Inhibition of phosphorus sorption to goethite, gibbsite and kaolin by fresh and decomposed organic matter. Biol Fertil Soils 44:277–288

Huntingford C, Fisher RA, Mercado L et al (2008) Towards quantifying uncertainty in predictions of Amazon 'dieback'. Philos Trans R Soc Lond B Biol Sci 363:1857–1864

Huntingford C, Zelazowski P, Galbraith D, Mercado LM, Sitch S, Fisher R, Lomas M, Walker AP, Jones CD, Booth BBB, Malhi Y, Hemming D, Kay G, Good P, Lewis SL, Phillips OL, Atkin OK, Lloyd J, Gloor E, Zaragoza-Castells J, Meir P, Betts R, Harris PP, Nobre C, Marengo J, Cox PM (2013) Simulated resilience of tropical rainforests to CO_2-induced climate change. Nature Geosci 6(4):268–273

Jacobs CMJ, Van den Hurk BMM, De Bruin HAR (1996) Stomatal behaviour and photosynthetic rate of unstressed grapevines in semi-arid conditions. Agric For Meteorol 80:111–134

Jarvis PG (1976) Interpretation of variations in leaf water potential and stomatal conductance found in canopies in field. Philos Trans R Soc Lond B Biol Sci 273:593–610

Jupp TE, Cox PM, Rammig A, Thonicke K, Lucht W, Cramer W (2010) Development of probability density functions for future South American rainfall. New Phytol 187(3):682–693

Kim Y, Knox RG, Longo M, Medvigy D, Hutyra LR, Pyle EH, Wofsy SC, Bras RL, Moorcroft PR (2012) Seasonal carbon dynamics and water fluxes in an Amazon rainforest. Glob Chang Biol 18:1322–1334

King DA (1993) A model analysis of the influence of root and foliage allocation on forest production and competition between trees. Tree Physiol 12:119–135

Körner C, Asshoff R, Bignucolo O, Hättenschwiler S, Keel SG, Peláez-Riedl S, Pepin S, Siegwolf RTW, Zotz G (2005) Carbon flux and growth in mature deciduous forest trees exposed to elevated CO_2. Science 309:1360–1362

Korning J, Balslev H (1994) Growth and mortality of trees in Amazonian tropical rain forest in Ecuador. J Veg Sci 4:77–86

Krinner G, Viovy N, de Noblet-Ducoudré N, Ogée J, Polcher J, Friedlingstein P, Ciais P, Sitch S, Prentice IC (2005) A dynamic global vegetation model for studies of the coupled atmosphere-biosphere system. Global Biogeochem Cycles 19, GB1015. doi:10.1029/2003GB002199

Kruijt B, Witte JPM, Jacobs CM, Kroon T (2008) Effects of rising atmospheric CO_2 on evapotranspiration and soil moisture: a practical approach for the Netherlands. J Hydrol 349 (3):257–267

Kucharik CJ, Foley JA, Delire C, Fisher VA, Coe MT, Lenters JD, Young-Molling C, Ramankutty N (2000) Testing the performance of a dynamic global ecosystem model: water balance, carbon balance, and vegetation structure. Global Biogeochem Cycles 4(3):795–825

Kull O, Kruijt B (2002) Leaf photosynthetic light response: a mechanistic model for scaling photosynthesis to leaves and canopies. Funct Ecol 12:767–777

Kursar TA, Engelbrecht BMJ, Burke A, Tyree MT, El Omari B, Giraldo JP (2009) Tolerance to low leaf water status of tropical tree seedlings is related to drought performance and distribution. Funct Ecol 23:93–102

Lapola DM, Oyama MD, Nobre CA (2009) Exploring the range of climate biome projections for tropical South America: the role of CO_2 fertilization and seasonality. Global Biogeochem Cycles 23:1–16

Leuning R (1995) A critical appraisal of a combined stomatal-photosynthesis model for C_3 plants. Plant Cell Environ 18:339–355

Levy PE, Cannell MGR, Friend AD (2004) Modelling the impact of future changes in climate, CO_2 concentration and land use on natural ecosystems and the terrestrial carbon sink. Glob Environ Chang 14:21–30

Lewis SL, Brando PM, Phillips OL, van der Heijden GM, Nepstad D (2011) The 2010 amazon drought. Science 331(6017):554

Li W, Fu R, Dickinson RE (2006) Rainfall and its seasonality over the Amazon in the 21st century as assessed by the coupled models for the IPCC AR4. J Geophys Res Atmos 111(D2)

Lieberman D, Lieberman M, Peralta R, Hartshorn GS (1985) Mortality patterns and stand turnover rates in a wet tropical forest in Costa Rica. J Ecol 73:915–924

Lines ER, Coomes DA, Purves DW (2010) Influences of forest structure, climate and species composition on tree mortality across the eastern US. PLoS One 5, e13212

Lloyd J, Farquhar GD (2008) Effects of rising temperatures and [CO_2] on the physiology of tropical forest trees. Philos Trans R Soc Lond B Biol Sci 363:1811–1817

Lloyd J, Bird MI, Veenendaal EM, Kruijt B (2001) Should phosphorus availability be constraining moist tropical forest responses to increasing CO_2 concentrations? In: Shulze ED, Harrison SP, Heimann M, Holland EA, Lloyd J, Prentice IC, Shimel D (eds) Global biogeochemical cycles in the climate system. Academic Press, New York, NY

Lloyd J, Patino S, Paiva RQ et al (2010) Optimisation of photosynthetic carbon gain and within-canopy gradients of associated foliar traits for Amazon forest trees. Biogeosciences 7:1833–1859

Lola da Costa AC, Galbraith D, Almeida S, Portela BTT, da Costa M, de Athaydes Silva Junior J, Braga AP, de Gonçalves PHL, de Oliveira AAR, Fisher R, Phillips OL, Metcalfe DB, Levy P, Meir P (2010) Effect of 7 yr of experimental drought on vegetation dynamics and biomass storage of an eastern Amazonian rainforest. New Phytol 187(3):579–591

Long SP, Hutchin PR (1991) Primary production in grasslands and coniferous forests with climate change—an overview. Ecol Appl 1:139–156

Loveys BR, Atkinson LJ, Sherlock DJ, Roberts RL, Fitter AH, Atkin OK (2003) Thermal acclimation of leaf and root respiration: an investigation comparing inherently fast- and slow-growing plant species. Glob Chang Biol 9(6):895–910

Malhi Y, Nobre AD, Grace J, Kruijt B, Pereira MGP, Culf AD, Scott S (1998) Carbon dioxide transfer over a Central Amazonian rain forest. J Geophys Res Atmos 103:31593–31612

Malhi Y, Phillips OL, Lloyd J et al (2002) An international network to monitor the structure, composition and dynamics of Amazonian forests (RAINFOR). J Veg Sci 13:439–450

Malhi Y, Aragao LEOC, Galbraith D, Huntingford C, Fisher R, Zelazowski P, Sitch S, McSweeney C, Meir P (2009a) Exploring the likelihood and mechanism of a climate-change-induced dieback of the Amazon rainforest. Proc Nat Acad Sci U S A 106(49):20610–20615

Malhi Y, Aragão LEOC, Metcalfe DB, Paiva R, Quesada CA, Almeida S, Anderson LO, Brando PM, Chambers JQ, COSTA ACL, Hutyra LR, Oliveira PJC, Patiño S, Pyle EH, Robertson AL, Teixeira LM (2009b) Comprehensive assessment of carbon productivity, allocation and storage in three Amazonian forests. Glob Chang Biol 15:1255–1274

Malhi Y, Doughty C, Galbraith D (2011) The allocation of ecosystem net primary productivity in tropical forests. Philos Trans R Soc Lond B Biol Sci 366:3225–3245

Malhi Y, Amezquita FF, Doughty CE, Silva-Espejo J, Girardin CAJ, Metcalfe DB, Aragão LEOC, Huaraca-Quispe LP, Almazora-Taype I, Eguiluz-Mora L et al (2014) The productivity, metabolism and carbon cycle of two lowland tropical forest plots in south-western Amazonia, Peru. Plant Ecol Divers 7(1–2):85–105

Marengo JA, Nobre CA, Tomasella J, Cardoso MF, Oyama MD (2008) Hydro-climate and ecological behaviour of the drought of Amazonia in 2005. Philos Trans R Soc Lond B Biol Sci 363:1773–1778

Markewitz D, Devine S, Davidson EA, Brando P, Nepstad DC (2010) Soil moisture depletion under simulated drought in the Amazon: impacts on deep root uptake. New Phytol 187:592–607

Marthews TR, Quesada CA, Galbraith DR, Malhi Y, Mullins CE, Hodnett MG, Dharssi I (2013) High-resolution hydraulic parameter maps for surface soils in tropical South America. Geosci Model Dev Discuss 6:6741–6774. doi:10.5194/gmdd-6-6741-2013

Mcdowell N, Pockman WT, Allen CD et al (2008) Mechanisms of plant survival and mortality during drought: why do some plants survive while others succumb to drought? New Phytol 178:719–739

Mcdowell NG, Beerling DJ, Breshears DD, Fisher RA, Raffa KF, Stitt M (2011) The interdependence of mechanisms underlying climate-driven vegetation mortality. Trends Ecol Evol 26:523–532

McMurtrie RE, Norby RJ, Medlyn BE, Dewar RC, Pepper DA, Reich PB, Barton CVM (2008) Why is plant-growth response to elevated CO_2 amplified when water is limiting, but reduced when nitrogen is limiting? A growth-optimisation hypothesis. Funct Plant Biol 35:521–534

Medlyn BE, Dreyer E, Ellsworth D, Fortsreuter M, Harley PC, Kirschbaum MUF, Le Roux X, Montpied P, Strassemeyer J, Walcroft A, Wang K, Lousteau D (2002) Temperature response of parameters of a biochemically based model of photosynthesis. II. A review of experimental data. Plant Cell Environ 25:1167–1179

Medlyn BE, Duursma RA, Eamus D, Ellsworth DS, Prentice IC, Barton CV, Wingate L (2011) Reconciling the optimal and empirical approaches to modelling stomatal conductance. Glob Chang Biol 17(6):2134–2144

Medvigy D, Wofsy SC, Munger JW, Hollinger DY, Moorcroft PR (2009) Mechanistic scaling of ecosystem function and dynamics in space and time: ecosystem demography model version 2. J Geophys Res Biogeosci 114, G01002

Meir P, Woodward FI (2010) Amazonian rain forests and drought: response and vulnerability. New Phytol 187:553–557

Meir P, Kruijt B, Broadmeadow M et al (2002) Acclimation of photosynthetic capacity to irradiance in tree canopies in relation to leaf nitrogen concentration and leaf mass per unit area. Plant Cell Environ 25:343–357

Meir P, Metcalfe DB, Costa ACL, Fisher RA (2008) The fate of assimilated carbon during drought: impacts on respiration in Amazon rainforests. Philos Trans R Soc Lond B 363 (1498):1849–1855

Mercado LM, Lloyd J, Dolman AJ, Sitch S, Patiño S (2009) Modelling basin-wide variations in Amazon forest productivity—Part 1: model calibration, evaluation and upscaling functions for canopy photosynthesis. Biogeosciences 6(7):1247–1272

Mercado LM et al (2011) Variations in Amazon forest productivity correlated with foliar nutrients and modelled rates of photosynthetic carbon supply. Philos Trans R Soc B Biol Sci 366 (1582):3316–3329

Metcalfe DB, Meir P, Aragao LEOC et al (2010) Shifts in plant respiration and carbon use efficiency at a large-scale drought experiment in the eastern Amazon. New Phytol 187:608–621

Mitchard ETA, Saatchi SS, Baccini A, Asner GP, Goetz SJ, Harris NL, Brown S (2013) Uncertainty in the spatial distribution of tropical forest biomass: a comparison of pan-tropical maps. Carbon Balance Manage 8:10. doi:10.1186/1750-0680-8-10

Monteith JL (1995) Accommodation between transpiring vegetation and the convective boundary-layer. J Hydrol 166:251–263

Moorcroft PR, Hurtt GC, Pacala SW (2001) A method for scaling vegetation dynamics: the ecosystem demography model (ED). Ecol Monogr 71(4):557–586

Morton DC, Nagol J, Carabajal CC, Rosette J, Palace M, Cook BD, Vermote EF, Harding DJ, North PRJ (2014) Amazon forests maintain consistent canopy structure and greenness during the dry season. Nature 506(7487):221–224

Nepstad DC, Carvalho CR, Davidson EA, Jipp P, Lefebvre P, Negreiros GHD, da Silva ED, Stone TA, Trumbore S, Vieira S (1994) The role of deep roots in the hydrological and carbon cycles of Amazonian forests and pastures. Nature 372:666–669

Nepstad DC, Moutinho P, Dias MB et al (2002) The effects of partial throughfall exclusion on canopy processes, aboveground production, and biogeochemistry of an Amazon forest. J Geophys Res Atmos 107:1–18

Nepstad DC, Marisa Tohver I, Ray D, Moutinho P, Cardinot G (2007) Mortality of large trees and lianas following experimental drought in an Amazon forest. Ecology 88:2259–2269

Nobre CA, Borma LS (2009) "Tipping points" for the Amazon forest. Curr Opin Environ Sustain 1:28–36

Norby RJ, Warren JM, Iversen CM, Medlyn BE, Mcmurtrie RE (2010) CO_2 enhancement of forest productivity constrained by limited nitrogen availability. Proc Nat Acad Sci U S A 107:19368–19373

O'Sullivan OS, Weerasinghe KWLK, Evans JR, Egerton JJG, Tjoelker MG, Atkin OK (2013) High-resolution temperature responses of leaf respiration in snow gum (*Eucalyptus pauciflora*) reveal high-temperature limits to respiratory function. Plant Cell Environ 36(7):1268–1284

Olander LP, Vitousek PM (2000) Regulation of soil phosphatase and chitinase activity by N and P availability. Biogeochemistry 49:175–190

Pasquel JA, Doughty CE, Metcalfe DB, Silva-Espejo J, Girardin AJ, Gutierrez JAC, Aguilar GEN, Quesada CA, Pizango JMR, Huaymacari JMR et al (2014) The seasonal cycle of productivity, metabolism and carbon dynamics in a wet seasonal forest in north-west Amazonia (Iquitos, Peru). Plant Ecol Divers 7(1–2):71–83

Phillips OL, Baker TR, Arroyo L, Higuchi N, Killeen TJ, Laurance WF, Lewis SL, Lloyd J, Malhi Y, Monteagudo A, Neill DA, Vargas PN, Silva JNM, Terborgh J, Martinez RV, Alexiades M, Almeida S, Brown S, Chave J, Comiskey JA, Czimczik CI, Di Fiore A, Erwin T, Kuebler C, Laurance SG, Nascimento HEM, Olivier J, Palacios W, Patino S, Pitman NCA, Quesada CA, Salidas M, Lezama AT, Vinceti B (2004) Pattern and process in Amazon tree turnover, 1976-2001. Philos Trans R Soc Lond B Biol Sci 359:381–407

Phillips OL, Aragão LEOC, Lewis SL et al (2009) Drought Sensitivity of the Amazon Rainforest. Sci Mag 323:1344–1347

Phillips OL, Van Der Heijden G, Lewis SL et al (2010) Drought-mortality relationships for tropical forests. New Phytol 187:631–646

Plaut JA, Yepez EA, Hill J, Pangle R, Sperry JS, Pockman WT, Mcdowell NG (2012) Hydraulic limits preceding mortality in a pinon-juniper woodland under experimental drought. Plant Cell Environ 35:1601–1617

Potter C, Klooster S, Genovese V (2012) Net primary production of terrestrial ecosystems from 2000 to 2009. Clim Change 115:365–378

Poulter B, Hattermann F, Hawkins E, Zaehle S, Sitch S, Restrepo-Coupe N, Heyder U, Cramer W (2010) Robust dynamics of Amazon dieback to climate change with perturbed ecosystem model parameters. Glob Chang Biol 16:2476–2495

Powell TL, Galbraith DR, Christoffersen BO, Harper A, Imbuzeiro H, Rowland L et al (2013) Confronting model predictions of carbon fluxes with measurements of Amazon forests subjected to experimental drought. New Phytol 200(2):350–365

Putz FE, Milton K (1982) Tree mortality rates on Barro Colorado Island. In: Leigh AS, Rand AS, Windsor DM (eds) Ecology of a tropical forest. Smithsonian Press, Washington, DC, pp 95–100

Quesada CA, Lloyd J, Schwarz M et al (2010) Variations in chemical and physical properties of Amazon forest soils in relation to their genesis. Biogeosciences 7:1515–1541

Quesada CA, Lloyd J, Anderson LO, Fyllas NM, Schwarz M, Czimczik CI (2011) Soils of Amazonia with particular reference to the RAINFOR sites. Biogeosciences 8:1415–1440

Quesada CA, Phillips OL, Schwarz M, Czimczik CI, Baker TR, Patiño S, Fyllas NM, Hodnett M, Herrera R, Almeida S, Alvarez DE, Arneth A, Arroyo L, Chao K-J, Dezzeo N, Erwin T, di Fiore A, Higuchi N, Honorio CE, Jiménez EM, Killeen T, Lezama AT, Lloyd G, López-González G, Luizão FJ, Malhi Y, Monteagudo A, Neill DA, Núñez VP, Paiva R, Peacock J, Peñuela MC, Peña CA, Pitman NCA, Priante FN, Prieto A, Ramírez H, Rudas A, Salomão R, Santos AJB, Schmerler J, Silva N, Silveira M, Vásquez R, Vieira I, Terborgh J, Lloyd J (2012) Basin-wide variations in Amazon forest structure and function are mediated by both soils and climate. Biogeosciences 9:2203–2246

Rammig A, Jupp T, Thonicke K, Tietjen B, Heinke J, Ostberg S, Lucht W, Cramer W, Cox P (2010) Estimating the risk of Amazonian forest dieback. New Phytol 187(3):694–706

Rocha W, Metcalfe DB, Doughty CE, Brando P, Silvério D, Halladay K, Nepstad DC, Balch JK, Malhi Y (2014) Ecosystem productivity and carbon cycling in intact and annually burnt forest at the dry southern limit of the Amazon rainforest (Mato Grosso, Brazil). Plant Ecol Divers 7 (1–2):25–40

Saleska SR, Miller SD, Matross DM, Goulden ML, Wofsy SC, Rocha HR, Camargo PB, Crill P, Daube BC, Freitas HC, Hutyra LR, Keller M, Kirchhoff V, Menton MC, Munger JW, Pyle EH, Rice AH, Silva H (2003) Carbon in Amazon forests: unexpected seasonal fluxes and disturbance-induced losses. Science 302:1554–1557

Saleska SR, Didan K, Huete AR, da Rocha HR (2007) Amazon forests green-up during 2005 drought. Science 318(5850):612

Samanta A, Ganguly S, Hashimoto H, Devadiga S, Vermote E, Knyazikhin Y, Nemani RR, Myneni RB (2010) Amazon forests did not green-up during the 2005 drought. Geophys Res Lett 37(5), L05401

Sampaio G, Nobre CA, Costa MH, Satyamurty P, Filho BSS, Cardoso MF (2007) Regional climate change over eastern Amazonia caused by pasture and soybean cropland expansion. Geophys Res Lett 34:1–7

Scheiter S, Langan L, Higgins SI (2013) Next-generation dynamic global vegetation models: learning from community ecology. New Phytol 198(3):957–969

Seiler C, Hutjes RWA, Kruijt B, Quispe J, Añez S, Arora VK et al (2014) Modeling forest dynamics along climate gradients in Bolivia. J Geophys Res Biogeosci. doi:10.1002/2013JG002509

Sharkey TD, Schrader SM (2006) High temperature sense. In: Rao KVM, Raghavendra AS, Reddy KJ (eds) Physiology and molecular biology of stress tolerance in plants. Springer, Dordrecht, pp 101–129

Silvestrini RA et al (2011) Simulating fire regimes in the Amazon in response to climate change and deforestation. Ecol Appl 21:1573–1590

Sitch S, Smith B, Prentice IC et al (2003) Evaluation of ecosystem dynamics, plant geography and terrestrial carbon cycling in the LPJ dynamic global vegetation model. Glob Chang Biol 9:161–185

Slot M, Rey-Sánchez C, Gerber S, Lichstein JW, Winter K, Kitajima K (2014) Thermal acclimation of leaf respiration of tropical trees and lianas: response to experimental canopy warming, and consequences for tropical forest carbon balance. Glob Chang Biol. doi:10.1111/gcb.12563

Smith NG, Dukes JS (2013) Plant respiration and photosynthesis in global-scale models: incorporating acclimation to temperature and CO_2. Glob Chang Biol 19:45–63

Smith B, Prentice IC, Sykes MT (2001) Representation of vegetation dynamics in the modelling of terrestrial ecosystems: comparing two contrasting approaches within European climate space. Glob Ecol Biogeogr 10:621–637

Smith B, Wårlind D, Arneth A, Hickler T, Leadley P, Siltberg J, Zaehle S (2014) Implications of incorporating N cycling and N limitations on primary production in an individual-based dynamic vegetation model. Biogeosciences 11:2027–2054. doi:10.5194/bg-11-2027-2014

Spracklen DV, Arnold SR, Taylor CM (2012) Observations of increased tropical rainfall preceded by air passage over forests. Nature 489(7415):282–285

Stewart JWB (1988) Modelling surface conductance of Pine forest. Agric For Meteorol 43:19–35

Stockmann U, Adams MA, Crawford JW et al (2013) The knowns, known unknowns and unknowns of sequestration of soil organic carbon. Agric Ecosyst Environ 164:80–99

Subke J-A, Inglima I, Francesca Cotrufo M (2006) Trends and methodological impacts in soil CO_2 efflux partitioning: a metaanalytical review. Glob Chang Biol 12(6):921–943

Tardieu F, Simonneau T (1998) Variability of species among stomatal control under fluctuating soil water status and evaporative demand: modeling isohydric and anisohydric behaviours. J Exp Bot 49:419–432

Thonicke K, Spessa A, Prentice IC, Harrison SP, Dong L, Carmona-Moreno C (2010) The influence of vegetation, fire spread and fire behaviour on biomass burning and trace gas emissions: results from a process based model. Biogeosciences 7(6):1991–2011

Thornton PE, Doney SC, Lindsay K, Moore JK, Mahowald N, Randerson JT, Fung I, Lamarque J-F, Feddema JJ, Lee Y-H (2009) Carbon-nitrogen interactions regulate climate-carbon cycle feedbacks: results from an atmosphere-ocean general circulation model. Biogeosciences 6:2099–2120

Toledo JJ, Magnusson WE, Castilho CV, Nascimento HEM (2012) Tree mode of death in Central Amazonia: effects of soil and topography on tree mortality associated with storm disturbances. For Ecol Manage 263:253–261

Tollefson J (2013) Experiment aims to steep rainforest in carbon dioxide. Nature 496:405–406

Tomasella J, Hodnett M (2004) Pedotransfer functions for tropical soils. Dev Soil Sci 30:415–435

Turner BL, Blackwell MSA (2013) Isolating the influence of pH on the amounts and forms of soil organic phosphorus. Eur J Soil Sci 64:249–259

Turner BL, Engelbrecht BMJ (2011) Soil organic phosphorus in lowland tropical rain forests. Biogeochemistry 103:297–315

Uhl C, Kauffman JB (1990) Deforestation, fire susceptibility, and potential tree responses to fire in the eastern Amazon. Ecology 71(2):437–449

Van De Weg MJ, Meir P, Grace J, Ramos GD (2012) Photosynthetic parameters, dark respiration and leaf traits in the canopy of a Peruvian tropical montane cloud forest. Oecologia 168:23–34

Van der Molen MK, Dolman AJ, Ciais P, Eglin T, Gobron N, Law BE, Meir P, Peters W, Phillips OL, Reichstein M, Chen T, Dekker SC, Doubková M, Friedl MA, Jung M, van den Hurk BJJM, de Jeu RAM, Kruijt B, Ohta T, Rebel KT, Plummer S, Seneviratne SI, Sitch S, Teuling AJ, van der Werf GR, Wang G (2011) Drought and ecosystem carbon cycling. Agric For Meteorol 151 (7):765–773

Verbeeck H et al (2011) Seasonal patterns of CO_2 fluxes in Amazon forests: fusion of eddy covariance data and the ORCHIDEE model. J Geophys Res 116(G2), G02018

Vergara W, Sholz SM (2010) Assessment of the risk of Amazon dieback. World Bank report. http://publications.worldbank.org/index.php?main_page=product_info&products_id=23931

Von Caemmerer S, Evans JR, Hudson GS, Andrews TJ (1994) The kinetics of ribulose-1,5-bisphosphate carboxylase/oxygenase in vivo inferred from measurements of photosynthesis in leaves of transgenic tobacco. Planta 195:88–97

Vermeulen MH, Kruijt BJ, Hickler T, Kabat, P (2015) Modelling short-term variability in carbon and water exchange in a temperate Scots pine forest. Ear Sys Dyna 6(2):485–503

Von Randow C, Zeri M, Restrepo-Coupe N, Muza MN, de Gonçalves LGG, Costa MH et al (2013) Inter-annual variability of carbon and water fluxes in Amazonian forest, Cerrado and pasture sites, as simulated by terrestrial biosphere models. Agric For Meteorol 182:145–155

Wang YP, Law RM, Pak B (2010) A global model of carbon, nitrogen and phosphorus cycles for the terrestrial biosphere. Biogeosciences 7:2261–2282

Wang X, Piao S, Ciais P, Friedlingstein P, Myneni RB, Cox P et al (2014) A two-fold increase of carbon cycle sensitivity to tropical temperature variations. Nature. doi:10.1038/nature12915

Waring RH, Landsberg JJ, Williams M (1998) Net primary production of forests: a constant fraction of gross primary production? Tree Physiol 18(2):129–134

Whitaker J, Ostle N, Nottingham AT, Ccahuana A, Salinas N, Bardgett RD et al (2014) Microbial community composition explains soil respiration responses to changing carbon inputs along an Andes-to-Amazon elevation gradient. J Ecol. doi:10.1111/1365-2745.12247

White A, Cannell MGR, Friend AD (1999) Climate change impacts on ecosystems and the terrestrial carbon sink: a new assessment. Glob Environ Chang 9:S21–S30

Williams M, Malhi Y, Nobre AD, Rastetter EB, Grace J, Pereira MGP (1998) Seasonal variation in net carbon exchange and evapotranspiration in a Brazilian rainforest: a modelling analysis. Plant Cell Environ 21:953–968

Woodrow IE, Berry JA (1988) Enzymatic regulation of photosynthetic CO2, fixation in C3 plants. Annu Rev Plant Physiol Plant Mol Biol 39(1):533–594

Yang X, Post WM, Thornton PE, Jain A (2013) The distribution of soil phosphorus for global biogeochemical modelling. Biogeosciences 10:2525–2537

Yang X, Thornton PE, Ricciuto DM, Post WM (2014) The role of phosphorus dynamics in tropical forests – a modelling study using CLM-CNP. Biogeosciences 11:1667–1681

Zaehle S, Friend AD (2010) Carbon and nitrogen cycle dynamics in the O-CN land surface model: 1. model description, site-scale evaluation, and sensitivity to parameter estimates. Global Biogeochem Cycles 24, GB1005

Zaehle S et al (2010) Carbon and nitrogen cycle dynamics in the O-CN land surface model: 2. Role of the nitrogen cycle in historical terrestrial carbon balance. Global Biogeochem Cycles 24, GB1006

Zimmermann M, Leifeld J, Conen F, Bird MI, Meir P (2012) Can composition and physical protection of soil organic matter explain soil respiration temperature sensitivity? Biogeochemistry 107:423–436

Part V
Integrating Considerations Between Biophysical and Social Aspects

Chapter 15
Land Use, Land Cover and Land Use Change in the Brazilian Amazon (1960–2013)

Jean P. Ometto, Eráclito R. Sousa-Neto, and Graciela Tejada

15.1 Introduction

The Amazon Basin in South America is home to the largest continuous remaining tropical rainforest, representing half the world's rainforest area, and is home to one-third of Earth's species (Tollefson 2008). Along with their rich biodiversity, the forests of Amazonia deliver important ecosystem services. For example, in Brazil, the forests of Amazonia alone contain more carbon stored than the amount of global human-induced fossil fuel CO_2 emissions of an entire decade (Öborn et al. 2011); therefore, they play an important role in the global carbon budget (Chambers et al. 2001; Loarie et al. 2009; Le Quere et al. 2009). In addition, the vegetation acts as an efficient 'pump water' in recycling water over the extension of the forest, and thus it is an important driver of the hydrological cycle and possibly a major contributor to regulating regional climate (Spracklen et al. 2012; Werth and Avissar 2002).

Despite recent reductions in the relative rates of deforestation[1] in Amazonia, deforestation continues at a high rate, and this process is leading to changes in the environment and society. In the past 40 years, the region has experienced drastic changes in its land use and land cover (LULC). Fostered chiefly by the replacement

[1] Deforestation is a process that begins with the intact forest and ends with the complete conversion of the original forest to other coverages. The first step is the removal of the noblest woods, and then the timber for the construction and, finally, the remaining softwoods are harvested for the production of plywood and boards. This process may take several years because the exploration of the forest is made generally by different enterprises, each one specialised in one phase (INPE 2008).

J.P. Ometto (✉) • E.R. Sousa-Neto • G. Tejada
Earth System Science Centre (CCST)—National Institute for Space Research (INPE), São José dos Campos, SP, Brazil
e-mail: jean.ometto@inpe.br

© Springer-Verlag Berlin Heidelberg 2016

L. Nagy et al. (eds.), *Interactions Between Biosphere, Atmosphere and Human Land Use in the Amazon Basin*, Ecological Studies 227,
DOI 10.1007/978-3-662-49902-3_15

of native vegetation by grazing land, sown with African grasses for cattle ranching and subsistence/family agriculture, more recently, large-scale agriculture such as soybean cultivation has become a major contributor to LULC change (Ometto et al. 2011). In general, deforestation and land conversion lead to the destruction of habitats and depletion of species, cause soil erosion and contribute to global climate change through the emissions of greenhouse gases (GHG).

Several studies have considered the future of the Amazon (Soares-Filho et al. 2010; Lapola et al. 2010; Gómez and Nagatani 2009; Malhi et al. 2008; Aguiar 2006; Soares-Filho et al. 2006; Laurance et al. 2001), following global concerns about biodiversity loss, deforestation-driven CO_2 emissions through the intensification of droughts and vulnerability to forest fires and major LULC changes. It appears that deforestation and global warming, acting synergistically, could lead to profound changes in the Amazon biome, and beyond. The potential shift in the energy and water cycles can cause changes in ecosystem structure (including biodiversity) and functioning, reducing the capacity of the forest to retain carbon and thereby increasing soil temperature and eventually affecting the regional hydrological cycle (Ometto et al. 2011).

Most of the above studies focused on Brazilian Legal Amazon (c. 5 million km^2), a legally designated entity that extends over nine federal states of Brazil (Fig. 15.1), whose inclusion in the designation in 1953 was underpinned by the similarities in their ecological structure, economic, political and social conditions. Currently, Brazilian Legal Amazon comprises the states of Acre, Amapá, Amazonas, Mato Grosso, Pará, Rondônia, Roraima and Tocantins, including a part of Maranhão (west of the 44° west).

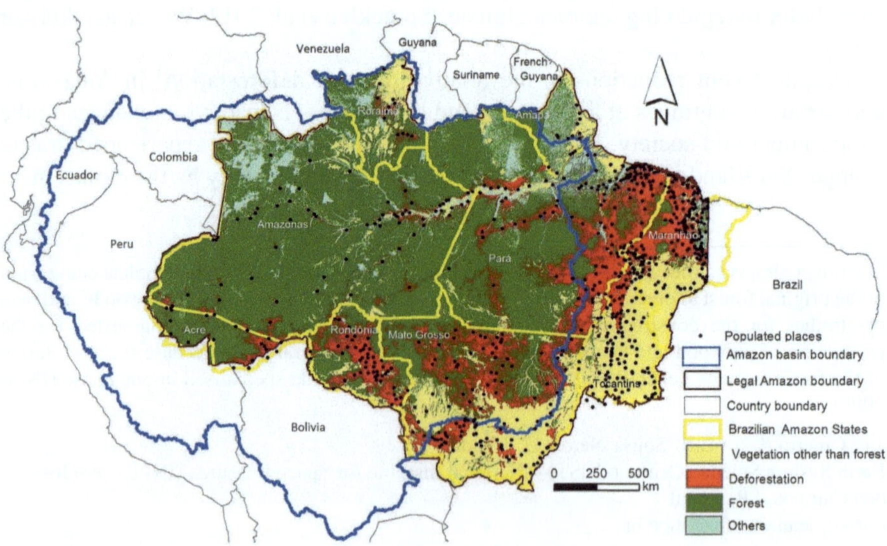

Fig. 15.1 The map of Brazilian Legal Amazon and its administrative division

In this chapter, we provide an overview of the extent and dynamics of land use and land use change of the Amazon Basin, with a focus on Brazilian Amazon, which occurred from 1960 to 2010, taking into account the environmental and social aspects related to the deforestation process. We provide some general information about deforestation rates and data sources available for their evaluation in all countries within the Amazon Basin.

15.2 Data and Information Sources Available on Land Use and Land Cover (LULC) Change

Until 1988, data on LULC change in Brazilian Amazon could be obtained from the agricultural census developed by the Brazilian Institute of Geography and Statistics (IBGE). Data were published in 1960, 1970 and every 5 years since then, 2006 being the latest census data released (IBGE 2006). The census data provided important information on land use (classes) at the municipality level but not in a spatially explicit format (Table 15.1).

Since 1988, the National Institute for Space Research (INPE) has been generating and compiling satellite data for monitoring LULC in Brazilian Amazon, as part of the Amazon Deforestation Calculation Program (PRODES). This system has provided a consistent and unique historical mapping of deforestation (INPE 2014a; Ometto et al. 2011).

PRODES has produced deforestation reports since 1978, using a mean deforestation rate from other data sources, such as IBGE until 1988; since 1997, the results have been presented in a spatially explicit format. From 2003 to the present, PRODES data and products (images, annual deforestation maps and deforestation statistics) are available on the Internet (INPE 2014a). Official statistics on deforestation rates for Brazilian Amazon are based on these data.

In addition to PRODES, other remote sensing products complement the deforestation and land use change-monitoring portfolio in Amazonia. DETER (INPE 2013) is an alert system developed by INPE that has been monitoring deforestation at a monthly basis since 2004, providing a reliable information source for decision-makers to implement rapid action on the ground. Based on indications of forest degradation obtained from DETER data, INPE has developed the DEGRAD system (mapping forest degradation in the Brazilian Amazon). The system uses LANDSAT and CBERS satellite images to map annually areas of degraded forest that are likely to be converted to clearcutting (INPE 2014b). TerraClass (INPE 2011) is a joint Project by INPE and the Brazilian Agricultural Research Enterprise (Embrapa) that uses the PRODES data for generating a LULC map of Brazilian Amazon every 2 years (INPE 2011). The data of TerraClass account for all the actual land use classes of converted land (see Table 15.1 for mapped land use classes).

Table 15.1 Land use classes considered in the deforested areas in Brazilian Amazon generated by TerraClass (2008)

Land use class	Definition
Annual crop	Extensive areas with a predominance of annual cycle crops, especially grains, with use of high technological standards, such as use of certified seeds, inputs, pesticides and mechanisation
Mosaic	Areas represented by an association of various types of land use and due to the spatial resolution of the satellite images, a discrimination between their components is not possible. In this class, family farming and subsystem of traditional pastures for livestock are carried out in conjugated form
Urban area	Areas resulting from population concentration forming villages, towns or cities with differentiated infrastructure in relation to rural areas and presenting density of roads, houses, buildings and other public facilities
Mining	Areas for extraction of valuable minerals or other geological materials with the presence of clearings and exposed soils, involving deforestation near superficial water bodies
Pasture	Pasture areas in the production process with a predominance of herbaceous vegetation and coverage of grass species between 90 % and 100 %
Pasture with shrubs	Pasture areas in the production process with a predominance of herbaceous vegetation and grass species coverage between 50 % and 80 %, associated with the presence of shrub with sparse vegetation with coverage between 20 % and 50 %
Pasture with areas of secondary regrowth	Areas that after clearcutting of natural vegetation and the development of some agro-pastoral activity are at the start of regeneration process of native vegetation, with dominance of shrubs and pioneer tree species. Areas characterized by high diversity of plant species
Eroded pasture (bare soil > 50 %)	Areas that after clearcutting of forests and the development of some agro-pastoral activity have a coverage of at least 50 % of exposed soil
Secondary vegetation	Areas that after the complete cut of forest vegetation are in advanced process of regeneration of shrub and/or trees or have been used for practising forestry or permanent agriculture with the use of native or exotic species
Other	These are areas that did not fit in the previous categories and that showed a differentiated coverage pattern such as rock outcrops, river beaches, sandbars and others
Areas with no data	Areas that have had their interpretation impossible by the presence of clouds or cloud shadow, at the time of passage for satellite image acquisition, in addition to areas recently burned

Other LULC datasets are available for the entire Amazon Basin (Table 15.2), including relevant parts of Bolivia, Ecuador, Peru, Colombia, Venezuela, Guyana, Suriname and French Guyana. The Terra-i system detects land cover changes resulting from human activities at 16-day intervals (Terra-i 2012). A regional initiative from the Amazon Geo-referenced Socio-environmental Information Network (RAISG) has used a standardised methodology for the whole Basin to produce

Table 15.2 Sources of land use and land cover change data for the Amazon Basin

Level	LUCC data	Description	Spatial/temporal resolution	Website	Source
Global	GLC2000	Vegetation map of South America (Global Land Cover 2000)	1 km/2000	http://www.gvm.jrc.it/glc2000	GLC (2003)
	GlobCover	Global composites and land cover map	300 m/2005–2006; 2009	http://due.esrin.esa.int/globcover/	ESA (2010)
Amazon Basin	Terra-i	Detects land cover changes resulting from human activities in near real time	250 m/2004 to 2011 update every 16 days	http://www.terra-i.org/terra-i.html	Terra-I (2012)
	RAISG	Deforestation map of the Amazon Basin	30 m/2000–2005 and 2010	www.raisg.socioambiental.org	RAISG (2012)
Brazilian Amazon	PRODES	Yearly deforestation map	60 m/yearly from 1988 to 2012	www.obt.inpe.br/prodes/	INPE (2013)
	DETER	Monthly deforestation alerts	250 m/monthly from	www.obt.inpe.br/deter/	INPE (2013)
	IBGE	Agricultural census data	Municipal level (not spatial data) every 5 years since 1960	www.ibge.gov.br	IBGE (2006)
	TerraClass	Land use map	30 m, 2008 available and 2010 only the report (missing spatially explicit data)	www.inpe.br/CRA	INPE (2011)
	DEGRAD	Forest degradation map	30 m/yearly since 2007	www.obt.inpe.br/degrade/	INPE (2014b)

deforestation maps for the years 2000, 2005 and 2010 (RAISG 2012). On the global scale, the GLC 2000 (https://ec.europa.eu/jrc/en/scientific-tool/global-land-cover) and the GlobCover (http://due.esrin.esa.int/page_globcover.php) are the most used map sources.

15.3 Occupation of Brazilian Amazon: Drivers and Trends in Deforestation

Apart from the impacts associated with indigenous settlements in Amazonia, dating back thousands of years, only in the past 40 years that the region experienced major changes in LULC. Nowadays, most of the deforestation is undertaken for cattle ranching, agriculture and creation/expansion of urban areas. Until the 1950s, the occupation of Brazilian Amazon was limited to the coastal region and the margins of the main rivers (Escada and Alves 2001), causing imperceptible deforestation at the regional scale. Economic activity was mostly related to the extraction of non-timber products, mainly rubber tapping (Costa 1997). Rubber became the first commodity produced in the region and had its golden years at the turn of the nineteenth and twentieth century and reached its decline by around 1920 (UICN 1995).

Government incentives to settle the region, underpinned by the construction of the Belém–Brasília highway in eastern Amazonia, caused the population to grow from 1 m to 5 m from 1955 to 1965 (Becker 1997). As part of the National Plan for Economic and Social Development (NPD), the main government strategies for the occupation of Amazonia included (1) infrastructure development (construction of roads, telecommunication, hydropower and urban areas) for spatial integration, (2) expropriation of land for implementation of mining and settlement projects and (3) subsidies to flow of capital and immigration (Becker 1997; Machado 1997). In the 1970s, the Trans-Amazônica and Cuiabá–Santarém highways—built alongside existing highways—formed the basic structure of road transportation within the National Integration Project (PIN) (Escada and Alves 2001).

While promoting the integration and the connectivity to regional and national markets, the construction of roads has led to high deforestation rates (Almeida 2009; Fearnside et al. 2009). The depletion of native vegetation during 1970s predominantly occurred along major roads and around new areas of human settlements (Skole and Tucker 1993; Machado 1997; Alves 2002; Fearnside 2005), primarily in south-eastern Amazon (Fig. 15.3), a region commonly known as the 'arc of deforestation' (Becker 2005; Ometto et al. 2011). Following opening the roads, logging, cattle ranching and small- and large-scale agriculture were the most common activities that have led to increasing deforestation (Aguiar 2006; Aguiar et al. 2012), reaching a total area of 152,000 km^2 deforested by the end of the 1970s (INPE 2002).

In the 1980s, the process of occupying of Amazonia included the expansion of agribusiness, mining and several settlement projects (Kitamura 1994). Tax incentives were a strong driver of deforestation (Fearnside 2005). Between 1978 and 1988, net deforestation in the Amazon region reached 360,889 km^2, a significant increase compared with the decade before (Fig. 15.4; INPE 2002).

The initial expansion of large-scale agriculture started in southern Brazilian Amazon, affecting the areas of the Cerrado 'biome' in the 1990s (Aguiar 2006); it changed the patterns of land use and the regional economy (Carvalho et al. 2002). The expansion of world markets improved access to local credit and government incentives, such as tax exemptions and funding for agricultural research. The improvement of market channels and infrastructure rapidly encouraged the expansion of mechanized agriculture with cash crops for export (Valdes 2006; Brown et al. 2004; Barbier 2004; Madi 2004). From the late 1990s to 2004, there was a significant increase in deforestation rates. This trend reflected the large-scale agriculture boom, especially in the states of Mato Grosso, Pará and Rondônia,

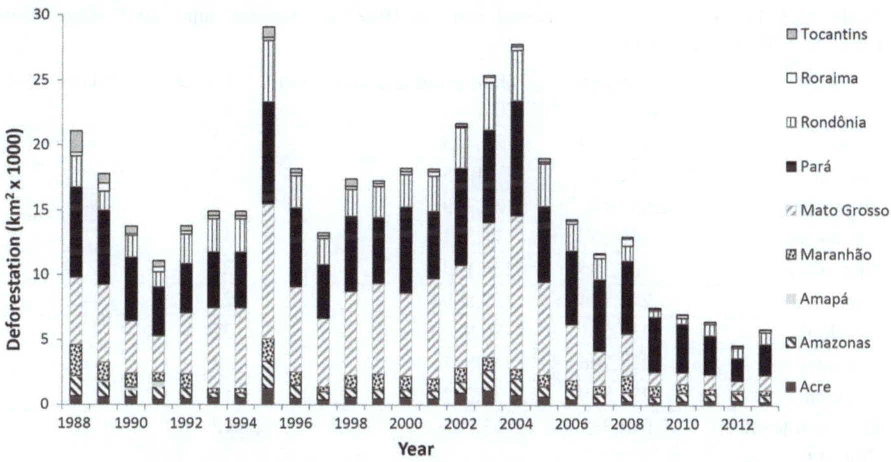

Fig. 15.2 Total deforested area per federal state in Brazilian Legal Amazon, from 1988 to 2013. Data from TerraClass 2010 (INPE 2011) and PRODES (INPE 2014a)

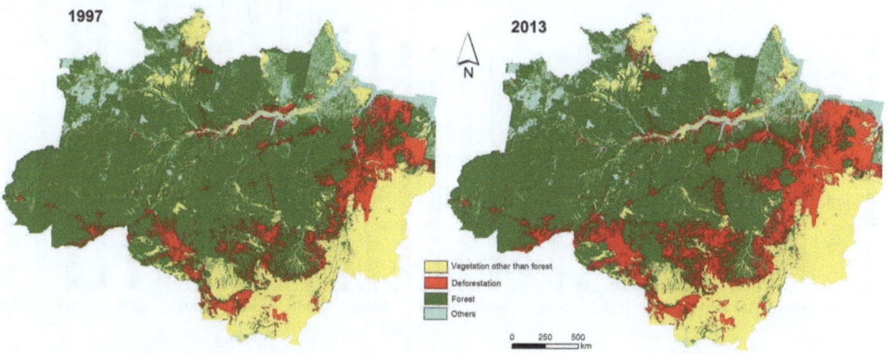

Fig. 15.3 Land use and land cover change from 1997 to 2013 in Brazilian Amazon. Data from PRODES (INPE 2014a)

leading to a deforestation rate of ca. 18,161 km^2 $year^{-1}$ for 1995–1996 and 27,772 km^2 $year^{-1}$ for 2004 (Fig. 15.2; INPE 2014a).

Due to several factors, such as policy formulations and pressure from the international community, since 2005 there has been a significant reduction in the annual deforestation rate in Brazilian Amazon, with 12,911 km^2 in 2008, 7464 km^2 in 2009, 4571 km^2 in 2012 (the lowest deforestation rate since 1988) and 5843 km^2 in 2013 (Fig. 15.2; INPE 2014a). The reduction was observed in all states, although Pará continued to be a state with a high absolute rate of deforestation until 2010 (Fig. 15.3, Table 15.2). Nonetheless, Maranhão has the highest accumulated deforestation on an area basis (Table 15.3). In total, 18.8 % of Brazilian Amazon has been converted from its natural vegetation (mainly tropical rainforest and cerrado)

Table 15.3 Deforested area per federal state in Brazilian Amazon until 2010. Data from TerraClass 2010 (INPE 2011)

State	Total area (km^2)	Deforested area (km^2) until 2013	Percentage of deforested area
Acre	164,170	20,455	12
Amapá	142,814	4925	3
Amazonas	1,559,160	32,799	2
Maranhão	262,297	111,351	42
Mato Grosso	903,385	209,143	23
Pará	1,247,794	257,869	21
Rondônia	237,581	86,821	37
Roraima	224,296	9871	4
Tocantins	271,849	30,271	11
Brazilian legal Amazon	5,013,347	763,505	15

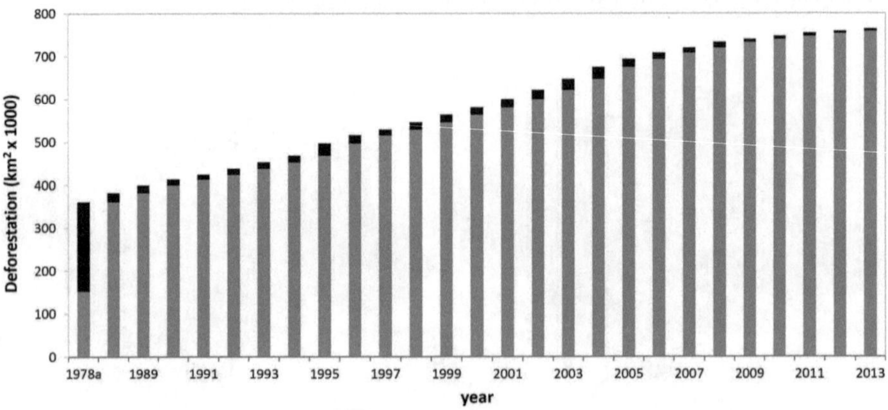

Fig. 15.4 Accumulated deforestation from 1978 to 2013 (*grey*) and annual deforestation (*black*) in Brazilian Amazon. Data from PRODES (INPE 2011) and TerraClass 2010 (INPE 2011). (**a**) Data for 1978 is from PRODES (INPE 2002). It is assumed that the deforestation reached 152,200 km^2 until 1978

to another land cover type through land use by 2013 (INPE 2013), of which 60 % occurred in the period from 1990 to 2010 (Fig. 15.4).

Some important causes of deforestation were associated with the demand for new land for agriculture and cattle ranching (Carvalho et al. C; Bickel and Dros 2003; Fearnside 2005; Baccini et al. 2012; Barona et al. 2010) (Fig. 15.5). In some cases, however, the area of cropland expanded at the expense of pastureland. As a result of international market pressure on curtailing soy produced on recently cleared land, the Brazilian Association of Vegetable Oil Industries (ABIOVE) and National Association of Cereal Exporters (ANEC) proposed the refusal of soy derived from land deforested in Brazilian Amazon after 2006 (known as the

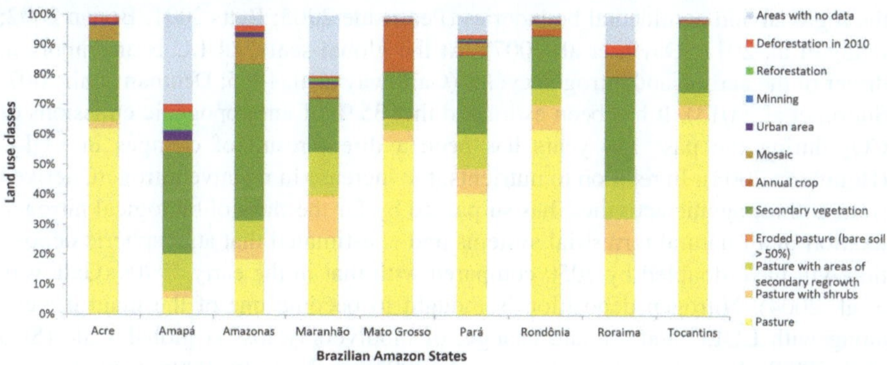

Fig. 15.5 Percentage of thematic land use classes per state in deforested areas of Brazilian Amazon. Data from TerraClass 2010 (INPE 2011)

'soy moratorium'). It has been suggested that the increase in soybean production in south-eastern Amazonia has, to some extent, displaced animal husbandry further to the north, where it subsequently has caused deforestation (Barona et al. 2010). Not all agree that this indirect land use change has been the case (Mueller 2003; Brandao et al. 2005).

Along with a reduction in deforestation rates, there has been a decrease in observed forest degradation recently, as shown by the results published for the years 2007, 2008, 2009 and 2010 by the DEGRAD and PRODES systems. The extent of degraded areas were 15,987 km^2 in 2007, 27,417 km^2 in 2008, 13,301 km^2 in 2009 and 7508 km^2 in 2010 (INPE 2014a, b). The states with higher accumulated deforestation according to INPE (2014a) are Maranhão (42 % considering only the area of the state within the Legal Amazon), Rondônia (37 %), Mato Grosso (23 %) and Pará (21 %) (Table 15.2).

Land use of the deforested land in Brazilian Amazon in 2010 was pasture (45.8 %), pasture with regeneration of woody vegetation (8.2 %), secondary woody vegetation (22.3 %) and cropland (5.4 %) (INPE 2011). The sequence of land use established after deforestation (e.g. forest to pasture, crops to pasture or pasture to secondary vegetation) and the time lag among the land uses are critical for land planning and development strategies. As well, this information is fundamental to deepen the understanding of the deforestation process, its drivers and the policies that can contribute to a sustainable use of the soil and Amazon conservation.

15.4 The Impacts of Land Use Change

The direct and indirect consequences of major changes in land use can affect human societies within Amazonia and beyond. Changes in ecosystem productivity, hydrological regime and climate are some of the impacts of deforestation that go beyond the regional and continental boundaries (Fearnside 2005; Betts 2001; Bonan 2002; Foley et al. 2011; Davin et al. 2007). At the global scale, LULC is an important driver of the carbon and nitrogen cycles (Galloway et al. 1995; Denman et al. 2007; Sutton et al. 2013). It has been estimated that 35 % of anthropogenic emissions of CO_2 during the past 150 years has been a direct result of changes in LULC (Houghton 2003). In relation to nutrients, the increase in reactive nitrogen, derived from anthropogenic activities, has surpassed by far the rates of biological nitrogen fixation in all natural terrestrial systems and is estimated that atmospheric deposition will have doubled by 2050 compared with that in the early 1990s (Galloway et al. 2004). Nitrogen deposition is thought to become one of the main drivers, along with LULC and climate change, of biodiversity loss at global scale (Sala et al. 2000). According to Galloway et al. (2004), in the early 1990s, a small region of south-east South America received inorganic N deposition over 1000 mg N m^2 $year^{-1}$. By 2050, this area is expected to grow significantly, and there will be a large region receiving >2000 mg N m^2 $year^{-1}$.

Deforestation, at the local scale, also causes loss of biodiversity, soil erosion, nutrient depletion and soil compaction. The degradation of soil quality results in low agricultural productivity (Martinelli et al. 2012) and increases the risk of further land clearing for extensive agriculture at the expense of native forest ecosystems (Sutton et al. 2013). Alterations to the hydrological regime have also been observed after deforestation. Conversion of forest can heavily impact the hydrological dynamics by increasing run-off, creating flash floods that can be followed by periods of greatly reduced stream flow. Regular flooding patterns are important for natural freshwater ecosystem functioning, for the riparian ecosystems as well as for floodplain agriculture (Fearnside 2005).

Deforestation reduces the options for sustainable forest management for timber or presently little-valued genetic or pharmacological resources (Fearnside 2005). The Amazon forest 'biome' is rich in biodiversity, for instance, comprising more than 50,000 vascular plant species, of which 30,000 are endemic (Vieira et al. 2008). Habitat fragmentation may directly drive the loss of fauna and flora, unbalancing ecological productivity (Tollefson 2013). Furthermore, biodiversity has an inherent value beyond the market value of diverse forest products (Fearnside 1999). The impact of continued deforestation on biodiversity is much greater in areas with little remaining forest, fragmented landscapes and high levels of endemism. According to Fearnside (2005), if Amazonian deforestation were allowed to continue unbridled, the same levels of risk to biodiversity would apply that had already happened to the Atlantic forest (see, e.g. Tabarelli et al. 2012).

Changes in vegetation canopy height alter the temperature and humidity balance leading to different patterns of precipitation that can feedback negatively to agricultural production (Ometto et al. 2011). Werth and Avissar (2002) found that deforestation effects in the Amazon were strong, with reductions in precipitation, evapotranspiration and cloudiness.

In addition, deforestation promotes the production of GHG from soils previously covered by native forests (Houghton 1999; McGuire et al. 2001; DeFries et al. 2002; Achard et al. 2004; Potter et al. 2008; Ometto et al. 2011). Along with C emissions, other GHGs are emitted from deforestation. Steudler et al. (1996)

showed that forest-to-pasture conversion resulted in a net source of CH_4 from soil of about 10 kg CH_4 ha^{-1} $year^{-1}$. According to Hao and Ward (2012), about 85 % of the total anthropogenic CH_4 emitted originates in the tropics, mainly resulting from agriculture, cattle husbandry, fuel wood use and deforestation. The Amazon is not a major contributor of nitrous oxide (N_2O) to the atmosphere, the exception being forest fires. However, the upland ('terra firme') Amazonian forest soils are estimated to emit about 15 % of global non-anthropogenic emissions (Davidson and Artaxo 2004).

Since early 1980s, Brazil has been one of the top GHG emitters, overall, due to the intense rate of deforestation at the contact zone between Amazon forest and Cerrado 'biomes'. The recent decreasing trend in deforestation rates in Brazilian Amazon has resulted from important commitments that the country has made during international climate change negotiations. The proposed reduction of 80 % in emissions derived from deforestation, in relation to a decadal mean (MMA 2009), is about to be achieved. The implications of these actions point to the necessity of profound changes and an alternative development plan for Brazilian Amazon, including capacity building, education and opportunities for economic activities, in particular at community level. At a more immediate timescale, incentives to aid the regrowth of secondary vegetation in areas illegally deforested could have the potential for the region to become a carbon sink.

15.5 Conclusions

Despite the recent decrease in deforestation rates in Brazilian Amazon, sustained efforts towards better land management are required to maintain efforts to harmonise economic development, social expectations and environmental conservation. Together, deforestation and global warming lead to profound changes in the forest structure with effects not only on the local environment, but beyond, potentially affecting human societies. Furthermore, deforestation causes changes in ecosystem productivity, hydrological and the climate regime. Investments in satellite monitoring together with studies on the impacts of deforestation are critical tools to understand and manage important processes for maintaining critical ecosystem services provided by the ecosystems of Brazilian Amazon.

References

Achard F, Eva HD, Mayaux P et al (2004) Improved estimates of net carbon emissions from land cover change in the tropics for the 1990s. Global Biogeochem Cycles 18, GB2008, doi:2010.1029/2003GB002142

Aguiar APD (2006) Modelagem de mudança do uso da terra na Amazônia: Explorando a heterogeneidade intra-regional. Ph.D. Thesis. National Institute for Space Research São José dos Campos, Brazil

Aguiar APD, Ometto JP, Nobre C, Lapola DM, Almeida C, Vieira IC, Soares JV, Alvala R, Saatchi S, Valeriano D, Castilla-Rubio JC (2012) Modeling the spatial and temporal heterogeneity of deforestation-driven carbon emissions: the INPE-EM framework applied to the Brazilian Amazon. Glob Chang Biol 18(11):3346–3366

Almeida C (2009) Estimativa da área e do tempo de permanência da vegetação secundária na Amazônia legal por meio de imagens Landsat/TM. Doctorate Thesis. National Institute for Space Research São José dos Campos, Brazil

Alves DS (2002) Space-time dynamics of deforestation in Brazilian Amazônia. Int J Remote Sens 23:2903–2908

RAISG. Amazon Geo-referenced Socio-environmental Information Network (2012) Amazonia 2012—protected areas and indigenous territories. www.raisg.socioambiental.org. Cited 25 Jun 2013

Baccini A et al (2012) Estimated carbon dioxide emissions from tropical deforestation improved by carbon-density maps. Nat Clim Change 2:182–185

Barbier EB (2004) Agricultural expansion, resource booms and growth in Latin America: implications for long-run economic development. World Dev 32:137–57

Barona E, Ramankutty N, Hyman G, Coomes OT (2010) The role of pasture and soybean in deforestation of the Brazilian Amazon. Environ Res Lett 5, 024002. doi:10.1088/1748-9326/5/2/024002

Becker BK (1997) Amazônia. Ática, São Paulo

Becker B (2005) Geopolítica da Amazônia. Estudos Avançados—Dossiê Amazônia Brasileira I 19 (53):71–86

Betts RA (2001) Biogeophysical impacts of land use on present-day climate: near-surface temperature change and radiative forcing. Atmos Sci Lett 1. doi: 10.1006/asle.2001.0023

Bickel U, Dros JM (2003) The impacts of soybean cultivation on Brazilian Ecoystems. AIDEnvironment-WWF, Bonn. http://assets.panda.org/downloads/impactsofsoybean.pdf. Cited 25 Jun 2013

Bonan GB (2002) Ecological climatology: concepts and applications. Cambridge University Press, Cambridge

Brandao ASP, Rezende GC, Da Costa Marques RW (2005) Agricultural growth in the period 1999–2004, outburst in soybeans area and environmental impacts in Brazil. IPEA Discussion paper no. 1062. Available via SSRN. http://ssrn.com/abstract=660442. Cited 25 Jun 2013

Brown JC, Jepson W, Price KP (2004) Expansion of mechanized agriculture and land-cover change in Southern Rondônia, Brazil. J Latin Am Geogr 3:96–102

Carvalho GO, Nepstad D, McGrath D, Vera Diaz MC, Barros AC (2002) Frontier expansion in the Amazon. Balancing development and sustainability. Environ Sci Policy Sustain Dev 44:32–42

Chambers JQ, Higuchi N, Tribuzy ES, Trumbore SE (2001) Sink for a century: carbon sequestration in the Amazon. Nature 410(6827):429–429

União Mundial para a Conservação–UICN (1995) Reservas Extrativistas. Murrieta JR, Rueda RF (eds), p 133

Costa WM (1997) O Estado e as políticas territoriais no Brasil. Contexto, São Paulo

Davidson EA, Artaxo P (2004) Globally significant changes in biological processes of the Amazon Basin: results of the Large-scale Biosphere-Atmosphere Experiment. Glob Change Biol 10:519–529

Davin EL, De Noblet-Ducoudre N, Friedlingstein P (2007) Impact of land cover change on surface climate: relevance of the radiative forcing concept. Geophys Res Lett 34. doi: 10.1029/2007GL029678

Instituto Nacional de Pesquisas Espaciais–INPE (2002) PRODES: deforestation in Brazilian Amazon. http://mtc-m18.sid.inpe.br/col/dpi.inpe.br/lise/2002/06.12.13.16/doc/Capa.htm. Cited 25 Jun 2013

Instituto Nacional de Pesquisas Espaciais–INPE (2008) Monitoramento da cobertura florestal da Amazônia por satélites sistemas PRODES, DETER, DEGRAD e Queimadas 2007-2008. Disponível via http://www.obt.inpe.br/prodes/Relatorio_Prodes2008.pdf. Cited 30 Apr 2015

Instituto Nacional de Pesquisas Espaciais–INPE (2013) Amazon deforestation detection in real time (DETER). www.obt.inpe.br/deter. Cited 10 Jul 2013

Instituto Nacional de Pesquisas Espaciais–INPE (2014a) PRODES: assessment of deforestation in Brazilian Amazonia. www.obt.inpe.br/prodes. Cited 28 May 2014

Instituto Nacional de Pesquisas Espaciais–INPE (2014b) DEGRAD: Mapeamento da Degradação Florestal na Amazônia Brasileira. www.obt.inpe.br/degrad. Cited 10 Jun 2014

DeFries RS, Houghton RA, Hansen MC et al (2002) Carbon emissions from tropical deforestation and regrowth based on satellite observations for the 1980s and 1990s. Proc Natl Acad Sci USA 99:14256–14261

Denman KL, Brasseur G, Chidthaisong A, Ciais P, Cox P, Dickinson RE, Haugustaine D, Heinze C, Holland E, Jacob D, Lohmann U, Ramachandran S, da Silva Dias PS, Wofsy SC, Zhang X (2007) Couplings between changes in the climate system and biogeochemistry. In Solomon S, Qin D, Manning M, Chen Z, Marquis M, Averyt KB, Tignor M, Miller HL (eds) Climate change 2007: the physical science basis. Contribution of working group I to the fourth assessment report of the intergovernmental panel on climate change. Cambridge University Press, Cambridge, pp 499–587

European Space Agency ESA (2010) GlobCover land cover map produced for the year 2009. Available at: http://www.esa.int

Escada MIS, Alves DS (2001) Mudanças de Uso e Cobertura do Solo na Amazônia: Impactos Sócio-Ambientais na Ocupação de Regiões de Fronteira Agrícola. Relatório Técnico Parcial. www.dpi.inpe.br/geopro/.../relatorio_amz_dinamica_ocupacao.pdf. Cited 25 Jun 2013

Fearnside PM (1999) Biodiversity as an environmental service in Brazil's Amazonian forests: risks, value and conservation. Environ Conserv 26(4):305–321

Fearnside PM (2005) Deforestation in Brazilian Amazonia: history, rates and consequences. Conserv Biol 19(3):680–688

Fearnside PM, Graça PMLA, Keizer EWH, Maldonado FD, Barbosa RI, Nogueira EM (2009) Modelagem de desmatamento e emissões de gases de efeito estufa na região sob influência da Rodovia Manaus-Porto Velho (BR-319). Revista Brasileira de Meteorologia 24(2):208–233

Foley JA, Ramankutty N, Brauman KA, Cassidy ES, Gerber JS, Johnston M, Mueller ND, O'Connell C, Ray DK, West PC, Balzer C, Bennett EM, Carpenter SR, Hill J, Monfreda C, Polasky S, Rockstrom J, Sheehan J, Siebert S, Tilman D, Zaks DPM (2011) Solutions for a cultivated planet. Nature 478:337–342

Galloway JN, Schlesinger WH, Levy H II, Michaels A, Schnoor JL (1995) Nitrogen fixation: anthropogenic enhancement-environmental response. Global Biogeochem Cycles 9:235–252

Galloway JN, Dentener FJ, Capone DG, Boyer EW, Howarth RW, Seitzinger SP, Asner GP, Cleveland CC, Green PA, Holland EA, Karl DM, Michaels AF, Porter JH, Townsend AR, Vorosmarty CJ (2004) Nitrogen cycles: past, present, and future. Biogeochemistry 70:153–226

Global Land Cover GLC (2003) Global Land Cover 2000 database. European Commission, Joint Research Centre. Available at: http://www.gvm.jrc.it/glc2000

Gómez R, Nagatani T et al. (2009) The future of Amazonia. In Geo Amazonia, 255–276

Hao WM, Ward DE (2012) Methane production from global biomass burning. J Geophys Res Atmos 98(D11):20657–20661

Houghton RA (1999) The annual net flux of carbon to the atmosphere from changes in land use 1850–1990. Tellus 51(B):298–313

Houghton RA (2003) Revised estimates of the annual net flux of carbon to the atmosphere from changes in land use 1850–2000. Tellus 55(B):378–390

Instituto Nacional de Pesquisas Espaciais–INPE/Empresa Brasileira de Pesquisa Agropecuária (EMBRAPA) (2011) TerraClass Project. www.inpe.br/cra/projetos_pesquisas/terraclass2010.php. Cited 25 Jun 2013

Instituto Brasileiro de Geografia e Estatística, IBGE (2006) censo agropecuário: 2006: resultados preliminares. Rio de Janeiro. ISSN: 01036157

Kitamura PC (1994) A Amazônia e o Desenvolvimento Sustentável. Embrapa, pp 30–48, 82–94 e pp 131–158

Lapola DM, Schaldacha R, Alcamo J, Bondeau A, Koch J, Koelking C, Priess JA (2010) Indirect land-use changes can overcome carbon savings from biofuels in Brazil. Proc Natl Acad Sci USA 107(8):3388–3393

Laurance WF, Cochrane MA, Bergen S, Fearnside PM, Delamônica P, Barber C, D'Angelo S, Fernandes T (2001) The future of the Brazilian Amazon. Science 291:438–439

Le Quere C, Raupach MR, Canadell JG et al (2009) Trends in the sources and sinks of carbon dioxide. Nat Geosci 2:831–836

Loarie SR, Asner GP, Field CB (2009) Boosted carbon emissions from Amazon deforestation. Geophys Res Lett 36(14):1–5

Machado LA (1997) Fronteira Agrícola na Amazônia. in: Becker BK, Christofoletti A, Davidovich FR, Geiger PP (eds) Geografia e Meio Ambiente no Brasil, pp 181–217

Madi MAC (2004) Financial liberalization and macroeconomic policy options: Brazil, 1994-2003. Instituto de Economia da UNICAMP, Campinas, p 32. www.eco.unicamp.br/Downloads/Publicacoes/TextosDiscussao/texto117.pdf. Cited 25 Jun 2013

Malhi Y, Roberts JT, Betts RA, Killeen TJ, Li W, Nobre CA (2008) Climate change, deforestation, and the fate of the Amazon. Science 319:169–172

Martinelli LA, Pinto AS, Nardoto GB, Ometto JPHB, Filoso S, Coletta LD, Ravagnani EC (2012) Nitrogen mass balance in the Brazilian Amazon: an update. Braz J Biol 72(3):683–690

McGuire AD, Sitch S, Clein JS et al (2001) Carbon balance of the terrestrial biosphere in the twentieth century: analyses of CO_2, climate and land use effects with four process-based ecosystem models. Global Biogeochem Cycles 15:183–206

MMA 2009 and add Ministério do Meio Ambiente MMA (2009) Plano Nacional sobre Mudança do Clima. Available at http://www.mma.gov.br/clima/politica-nacional-sobre-mudanca-do-clima/planonacional-sobre-mudanca-do-clima

Mueller CC (2003) Expansion and modernization of agriculture in the Cerrado—the case of soybeans in Brazil's Center-West. University of Brasilia, Department of Economics, Brasilia. http://e-groups.unb.br/face/eco/cpe/TD/306nov2003CMueller.pdf. Cited 25 Jun 2013

Öborn I, Magnusson U, Bengtsson J, Vrede K, Fahlbeck E, Jensen ES, Westin C, Jansson T, Hedenus F, Lindholm Schulz H, Stenström M, Jansson B, Rydhmer L (2011) Five scenarios for 2050—conditions for agriculture and land use. Swedish University of Agricultural Sciences, Uppsala. ISBN: 978-91-576-9032-6. www.slu.se/framtidenslantbruk. Cited 25 Jun 2013

Ometto JPHB, Aguiar AP, Martinelli LA (2011) Amazon deforestation in Brazil: effects, drivers and challenges. Carbon Manage 2(5):575–585

Potter C, Klooster S, Genovese V (2008) Carbon emissions from deforestation in the Brazilian Amazon Region. Biogeosciences 6:2369–2381

Sala OE, Chapin FS III, Armesto JJ et al (2000) Global biodiversity scenarios for the year 2100. Science 287:1770–1774

Skole DL, Tucker CJ (1993) Tropical deforestation and habitat fragmentation in the Amazon: satellite data from 1978 to 1988. Science 260:1905–1910

Soares-Filho BS, Nepstad DC, Curran LM, Cerqueira GC, Garcia RA, Azevedo Ramos C, Voll E, Macdonald A, Lefebvre P, Schlesinger P (2006) Modelling conservation in the Amazon basin. Nature 440:520–523

Soares-Filho B et al (2010) Role of Brazilian Amazon protected areas in climate change mitigation. Proc Natl Acad Sci USA 107(24):10821–10826

Spracklen DV, Arnold SR, Taylor CM (2012) Observations of increased tropical rainfall preceded by air passage over forests. Nature 489:282–285. doi:10.1038/nature11390

Steudler PA, Melillo JM, Feigl BJ, Neill C, Piccolo MC, Cerri CC (1996) Consequence of forest-to-pasture conversion on CH_4 fluxes in the Brazilian Amazon Basin. J Geophys Res Atmos 101 (D13):18547–18554

Sutton MA, Howard CM, Bleeker A, Datta A (2013) The global nutrient challenge: from science to public engagement. Environ Dev 6:80–85

Tabarelli M, Peres CA, Melo FPL (2012) The 'few winners and many losers' paradigm revisited: emerging prospects for tropical forest biodiversity. Biol Conserv 155:136–140. doi:10.1016/j.biocon.2012.06.020

Terra-i (2012) Near real-time monitoring of habitat change at continental scales using MODIS-NDVI and TRMM. http://www.terra-i.org. Cited 25 Jun 2013

Tollefson J (2008) Brazil goes to war against logging. Nature 452:134–135

Tollefson J (2013) Forest ecology: splinters of the Amazon. Nature 496:286–289

Valdes C (2006) Brazil's booming agriculture faces obstacles. Amber Waves Economic Research Service/USDA 4:28–35. www.ers.usda.gov/AmberWaves/November06/. Cited 25 Jun 2013

Vieira ICG, Toledo PM, Silva JMC, Higuchi H (2008) Deforestation and threats to the biodiversity of Amazonia. Braz J Biol 68:949–956

Werth D, Avissar R (2002) The local and global effects of Amazon deforestation. J Geophys Res Atmos 107(D20):551–558

Chapter 16
The Impact of Land Use on Carbon Stocks and Fluxes in Brazilian Amazonia: Implications for Policy

Philip Fearnside

16.1 Introduction

Assessing impact of land use on carbon stocks and fluxes depends on quantification of the magnitudes and understanding the processes operating in all three of these sectors. This requires estimates of biomass and carbon stocks, the carbon fluxes from transformations such as deforestation and abandonment to secondary succession and the associated models of land-use change needed to estimate the areas affected. Although significant progress has been made in these three areas in the past few years, much remains to be done. The social and physical changes themselves have been evolving, as have expectations regarding future changes in Brazilian Amazonia, or Brazilian Legal Amazon (BLA) (Fig. 16.1). This chapter reviews progress on emissions estimates for deforestation and logging and for the magnification of these effects by extreme events. Future prospects for monitoring these processes are also reviewed. Finally, the chapter reviews land-use change, its modelling, recent developments and probable future drivers. This includes both the forces driving increased biomass loss and consequent carbon emission and efforts such as Reducing Emissions from Deforestation and Forest Degradation (REDD) that are aimed at restraining forest destruction by using the value that avoiding these emissions has for global efforts to mitigate climate change.

P. Fearnside (✉)
Instituto Nacional de Pesquisas da Amazônia, Manaus, Brazil
e-mail: philip.fearnside@gmail.com

© Springer-Verlag Berlin Heidelberg 2016

L. Nagy et al. (eds.), *Interactions Between Biosphere, Atmosphere and Human Land Use in the Amazon Basin*, Ecological Studies 227,
DOI 10.1007/978-3-662-49902-3_16

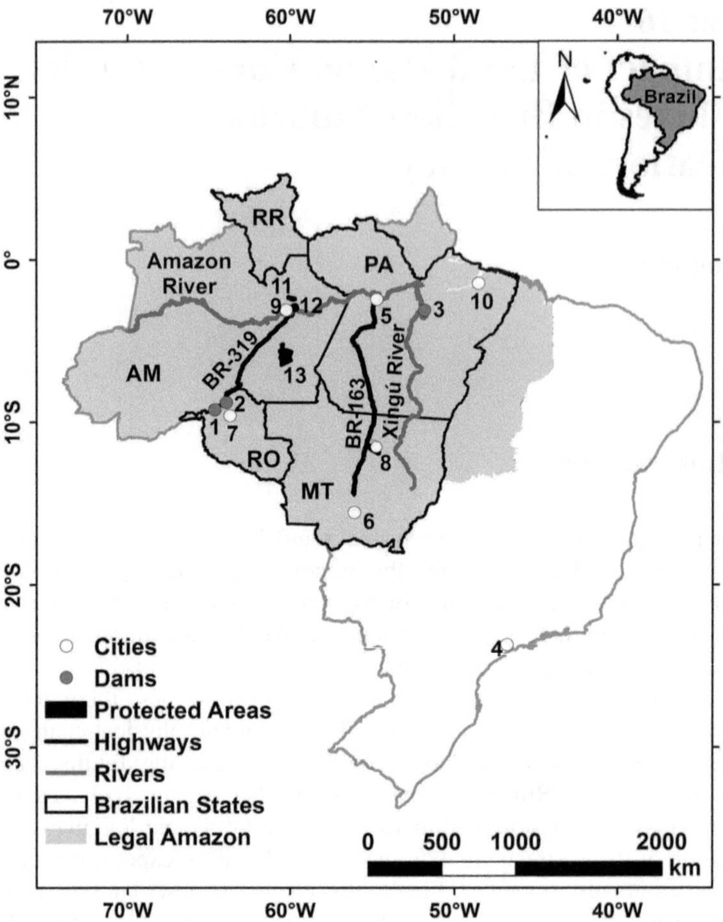

Fig. 16.1 Brazilian Amazonia and locations mentioned in the text. Dams: (1) Jirau, (2) Santo Antônio, (3) Belo Monte; Cities: (4) São Paulo, (5) Santarém, (6) Cuiabá, (7) Porto Velho, (8) Cláudia, (9) Manaus, (10) Belém; Other: (11) PDBFF, (12) Ducke Reserve, (13) RDS Juma

16.2 Emissions Estimates

16.2.1 Deforestation

There have been a number of advances in quantifying emissions from deforestation, but this is still the most uncertain portion of global emissions estimates. Discrepancies are large among recent estimates of emissions (but see Chaps. 5 and 6). Harris et al. (2012) estimated average annual gross emission from tropical deforestation over the 2000–2005 period at 0.81 Pg C (0.57–1.22 Pg C 90 % confidence interval). This contrasts with an estimate of 2.2 Pg C by Baccini et al. (2012) for the 2000–2010 period as well as estimates by Houghton (2003) of 1.93 Pg C for

1980–1989 and 2.2 Pg C for 1990–1999. It also differs from the ‘classic’ value of 1.6 Pg C for the annual net emission from land-use change that persisted through a series of the Intergovernmental Panel on Climate Change (IPCC) reports on the basis of an evolving series of rationales (see Fearnside 2000b).

The study by Harris et al. (2012) has claimed as positive points the fact that the study was limited to gross emissions (i.e. ignoring regrowth in the deforestation emission) and that it omitted soil carbon. The study also omitted the trace-gas emissions that, considering the global warming potentials used to express CO_2 equivalents under the 1997 Kyoto Protocol of the United Nations Framework Convention on Climate Change (UNFCCC), increased the impact of global warming of each ton of carbon emitted by deforestation by c. $15.3\,\% \pm 9.7\,\%$ (depending on the emission factors used) as compared to a ton of carbon emitted by fossil fuels, which emit almost all of their carbon as CO_2 (Fearnside 2000c, pp. 143–145). This author holds the view that all components must be included based on the best available data, even if the estimates have substantial uncertainty. Excluding uncertain components does not increase the utility of the result for assessing the impact of land use on carbon stocks and fluxes—it only makes the result less relevant.

The trace-gas emissions depend on how much of the biomass is oxidized through combustion and how much through decay (e.g. Fearnside 2000c). For the portion oxidized through combustion, the amount emitted through flaming versus smouldering combustion is important, more trace gases being emitted by smouldering. Burn quality is an important factor in determining the biomass exposed to burning that is actually oxidized, the unburned biomass that will be subject to decay or to subsequent burning, and how much is converted to charcoal. Studies include those of Soares Neto et al. (2009), who found 23.9 % burning efficiency in Alta Floresta, Mato Grosso, similar to other results from Amazonian burns. A study by Righi et al. (2009) in a transition forest in Feliz Natal, Mato Grosso, in 2007 (a dry year) found a burning efficiency of 65 %. Higher burning efficiencies in forests with lower biomass in the ‘arc of deforestation’ imply additional trace-gas emissions (Fearnside et al. 2009).

16.2.2 Logging

16.2.2.1 Logging and Emissions

Logging is a major factor of forest disturbance that has received much less research attention than deforestation, in part, because of the greater difficulty of quantification by remote sensing. Merry et al. (2009) have simulated the future advance of logging, and Ahmed and Ewers (2012) have produced maps of remaining timber resources.

The question of how much carbon is emitted by logging is an important one, and it has not been adequately included in global emissions estimates and in national accounts. Brazil's first inventory included an estimate (not included in the inventory's accounting of national emissions) of only 2.4×10^6 Mg C per year (MCT 2004, p. 149), and logging emissions were completely omitted in the second inventory (MCT 2010). By contrast, Asner et al. (2005) estimated annual emissions from logging at 80×10^6 Mg C. An unpublished critique by researchers at the National Institute for Space Research (INPE) of Brazil (Câmara et al. 2005) pointed out problems in interpreting the satellite imagery and suggested that the annual emission will be about half as much, or 40×10^6 Mg C. Both results provide confirmation that emissions are substantial, and they bracket the earlier estimate of 62×10^6 Mg C (Fearnside 2000a). In 1990 (the standard year for initial inventories under the UNFCCC), this represented 15 % of the annual balance of emission from Amazonian deforestation and logging in Brazil (Fearnside 2000a); the percentage today would be higher, since the deforestation rate in 2013 had declined to roughly half the rate in 1990, but logging activity has not declined.

16.2.2.2 Future Prospects for Monitoring Logging

Progress has been disappointing in automatic interpretation of satellite imagery. One algorithm (Souza et al. 2005) tried and failed to win approval by the Brazilian Forest Service (SFB) for use in application on a regional scale. Graça et al. (2005) developed an algorithm that performed well in a limited area near Cláudia, Mato Grosso, but has not yet been translated into an automated form that can be easily applied on a regional scale. The CLASS algorithm (Asner et al. 2005) has a better computational implementation for large-scale 'operational' use. The algorithm needs to be complemented with local expertise to clean the results of areas of rock outcrops, hilltops and some seasonally flooded ecosystems (*várzea*) that were mistakenly identified as logging in the study by Asner et al. (2005) as pointed out by Câmara et al. (2005). A subsequent 'light' adaptation (ClassLite: Asner et al. 2009) is not designed specifically for logging, but rather for detecting biomass loss from any cause, including fire.

A key to identifying logging activity is the correct identification of the logging decks (small clearings where logs are stockpiled for loading on trucks). This has been hindered by the effect of shadows that often impede the algorithms from correctly identifying the decks. An important advance has been made by Maldonado et al. (2009), who developed an algorithm that eliminates the effect of shading, so that the logging decks stand out clearly.

Unfortunately, none of these algorithms has yet advanced to the point where regular estimates of degradation (biomass loss) from logging are available similar to those produced yearly for deforestation. The DETEX programme of INPE is intended to monitor areas of logging, but, so far, the results of this effort are not posted on INPE's website, in contrast to the annual deforestation data from the PRODES programme. It should be noted that logging interests are adamant that logged areas not be characterized as 'degraded', but rather as 'managed' areas. However, considering the definition of degradation as a reduction in biomass and carbon stock, logged areas are greatly degraded (e.g. Keller et al. 2004). They are also damaged in other ways as compared to undisturbed forest. In practice, logging often serves as a prelude to conversion to other uses, despite the discourse regarding sustainable management (e.g. Fearnside 2003).

16.2.3 Extreme Events

16.2.3.1 Extreme Events and Emissions

The impact of land-use change on carbon stocks and flows is aggravated by extreme events, especially droughts. Deforestation creates forest edges, where the microclimate is hotter and drier than in continuous forest, causing increased tree mortality due to water stress (Nascimento and Laurance 2004). Droughts magnify this source of mortality. The edges are also the main entry point for forest fires (Cochrane and Laurance 2002), which are also directly related to rainfall (Vasconcelos et al. 2013a; Nepstad et al. 2004). Amazonian droughts affect biomass both through mortality and through impeding growth (Gatti et al. 2014; Phillips et al. 1998, 2009, 2010). Tree mortality from forest-fire events has been estimated in on-the-ground studies by various authors (Table 16.1).

Almost all Amazonian forest fires are at least partially the result of human activities, including both the ignition sources and, in many cases, the increased flammability of standing forest due to the impact of logging in increasing necromass and in opening the canopy (e.g. Gerwing 2002; Berenguer et al. 2014). These factors make it possible for forest fires to develop whenever a major drought episode occurs, such as the El Niño events of 1997–1998 and 2003 and the Atlantic dipole events of 2005 and 2010 (Lewis et al. 2011; Marengo et al. 2008, 2011, and Chap. 4).

The losses are more severe when logging and fire are combined (e.g. Barlow and Peres 2006). In a study in Paragominas, Pará, Gerwing (2002, p. 136) found, as compared to 'intact' forest, 11.8 % less total above-ground biomass (live + dead) in forest that had been moderately logged, 12.9 % less in forest that had been heavily logged, 23.4 % less in forest that had been logged and lightly burned and 51.1 % less in forest that had been logged and heavily burned. Berenguer et al. (2014, p. 6) found forest that had been logged but not burned to have c. 24.5 % less carbon in above-ground biomass (live + dead), on average, than 'undisturbed' forest in Paragominas, while forest that had been both logged and burned had 48.2 % less. In Santarém, the same study found that forests that had been logged but not burned had above-ground biomass carbon reduced by only 2.2 %, while those that had only been burned had 5.6 % less and those both logged and burned 22.2 % less than 'undisturbed' forest.

The occurrence of fire is related to rainfall and soil water (Alencar et al. 2004; Aragão et al. 2008; Chap. 13; Nepstad et al. 2004; Silvestrini et al. 2011). This indicates the likelihood of increased fires if Amazonia experiences the projected by

Table 16.1 Tree mortality from Amazonian forest fires

Initial forest condition	Year of fire	Measurement (years after fire)	Location	Mortality (% individuals)	Mortality (% biomass)	Reference
"Undisturbed" forest						
		0	Tapajós-Arapiuns Extractive Reserve, Pará	11 %		Peres (1999)
		9–15	Tapajós-Arapiuns Extractive Reserve, Pará	36 %		Haugaasen et al. (2003)
	1997	1	Tapajós-Arapiuns Extractive Reserve, Pará	34 %	28 %	Barlow et al. (2003)
	1997	3	Tapajós-Arapiuns Extractive Reserve, Pará	48 %	51 %	Barlow et al. (2003)
		~36	Tapajós-Arapiuns Extractive Reserve, Pará	42 %		Barlow and Peres (2004)
	1997	9	Tapajós-Arapiuns Extractive Reserve, Pará	No significant change since year 3	No significant change since year 3	Barlow and Peres (2008)
	2001	2	Alta Floresta, Mato Grosso	29.5 %	–	Carvalho Júnior et al. (2010)
	Various	Various	Paragominas, Pará			Gerwing (2002)
	Various	Various	Santarém, Pará	–	~5.6%[a]	Berenguer et al. (2014), p. 6.
	2005	3	Chico Mendes Extractive Reserve, Acre	ns	ns	Barlow et al. (2012)
	2005	1	Embrapa plots, Senador Guiomard and Acrelândia, Acre	17.3 %	5.3 %	Vasconcelos et al. (2013b)
	2005	4	Embrapa plots, Senador Guiomard and Acrelândia, Acre	7.6 %	14.4 %	Vasconcelos et al. (2013b)
	1998	0.3	Roraima	8 %		Barbosa and Fearnside (1999)
	1998	<0.5	Roraima		8.9 %	dos Santos et al. (1998), p. 102

Previously logged						
		1	Tailândia, Pará	38 %		Cochrane and Schulze (1999)
		0.1–1.5	Paragominas, Pará	41 %		Holdsworth and Uhl (1997)
		–	Paragominas, Pará	36–54 %		Kauffman (1991)
	Various	Various	Paragominas, Pará			Gerwing (2002)

[a]Approximate percent above-ground biomass carbon loss (live + dead) from all causes

dry-season rainfall decreases expected to result from continued global warming (Justino et al. 2011; Malhi et al. 2008, 2009; Nepstad et al. 2008; Nobre and Borma 2009). Among the consequences of this would be reduced capacity for REDD to provide benefits, both for climate and for local populations (Aragão and Shimabukuro 2010; Barlow et al. 2012).

Emissions from the major forest fires that occurred during the El Niño of 1997–1998 were estimated in Pará (Alencar et al. 2006) and in Roraima (Barbosa and Fearnside 1999). Potential emissions from the fires in Southwestern Amazonia during the 2005 Atlantic dipole drought were estimated by Vasconcelos et al. (2011, 2013b). All of these studies indicate major emissions.

Because forest fires represent a threat to Amazonian forest, it is important to understand the likely changes in fire frequency and area under climate regimes altered by global warming (Chap. 13). The distribution of fires of different sizes is important in helping to define the levels of atmospheric greenhouse gases that correspond to 'dangerous' interference with the global climate system, as required by Article 2 of the climate convention (UNFCCC 1992). A study by Pueyo et al. (2010) has contributed to this in improving the mathematical characterization of fire-size distributions in Amazonian forest and savanna ecosystems under drought conditions. The study finds evidence of a critical transition to a megafire regime under extreme drought in rainforests.

The effect of increased forest fires under climatic regimes altered by projected global warming has been omitted from most modelled estimates of future emissions from Amazonia (e.g. Cox et al. 2004, 2008). Forest fires would both increase emissions and speed the demise of the forest, as compared to a scenario without fires. The effect of fire is not included in recent models that indicate resistance of Amazon forest to climates with as much as four times the preindustrial atmospheric CO_2 concentration, based on CO_2 fertilization increasing tree growth and reducing water loss from transpiration (Cox et al. 2013; Good et al. 2013; Huntingford et al. 2013). Fires can kill trees irrespective of their ability to survive drought alone (e.g. Fearnside 2013a).

Emissions from forest fires are not included in national accounts under the UNFCCC (IPCC 2006). However, if global warming is to be contained, it is necessary to have estimates of all emissions sources, including those that are wholly or partially the result of natural events. Only emissions that are 'directly human induced' are covered under the Kyoto Protocol (UNFCCC 1997) and are considered to be the responsibility of the country where the emission occurred. The objective of the UNFCCC is to avoid 'dangerous' concentrations of greenhouse gases (UNFCCC 1992, Art. 2). To keep concentrations within this limit, it is necessary to know the total that is being emitted in the world, so that the quotas ('assigned amounts') negotiated for the different countries will be sufficient to limit the total increase, not just the increase that is deliberately emitted by society. If emissions, such as those of anthropogenic forest fires are not counted, then the quotas negotiated may be insufficient to contain global warming.

16.2.3.2 Monitoring Fires

The interpretation of satellite imagery to detect and quantify damage from Amazonian fires has advanced using LANDSAT-TM imagery with 30-m resolution (e.g. Graça et al. 2012; Chap. 13). INPE's DEGRAD programme uses 250-m resolution MODIS imagery to measure fire scars at least 25 ha in area (INPE 2014a). These results are not yet posted on INPE's website, but they are communicated to the Ministry of the Environment.

Canopy damage has been mapped by Morton et al. (2011) for one LANDSAT scene in southern Amazonia using both LANDSAT and MODIS imagery. The algorithm that these authors developed is able to differentiate canopy damage from fires and from logging for areas above 1.5 ha using LANDSAT or 10 ha using MODIS, using a 4-year moving window to characterize the changes in each group of pixels over time. These developments offer hope that the spatial extent of fire damage can be quantified on a regional basis, thereby addressing an important source of uncertainty regarding Amazonian emissions from forest degradation (Chap. 13).

16.3 Land-Use Change

16.3.1 Modelling Land-Use Change

Difficult as it is, the ability to understand and model land-use change is essential if effective governance measures are to be implemented to bring the process under control. The causes of Amazonian deforestation are many, and the relative importance of each varies among locations and over time in any given location (e.g. Fearnside 2005, 2008a). The dynamics of clearing by small farmers depends on a variety of economic and demographic factors (Caldas et al. 2010; Perz and Walker 2002). Roads are important factors for actors of all sizes, but the tens of thousands of kilometres of clandestine or 'endogenous' roads that have been built in the forest are particularly important for entry of small farmers (Brandão Júnior et al. 2007). Roads speed deforestation not only though clearing spreading laterally from the roadside, but also by allowing migration flows to areas far beyond the end of the road in question, as in the case of the proposed reopening of the BR-319 Highway that would reconnect Manaus and Porto Velho (Barni et al. 2015). Land speculation, which is also stimulated by roads, is a significant force in many parts of the region, including among small farmers in settlement areas established by the government (Carrero and Fearnside 2011).

Cattle pasture is still the main replacement for forest in Brazilian Amazonia as a whole (McAlpine et al. 2010). However, soybeans are the major force driving land-use change in much of Mato Grosso and in smaller areas in eastern Rondônia and in the Santarém area of Pará (Fargione et al. 2008; Gibbs et al. 2008; Fearnside 2001;

Morton et al. 2006). The role of China has recently become a dominant factor in the advance of soy (Fearnside et al. 2013; Fearnside and Figueiredo 2015). Even when soy is planted in former cattle pastures rather than in freshly felled forest, it has an indirect effect on deforestation by displacing ranching activity into rainforest areas, as has been shown statistically for movement of this activity from Mato Grosso to Pará (Arima et al. 2011). Note that Brazilian diplomats do not accept this effect and were successful in getting mention of it deleted from the summary for policymakers for the IPCC's Fifth Assessment Report (Garcia 2014).

Quantifying the effect of protected areas on deforestation is particularly important as a guide to policy in this area. The conservation units created and strengthened under the Amazon Region Protected Areas (ARPA) programme have been shown to have a significant effect in slowing deforestation (Soares-Filho et al. 2009, 2010). Differences in effectiveness in resisting deforestation have been found for the various types of reserves, such as those under federal versus state-level control and 'integral protection' versus 'sustainable use' categorization in Brazil's National System of Conservation Units (SNUC) (Vitel et al. 2009). Indigenous areas have consistently been found to be the most resistant to deforestation, and in many areas in the arc of deforestation, indigenous areas represent the only forest that remains standing (Nepstad et al. 2006a).

The SimAmazonia model in the DINAMICA software by Soares-Filho et al. (2006) has been an important tool for visualizing likely trends over the 2000–2050 period. However, for assessing the impact of specific development projects, such as opening a highway or creating a reserve, a number of additional features are needed. In the case of highways, these projects act as positive forces in increasing the total amount of deforestation that takes place, not merely in redistributing the location of a given amount of deforestation activity that has been calculated separately based on macroeconomic indicators such as expected growth in gross domestic product (GDP). Simulations using DINAMICA indicate a substantial impact from currently planned highway projects, such as the reopening of the BR-319 Highway that would connect the arc of deforestation in Rondônia with Manaus in central Amazonia (Fearnside et al. 2009; see also Fearnside and Graça 2006). In the case of reserves, there can be a significant distortion in the expected amount of deforestation in the reserve area. When the baseline deforestation is calculated by multiplying a deforestation rate expressed as a proportion of the forest area in a large subregion (in one case representing about one-third of Brazilian Amazonia), a large total area to be deforested each year results; when this is spatially allocated based on attractive features such as presence of roads and of previous clearings, the large area to be deforested is placed in a single corner of the subregion, producing unrealistically high-clearing rates for this particular location. This is the case for the Juma Sustainable Development Reserve (RDS Juma) in the state of Amazonas, which is the location of the first Reducing Emissions from Deforestation and Forest Degradation (REDD) project (Yanai et al. 2012).

16.3.2 Monitoring Deforestation

There has been a trend to smaller clearings detected by INPE's PRODES programme (using LANDSAT-TM with 30-m resolution), with increasing percentages of the area deforested each year being in clearings with smaller areas. Deforestation estimates with higher-resolution sensors will be needed to capture small clearings (LANDSAT-TM has a detection limit of 6.25 ha for clearings). High-resolution sensors, such as IKONOS or QuickBird, will also be needed to monitor the narrow strips of forest along watercourses that are still considered as areas of environmental protection (APPs) under the Forest Code as revised in 2012. The Ministry of the Environment and INPE have plans for compiling a higher-resolution mosaic that would at least be able to detect the 30-m wide APPs of large properties, but not the 5-m wide APPs of small properties. Satellite monitoring linked to Google Earth, aided by on-the-ground input from civil society via the internet, is expected to increase the speed and accuracy of deforestation monitoring (Tollefson 2009).

Information on land uses in deforested areas has long been a limitation in quantifying net emissions of greenhouse gases. The limitation of deforestation monitoring to just two classes, forest and non-forest, does not allow quantification of the stock and uptake of carbon in the deforested landscape, forcing calculations of emissions to rely on extrapolations from samples in small areas. Especially important are estimates of areas of secondary forest and of degraded pasture. An important improvement is the Terra Class data set (EMBRAPA and INPE 2011; INPE 2014b; Chap. 15).

16.3.3 Recent Developments in the Region

Changes in deforestation rates have a direct relation with carbon emissions and the rate of depletion of carbon stocks. Deforestation rates in Brazilian Amazonia declined markedly from 2004 (when 27.8×10^3 km^2 were cleared) to 2014 (when 4.8×10^3 km^2 were cleared) (INPE 2015). Note, however, that deforestation soared in the 3 months following the July cut-off of the official annual deforestation estimate for 2014 (Fearnside 2015). This was followed by the rainy season in Amazonia, when, although deforestation rates were also much higher than in previous years, the effect is insignificant in terms of the annual total. Whether or not the upturn of deforestation beginning in August 2014 signals a sustained rebound, the basic forces driving deforestation continue to grow on the long term.

The decline in deforestation rates after 2004 was the result of a variety of different forces, with significant differences depending on the location and the year in question. From 2004 to 2008, the slowdown can largely be explained by the decline in international prices of commodities such as soy and beef and a rise in the value of the Brazilian real relative to other currencies, thus decreasing the profitability of exporting commodities that drive Amazonian deforestation (Fearnside 2009a). From 2009 onwards, the trend in deforestation rates diverged from those of commodity indicators, indicating that government regulatory measures were having a significant effect (Assunção et al. 2012; Hargrave and Kis-Katos 2011).

Despite the decreases in deforestation rates since 2004, a number of changes point in the direction of greater future deforestation. These include the continuing increases of the Amazonian population and of investment in the region, the planning and construction of ever more highways, dams and other infrastructure projects and some notable changes weakening environmental protections. One is the revision of the Forest Code (*Codigo Florestal*), as finally passed on 25 September 2012 (Law No. 12.651/12). This greatly reduces (or eliminates) requirements for maintaining forest along watercourses and on steep hillsides and, by effectively pardoning most of the past violations, creates the expectation that deforestation in violation of the present regulations will eventually be pardoned in future 'amnesties' (Fearnside 2010; Metzger et al. 2010; Vieira and Becker 2010; Sparovek et al. 2012). Another serious setback for environmental protection is the weakening of the environmental impact statement and licensing process by the precedents set in 2011 and 2012 in the cases of the Santo Antônio and Jirau dams on the Madeira River and the Belo Monte Dam on the Xingu River (Fearnside 2012a, 2013b, 2014). By allowing infrastructure projects to be approved over the objections of the technical staff of the licensing agencies, and by granting licences without having satisfied the 'conditionalities' that had been established as preconditions, the door is opened to approving any project no matter how great its impacts may be. Other setbacks include a virtually complete halt to creation of new protected areas after 2010 (Alencastro 2014), continued reduction or rescinding of existing protected areas (Bernard et al. 2014) and a 72 % cut in government funds for controlling deforestation in 2015 (Leite 2015).

Of great concern is proposed legislation limiting the authority of the executive branch of the federal government to enforce environmental regulations and to create new indigenous areas and conservation units. Requiring congressional approval would effectively make it impossible to create more protected areas in the foreseeable future. As demonstrated by the recent weakening of the Forest Code, the national congress is currently dominated by the 'ruralist block' (representatives of large land-holders).

16.3.4 Future Forces in Land-Use Change

A variety of forces can be expected to affect future trends in Amazonian land-use change. Brazil's National Plan on Climate Change (PNMC), and the 'voluntary objectives' Brazil announced at the 2009 Conference of the Parties of the climate convention held in Copenhagen, call for a reduction of the annual deforestation rate to 5×10^3 km^2 by 2020 (CIMC 2008). This reduction is substantial as compared to the 19.5×10^3 km^2 annual deforestation rate used as the baseline for the plan, but is much less so as compared to recent rates (5.8×10^3 km^2 in 2013). Nevertheless, achieving this will require significant governance efforts given the likely forces acting to increase deforestation in the coming years. These include the effects of planned reconstruction of key highways (together with the opening of side roads): the Santarém-Cuiabá (BR-163) and the Manaus-Porto Velho (BR-319) Highways (Fearnside 2007; Fearnside and Graça 2006). Roads are generally the key drivers in Amazonian deforestation (e.g. Arima et al. 2008; Perz et al. 2008; Southworth et al. 2011). The effect of planned dams is already being felt (Barreto et al. 2011). Planned waterways for transport of soybeans can be expected to strengthen this deforestation force (Fearnside 2002). The effect of biofuels, including oil palm, may be significant (Fearnside 2009b). Increasing Brazilian exports of beef, along with investments in both deforestation and in pasture intensification, represents another significant trend (McAlpine et al. 2010).

Various possible forces have been suggested as acting to reduce deforestation pressure in the future. These include increasing urbanization (Wright and Muller-Landau 2006); however, various factors make this effect much less than claimed, especially the fact that most of the people moving to cities are not from the major groups of actors in Amazonian deforestation (Fearnside 2008b). Another is the effect of a moratorium on soy purchases from land deforested for this crop (Gibbs et al. 2015a; Nepstad et al. 2014). Certification of cattle ranches and slaughter-houses is also being promoted as a means of decreasing deforestation pressure (Gibbs et al. 2015b; Nepstad et al. 2006b, 2014; Newton et al. 2014). Note, however, that a variety of practices allowing 'leakage' and 'laundering' reduce the effectiveness of these agreements at present (Gibbs et al. 2015b). Similar challenges face timber certification that is promoted both by the government and by NGOs as a means of encouraging sustainable forest management (e.g. Barreto et al. 1998). The net effect of the spread of sustainable forest management is much more complicated than is often portrayed because of economic contradictions and regulatory loopholes that can make the management plans a mere front for obtaining authorization for harvesting and transporting the logs, but with future conversion to deforested land uses as the ultimate result (Fearnside 2003).

One of the most controversial topics regarding future deforestation is the potential role of reducing emissions from deforestation and forest degradation (REDD). Potential benefits include reducing clearing in private properties (Stickler et al. 2009), creating protected areas (Nepstad et al. 2011) and implementing a variety of policy changes for reducing deforestation (Moutinho et al. 2011a, b). Challenges include the proper accounting for leakage (Fearnside 2009c; Yanai et al. 2012) and a series of unresolved controversies ranging from how the carbon accounting is done to how the resulting funds are used (Fearnside 2012b, c). Many have strong opinions on REDD, favouring either throwing it out altogether or working to fix its problems. Strong reasons to solve the very real problems that face REDD include the still significant amounts of carbon emitted annually by Amazonian deforestation, the very large stocks of carbon in the remaining forest at risk of future emission, the lower cost and greater speed of avoiding deforestation emissions as compared to many other mitigation options and the substantial non-carbon environmental benefits and social gains from maintaining Amazonian rainforest.

Acknowledgements I thank Conselho Nacional de Desenvolvimento Científico e Tecnológico (CNPq: Proc. 304020/2010-9; 573810/2008-7), Fundação de Amparo à Pesquisa do Estado do Amazonas (FAPEAM: Proc. 708565) and Instituto Nacional de Pesquisas da Amazônia (INPA: PRJ15.125) for financial support. M.A. dos Santos Jr. prepared Fig. 16.1. I thank L. Nagy and two referees for comments.

References

Ahmed SE, Ewers RM (2012) Spatial pattern of standing timber value across the Brazilian Amazon. PLoS One 7(5), e36099

Alencar AC, Solórzano LA, Nepstad DC (2004) Modeling forest understory fires in an eastern Amazonian landscape. Ecol Appl 14(4):S139–S149

Alencar AC, Nepstad D, Diaz MCV (2006) Forest understory fire in the Brazilian Amazon in ENSO and non-ENSO years: area burned and committed carbon emissions. Earth Interact 10 (6):1–17

Alencastro C (2014) Dilma não criou nenhuma nova unidade de conservação na Amazônia. O Globo. Accessed 4 Aug 2014

Aragão LEOC, Shimabukuro YE (2010) The incidence of fire in Amazonian forests with implications for REDD. Science 328:1275–1278. doi:10.1126/science.1186925

Aragão LEOC, Malhi Y, Barbier N, Lima A, Shimabukuro Y, Anderson L, Saatchi S (2008) Interactions between rainfall, deforestation and fires during recent years in the Brazilian Amazonia. Philos Trans R Soc B 363:1779–1785. doi:10.1098/rstb.2007.0026

Arima EY, Walker RT, Sales M, Souza C Jr, Perz SG (2008) The fragmentation of space in the Amazon basin: emergent road networks. Photogramm Eng Remote Sens 74:699–709

Arima EY, Richards P, Walker R, Caldas MM (2011) Statistical confirmation of indirect land use change in the Brazilian Amazon. Environ Res Lett 6:024010. doi:10.1088/1748-9326/6/2/024010

Asner GP, Knapp DE, Broadbent EN, Oliveira PJC, Keller M, Silva JN (2005) Selective logging in the Brazilian Amazon. Science 310:480–482. doi:10.1126/science.1118051

Asner GP, Knapp DE, Balaji A, Paez-Acosta G (2009) Automated mapping of tropical deforestation and forest degradation: CLASlite. J Appl Remote Sens 3(3):35–43. doi:10.1117/1.3223675

Assunção J, Gandour CC, Rocha R (2012) Deforestation slowdown in the legal Amazon: prices or policies? Climate Policy Initiative (CPI) Working paper. Pontífica Universidade Católica (PUC), Rio de Janeiro, Brazil, 37 pp. http://climatepolicyinitiative.org/publication/deforestation-slowdown-in-the-legal-amazon-prices-or-policie/

Baccini A, Goetz SJ, Walker WS, Laporte NT, Sun M, Sulla-Menashe D, Hackler J, Beck PSA, Dubayah R, Friedl MA, Samanta S, Houghton RA (2012) Estimated carbon dioxide emissions from tropical deforestation improved by carbon-density maps. Nat Clim Chang 2:182–185. doi:10.1038/nclimate1354

Barbosa RI, Fearnside PM (1999) Incêndios na Amazônia brasileira: Estimativa da emissão de gases do efeito estufa pela queima de diferentes ecossistemas de Roraima na passagem do evento "El Niño" (1997/98). Acta Amazonica 29(4): 513–534. English version available at: http://philip.inpa.gov.br/publ_livres/mss%20and%20in%20press/RR-Fire-Acta-engl.pdf

Barlow J, Peres CA (2004) Ecological response to El Niño-induced surface fires in central Amazonia: management implications for flammable tropical forests. Philos Trans R Soc Lond B 359:367–380

Barlow J, Peres CA (2006) Consequences of cryptic and recurring fire disturbances for ecosystem structure and biodiversity in Amazonian forests. In: Laurance WF, Peres CA (eds) Emerging threats to tropical forests. University of Chicago Press, Chicago, IL, pp 225–240, 563 pp

Barlow J, Peres CA (2008) Fire-mediated dieback and compositional cascade in an Amazonian Forest. Philos Trans R Soc B 363:1787–1794. doi:10.1098/rstb.2007.0013

Barlow J, Peres CA, Lagan BO, Haugaasen T (2003) Large tree mortality and the decline of forest biomass following Amazonian wildfires. Ecol Lett 6:6–8

Barlow J, Parry L, Gardner TA, Ferreira J, Aragao LEOC, Carmenta R, Berenguer E, Vieira ICG, Souza C, Cochrane MA (2012) The critical importance of considering fire in REDD+ programs. Biol Conserv 154:1–8

Barni PE, Fearnside PM, Graça PMLA (2015) Simulating deforestation and carbon loss in Amazonia: impacts in Brazil's Roraima state from reconstructing highway BR-319 (Manaus-Porto Velho). Environ Manage 55(2):259–278. doi:10.1007/s00267-014-0408-6

Barreto P, Amaral P, Vidal E, Uhl C (1998) Custos e benefícios de manejo florestal para produção de madeira na Amazônia oriental. Serie Amazonia 10. Instituto do Homem e Ambiente na Amazônia (IMAZON), Belém, Pará, Brazil

Barreto P, Brandão A Jr, Martins H, Silva D, Souza C Jr, Sales M, Feitosa T (2011) Risco de Desmatamento Associado à Hidrelétrica de Belo Monte. Instituto do Homem e Meio Ambiente da Amazônia (IMAZON), Belém, Pará, Brazil, 98 pp. http://www.imazon.org.br/publicacoes/livros/risco-de-desmatamento-associado-a-hidreletrica-de-belo-monte/at_download/file

Berenguer E, Ferreira J, Gardner TA, Aragão LEOC, de Camargo PB, Cerri CE, Durigan M, de Oliveira Junior RC, Vieira ICG, Barlow J (2014) A large-scale field assessment of carbon stocks in human-modified tropical forests. Glob Chang Biol. doi:10.1111/gcb.12627

Bernard E, Penna LAO, Araújo E (2014) Downgrading, downsizing, degazettement, and reclassification of protected areas in Brazil. Conserv Biol 28:939–950. doi:10.1111/cobi.12298

Brandão Júnior AO, Souza CM Jr, Ribeiro JGF, Sales MHR (2007) Desmatamento e estradas não-oficiais da Amazônia. In: Anais do XIII Simpósio Brasileiro de Sensoriamento Remoto (Florianópolis, 2007). Florianópolis, Santa Catarina, Brazil, pp 2357–2364 [online]. http://urlib.net/dpi.inpe.br/sbsr@80/2006/11.15.19.47?languagebutton=pt-BR

Caldas MM, Simmons C, Walker R, Perz S, Aldrich S, Pereira R, Leite F, Arima E (2010) Settlement formation and land cover use change: a case study in the Brazilian Amazon. J Lat Am Geogr 9(1):125–144

Câmara G, Valeriano D, Soares JV, Souza C Jr (2005) Nota técnica sobre o artigo "Selective Logging in the Brazilian Amazon". Instituto Nacional de Pesquisas Espaciais (INPE), São José dos Campos, SP, Brazil, 3 pp. http://lba.cptec.inpe.br/lba/site/?p=noticia&t=0&op=227

Carrero GC, Fearnside PM (2011) Forest clearing dynamics and the expansion of land holdings in Apuí, a deforestation hotspot on Brazil's Transamazon highway. Ecol Soc 16(2):26. [online] http://www.ecologyandsociety.org/vol16/iss2/art26/

Carvalho Júnior JA, Veras CAG, Alvarado EC, Sandberg DV, Leite SJ, Gielow R, Rabelo ERC, Santos JC (2010) Understorey fire propagation and tree mortality on adjacent areas to an Amazonian deforestation fire. Int J Wildland Fire 19:1–5

CIMC (Comitê Interministerial sobre Mudança do Clima) (2008) Plano Nacional sobre Mudança do Clima—PNMC—Brasil. Ministério do Meio Ambiente, Brasília, DF, Brazil, 129 pp. http://www.mma.gov.br/estruturas/imprensa/_arquivos/96_01122008060233.pdf

Cochrane MA, Laurance WF (2002) Fire as a large-scale edge effect in Amazonian forests. J Trop Ecol 18:311–325

Cochrane MA, Schulze MD (1999) Fire as a recurrent event in tropical forests of the eastern Amazon: effects on forest structure, biomass, and species composition. Biotropica 31:2–16

Cox PM, Betts RA, Collins M, Harris PP, Huntingford C, Jones CD (2004) Amazonian forest dieback under climate-carbon cycle projections for the 21st century. Theor Appl Climatol 78:137–156
Cox PM, Harris PP, Huntingford C, Betts RA, Collins M, Jones CD, Jupp TE, Marengo JA, Nobre CA (2008) Increasing risk of Amazonian drought due to decreasing aerosol pollution. Nature 453:212–215. doi:10.1038/nature06960
Cox PM, Pearson D, Booth BB, Friedlingstein P, Huntingford C, Jones CD, Luke CM (2013) Sensitivity of tropical carbon to climate change constrained by carbon dioxide variability. Nature 494:341–344
dos Santos JR, Lacruz MSP, de Araújo LS, Xaud HAM (1998) El proceso de quema de biomasa de bosque tropical y de sabanas en la Amazonia brasilera: experiencias de monitoreo con datos ópticos y de microondas. Serie Geografica 7:97–108
EMBRAPA (Empresa Brasileira de Pesquisa Agropecuária) and INPE (Instituto Nacional de Pesquisas Espaciais) (2011) Terra Class: Levantamento de informações de uso e cobertura da terra na Amazônia: Sumário Executivo. INPE, São José dos Campos, SP, Brazil, 20 pp. http://www.inpe.br%2Fcra%2Fprojetos_pesquisas%2Fsumario_executivo_terraclass_2008.pdf
Fargione J, Hill J, Tilman D, Polasky S, Hawthorne P (2008) Land clearing and the biofuel carbon debt. Science 319:1235–1238
Fearnside PM (2000a) Greenhouse gas emissions from land-use change in Brazil's Amazon region. In: Lal R, Kimble JM, Stewart BA (eds) Global climate change and tropical ecosystems, Advances in Soil Science. CRC Press, Boca Raton, FL, pp 231–249, 438 pp
Fearnside PM (2000b) Effects of land use and forest management on the carbon cycle in the Brazilian Amazon. J Sustain For 12(1–2):79–97
Fearnside PM (2000c) Global warming and tropical land-use change: greenhouse gas emissions from biomass burning, decomposition and soils in forest conversion, shifting cultivation and secondary vegetation. Climatic Chang 46(1–2):115–158. doi:10.1023/A:1005569915357
Fearnside PM (2001) Soybean cultivation as a threat to the environment in Brazil. Environ Conserv 28(1):23–38. doi:10.1017/S0376892901000030
Fearnside PM (2002) Avança Brasil: environmental and social consequences of Brazil's planned infrastructure in Amazonia. Environ Manage 30(6):748–763. doi:10.1007/s00267-002-2788-2
Fearnside PM (2003) Conservation policy in Brazilian Amazonia: understanding the dilemmas. World Dev 31(5):757–779. doi:10.1016/S0305-750X(0)00011-13
Fearnside PM (2005) Deforestation in Brazilian Amazonia: history, rates and consequences. Conserv Biol 19(3):680–688. doi:10.1111/j.1523-1739.2005.00697.x
Fearnside PM (2007) Brazil's Cuiabá-Santarém (BR-163) Highway: The environmental cost of paving a soybean corridor through the Amazon. Environ Manage 39(5):601–614. doi:10.1007/s00267-006-0149-2
Fearnside PM (2008a) The roles and movements of actors in the deforestation of Brazilian Amazonia. Ecol Soc 13(1):23. [online] http://www.ecologyandsociety.org/vol13/iss1/art23/
Fearnside PM (2008b) Will urbanization cause deforested areas to be abandoned in Brazilian Amazonia? Environ Conserv 35(3):197–199. doi:10.1017/S0376892908004906
Fearnside PM (2009a) Brazil's evolving proposal to control deforestation: Amazon still at risk. Environ Conserv 36(3):176–179. doi:10.1017/S0376892909990294
Fearnside PM (2009b) Potential benefits and impacts of biofuel production in the Brazilian Amazon, pp. 29–36. In: Garen EJ, Mateo-Vega J (eds) Biofuels and neotropical forests: trends, implications, and emerging alternatives. Proceedings of the ELTI/PRORENA conference, December 5, 2008, Environmental Leadership & Training Initiative (ELTI)/Proyecto de Reforestación con Especies Nativas (PRORENA), Vol. 3, No. 1. Smithsonian Tropical Research Institute (STRI), Panama City, Panama, 101 pp. http://environment.yale.edu/elti/files/Biocombustibles.pdf

Fearnside PM (2009c) Carbon benefits from Amazonian forest reserves: leakage accounting and the value of time. Mitig Adapt Strateg Glob Chang 14(6):557–567. doi:10.1007/s11027-009-9174-9

Fearnside PM (2010) Código florestal: as perdas invisíveis. Ciência Hoje 46(273):66–67

Fearnside PM (2012a) Belo Monte Dam: a spearhead for Brazil's dam building attack on Amazonia? GWF discussion paper 1210. Global water forum, Canberra, Australia. http://www.globalwaterforum.org/wp-content/uploads/2012/04/Belo-Monte-Dam-A-spearhead-for-Brazils-dam-building-attack-on-Amazonia_-GWF-1210.pdf

Fearnside PM (2012b) The theoretical battlefield: accounting for the climate benefits of maintaining Brazil's Amazon forest. Carbon Manage 3(2):145–148. doi:10.4155/CMT.12.9

Fearnside PM (2012c) Brazil's Amazon Forest in mitigating global warming: unresolved controversies. Clim Pol 12(1):70–81. doi:10.1080/14693062.2011.581571

Fearnside PM (2013a) Vines, CO_2 and Amazon forest dieback. Nature. [online] http://www.nature.com/nature/journal/vaop/ncurrent/full/nature11882.html

Fearnside PM (2013b) Decision-making on Amazon dams: politics trumps uncertainty in the Madeira River sediments controversy. Water Altern 6(2):313–325

Fearnside PM (2014) Brazil's Madeira River dams: a setback for environmental policy in Amazonian development. Water Altern 7(1):156–169, http://www.water-alternatives.org/index.php/alldoc/articles/vol7/v7issue1/244-a7-1-15/file

Fearnside PM (2015) Deforestation soars in the Amazon. Nature 521:423. doi:10.1038/521423b

Fearnside PM, Figueiredo AMR (2015) China's influence on deforestation in Brazilian Amazonia: a growing force in the state of Mato Grosso. BU Global economic governance initiative discussion papers 2015-3. Boston University, Boston, MA, 51 pp. http://www.bu.edu/pardeeschool/files/2014/12/Brazil1.pdf

Fearnside PM, Graça PMLA (2006) BR-319: Brazil's Manaus-Porto Velho highway and the potential impact of linking the arc of deforestation to central Amazonia. Environ Manage 38 (5):705–716. doi:10.1007/s00267-005-0295-y

Fearnside PM, Righi CA, Graça PMLA, Keizer EWH, Cerri CC, Nogueira EM, Barbosa RI (2009) Biomass and greenhouse gas emissions from land-use change in Brazil's Amazonian "arc of deforestation": the states of Mato Grosso and Rondônia. For Ecol Manage 258:1968–1978. doi:10.1016/j.foreco.2009.07.042

Fearnside PM, Figueiredo AMR, Bonjour SCM (2013) Amazonian forest loss and the long reach of China's influence. Environ Dev Sustain 15(2):325–338. doi:10.1007/s10668-012-9412-2

Garcia R (2014) Impacto do clima será mais amplo, porém mais incerto. Folha de São Paulo. 31 Mar 2014, pp C-5

Gatti LV et al (2014) Drought sensitivity of Amazonian carbon balance revealed by atmospheric measurements. Nature 506:76–80. doi:10.1038/nature12957

Gerwing JJ (2002) Degradation of forests through logging and fire in the eastern Brazilian Amazon. For Ecol Manage 157:131–141

Gibbs HK, Johnston M, Foley JA, Holloway T, Monfreda C, Ramankutty N, Zaks D (2008) Carbon payback times for crop-based biofuel expansion in the tropics: the effects of changing yield and technology. Environ Res Lett 3, 034001. doi:10.1088/1748-9326/3/3/034001

Gibbs HK, Munger J, L'Roe J, Barreto P, Pereira R, Christie M, Amaral T, Walker NF (2015a) Did ranchers and slaughterhouses respond to zero-deforestation agreements in the Brazilian Amazon? Conserv Lett. doi:10.1111/conl.12175

Gibbs HK, Rausch L, Munge J, Schelly I, Morton DC, Noojipady P, Soares-Filho B, Barreto P, Micol L, Walker NF (2015b) Brazil's soy moratorium. Science 347:377–378. doi:10.1126/science.aaa0181

Good P, Jones C, Lowe J, Betts R, Gedney N (2013) Comparing tropical forest projections from two generations of Hadley Centre Earth System models, HadGEM2-ES and HadCM3LC. J Climate 26(2):495–511

Graça PMLA, Santos JR, Soares JV, Souza PEU (2005) Desenvolvimento metodológico para detecção e mapeamento de áreas florestais sob exploração madeireira: estudo de caso, região norte do Mato Grosso. In: Simpósio Brasileiro de Sensoriamento Remoto, 12, 2005. Goiânia. Anais do XII Simpósio Brasileiro de Sensoriamento Remoto, Goiânia, INPE, São José dos Campos, São Paulo, Brazil, pp 1555–1562

Graça PMLA, Vitel CSMN, Fearnside PM (2012) Detecção de cicatrizes de incêndios florestais utilizando a técnica de análise por vetor de mudança na Terra Indígena Sete de Setembro—Rondonia. Ambiência 8:511–521. doi:10.5777/ambiencia.2012.04.06

Hargrave J, Kis-Katos K (2011) Economic causes of deforestation in the Brazilian Amazon: a panel data analysis for the 2000s. Discussion paper series n. 17, University of Freiburg, Freiburg, Germany, 30 pp. http://www.vwl.uni-freiburg.de/iwipol/discussion_papers/DP17_Hargrave_Kis-Katos%20-%20Economic%20Causes%20of%20Deforestation%20in%20the%20Brazilian%20Amazon.pdf

Harris NL, Brown S, Hagen SC, Saatchi SS, Petrova S, Salas W, Hansen MC, Potapov PV, Lotsch A (2012) Baseline map of carbon emissions from deforestation in tropical regions. Science 336:1573–1576. doi:10.1126/science.1217962

Haugaasen T, Barlow J, Peres CA (2003) Surface wildfires in central Amazonia: short-term impact on forest structure and carbon loss. For Ecol Manage 179:321–331

Holdsworth AR, Uhl C (1997) Fire in Amazonian selectively logged rain forest and the potential for fire reduction. Ecol Appl 7:713–725

Houghton RA (2003) Revised estimates of the annual net flux of carbon to the atmosphere from changes in land use and land management 1850-2000. Tellus B 55:378–390

Huntingford C et al (2013) Simulated resilience of tropical rainforests to CO_2-induced climate change. Nat Geosci 6:268–273. doi:10.1038/ngeo1741

INPE (Instituto Nacional de Pesquisas Espaciais) (2014a) Sistema DEGRAD. INPE, São José dos Campos, São Paulo, Brazil. http://www.obt.inpe.br/degrad

INPE (Instituto Nacional de Pesquisas Espaciais) (2014b) Terra Class. INPE, São José dos Campos, São Paulo, Brazil. http://www.inpe.br/cra/projetos_pesquisas/terraclass.php

INPE (Instituto Nacional de Pesquisas Espaciais) (2015) Projeto PRODES: Monitoramento da Floresta Amazônica Brasileira por Satélite. INPE, São José dos Campos, São Paulo, Brazil. http://www.obt.inpe.br/prodes/

IPCC (2006) 2006 IPCC Guidelines for National Greenhouse Gas Inventories. In: Eggleston HS, Buendia L, Miwa K, Ngara T, Tanabe K (eds) National Greenhouse Gas Inventories Programme. Intergovernmental Panel on Climate Change (IPCC) and Institute for Global Environmental Strategies (IGES), Kanagawa, Japan

Justino F, de Melo AS, Setzer A, Sismanoglu R, Sediyama GC, Ribeiro GA, Machado JP, Sterl A (2011) Greenhouse gas induced changes in the fire risk in Brazil in ECHAM5/MPI-OM coupled climate model. Clim Change 106(2):285–302. doi:10.1007/s10584-010-9902-x

Kauffman JB (1991) Survival by sprouting following fire in tropical forests of the Eastern Amazon. Biotropica 23(3):219–224

Keller M, Palace M, Asner GP, Pereira R, Silva JNM (2004) Coarse woody debris in undisturbed and logged forests in the eastern Brazilian Amazon. Glob Chang Biol 10:784–795. doi:10.1111/j.1529-8817.2003.00770.x

Leite M (2015) Dilma corta 72% da verba contra desmatamento na Amazônia. Folha de São Paulo. http://www1.folha.uol.com.br/ambiente/2015/03/1610479-dilma-corta-72-da-verba-contra-desmatamento-na-amazonia.shtml. Accessed 31 Mar 2015

Lewis SL, Brando PM, Phillips OL, van der Heijden GMF, Nepstad D (2011) The 2010 Amazon drought. Science 331:554. doi:10.1126/science.1200807

Maldonado FD, Graça PM, dos Santos JR (2009) Aplicação da fração solo por rotação radiométrica para detecção do corte seletivo em dominio da floresta amazonica. In: XIV Simposio Brasileiro de Sensoriamento Remoto, 2009, Natal-RN. Anais. Instituto Nacional de Pesquisas Espaciais (INPE), São José dos Campos, SP, Brazil. pp 2823–2830

Malhi Y, Roberts JT, Betts RA, Killeen TJ, Li W, Nobre CA (2008) Climate change, deforestation, and the fate of the Amazon. Science 319:169–172

Malhi Y, Aragão LEOC, Galbraith D, Huntingford C, Fisher R, Zelazowski P, Sitch S, McSweeney C, Meir P (2009) Exploring the likelihood and mechanism of a climate-change-induced dieback of the Amazon rainforest. Proc Natl Acad Sci USA 106(49):20610–20615. doi:10.1073/pnas.0804619106

Marengo JA, Nobre CA, Tomasella J, Oyama MD, Sampaio de Oliveira G, de Oliveira R, Camargo H, Alves LM, Brown IF (2008) The drought of Amazonia in 2005. J Climate 21:495–516

Marengo JA, Tomasella J, Alves LM, Soares WR, Rodriguez DA (2011) The drought of 2010 in the context of historical droughts in the Amazon region. Geophys Res Lett 38, L12703. doi:10.1029/2011Gl047436

McAlpine CA, Ryan JG, Seabrook L, Thomas S, Dargusch PJ, Syktus JI, Pielke RA Sr, Etter AE, Fearnside PM, Laurance WF (2010) More than CO_2: a broader paradigm for managing climate change and variability to avoid ecosystem collapse. Curr Opin Environ Sustain 2:334–346. doi:10.1016/j.cosust.2010.10.001

MCT (Ministério de Ciência e Tecnologia) (2004) Brazil's initial National communication to the United Nations Framework Convention on Climate Change. MCT, Brasília, DF, Brazil, 271 pp. http://www.mct.gov.br/upd_blob/0005/5142.pdf

MCT (Ministério da Ciência e Tecnologia) (2010) Second National communication of Brazil to the United Nations Framework Convention on Climate Change, vol 2. MCT, Brasília, DF, Brazil, 493 pp. http://www.mct.gov.br/upd_blob/0214/214078.pdf and http://www.mct.gov.br/upd_blob/0214/214079.pdf

Merry F, Soares-Filho B, Nepstad D, Amacher G, Rodrigues H (2009) Balancing conservation and economic sustainability: the future of the Amazon timber Industry. Environ Manage 44:395–407

Metzger J-P, Lewinsohn T, Joly CA, Verdade LM, Rodrigues RR (2010) Brazilian law: full speed in reverse. Science 329:276–277

Morton DC, DeFries RS, Shimabukuro YE, Anderson LO, Arai E, del Bon Espirito-Santo F, Freitas R, Morisette J (2006) Cropland expansion changes deforestation dynamics in the southern Brazilian Amazon. Proc Natl Acad Sci U S A 103(39):14637–14641

Morton DC, DeFries RS, Nagol J, Souza CM, Kasischke ES, Hurtt GC (2011) Mapping canopy damage from understory fires in Amazon forests using annual time series of Landsat and MODIS data. Remote Sens Environ 115(7):1706–1720. doi:10.1016/j.rse.2011.03.002

Moutinho P, Martins OS, Christovam M, Lima A, Nepstad D, Crisostomo AC (2011a) The emerging REDD+ regime of Brazil. Carbon Manage 2:587–602

Moutinho P, Stella O, Lima A, Christovam M, Alencar A, Castro I, Nepstad D (2011b) REDD in Brazil: a focus on the Amazon. Principles, criteria, and institutional structures for a national program for Reducing Emissions from Deforestation and Forest Degradation—REDD. Center for Strategic Studies and Management, Brasília, DF, Brazil, 148 pp. http://ipam.org.br/download/livro/REDD-in-Brazil-A-focus-on-the-Amazon-Principles-criteria-and-institutional-structures-for-a-national-program-for-Reducing-Emissions-from-Deforestation-and-Forest-Degradation-REDD/583

Nascimento HEM, Laurance WF (2004) Biomass dynamics in Amazonian forest fragments. Ecol Appl 14(4) Supplement: S127–S138

Nepstad DC, Lefebvre WP, Silva UL, Tomasella J, Schlesinger P, Solórzano L, Moutinho P, Ray R, Benito JG (2004) Amazon drought and its implications for forest flammability and tree growth: a basin-wide analysis. Glob Chang Biol 10:704–717. doi:10.1111/j.1529-8817.2003.00772.x

Nepstad DC, Schwartzman S, Bamberger B, Santilli M, Ray D, Schlesinger P, Lefebvre P, Alencar A, Prinz E, Fiske G, Rolla A (2006a) Inhibition of Amazon deforestation and fire parks and indigenous lands. Conserv Biol 20(1):65–73

Nepstad DC, Stickler CM, Almeida OT (2006b) Globalization of the Amazon soy and beef industries: opportunities for conservation. Conserv Biol 20(6):1595–1603

Nepstad DC, Stickler C, Soares-Filho BS, Merry F (2008) Interactions among Amazon land use, forests, and climate: prospects for a near-term forest tipping point. Philos Trans R Soc Lond B 363:1737–1746

Nepstad DC, McGrath DG, Soares-Filho B (2011) Systemic conservation, REDD, and the future of the Amazon Basin. Conserv Biol 25(6):1113–1116

Nepstad DC, McGrath DG, Stickler C, Alencar A, Azevedo A, Swette B, Bezerra T, DiGiano M, Shimada J, da Motta RS, Armijo E, Castello L, Brando P, Hansen MC, McGrath-Horn M, Carvalho O, Hess L (2014) Slowing Amazon deforestation through public policy and interventions in beef and soy supply chains. Science 344:1118–1123. doi:10.1126/science.1248525

Newton P, Alves-Pinto HN, Pinto LFG (2014) Certification, forest conservation, and cattle: theories and evidence of change in Brazil. Conserv Lett. doi:10.1111/conl.12116

Nobre CA, Borma LS (2009) Tipping points for the Amazon forest. Curr Opin Environ Sustain 1:28–36

Peres CA (1999) Ground fires as agents of mortality in a central Amazonian forest. J Trop Ecol 15:535–541

Perz SG, Walker RT (2002) Household life cycles and secondary forest cover among small farm colonists in the Amazon. World Dev 30(6):1009–1027

Perz SG, Caldas M, Walker R, Arima E (2008) Road networks and forest fragmentation in the amazon: explanations for local differences with implications for conservation and development. J Lat Am Geogr 7(2):85–104. doi:10.1353/lag.0.0004

Phillips OL, Malhi Y, Higuchi N, Laurance WF, Núñez PV, Vásquez RM, Laurance SG, Ferreira LV, Stern M, Brown S, Grace J (1998) Changes in the carbon balance of tropical forests: evidence from long-term plots. Science 282:439–442

Phillips OL et al (2009) Drought sensitivity of the Amazon rainforest. Science 323:1344–1347

Phillips OL, Lewis SL, López-González G, Aragão LEC, Lloyd J, Malhi Y, Almeida S, Arroyo L, Baker TR, Feldpausch TR, Gloor E, Higuchi N (2010) Drought mortality relationships for tropical forests. New Phytol 187:631–646

Pueyo S, Graça PMLA, Barbosa RI, Cots R, Cardona E, Fearnside PM (2010) Testing for criticality in ecosystem dynamics: the case of Amazonian rainforest and savanna fire. Ecol Lett 13:793–802. doi:10.1111/j.1461-0248.2010.01497.x

Righi CA, Graça PMLA, Cerri CC, Feigl BJ, Fearnside PM (2009) Biomass burning in Brazil's Amazonian "arc of deforestation": burning efficiency and charcoal formation in a fire after mechanized clearing at Feliz Natal, Mato Grosso. For Ecol Manage 258:2535–2546. doi:10.1016/j.foreco.2009.09.010

Silvestrini RA, Soares-Filho BS, Nepstad D, Coe M, Rodrigues H, Assunção R (2011) Simulating fire regimes in the Amazon in response to climate change and deforestation. Ecol Appl 21:1573–1590. doi:10.1890/10-0827.1

Soares Neto T, Carvalho J Jr, Veras C, Alvarado E, Goelo R, Lincoln E, Christian T, Yokelson R, Santos J (2009) Biomass consumption and CO_2, CO and main hydrocarbon gas emissions in an Amazonian forest clearing fire. Atmos Environ 43:438–446

Soares-Filho BS, Nepstad D, Curran LM, Cerqueira GC, Garcia RA, Ramos CA, Voll E, Mcdonald A, Lefebvre P, Schlesinger P (2006) Modeling conservation in the Amazon basin. Nature 440(23):520–523

Soares-Filho BS, Dietzsch L, Moutinho P, Falieri A, Rodrigues H, Pinto E, Maretti CC, Suassuna K, Scaramuzza CAM, de Araújo FV (2009) Reduction of carbon emissions associated with deforestation in Brazil: the role of the Amazon Region Protected Areas Program. Instituto de Pesquisa Ambiental da Amazônia (IPAM), Belém, Pará, Brazil, 32 pp. http://www.ipam.org.br/download/livro/Reduction-of-Carbon-Emissions-Associated-With-Deforestation-in-Brazil-The-Role-of-The-Amazon-Region-Protected-Areas-Program-ARPA-/184

Soares-Filho BS, Moutinho P, Nepstad D, Anderson A, Rodrigues H, Garcia R, Dietzsch L, Merry F, Bowman M, Hissa L, Silvestrini R, Maretti C (2010) Role of Brazilian Amazon protected areas in climate change mitigation. Proc Natl Acad Sci U S A 107(24):10821–10826. doi:10.1073/pnas.0913048107

Southworth J, Marsik M, Qiu Y, Perz S, Cumming G, Stevens F, Rocha K, Duchelle A, Barnes G (2011) Roads as drivers of change: trajectories across the Tri-national frontier in MAP, the Southwestern Amazon. Remote Sens 3:1047–1066

Souza C, Roberts DA, Cochrane MA (2005) Combining spectral and spatial information to map canopy damages from selective logging and forest fires. Remote Sens Environ 98:329–343. doi:10.1016/j.rse.2005.07.013

Sparovek G, Berndes G, Barretto AGOP, Klug ILF (2012) The revision of the Brazilian Forest Act: increased deforestation or a historic step towards balancing agricultural development and nature conservation? Environ Sci Policy 16:65–71. doi:10.1016/j.envsci.2011.10.008

Stickler CM, Nepstad D, Coe MT, McGrath DG, Rodrigues HO, Walker WS, Soares-Filho BS, Davidson EA (2009) The potential ecological costs and cobenefits of REDD: a critical review and case study from the Amazon region. Glob Chang Biol 15(12):2803–2824, http://onlinelibrary.wiley.com/doi/10.1111/j.1365-2486.2009.02109.x/full

Tollefson J (2009) Counting carbon in the Amazon. Nature 461:1048–1052

UNFCCC (United Nations Framework Convention on Climate Change) (1992) United Nations Framework Convention on Climate Change. UNFCCC, Bonn, Germany, http://www.unfccc.de

UNFCCC (United Nations Framework Convention on Climate Change) (1997) Kyoto Protocol to the United Nations Framework Convention on Climate Change. Document FCCC/CP/1997;7/Add1 UNFCCC, Bonn, Germany. http://www.unfccc.de

Vasconcelos SS, Fearnside PM, Graça PMLA, Nogueira EM (2011) Mapeamento das áreas afetadas por incêndios florestais no sul do amazonas e estimativas das emissões potenciais de carbono, pp 8059–8066. In: Epiphanio JCN, Galvão LS (eds) Anais XV Simpósio Brasileiro de Sensoriamento Remoto, Curitiba, Brasil 2011. Instituto Nacional de Pesquisas Espaciais (INPE), São José dos Campos, São Paulo, Brazil, 9268 pp. http://www.dsr.inpe.br/sbsr2011

Vasconcelos SS, Fearnside PM, Graça PMLA, Dias DV, Correia FWS (2013a) Variability of vegetation fires with rain and deforestation in Brazil's state of Amazonas. Remote Sen Environ 136:199–209. doi:10.1016/j.rse.2013.05.005

Vasconcelos SS, Fearnside PM, Graça PLMA, Nogueira EM, de Oliveira LC, Figueiredo EO (2013b) Forest fires in southwestern Brazilian Amazonia: estimates of area and potential carbon emissions. For Ecol Manage 291:199–208. doi:10.1016/j.foreco.2012.11.044

Vieira ICG, Becker BK (2010) A revisão do Código Florestal e o desenvolvimento do país. Ciência Hoje 46(274):64–67

Vitel CSMN, Fearnside PM, Graça PMLA (2009) Análise da inibição do desmatamento pelas áreas protegidas na parte Sudoeste do Arco de desmatamento. In: Epiphanio JCN, Galvão LS (eds) Anais XIV Simpósio Brasileiro de Sensoriamento Remoto, Natal, Brasil 2009. Instituto Nacional de Pesquisas Espaciais (INPE), São José dos Campos, São Paulo, Brazil, pp 6377–6384. http://sbsr.dpi.inpe.br/col/dpi.inpe.br/sbsr@80/2008/11.13.14.42/doc/6377-6384.pdf

Wright SJ, Muller-Landau HC (2006) The future of tropical species. Biotropica 38(3):287–301

Yanai AM, Fearnside PM, Graça PMLA, Nogueira EM (2012) Avoided deforestation in Brazilian Amazonia: simulating the effect of the Juma Sustainable Development Reserve. For Ecol Manage 282:78–91. doi:10.1016/j.foreco.2012.06.029

Chapter 17
An Amazonian Forest and Its Fragments as a Laboratory of Global Change

William F. Laurance, José L.C. Camargo, Philip M. Fearnside, Thomas E. Lovejoy, G. Bruce Williamson, Rita C.G. Mesquita, Christoph F.J. Meyer, Paulo E.D. Bobrowiec, and Susan G.W. Laurance

17.1 Introduction

The Biological Dynamics of Forest Fragments Project (BDFFP) is the world's largest and longest-running experimental study of habitat fragmentation (Box 17.1). Located in central Amazonia near the city of Manaus, the BDFFP has evolved since its inception in 1979 into a major epicentre for long-term research. The BDFFP's research mission has gradually broadened to include not only long-

W.F. Laurance (✉) • S.G.W. Laurance (✉)
Centre for Tropical Environmental and Sustainability Science (TESS) and College of Science and Engineering, James Cook University, Cairns, QLD 4878, Australia
e-mail: bill.laurance@jcu.edu.au; susan.laurance@jcu.edu.au

J.L.C. Camargo
Biological Dynamics of Forest Fragments Project, National Institute for Amazonian Research (INPA) and Smithsonian Tropical Research Institute, Manaus, AM 69067-375, Brazil

P.M. Fearnside
Department of Environmental Dynamics, National Institute for Amazonian Research (INPA), Manaus, AM 69067-375, Brazil

T.E. Lovejoy
Biological Dynamics of Forest Fragments Project, National Institute for Amazonian Research (INPA) and Smithsonian Tropical Research Institute, Manaus, AM 69067-375, Brazil

Department of Environmental Science and Policy, George Mason University, Fairfax, VA 22030, USA

G.B. Williamson
Department of Biological Sciences, Louisiana State University, Baton Rouge, LA 70803, USA

R.C.G. Mesquita
Long-term Ecological Research (PELD), National Institute for Amazonian Research (INPA), Manaus, AM 69067-375, Brazil

© Springer-Verlag Berlin Heidelberg 2016

L. Nagy et al. (eds.), *Interactions Between Biosphere, Atmosphere and Human Land Use in the Amazon Basin*, Ecological Studies 227,
DOI 10.1007/978-3-662-49902-3_17

term studies of forest fragmentation but also important work on global-change phenomena and a variety of basic research topics.

Here we highlight some key contributions of this singular project to the study of land-use change and regional- and global-scale drivers in central Amazonia, at the heart of the world's largest tropical forest.

17.1.1 Amazonia and Global Change

Amazonia stands at the intersection of several key questions for global change, both for study and for action. It is believed to be one of the regions that will be most impacted by projected climate changes (Salazar et al. 2007; Dai 2012; IPCC 2013). It has the potential to contribute significantly to efforts to mitigate climate change during the narrow window of time that we have to avert 'dangerous' global warming (Fearnside 2000, 2012). It is also one of the places where sharply reducing greenhouse gas emissions—by limiting forest loss and degradation—could deliver the greatest global benefits for humankind (Stickler et al. 2009).

The rapid loss and fragmentation of old-growth forests are among the greatest threats to tropical biodiversity (Lovejoy et al. 1986; Sodhi et al. 2004; Laurance and Peres 2006; Gibson et al. 2011). More than half of all remaining tropical forest occurs in the Amazon basin, which is being seriously altered by large-scale agriculture (Fearnside 2001a; Gibbs et al. 2010), industrial logging (Asner et al. 2005), proliferating roads and energy infrastructure (Laurance et al. 2001a; Fearnside 2002, 2007; Killeen 2007), increasing biofuel production (Butler and Laurance 2009) and oil, gas and mining developments (Finer et al. 2008).

The exploitation of Amazonia is driving forest fragmentation on a vast spatial scale. By the early 1990s, the area of Amazonian forest that was fragmented (<100 km^2) or vulnerable to edge effects (<1 km from the edge) was over 150 %

C.F.J. Meyer
Biological Dynamics of Forest Fragments Project, National Institute for Amazonian Research (INPA) and Smithsonian Tropical Research Institute, Manaus, AM 69067-375, Brazil

Center for Environmental Biology, University of Lisbon, Campo Grande C2, 1749-016 Lisbon, Portugal

P.E.D. Bobrowiec
Biological Dynamics of Forest Fragments Project, National Institute for Amazonian Research (INPA) and Smithsonian Tropical Research Institute, Manaus, AM 69067-375, Brazil

Coordination of Biodiversity, National Institute for Amazonian Research (INPA), Manaus, AM 69067-375, Brazil greater than the area that had been deforested (Skole and Tucker 1993). From 1999 to 2002, deforestation and logging in Brazilian Amazonia, respectively, created ~32,000 and ~38,000 km of new forest edge annually (Broadbent et al. 2008). Prevailing land uses in Amazonia, such as cattle ranching and small-scale farming, produce landscapes dominated by small (<400 ha) and irregularly shaped forest fragments (Cochrane and Laurance 2002; Broadbent et al. 2008). Such fragments are highly vulnerable to edge effects, fires and other deleterious consequences of forest fragmentation (Laurance et al. 2002; Barlow et al. 2006; Cochrane and Laurance 2008). While model predictions for future climate in Amazonia vary considerably, it is expected that parts of the region will be hotter and drier under expected global warming (Dai 2012; IPCC 2013). What this warming portends for affected areas of Amazonian forest is a matter of some controversy. Disastrous die-offs projected by the UK Meteorological Office Hadley Centre at atmospheric CO_2 concentrations about twice those in the pre-industrial atmosphere (Cox et al. 2000, 2004) have now been countered by a new model version from the same group indicating the Amazon forest is likely to remain almost entirely intact, even with up to four times the pre-industrial CO_2 concentration (Cox et al. 2013; Good et al. 2013; Huntingford et al. 2013). The main difference is inclusion of CO_2 fertilization effects, making the trees grow faster, resist stress better and close their stomata more frequently such that they use and need less water.

17.1.2 Contributions of the BDFFP to Global-Change Research

The BDFFP, with 37 years of research in fragmented and continuous forest in central Amazonia (Box 17.1, Fig. 17.1), has been contributing to quantifying the interactions of land use and global climate change. BDFFP studies have assessed the vulnerability of the forest to changes in meteorological parameters (Laurance et al. 2009a), including those that are aggravated by fragmentation (Laurance 2004a, b). The long-term monitoring of thousands of individual forest trees, and of populations of various other plant and animal species in the same locations, provides the potential for early detection of global environmental changes.

The BDFFP is a source of invaluable long-term datasets. These include high-quality estimates of Amazon forest biomass and carbon stocks (Phillips et al. 1998; Baker et al. 2004). The project also contributes greatly to knowledge of the diversity of species and their relationships in an Amazon forest ecosystem (Laurance et al. 2010a; Ter Steege et al. 2013). Biodiversity and ecosystem processes represent part of what is lost when the forest is destroyed or degraded, whether by direct human action, by climate change or by the interaction of both together. Understanding these processes is essential for assessing not only the

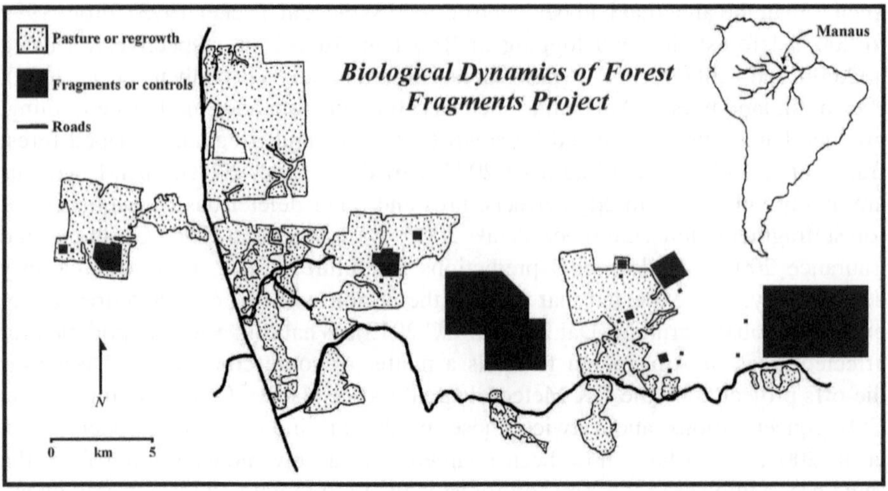

Fig. 17.1 Map of the Biological Dynamics of Forest Fragments Project in central Brazilian Amazonia

vulnerability of forests but also their potential resilience in the face of global change and the rate of recovery following perturbation (Williamson et al. 2014).

The BDFFP has made a substantial contribution to debate over climatic influences on the Amazon via its monitoring of lianas and forest dynamics (Laurance et al. 2014a, b). Lianas evidently make better use of rising CO_2 than do trees (e.g. Condon et al. 1992) and contribute significantly to tree damage and mortality (Ingwell et al. 2010). They also form heavily vine-dominated 'liana forests' in drier parts of Amazonia (Fearnside 2013). BDFFP plots show a marked increase in liana abundance and biomass between censuses in 1997–1999 and 2012 (Laurance et al. 2014a, b). Liana increases have also been found in tropical forests in western Amazonia, Central America, the Guianas and elsewhere, with rising CO_2 levels being one of the more likely explanations (see Laurance et al. 2014a, b and references therein). This negative effect of CO_2 enrichment is not included in the Hadley Centre models and would likely cancel out some of the benefits indicated in a high-CO_2 future.

BDFFP data have helped to identify the direct effects of a warmer, drier climate on the forest. The microclimate at forest edges is significantly hotter and drier than that in the continuous forest (Kapos 1989; Kapos et al. 1993; Camargo and Kapos 1995). Canopy trees are vulnerable to changing microclimates on forest edges during the dry season, with desiccation detected up to 2 km from clearings (Briant et al. 2010). At the BDFFP, edge-associated tree mortality and 'biomass collapse' have been extensively documented (Laurance et al. 1997, 1998a, b, c, 2000; Nascimento and Laurance 2004). Because the entire forest can be expected to face comparable conditions under projected climate change, the dead trees in the BDFFP fragment edges stand as a clear warning of the power of these changes.

Better estimates of how the forest will fare under changed climate are essential for many reasons, including providing the scientific basis needed to convince both world leaders and the general public that containing climate change is worth the cost. But just as basic is the question: what should we do about climate change once the world finally decides to act? The role of tropical forests is critical to this debate, as they contain a large stock of carbon that could either be substantially released by deforestation, logging and fire or conserved for their crucial environmental values. The ways how avoiding these emissions could be incorporated into global mitigation efforts, how carbon benefits would be rewarded and how they should be calculated have been the subject of long-standing controversy, dividing environmental groups, national governments and scientists (see Fearnside 2001b, 2012).

One aspect of this discussion to which the BDFFP makes an important contribution is in reducing the uncertainty surrounding biomass and carbon stock estimates for Amazon forest. The BDFFP tree survey is much more complete than, for example, the 3000 1-ha plots surveyed by the RADAMBRASIL project (Nogueira et al. 2008). The BDFFP has much better species identifications and includes data on other forest components, such as palms, lianas, strangler figs, understory vegetation and dead vegetation (necromass). Correct species identification allows better matching with plant functional traits such as wood density and tree form (e.g. Fearnside 1997; Nogueira et al. 2005, 2007; Chave et al. 2006).

Crucially, the BDFFP forest data allow one to see the variability in biomass from one hectare to another. The mean above-ground biomass of live trees across 69 1-ha plots was 356 ± 47 Mg ha^{-1} (Laurance et al. 1999). This great variability indicates the need for many plots, rather than relying on only a few plots of 1 ha or less scattered around the region as the basis for calibrating satellite imagery for biomass mapping and for estimating greenhouse gas emissions from deforestation (see Chap. 16).

17.2 Long-Term Studies of Forest Fragmentation

The BDFFP's original mission focuses on assessing the effects of forest fragmentation on Amazonian forests and fauna and on important ecological and ecosystem processes. Here we summarize some key conservation lessons that have been gleaned to date.

17.2.1 Sample Effects Are Important in Amazonia

Many species in Amazonian forests are rare or patchily distributed. This phenomenon is especially pronounced in the large expanses of the basin that overlay heavily weathered, nutrient-poor soils (e.g. Radtke et al. 2008), where resources such as fruits, flowers and nectars are scarce and plants are heavily defended against herbivore attack (Laurance 2001). Herein lies a key implication for understanding forest fragmentation: given their rarity, many species may be absent from fragments not because their populations have vanished, but because they were simply not present at the time of fragment creation—a phenomenon termed the 'sample effect' (Wilcox and Murphy 1985). Such sample effects are the hypothesized explanation for the absence of many rare understory bird species from fragments (Ferraz et al. 2007). In addition, many beetles (Didham et al. 1998a), bats (Sampaio et al. 2003), ant-defended plants (Bruna et al. 2005) and trees (Bohlman et al. 2008; Laurance et al. 2010b) at the BDFFP exhibit high levels of habitat specialization or patchiness. In a region where rarity and patchy distributions of species are the norm, sample effects appear to play a major role in structuring fragmented communities. Given these sample effects, nature reserves will have to be especially large to sustain viable populations of rare species (Lovejoy and Oren 1981; Laurance 2005; Peres 2005; Radtke et al. 2008).

17.2.2 Fragment Size Is Vital

Although fragments range from just 1 ha to 100 ha in the BDFFP study area, understanding fragment-area effects has long been a central goal of the project (Lovejoy and Oren 1981; Lovejoy et al. 1984, 1986). The species richness of many organisms declines with fragment area, even with constant sampling effort across all fragments. Such declines are evident in leaf bryophytes (Zartman 2003), tree seedlings (Benítez-Malvido and Martinez-Ramos 2003a), palms (Scariot 1999), understory insectivorous birds (Stratford and Stouffer 1999; Ferraz et al. 2007), gleaning animal-eating bats (Sampaio 2000; Rocha et al. 2013), primates (Gilbert and Setz 2001; Boyle and Smith 2010a) and larger herbivorous mammals (Timo 2003), among others. For these groups, smaller fragments are often unable to support viable populations, and deleterious edge effects—ecological changes associated with the abrupt, artificial edges of forest fragments—can also rise sharply in intensity (Didham et al. 1998a). A few groups, such as ant-defended plants and their ant mutualists, show no significant decline in diversity with fragment area (Bruna et al. 2005).

Fragment size also influences the rate of species losses, with smaller fragments losing species more quickly (Lovejoy et al. 1986; Stouffer et al. 2008). Assuming the surrounding matrix is hostile to bird movements and precludes colonization, Ferraz et al. (2003) estimated that a 1000-fold increase in fragment area would be needed to slow the rate of local species extinctions by tenfold. Even a fragment of 10,000 ha in area would be expected to lose a substantial part of its bird fauna within one century (Ferraz et al. 2003). Similarly, mark–recapture data suggest that very large fragments will be needed to maintain fully intact assemblages of some faunal groups, such as ant-following birds, which forage over large areas of the forest (Van Houtan et al. 2007).

17.3 Edge Effects

17.3.1 Forest Hydrology Is Disrupted

The hydrological regimes of fragmented landscapes differ markedly from those of intact forest (Kapos 1989). Pastures or crops surrounding fragments have much lower rates of evapotranspiration than do forests because they have far lower leaf area and thus less rooting depth. Additionally, such clearings are hotter and drier than forests (Camargo and Kapos 1995). Field observations and heat flux simulations suggest that desiccating conditions can penetrate up to 100–200 m into fragments from adjoining clearings (Malcolm 1998; Didham and Lawton 1999). Further, streams in fragmented landscapes experience greater temporal variation in flows than do those in forests, because clearings surrounding fragments have less evapotranspiration and rainfall interception by vegetation (Trancoso 2008). Free run-off promotes localized flooding in the wet season and stream failure in the dry season, with potentially important impacts on aquatic invertebrates (Nessimian et al. 2008) and other organisms.

Forest fragmentation also can alter low-level atmospheric circulation, which in turn affects local cloudiness and rainfall. The warm, dry air over clearings tends to rise, creating zones of low air pressure. The relatively cool, moist air over the forests is drawn into this vacuum (Avissar and Schmidt 1998). As it warms, it also rises and forms convectional clouds over the clearing, which can lead to localized thunderstorms (Avissar and Liu 1996). In this way, clearings of a few hundred hectares or more can draw moisture away from nearby forests (Laurance 2004a; Cochrane and Laurance 2008). In eastern Amazonia, satellite observations of canopy–water content suggest that such desiccating effects typically penetrate 1.0–2.7 km into fragmented forests (Briant et al. 2010). This moisture-robbing function of clearings, in concert with frequent burning in adjoining pastures, could help explain why fragmented forests are so vulnerable to destructive, edge-related fires (Cochrane and Laurance 2002, 2008).

17.3.2 Edge Effects Often Dominate Fragment Dynamics

Edge effects are among the most important drivers of ecological change in the BDFFP fragments. The distance to which different edge effects penetrate into fragments varies widely, ranging from <10 m to 300 m at the BDFFP (Laurance et al. 2002) and considerably further (at least 2–3 km) in areas of the Amazon where edge-related fires are common (Cochrane and Laurance 2002, 2008; Briant et al. 2010).

Edge phenomena are remarkably diverse (Fig. 17.2). They include increased desiccation stress, wind shear and wind turbulence that sharply elevate rates of tree mortality and damage (Laurance et al. 1997, 1998a). These in turn cause wide-

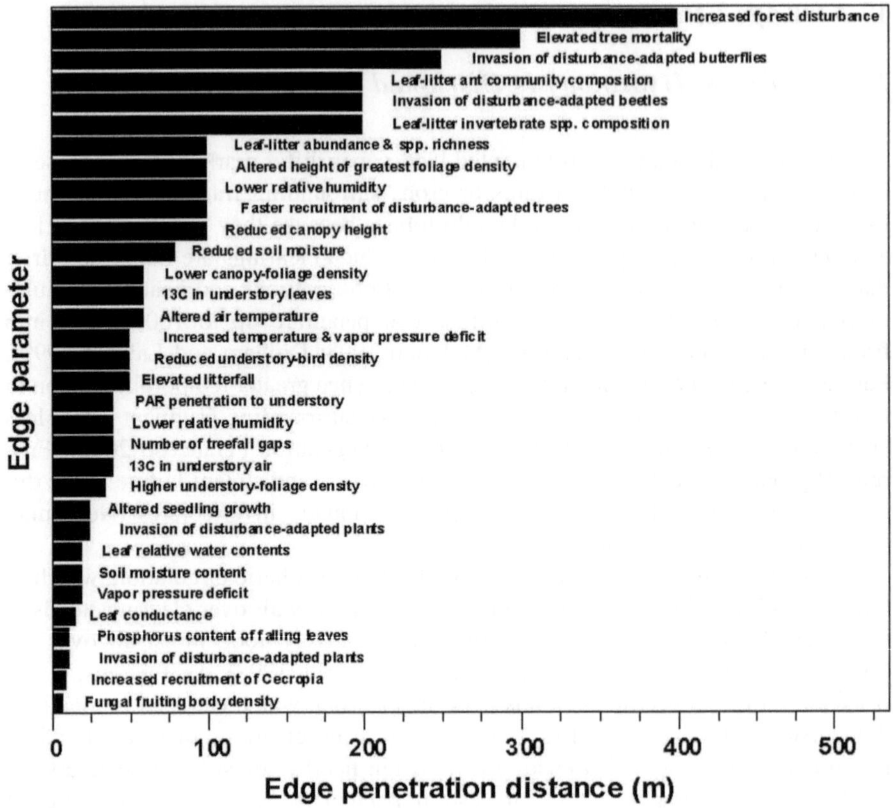

Fig. 17.2 The diversity of edge-effect phenomena studied at the BDFFP and the distance to which each was found to penetrate into fragment interiors (after Laurance et al. 2002)

ranging alterations in the community composition of trees (Laurance et al. 2000, 2006a, b) and lianas (Laurance et al. 2001b). Such stresses may also reduce germination (Bruna 1999) and establishment (Uriarte et al. 2010) of shade-tolerant plant species in fragments, leading to dramatic changes in the composition and abundance of tree seedlings (Benítez-Malvido 1998; Benítez-Malvido and Martinez-Ramos 2003a).

Many animal groups, such as numerous bees, wasps, flies (Fowler et al. 1993), beetles (Didham et al. 1998a, b), ants (Carvalho and Vasconcelos 1999), butterflies (Brown and Hutchings 1997), understory birds (Quintela 1985; Laurance 2004b) and gleaning animal-eating bats (Rocha et al. 2013), decline in abundance near fragment edges. Negative edge effects are apparent even along forest roads (20–30-m width) in large forest tracts. Among understory birds, for example, five of eight foraging guilds declined significantly in abundance within 70 m of roads, whereas tree mortality increased and canopy cover declined (Laurance 2004b).

Some groups of organisms remain stable or even increase in abundance near edges. Leaf bryophytes (Zartman and Nascimento 2006), wandering spiders

(*Ctenus* spp.; Rego et al. 2007; Mestre and Gasnier 2008) and many frogs (Gascon 1993) show no significant response to edges. Organisms that favour forest ecotones or disturbances, such as many species of gap-favouring and frugivorous birds (Laurance 2004b), hummingbirds (Stouffer and Bierregaard 1995a), frugivorous bats that exploit early successional plants (Sampaio 2000), light-loving butterflies (Leidner et al. 2010) and fast-growing lianas (Laurance et al. 2001b), increase in abundance near edges, sometimes dramatically.

17.3.3 Edge Effects Are Cumulative

BDFFP research provides strong support for the idea that two or more nearby edges create more severe edge effects than does just one (Fig. 17.3). This conclusion is supported by studies of edge-related changes in forest microclimate (Kapos 1989; Malcolm 1998), vegetation structure (Malcolm 1994), tree mortality (Laurance et al. 2006a), abundance and species richness of tree seedlings (Benítez-Malvido 1998; Benítez-Malvido and Martinez-Ramos 2003a), liana abundance (Laurance et al. 2001b) and the density and diversity of disturbance-loving pioneer trees (Laurance et al. 2006a, b, 2007). The additive effects of nearby edges could help to explain why small (<10 ha) or irregularly shaped forest remnants are often so severely altered by forest fragmentation (Zartman 2003; Laurance et al. 2006a).

17.3.4 Edge Age, Structure and Adjoining Vegetation Influence Edge Effects

When a forest edge is newly created, it is open to fluxes of wind, heat and light, creating sharp edge–interior gradients in forest microclimate that stress or kill many rainforest trees (Lovejoy et al. 1986; Sizer and Tanner 1999). As the edge ages, however, proliferating vines and lateral branch growth tend to 'seal' the edge, making it less permeable to microclimatic changes (Camargo and Kapos 1995; Didham and Lawton 1999). Tree death from microclimatic stress is likely to decline over the first few years after edge creation (D'Angelo et al. 2004) because the edge becomes less permeable, because many drought-sensitive individuals die immediately and because surviving trees may acclimate to drier, hotter conditions near the edge (Laurance et al. 2006a). Tree mortality from wind turbulence, however, probably increases as the edge ages and becomes more closed because, as suggested by wind tunnel models, downwind turbulence increases when edges are less permeable (Laurance 2004a).

Regrowth forest adjoining fragment edges can also lessen edge-effect intensity. Microclimatic alterations (Didham and Lawton 1999), tree mortality (Mesquita et al. 1999) and edge avoidance by understory birds (Develey and Stouffer 2001;

Fig. 17.3 Forest plots affected by two or more nearby edges (plot centre <100 m from the edge) suffer greater tree mortality (A) and have a higher density (B) and species richness (C) of disturbance-loving pioneer trees than do plots with just one nearby edge. Values shown are the mean ± SD (after Laurance et al. 2006a)

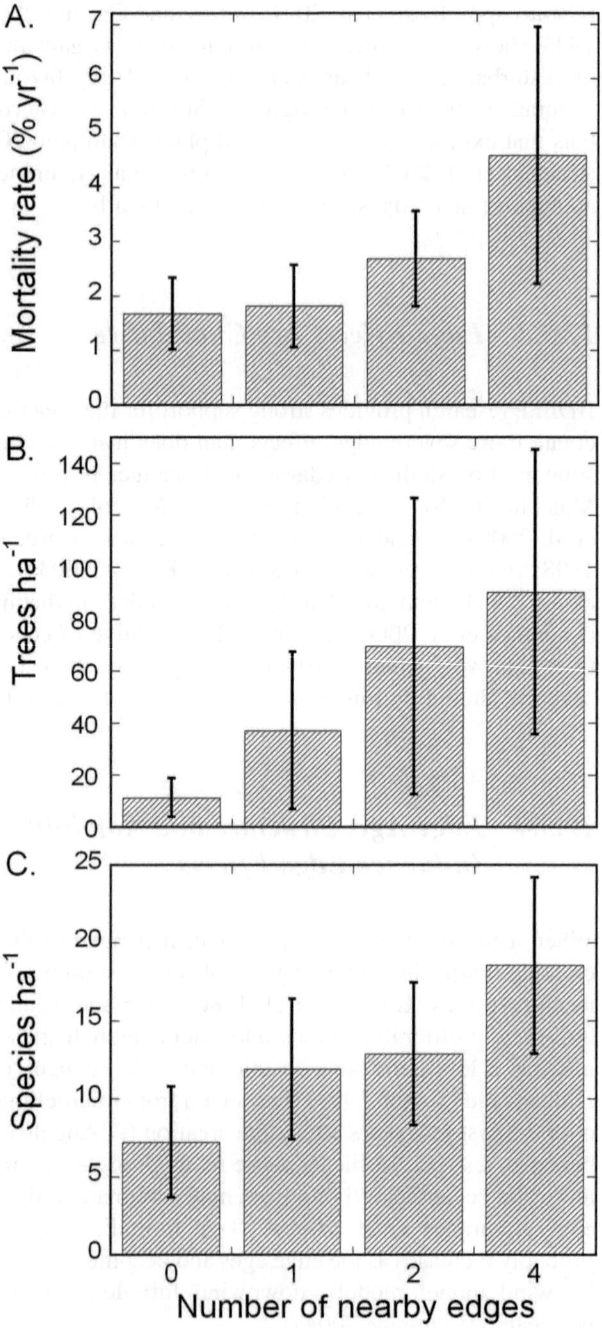

Laurance 2004b; Laurance et al. 2004a, b) and gleaning bats that feed on invertebrates and small vertebrates (Rocha et al. 2013) are all reduced substantially when forest edges are buffered by adjoining regrowth forest, relative to edges adjoined by cattle pastures.

17.4 Isolation and Matrix Effects

17.4.1 Matrix Structure and Composition Affect Fragments

Secondary forests have gradually overtaken most pastures in the BDFFP landscape. This regrowth lessens the effects of fragmentation for some taxa as the matrix becomes less hostile to faunal use and movements. Several species of insectivorous birds that had formerly disappeared have recolonized fragments as the surrounding secondary forest grew back (Stouffer and Bierregaard 1995b). The rate of local extinctions of birds has also declined (Stouffer et al. 2008). Similarly, gleaning animal-eating bats, which occurred at low abundances in fragments (Sampaio 2000) and in secondary regrowth (Bobrowiec and Gribel 2010) 10–15 years ago, have since increased in response to matrix regeneration (Rocha et al. 2013). A number of other species, including certain forest spiders (Mestre and Gasnier 2008), dung beetles (Quintero and Roslin 2005), euglossine bees (Becker et al. 1991) and monkeys such as red howlers, bearded sakis and brown capuchins (Boyle and Smith 2010a), have recolonized some fragments.

The surrounding matrix also has a strong effect on plant communities in fragments by mediating certain edge effects (see above), influencing the movements of pollinators (Dick 2001; Dick et al. 2003) and seed dispersers (Jorge 2008; Bobrowiec and Gribel 2010; Boyle and Smith 2010a) and strongly affecting the seed rain that arrives in fragments. For instance, pioneer trees regenerating in fragments differed strikingly in composition between fragments surrounded by *Cecropia*-dominated regrowth and those encircled by *Vismia*-dominated regrowth (Nascimento et al. 2006). In this way plant and animal communities in fragments could come to mirror to some extent the composition of the surrounding matrix (Laurance et al. 2006a, b), a phenomenon observed elsewhere in the tropics (e.g. Janzen 1983; Diamond et al. 1987).

17.4.2 Matrix Is Affected by History and Forest Proximity

Land-use history is a primary driver of secondary succession in the central Amazon, resulting in the establishment of distinct trajectories differing in structure, composition, biomass and dynamics (Mesquita et al. 1999; Williamson et al. 2014). Intensive use with prescribed fire to maintain pastures compromises the regenerative potential of land which, once abandoned, is colonized by few species and dominated by the genus *Vismia*, resulting in secondary forests that are depauperate in richness and stalled in succession. Where land use has been less intensive, a more diverse vegetation, dominated by the genus *Cecropia*, colonizes, fostering relatively rapid plant succession.

Plant density and species diversity in secondary forests decrease with distance from forest edge and are significantly different between *Vismia*- and *Cecropia*-dominated secondary forests. These differences were initially attributed to differential seed dispersal limitations (Mesquita et al. 2001; Puerta 2002). Wieland et al. (2011), however, showed that the seed rain was similar for both types of second growth and dominated by pioneer species, with only the occasional presence of mature forest species, even very close to forest edges. These results point to other relevant processes affecting plant establishment, such as seed consumption, germination success and seedling herbivory (Wieland et al. 2011; Massoca et al. 2012).

17.4.3 Even Narrow Clearings Are Harmful

Many Amazonian species avoid clearings, and even a forest road can be an insurmountable barrier for some. A number of understory insectivorous birds exhibit depressed abundances near forest roads (20–40-m width) (Laurance 2004b) and strongly inhibited movements across those roads (Laurance et al. 2004a, b). Experimental translocations of resident adult birds reveal such bird species will cross a highway (50–75-m width) but not a small pasture (250-m width) to return to their territory (Laurance and Gomez 2005). Individuals of other vulnerable species, however, have traversed clearings to escape from small fragments to larger forest areas (Harper 1989; Van Houtan et al. 2007). Captures of understory birds declined dramatically in fragments when a 100-m wide swath of regrowth forest was cleared around them, suggesting that species willing to traverse regrowth would not cross clearings (Stouffer et al. 2006).

Aside from birds, clearings of just 100–200-m width can evidently reduce or halt the movements of many forest-dependent organisms (Laurance et al. 2009b), ranging from herbivorous insects (Fáveri et al. 2008), euglossine bees (Powell and Powell 1987) and dung beetles (Klein 1989) to the spores of epiphyllous lichens (Zartman and Nascimento 2006; Zartman and Shaw 2006). Narrow clearings can also provide invasion corridors into forests for exotic and nonforest species (Gascon et al. 1999; Laurance et al. 2009b).

17.5 Landscape Dynamics

17.5.1 Rare Disturbances Can Leave Lasting Legacies

Rare events such as windstorms and droughts have strongly influenced the ecology of fragments. Rates of tree mortality rose abruptly in fragmented (Laurance et al. 2001c) and intact forests (Williamson et al. 2000; Laurance et al. 2009a) in the year after the intense 1997 El Niño drought. Such pulses of tree death help drive changes in the floristic composition and carbon storage of fragments (Laurance et al. 2007). Leaf shedding by drought-stressed trees also increases markedly during droughts, especially within ~60 m of forest edges (Laurance and Williamson 2001). The additional litter increases the susceptibility of fragments to intrusion by surface fires (Cochrane and Laurance 2002, 2008).

Intense windblasts from convectional thunderstorms have occasionally strafed parts of the BDFFP landscape and caused intense forest damage and tree mortality, especially in the fragments. Fragments in the easternmost cattle ranch at the BDFFP have had substantially lower rates of tree mortality than did those in the other two ranches, because the former have so far escaped windstorms (Laurance et al. 2007). These differences have strongly influenced the rate and trajectory of change in tree-community composition in fragments (Laurance et al. 2006b). Hence, by altering forest dynamics, composition, structure and carbon storage, rare disturbances have left an enduring imprint on the ecology of fragmented forests.

17.5.2 Fragments Are Hyperdynamic

The BDFFP fragments experience exceptionally large variability in population and community dynamics, relative to intact forest, despite being largely protected from ancillary human threats such as fires, logging and overhunting. Being a small resource base, a habitat fragment is inherently vulnerable to stochastic effects and external vicissitudes. Species abundances can fluctuate dramatically in small communities, especially when immigration is low and disturbances are frequent (Hubbell 2001). Edge effects, reduced dispersal, external disturbances and changing herbivore or predation pressure can all elevate the dynamics of plant and animal populations in fragments (Laurance 2002, 2008).

Many examples of hyperdynamism have been observed in the BDFFP fragments. Some butterfly species have experienced dramatic population irruptions in response to a proliferation of their favoured host plants along fragment margins (Brown and Hutchings 1997), and butterfly communities in general are hyperdynamic in fragments (Leidner et al. 2010). Bat assemblages also show pronounced species turnover, particularly in 1-ha fragments (Rocha et al. 2013). Streamflows are far more variable in fragmented than forested watersheds (Trancoso 2008). Rates of tree mortality and recruitment are chronically elevated in fragments (Laurance et al. 1998a, b), with major pulses associated with rare disturbances (see above). Further, tree species disappear and turn over far more rapidly in fragments than intact forest, especially within ~100 m of forest margins (Laurance et al. 2006b). These and many other instabilities plague small, dwindling populations in the BDFFP fragments.

17.5.3 Fragments in Different Landscapes Diverge

An important insight is that different fragmented landscapes—even those as alike as the three large cattle ranches in the BDFFP, which have very similar forests, soils, climate, fragment ages and land-use histories—can diverge to a surprising degree in species composition and dynamics. Although spanning just a few dozen kilometres, the three ranches are following unexpectedly different trajectories of change.

At the outset, small initial differences among the ranches multiplied into much bigger differences. Parts of the western and eastern ranches were cleared in 1983, when an early wet season prevented burning of the felled forest. Tall, floristically diverse *Cecropia*-dominated regrowth quickly developed in these areas, whereas areas cleared in the years just before or after became cattle pastures or, eventually, scrubby *Vismia*-dominated regrowth (Williamson and Mesquita 2001). These different successional trajectories manifested, for instance, as distinct differences in bat assemblages, whereby *Cecropia*-dominated regrowth retained a considerable fraction of forest-specialist bat species found in continuous forest compared to *Vismia* regrowth (Bobrowiec and Gribel 2010). As discussed above, the differing matrix vegetation strongly affected the dynamics of plant and animal communities in the nearby fragments. These differences were magnified by subsequent windstorms, which heavily damaged most fragments in the central and western ranches, and yet left fragments in the eastern ranch unscathed. Even identically sized fragments in the three ranches have had remarkably different dynamics and vectors of compositional change (Laurance et al. 2007).

The apparently acute sensitivity of fragments to local landscape and weather dynamics—even within a study area as initially homogeneous as ours—prompted us to propose a 'landscape-divergence hypothesis' (Laurance et al. 2007). We argue that fragments within the same landscape tend to have similar dynamics and trajectories of change in species composition, which will often differ from those in other landscapes. Over time, this process will tend to homogenize fragments in the same landscape and promote ecological divergence among fragments in different landscapes. Evidence for this hypothesis is provided by tree communities in our fragments, which appear to be diverging in composition among the three cattle ranches (Fig. 17.4). Pioneer and weedy trees are increasing in all fragments, but the composition of these generalist plants and their rate of increase differ markedly among the three ranches (Scariot 2001; Laurance et al. 2006a, 2007; Nascimento et al. 2006). This same pattern of landscape homogenization within

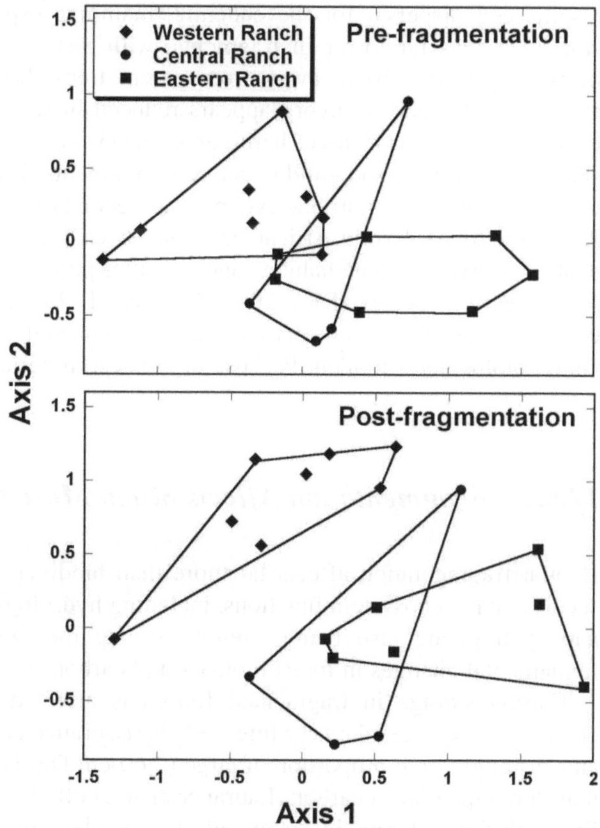

Fig. 17.4 Increasing divergence of tree-community composition in three fragmented Amazonian landscapes. Tree communities in forest-edge plots (<100 m from the nearest edge) are shown before forest fragmentation and 13–18 years after fragmentation, based on a single ordination of all plots and censuses in the study area. The ordination used importance values for all 267 tree genera found in the plots (after Laurance et al. 2007)

ranches can also be seen for bat assemblages in the secondary forest matrix (Bobrowiec and Gribel 2010).

17.6 Broader Consequences of Fragmentation

17.6.1 Ecological Distortions Are Common

Many ecological interactions are altered in fragmented forests. Fragmented communities can pass through unstable transitional states that may not otherwise occur in nature (Terborgh et al. 2001). Moreover, species at higher trophic levels, such as predators and parasites, are often more vulnerable to fragmentation than are herbivores, thereby altering the structure and functioning of food webs (Didham et al. 1998b; Terborgh et al. 2001).

BDFFP findings suggest that even forest fragments that are unhunted, unlogged and unburned have reduced densities of key mammalian seed dispersers. As a result, seed dispersal for the endemic, mammal-dispersed tree *Duckeodendron cestroides* was far lower in fragments, with just ~5 % of the number of seeds being dispersed >10 m away from parent trees than in intact forest (Cramer et al. 2007a). Leaf herbivory appears reduced in fragments, possibly because of lower immigration of insect herbivores (Fáveri et al. 2008). Dung beetles exhibit changes in biomass and guild structure in fragments (Radtke et al. 2008) that could alter rates of forest nutrient cycling and secondary seed dispersal (Klein 1989; Andresen 2003). Exotic Africanized honeybees, a generalist pollinator, are abundant in matrix and edge habitats and can alter pollination distances and gene flow for some tree species (Dick 2001; Dick et al. 2003). A bewildering variety of ecological distortions can pervade fragmented habitats, and a challenge for conservation biologists is to identify those of greatest importance and generality.

17.6.2 Fragmentation Affects Much More than Biodiversity

Habitat fragmentation affects far more than biodiversity and interactions among species; many ecosystem functions, including hydrology (see above) and biochemical cycling, are also being altered. Among the most important of these are fundamental changes in forest biomass and carbon storage.

Carbon storage in fragmented forests is affected by a suite of interrelated changes. Many trees die near forest edges (Laurance et al. 1997, 1998a), including an alarmingly high proportion of large (≥60-cm DBH) canopy and emergent trees that store much forest carbon (Laurance et al. 2000). Fast-growing pioneer trees and lianas that proliferate in fragments are smaller and have lower wood density (Fig. 17.5) and thereby sequester much less carbon, than do the mature-phase trees they replace (Laurance et al. 2001b, 2006a). Based on current rates of forest fragmentation, the edge-related loss of forest carbon storage might produce up to

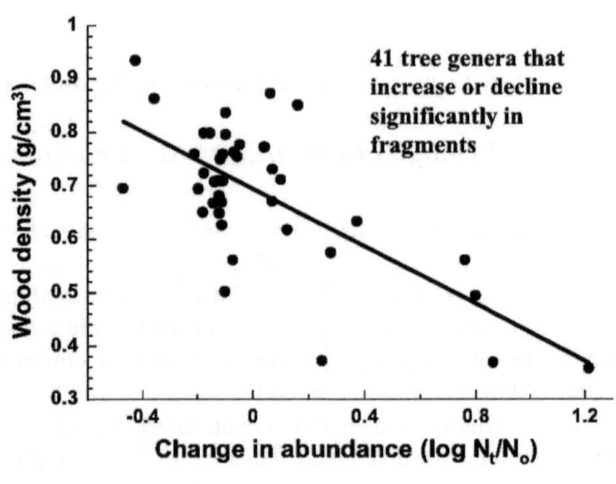

Fig. 17.5 Tree genera that increase in forest fragments (those with positive values on the X axis) tend to have lower wood density than do those that decline in fragments (those with negative values on the X axis) (after Laurance et al. 2006b)

150 million tons of atmospheric carbon emissions annually, above and beyond that from tropical deforestation per se (Laurance et al. 1998c). Such discharge would exceed the yearly carbon emissions of the entire United Kingdom. Note, however, that most of this emission is already counted in the existing estimates of the impact of Amazonian land-use change because the deforestation emission estimates use forest biomass values for undegraded forest (Fearnside 2000). Because most deforestation occurs by expansion of already-existing clearings, forest edges (with reduced biomass) are the first areas to be cleared. Only the annual increase in the total length of forest edges represents an addition. Improved emission estimates, accounting for degradation by logging, fire and fragmentation, are a high priority.

In addition, biomass is being redistributed in fragmented forests. Less biomass is stored in large, densely wooded old-growth trees and more in fast-growing pioneer trees, disturbance-loving lianas, woody debris and leaf litter (Sizer et al. 2000; Nascimento and Laurance 2004; Vasconcelos and Luizão 2004). Finally, carbon cycling accelerates. The large, mature-phase trees that predominate in intact forests can live for many centuries or even millennia (Chambers et al. 1998; Laurance et al. 2004a, b), sequestering carbon for long periods of time. However, the residence time of carbon in early successional trees, vines and necromass (wood debris, litter), which proliferate in fragments, is far shorter (Nascimento and Laurance 2004). Other biochemical cycles, such as those affecting key nutrients like phosphorus (Sizer et al. 2000) and calcium (Vasconcelos and Luizão 2004), may also be altered in fragmented forests, given the striking changes in biomass dynamics, hydrology and thermal regimes they experience.

17.7 Predicting Species Responses to Fragmentation

17.7.1 Species Losses Are Highly Nonrandom

Local extinctions of species in the BDFFP fragments have occurred in a largely predictable sequence, with certain species being consistently more vulnerable than others. Among birds, a number of understory insectivores, including army ant followers, solitary species, terrestrial foragers and obligate mixed-flock members, are most susceptible to fragmentation. Others, including edge/gap species, insectivores that use mixed flocks facultatively, hummingbirds and many frugivores, are far less vulnerable (Antongiovanni and Metzger 2005; Stouffer et al. 2006, 2008). In a similar vein, among bats, gleaning predators are consistently the most vulnerable species, whereas many frugivores respond positively to fragmentation and disturbance (Sampaio 2000; Bobrowiec and Gribel 2010; Rocha et al. 2013). Primates exhibit similarly predictable patterns of species loss, with wide-ranging frugivores, especially the black spider monkey, being the most vulnerable (Boyle and Smith 2010a). Local extinctions in fragments follow a foreseeable pattern, with species assemblages in smaller fragments rapidly forming a nested subset of those in larger fragments (Stouffer et al. 2008). Random demographic and genetic processes may help to drive tiny populations into oblivion, but the species that reach this precarious threshold are far from random.

17.7.2 *Fragmented Communities Are Not Neutral*

An important corollary of nonrandom species loss is that fragmented forests are not neutral. Neutral theory (Hubbell 2001) assumes that species in diverse, space-limited communities, such as tropical trees, are competitively equivalent in order to make predictions about phenomena such as the species–area curves, the relative abundances of species in communities and the rate of species turnover in space. Hubbell (2001) emphasizes the potential relevance of neutral theory for predicting community responses to habitat fragmentation: for isolated communities, locally abundant species should be least extinction prone, with rare species being lost more frequently from random demographic processes. Over time, fragments should become dominated by initially abundant species, with rare species gradually vanishing; other ecological traits of species are considered unimportant.

Gilbert et al. (2006) tested the efficacy of neutral theory for predicting changes in tree communities at the BDFFP. Neutral theory effectively predicted the rate of local extinctions of species from plots in fragmented and intact forest as a function of the local diversity and the mortality rate of trees. However, in most fragments, the observed rate of change in species composition was two to six times faster than predicted by the theory. Moreover, the theory was wildly erroneous in predicting which species are most prone to local extinction. Rather than becoming increasingly dominated by initially common species, fragments in the BDFFP landscape have experienced striking increases in disturbance-loving pioneer species (Fig. 17.6) (Laurance et al. 2006a), which were initially rare when the fragments were created. As a model for predicting community responses to habitat

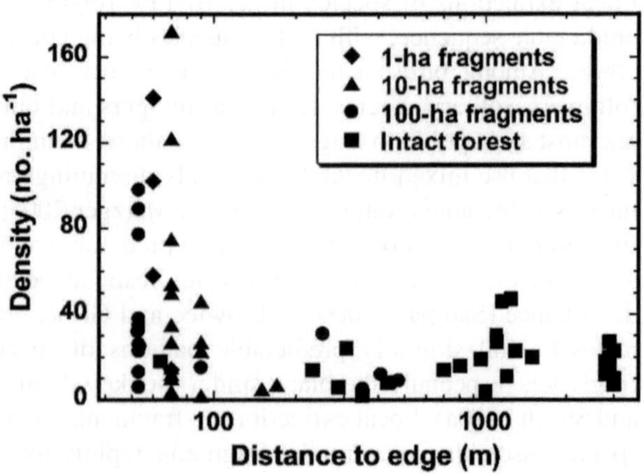

Fig. 17.6 Striking increases in the density of 52 species in nine pioneer or early successional genera (*Annona, Bellucia, Cecropia, Croton, Goupia, Jacaranda, Miconia, Pourouma, Vismia*) near forest edges in the BDFFP study area (after Laurance et al. 2006a)

fragmentation, neutral theory clearly failed, demonstrating that ecological differences among species strongly influence their responses to fragmentation.

17.7.3 Matrix Use and Area Needs Determine Animal Vulnerability

The responses of animal species to fragmentation appear largely governed by two key sets of traits. The first is their spatial requirements for forest habitat. In birds (Van Houtan et al. 2007) and mammals (Timo 2003), wide-ranging forest species are more vulnerable than are those with localized ranges and movements. Species with limited spatial needs, such as many small mammals (Malcolm 1997), hummingbirds (Stouffer et al. 2008), frogs (Tocher et al. 1997) and ants (Carvalho and Vasconcelos 1999), are generally less susceptible to fragmentation.

The second key trait for fauna is their tolerance of matrix habitats (Gascon et al. 1999), which comprises cattle pastures and regrowth forest in the BDFFP landscape. Populations of species that avoid the matrix will be entirely isolated in fragments and therefore vulnerable to local extinction, whereas those that tolerate or exploit the matrix often persist (Laurance 1991; Malcolm 1997; Antongiovanni and Metzger 2005; Ferraz et al. 2007; Bobrowiec and Gribel 2010). At least among terrestrial vertebrates, matrix use is positively associated with tolerance of edge habitats (Laurance 2004b; Farneda 2013), an ability to traverse small clearings (Laurance et al. 2004a, b; Laurance and Gomez 2005) and behavioural flexibility (Neckel-Oliveira and Gascon 2006; Stouffer et al. 2006; Van Houtan et al. 2006; Boyle and Smith 2010b). Within particular animal groups, such as beetles or small mammals, traits such as body size and natural abundance are poor or inconsistent predictors of vulnerability (Laurance 1991; Didham et al. 1998a; Jorge 2008; Boyle and Smith 2010a). Natural abundance, however, is an important predictor of sensitivity to fragmentation for bats at the BDFFP (Farneda 2013).

17.7.4 Disturbance Tolerance and Mutualisms Affect Plant Vulnerability

Among plants, a different suite of factors is associated with vulnerability to fragmentation. Because fragments suffer chronically elevated tree mortality, faster-growing pioneer trees and lianas that readily colonize treefall gaps are favoured at the expense of slower-growing mature-phase trees (Laurance et al. 2006a, b). Pioneer species often flourish in the matrix and produce abundant small fruits that are carried into fragments by frugivorous birds and bats that move between the matrix and nearby fragments (Sampaio 2000; Nascimento et al. 2006). Especially vulnerable in fragments are the diverse assemblages of smaller subcanopy trees that are physiologically specialized for growth and reproduction in dark, humid, forest-interior conditions (Laurance et al. 2006b). Tree species that have obligate outbreeding systems, rely on animal seed dispersers or have relatively large, mammal-dispersed seeds also appear vulnerable (Laurance et al. 2006b; Cramer et al. 2007b).

These combinations of traits suggest that plant communities in fragmented forests are structured primarily by chronic disturbances and microclimatic stresses and possibly also by alterations in animal pollinator and seed-disperser communities. For long-lived plants such as *Heliconia* species and many mature-phase trees, demographic models suggest that factors that reduce adult survival and growth—such as recurring wind disturbance and edge-related microclimatic stresses—exert a strong influence on population growth (Bruna 2003; Bruna and Oli 2005).

Differential tolerance to drought also seems to play a role on secondary forests. We find higher and significant mortality and lower biomass accumulation rates in *Cecropia*-dominated secondary forests, associated with drier years, whereas *Vismia*-dominated regrowth showed a non-significant, but similar, trend. It is likely that different species assemblages account for the differential ability of these successional pathways to tolerate extreme climate events (Mesquita et al. 2012).

17.8 Broad Perspectives

17.8.1 Long-Term Research Is Crucial

Many insights from the BDFFP would have been impossible in a shorter-term study. The exceptional vulnerability of large trees to fragmentation (Laurance et al. 2000) only became apparent after two decades of fragment isolation. Likewise, the importance of ephemeral events such as El Niño droughts (Williamson et al. 2000; Laurance et al. 2001c) and major windstorms (Laurance et al. 2007) would not have been captured in a less-enduring project. Many other key phenomena, such as the kinetics of species loss in fragments (Ferraz et al. 2003), the strong effects of matrix dynamics on fragmented bird and bat assemblages (Antongiovanni and Metzger 2005; Stouffer et al. 2006; Rocha et al. 2013), the divergence of fragments in different landscapes (Laurance et al. 2007) and the effects of fragmentation on rare or long-lived species (Benítez-Malvido and Martinez-Ramos 2003b; Ferraz et al. 2007), are only becoming understood after decades of effort.

Far more remains to be learned. For example, forest simulation models parameterized with BDFFP data suggest that even small ($<$10 ha) fragments will require a century or more to stabilize in floristic composition and carbon storage (Groeneveld et al. 2009), given the long-lived nature of many tropical trees. Eventually, these fragments might experience a fundamental reorganization of their plant communities, given striking shifts in the composition of their tree, palm, liana and herb seedlings (Scariot 2001; Benítez-Malvido and Martinez-Ramos 2003a; Brum et al. 2008). If these newly recruited plants represent the future of the forest, then the BDFFP fragments will eventually experience dramatic changes in floristic composition—comparable to those observed in some other long-fragmented ecosystems (e.g. da Silva and Tabarelli 2000; Girão et al. 2007; Santos et al. 2010).

17.8.2 The BDFFP Is a Best-Case Scenario

Although forest fragments in the BDFFP are experiencing a wide array of ecological changes, it is important to emphasize that it is a controlled experiment. The fragments are square, not irregular, in shape. They are isolated by distances of just 80–650 m from large tracts of surrounding mature forest. They are embedded in a relatively benign matrix increasingly dominated by regrowth forest. And they lack many of the ancillary threats, such as selective logging, wildfires and overhunting, that plague many fragmented landscapes and wildlife elsewhere in the tropics (e.g. Moura et al. 2014). Such threats can interact additively or synergistically with fragmentation, creating even greater perils for the rainforest biota (Laurance and Cochrane 2001; Michalski and Peres 2005; Brook et al. 2008). For these reasons, results from the BDFFP are clearly optimistic relative to many human-dominated landscapes elsewhere in the tropics.

17.9 Conservation Lessons from the BDFFP

17.9.1 Amazonian Reserves Should Be Large and Numerous

A key conclusion from BDFFP research is that nature reserves in Amazonia should ideally be very large—on the order of thousands to tens of thousands of square kilometres (Laurance 2005; Peres 2005). Only at this size will they be likely to maintain natural ecological processes and sustain viable populations of the many rare and patchily distributed species in the region (Ferraz et al. 2007; Radtke et al. 2008); provide resilience from rare calamities such as droughts and intense storms (Laurance et al. 2007); facilitate persistence of terrestrial and aquatic animals that migrate seasonally (Bührnheim and Fernandes 2003); buffer the reserve from large-scale edge effects including fires, forest desiccation and human encroachment (Cochrane and Laurance 2002; Briant et al. 2010); maximize forest carbon storage (Laurance et al. 1997, 1998c); and provide resilience from future climatic and atmospheric changes—the effects of which are difficult to predict for Amazonia (Laurance and Useche 2009). Further, on the ancient soils of central and eastern Amazonia, low plant productivity translates into low population densities of many animals up the food chain, so reserves must be proportionately larger to harbour viable populations (Radtke et al. 2008; Deichmann et al. 2011, 2013).

Nature reserves in Amazonia should also be numerous and stratified across major river basins and climatic and edaphic gradients, in order to preserve locally endemic species (Bierregaard et al. 2001; Laurance 2007). Further, the core areas of reserves should ideally be free of roads, which can promote human encroachment and hunting, internally fragment wildlife populations, and facilitate invasions of exotic species and fire (Laurance et al. 2009b).

17.9.2 Protect and Reconnect Fragments

Few landscapes are as intact as those in the Amazon. Around the world, biodiversity hotspots, which sustain the majority of species at risk of extinction, have, by definition, lost over 70 % of their natural vegetation, and what remains is typically in small fragments (Myers et al. 2000). The BDFFP makes recommendations here, too. Reconnecting isolated fragments by forest restoration will be an effective way of creating areas large enough to slow the rate of local species extinctions (Lima and Gascon 1999; Pimm and Jenkins 2005).

In such heavily fragmented landscapes, protecting remaining forest remnants is highly desirable, as they are likely to be key sources of plant propagules and animal seed dispersers and pollinators (Mesquita et al. 2001; Chazdon et al. 2008). They may also act as stepping stones for animal movements (Laurance and Bierregaard 1997; Dick et al. 2003). In regions where forest loss is severe, forest fragments could also sustain the last surviving populations of locally endemic species, thereby underscoring their potential value for nature conservation (Arroyo-Rodríguez et al. 2009).

17.9.3 Fragmented Landscapes Can Recover

A further lesson is that fragmented landscapes, if protected from fires and other major disturbances, can begin to recover in just a decade or two. Forest edges tend to ‘seal’ themselves, reducing the intensity of deleterious edge effects (Camargo and Kapos 1995; Didham and Lawton 1999; Mesquita et al. 1999). Secondary forests can develop quickly in the surrounding matrix (Mesquita et al. 2001), especially if soils and seedbanks are not depleted by overgrazing or repeated burning (Ribeiro et al. 2009; Norden et al. 2011). Secondary forests facilitate movements of many animal species (Gascon et al. 1999), allowing them to recolonize fragments from which they had formerly disappeared (Becker et al. 1991; Quintero and Roslin 2005; Stouffer et al. 2008; Bobrowiec and Gribel 2010; Boyle and Smith 2010a; Rocha et al. 2013). Species clinging to survival in fragments can also be rescued from local extinction via the genetic and demographic contributions of immigrants (Zartman and Nascimento 2006; Stouffer et al. 2008).

17.10 The Future of the BDFFP

The BDFFP is one of the most enduring and influential ecological research projects in existence today (Gardner et al. 2009; Peres et al. 2010). From the prism of understanding habitat fragmentation, there are vital justifications for continuing it. The project, moreover, is engaged in far more than fragmentation research: it plays a leading role in training Amazonian scientists and decision-makers and sustains long-term research on global-change phenomena, forest regeneration and basic ecological studies.

In its 37-year history, the BDFFP has faced myriad challenges. These include, among others, the continuing fluctuations in currencies, challenges in obtaining research visas for foreign students and scientists, inadequate core funding from its US and Brazilian sponsors and the vagaries of finding soft money for long-term research and to sustain a minimal number of workers to support infrastructure and logistics. Yet today the BDFFP faces a far more direct threat: encroachment from colonists and hunters. Since the late 1990s, the paving of the 1100-km-long Manaus–Venezuela (BR-174) highway has greatly accelerated forest colonization and logging north of the city. SUFRAMA, a Brazilian federal agency that controls an expanse of land north of Manaus that includes the BDFFP, has begun settling families in farming plots around the immediate periphery of the study area. At least six colonization projects involving 180 families are planned for the area (Laurance and Luizão 2007). These settlements could be the beginning of a dramatic influx into the area, especially if the proposed BR-319 highway between Manaus and Rondônia, a major deforestation hotspot in southern Amazonia, is completed as planned (Fearnside and Graça 2006).

To date, BDFFP staff and supporters have managed to stave off most of the colonization projects—which also threaten to bisect the Central Amazonian Conservation Corridor, a budding network of protected and indigenous lands that is one of the most important conservation areas in the entire Amazon basin (Laurance and Luizão 2007). Yet it is an uphill battle against a government bureaucracy that appears myopically determined to push ahead with colonization at any cost—despite the fact that colonists can barely eke out a living on the region's infamously poor soils (Fearnside and Leal Filho 2001). That such a globally important research project and conservation area could be lost seems unthinkable. That it could be lost for such a limited gain seems tragic.

Amazon forest is under stress from a variety of global changes that are expected to increase in the coming decades. Beyond the considerable contributions of the BDFFP to date in providing information relevant to understanding these changes, the project is uniquely well placed to track the impacts of these changes as they occur. The BDFFP must continue its role in contributing to the scientific basis for more serious global efforts to contain the current human destruction of the environment at both the global and regional level.

Box 17.1: The Biological Dynamics of Forest Fragments Project
Since its inception in 1979, the Biological Dynamics of Forest Fragments Project (BDFFP) has been assessing the impacts of fragmentation on the Amazon rainforest and biota (Lovejoy et al. 1986; Bierregaard et al. 1992; Pimm 1998; Laurance et al. 2002, 2011). Today, it is the world's largest and longest-running experimental study of habitat fragmentation, as well as one of the most highly cited ecological investigations ever conducted (Gardner et al. 2009; Peres et al. 2010; Pitman et al. 2011). The BDFFP has also been a global leader in research, training and capacity development, with over 640 publications (http://pdbff.inpa.gov.br), more than 180 student theses, over 700 graduate students and conservation professionals participating in sponsored courses and over 1000 student interns to date.

The BDFFP is located 80 km north of Manaus, Brazil, and spans ~1000 km^2. The topography is relatively flat (80–160-m elevation) but dissected by numerous stream gullies. The heavily weathered, nutrient-poor soils of the study area are typical of large expanses of the Amazon basin. Rainfall ranges from 1900 mm to 3500 mm annually with a moderately strong dry season from June to October. The forest canopy is 30–37 m tall, with emergent trees to 55 m. Species richness of trees ($\geq$10-cm DBH) often exceeds 280 species ha^{-1} (de Oliveira and Mori 1999; Laurance et al. 2010b) with a comparably high level of diversity also evident in many other plant and animal taxa.

The study area includes three large cattle ranges (~5000 ha each) containing 11 forest fragments (five of 1 ha, four of 10 ha and two of 100 ha) and expanses of nearby continuous forest that serve as experimental controls. In the early 1980s, the fragments were isolated from nearby intact forest by distances of 80–650 m by clearing and burning the surrounding forest. A key feature was that pre-fragmentation censuses were conducted for many animal and plant groups (e.g. trees, understory birds, small mammals, primates, frogs, many invertebrate taxa), thereby allowing long-term changes in these groups to be assessed far more confidently than in most other fragmentation studies.

Because of poor soils and low productivity, the ranches surrounding the BDFFP fragments were largely abandoned after government fiscal incentives dried up from 1984 onwards. Secondary forests (initially dominated by *Vismia* spp. in areas that were cleared and burned or by *Cecropia* spp. in areas that were cleared without fire) proliferated in many formerly forested areas (Mesquita et al. 2001). Some of the regenerating areas initially dominated by *Cecropia* spp. later developed into quite mature (>20 m tall), species-rich secondary forests. *Vismia*-dominated regrowth, which is

(continued)

Box 17.1 (continued)

relatively species poor, is changing far more slowly (Norden et al. 2011; Williamson et al. 2014). To help maintain isolation of the experimental fragments, 100-m-wide strips of regrowth were cleared and burned around each fragment on four to five occasions, most recently in 2013–2014. Additional human disturbances that harm many fragmented landscapes in the Amazon, such as major fires and logging, are largely prevented at the BDFFP. Hunting pressure has been very limited until recently, following a government decision to increase colonization in the general area (Laurance and Luizão 2007). Laurance and Bierregaard (1997) and Bierregaard et al. (2001) provide detailed descriptions of the study area and design.

Acknowledgements We thank Laszlo Nagy and an anonymous referee for their helpful comments on the manuscript. The National Institute for Amazonian Research (INPA), Smithsonian Institution, US National Science Foundation, Brazilian Science Foundation (CNPq), Amazonian State Science Foundation (FAPEAM), NASA-LBA programme, USAID, Mellon Foundation, Blue Moon Fund, Marisla Foundation and other organizations generously supported the BDFFP. Substantial parts of this text are updated from Laurance et al. (2011). This is publication number 640 in the BDFFP technical series.

References

Andresen E (2003) Effect of forest fragmentation on dung beetle communities and functional consequences for plant regeneration. Ecography 26:87–97

Antongiovanni M, Metzger JP (2005) Influence of matrix habitats on the occurrence of insectivorous bird species in Amazonian forest fragments. Biol Conserv 122:441–451

Arroyo-Rodríguez V, Pineda E, Escobar F, Benítez-Malvido J (2009) Conservation value of small patches to plant species diversity in highly fragmented landscapes. Conserv Biol 23:729–739

Asner GP, Knapp D, Broadbent E, Oliveira P, Keller M, Silva J (2005) Selective logging in the Brazilian Amazon. Science 310:480–482

Avissar R, Liu Y (1996) A three-dimensional numerical study of shallow convective clouds and precipitation induced by land-surface forcing. J Geophys Res 101:7499–7518

Avissar R, Schmidt T (1998) An evaluation of the scale at which ground-surface heat flux patchiness affects the convective boundary layer using a large-eddy simulation model. J Atmos Sci 55:2666–2689

Baker TR, Phillips OL, Malhi Y, Almeida S, Arroyo L, Di Fiore A et al (2004) Increasing biomass in Amazonian forest plots. Philos Trans R Soc B 359:353–365

Barlow J, Peres CA, Henriques L, Stouffer PC, Wunderle J (2006) The responses of understorey birds to forest fragmentation, logging and wildfires: an Amazonian synthesis. Biol Conserv 128:182–192

Becker P, Moure JB, Peralta F (1991) More about euglossine bees in Amazonian forest fragments. Biotropica 23:586–591

Benítez-Malvido J (1998) Impact of forest fragmentation on seedling abundance in a tropical rain forest. Conserv Biol 12:380–389

Benítez-Malvido J, Martinez-Ramos M (2003a) Influence of edge exposure on tree seedling species recruitment in tropical rain forest fragments. Biotropica 35:530–541

Benítez-Malvido J, Martinez-Ramos M (2003b) Impact of forest fragmentation on understory plant species richness in Amazonia. Conserv Biol 17:389–400

Bierregaard RO, Lovejoy TE, Kapos V, dos Santos AA, Hutchings RW (1992) The biological dynamics of tropical rainforest fragments. Bioscience 42:859–866

Bierregaard RO, Gascon C, Lovejoy TE, Mesquita R (eds) (2001) Lessons from Amazonia: ecology and conservation of a fragmented forest. Yale University Press, New Haven, CT

Bobrowiec PED, Gribel R (2010) Effects of different secondary vegetation types on bat community composition in Central Amazonia, Brazil. Anim Conserv 13:204–216

Bohlman S, Laurance WF, Laurance SG, Nascimento H, Fearnside PM, Andrade A (2008) Effects of soils, topography, and geographic distance in structuring central Amazonian tree communities. J Veg Sci 19:863–874

Boyle SA, Smith AT (2010a) Can landscape and species characteristics predict primate presence in forest fragments in the Brazilian Amazon? Biol Conserv 143:1134–1143

Boyle SA, Smith AT (2010b) Behavioral modifications in northern bearded saki monkeys (*Chiropotes satanas chiropotes*) in forest fragments of central Amazonia. Primates 51:43–51

Briant G, Gond V, Laurance SGW (2010) Habitat fragmentation and the desiccation of forest canopies: a case study from eastern Amazonia. Biol Conserv 143:2763–2769

Broadbent E, Asner GP, Keller M, Knapp D, Oliveira P, Silva J (2008) Forest fragmentation and edge effects from deforestation and selective logging in the Brazilian Amazon. Biol Conserv 140:142–155

Brook BW, Sodhi NS, Bradshaw CJA (2008) Synergisms among extinction drivers under global change. Trends Ecol Evol 23:453–460

Brown KS, Hutchings RW (1997) Disturbance, fragmentation, and the dynamics of diversity in Amazonian forest butterflies. In: Laurance WF, Bierregaard RO (eds) Tropical forest remnants: ecology, management, and conservation of fragmented communities. University of Chicago Press, Chicago, IL, pp 91–110

Brum HD, Nascimento H, Laurance WF, Andrade A, Laurance SGW, Luizão R (2008) Rainforest fragmentation and the demography of the economically important palm *Oenocarpus bacaba* in central Amazonia. Plant Ecol 199:209–215

Bruna EM (1999) Seed germination in rainforest fragments. Nature 402:139

Bruna EM (2003) Are plant populations in fragmented habitats recruitment limited? Tests with an Amazonian herb. Ecology 84:932–947

Bruna E, Oli M (2005) Demographic consequences of habitat fragmentation for an Amazonian understory plant: analysis of life-table response experiments. Ecology 86:1816–1824

Bruna EM, Vasconcelos HL, Heredia S (2005) The effect of habitat fragmentation on communities of mutualists: a test with Amazonian ants and their host plants. Biol Conserv 124:209–216

Bührnheim CM, Fernandes CC (2003) Structure of fish assemblages in Amazonian rainforest streams: effects of habitats and locality. Copeia 2003:255–262

Butler RA, Laurance WF (2009) Is oil palm the next emerging threat to the Amazon? Trop Conserv Sci 2:1–10

Camargo JLC, Kapos V (1995) Complex edge effects on soil moisture and microclimate in central Amazonian forests. J Trop Ecol 11:205–211

Carvalho KS, Vasconcelos HL (1999) Forest fragmentation in central Amazonia and its effects on litter-dwelling ants. Biol Conserv 91:151–158

Chambers JQ, Higuchi N, Schimel JP (1998) Ancient trees in Amazonia. Nature 391:135–136

Chave J, Muller-Landau HC, Baker TR, Easdale TA, ter Steege H, Webb CO (2006) Regional and phylogenetic variation of wood density across 2,456 neotropical tree species. Ecol Appl 16:2356–2367

Chazdon RL, Harvey CA, Komar O, Griffith DM, Ferguson BG, Martinez-Ramos M, Morales H, Nigh R, Soto-Pinto L, van Breugel M, Philpott SM (2008) Beyond reserves: a research agenda for conserving biodiversity in human-modified tropical landscapes. Biotropica 41:142–153

Cochrane MA, Laurance WF (2002) Fire as a large-scale edge effect in Amazonian forests. J Trop Ecol 18:311–325
Cochrane MA, Laurance WF (2008) Synergisms among fire, land use, and climate change in the Amazon. Ambio 37:522–527
Condon MA, Sasek TW, Strain BR (1992) Allocation patterns in two tropical vines in response to increased atmospheric CO_2. Funct Ecol 6:680–685
Cox PM, Betts RA, Jones CD, Spall SA, Totterdell IJ (2000) Acceleration of global warming due to carbon cycle feedbacks in a coupled climate model. Nature 408:184–187
Cox PM, Betts RA, Collins M, Harris PP, Huntingford C, Jones CD (2004) Amazonian forest dieback under climate-carbon cycle projections for the 21st century. Theor Appl Climatol 78:137–156
Cox PM, Pearson D, Booth BB, Friedlingstein P, Huntingford C, Jones CD, Luke CMD (2013) Sensitivity of tropical carbon to climate change constrained by carbon dioxide variability. Nature 494:341–344
Cramer JM, Mesquita R, Bentos T, Moser B, Williamson GB (2007a) Forest fragmentation reduces seed dispersal of *Duckeodendron cestroides*, a Central Amazon endemic. Biotropica 39:709–718
Cramer JM, Mesquita R, Williamson GB (2007b) Forest fragmentation differentially affects seed dispersal of large and small-seeded tropical trees. Biol Conserv 137:415–423
D'Angelo S, Andrade A, Laurance SG, Laurance WF, Mesquita R (2004) Inferred causes of tree mortality in fragmented and intact Amazonian forests. J Trop Ecol 20:243–246
da Silva JMC, Tabarelli M (2000) Tree species impoverishment and the future flora of the Atlantic forest of northeast Brazil. Nature 404:72–74
Dai A (2012) Increasing drought under global warming in observations and models. Nat Clim Chang 3:52–58
de Oliveira AA, Mori SA (1999) A central Amazonian terra firme forest. I. High tree species richness on poor soils. Biodivers Conserv 8:1219–1244
Deichmann JL, Lima A, Williamson GB (2011) Effects of geomorphology and primary productivity on Amazonian leaf litter herpetofauna. Biotropica 43:149–156
Deichmann JL, Toft CA, Deichmann PM, Lima AP, Williamson GB (2013) Neotropical primary productivity affects biomass of leaf-litter herpetofaunal assemblage. J Trop Ecol 28:427–435
Develey P, Stouffer PC (2001) Roads affect movements by understory mixed-species flocks in central Amazonian Brazil. Conserv Biol 15:1416–1422
Diamond JM, Bishop KD, Balen SV (1987) Bird survival in an isolated Javan woodland: island or mirror? Conserv Biol 1:132–142
Dick CW (2001) Genetic rescue of remnant tropical trees by an alien pollinator. Proc R Soc B Biol Sci 268:2391–2396
Dick CW, Etchelecu G, Austerlitz F (2003) Pollen dispersal of tropical trees (*Dinizia excelsa*: Fabaceae) by native insects and African honeybees in pristine and fragmented Amazonian rainforest. Mol Ecol 12:753–764
Didham RK, Lawton JH (1999) Edge structure determines the magnitude of changes in microclimate and vegetation structure in tropical forest fragments. Biotropica 31:17–30
Didham RK, Hammond PM, Lawton JH, Eggleton P, Stork NE (1998a) Beetle species responses to tropical forest fragmentation. Ecol Monogr 68:295–303
Didham RK, Lawton JH, Hammond PM, Eggleton P (1998b) Trophic structure stability and extinction dynamics of beetles (Coleoptera) in tropical forest fragments. Proc Trans R Soc Lond B Biol Sci 353:437–451
Farneda FZ (2013) Bat species vulnerability to forest fragmentation in the Central Amazon. M.Sc. Thesis. University of Lisbon, Portugal
Fáveri SB, Vasconcelos HL, Dirzo R (2008) Effects of Amazonian forest fragmentation on the interaction between plants, insect herbivores, and their natural enemies. J Trop Ecol 24:57–64
Fearnside PM (1997) Wood density for estimating forest biomass in Brazilian Amazonia. For Ecol Manage 90:59–89

Fearnside PM (2000) Global warming and tropical land-use change: greenhouse gas emissions from biomass burning, decomposition and soils in forest conversion, shifting cultivation and secondary vegetation. Clim Change 46:115–158

Fearnside PM (2001a) Soybean cultivation as a threat to the environment in Brazil. Environ Conserv 28:23–38

Fearnside PM (2001b) Saving tropical forests as a global warming countermeasure: an issue that divides the environmental movement. Ecol Econ 39:167–184

Fearnside PM (2002) Avança Brasil: environmental and social consequences of Brazil's planned infrastructure in Amazonia. Environ Manage 30:748–763

Fearnside PM (2007) Brazil's Cuiabá-Santarém (BR-163) highway: the environmental cost of paving a soybean corridor through the Amazon. Environ Manage 39:601–614

Fearnside PM (2012) Brazil's Amazon forest in mitigating global warming: unresolved controversies. Climate Policy 12:70–81

Fearnside PM (2013) Vines, CO_2 and Amazon forest dieback. Nature online comment. http://www.nature.com/nature/journal/vaop/ncurrent/full/nature11882.html

Fearnside PM, Graça PMLA (2006) BR-319: Brazil's Manaus–Porto Velho highway and the potential impact of linking the arc of deforestation to central Amazonia. Environ Manage 38:705–716

Fearnside PM, Leal Filho N (2001) Soil and development in Amazonia: lessons from the Biological Dynamics of Forest Fragments Project. In: Bierregaard RO, Gascon C, Lovejoy TE, Mesquita R (eds) Lessons from Amazonia: the ecology and conservation of a fragmented forest. Yale University Press, New Haven, CT, pp 291–312

Ferraz G, Russell GJ, Stouffer PC, Bierregaard RO, Pimm SL, Lovejoy TE (2003) Rates of species loss from Amazonian forest fragments. Proc Nat Acad Sci U S A 100:14069–14073

Ferraz G, Nichols JD, Hines J, Stouffer PC, Bierregaard RO, Lovejoy TE (2007) A large-scale deforestation experiment: effects of patch area and isolation on Amazon birds. Science 315:238–241

Finer M, Jenkins C, Pimm SL, Keane B, Ross C (2008) Oil and gas projects in the western Amazon: threats to wilderness, biodiversity, and indigenous peoples. PLoS ONE 3, e2932

Fowler HG, Silva CA, Ventincinque E (1993) Size, taxonomic and biomass distributions of flying insects in central Amazonia: forest edge vs. understory. Rev Biol Trop 41:755–760

Gardner TA, Barlow J, Chazdon R, Ewers R, Harvey C, Peres CA, Sodhi NS (2009) Prospects for tropical forest biodiversity in a human-modified world. Ecol Lett 12:561–582

Gascon C (1993) Breeding habitat use by Amazonian primary-forest frog species at the forest edge. Biodivers Conserv 2:438–444

Gascon C, Lovejoy TE, Bierregaard RO, Malcolm JR, Stouffer PC, Vasconcelos H, Laurance WF, Zimmerman B, Tocher M, Borges S (1999) Matrix habitat and species persistence in tropical forest remnants. Biol Conserv 91:223–229

Gibbs HK, Reusch AS, Achard F, Clayton MK, Holmgren P, Ramankutty N, Foley JA (2010) Tropical forests were the primary sources of new agricultural lands in the 1980s and 1990s. Proc Nat Acad Sci U S A 107:16732–16737

Gibson L, Lee TM, Koh LP, Brook BW, Gardner TA, Barlow J, Peres CA, Bradshaw CJA, Laurance WF, Lovejoy TE, Sodhi NS (2011) Primary forests are irreplaceable for sustaining tropical biodiversity. Nature 478:378–381

Gilbert KA, Setz EZF (2001) Primates in a fragmented landscape: six species in central Amazonia. In: Laurance WF, Bierregaard RO (eds) Tropical forest remnants: ecology, management and conservation of fragmented communities. University of Chicago Press, Chicago, IL, pp 207–221

Gilbert B, Laurance WF, Leigh EG, Nascimento H (2006) Can neutral theory predict the responses of Amazonian tree communities to forest fragmentation? Am Nat 168:304–317

Girão LC, Lopes AV, Tabarelli M, Bruna EM (2007) Changes in tree reproductive traits reduce functional diversity in a fragmented Atlantic forest landscape. PLoS ONE 2, e908

Good P, Jones CD, Lowe JA, Betts RA, Gedney N (2013) Comparing tropical forest projections from two generations of Hadley Centre Earth System models, HadGEM2-ES and HadCM3LC. J Climate 26:495–511

Groeneveld J, Alves L, Bernacci L, Catharino E, Knogge C, Metzger J, Pütz S, Huth A (2009) The impact of fragmentation and density regulation on forest succession in the Atlantic rain forest. Ecol Model 220:2450–2459

Harper LH (1989) The persistence of ant-following birds in small Amazonian forest fragments. Acta Amazonica 19:249–263

Hubbell SP (2001) The neutral theory of biodiversity and biogeography. Princeton University Press, Princeton, NJ

Huntingford C, Zelazowski P, Galbraith D, Mercado LM, Sitch S, Fisher R, Lomas M, Walker AP, Jones CD, Booth BB, Malhi Y, Hemming D, Kay G, Good P, Lewis SL, Phillips OL, Atkin OK, Lloyd J, Gloor E, Zaragoza-Castells J, Meir P, Betts R, Harris PP, Nobre C, Marengo J, Cox PM (2013) Simulated resilience of tropical rainforests to CO_2-induced climate change. Nat Geosci 6:268–273

Ingwell LL, Wright SJ, Becklund KK, Hubbell SP, Schnitzer SA (2010) The impact of lianas on 10 years of tree growth and mortality on Barro Colorado Island, Panama. J Ecol 98:879–887

IPCC (2013) Summary for policymakers. In: Stocker TF et al (eds) Climate change 2013: the physical science basis. Intergovernmental Panel on Climate Change, Cambridge University Press, Cambridge

Janzen DH (1983) No park is an island: increase in interference from outside as park size decreases. Oikos 41:402–410

Jorge ML (2008) Effects of forest fragmentation on two sister genera of Amazonian rodents (*Myoprocta acouchy* and *Dasyprocta leporina*). Biol Conserv 141:617–623

Kapos V (1989) Effects of isolation on the water status of forest patches in the Brazilian Amazon. J Trop Ecol 5:173–185

Kapos V, Ganade G, Matusi E, Victoria RL (1993) Delta 13C as an indicator of edge effects in tropical rainforest reserves. J Ecol 81:425–432

Killeen TJ (2007) A perfect storm in the Amazon wilderness: development and conservation in the context of the initiative for the Integration of the Regional Infrastructure of South America (IIRSA). Conservation International, Washington, DC

Klein BC (1989) Effects of forest fragmentation on dung and carrion beetle communities in central Amazonia. Ecology 70:1715–1725

Laurance SGW (2004a) Responses of understory rain forest birds to road edges in central Amazonia. Ecol Appl 14:1344–1357

Laurance SGW, Stouffer PC, Laurance WF (2004b) Effects of road clearings on movement patterns of understory rainforest birds in central Amazonia. Conserv Biol 18:1099–1109

Laurance SGW, Gomez MS (2005) Clearing width and movements of understory rainforest birds. Biotropica 37:149–152

Laurance SGW, Laurance WF, Nascimento H, Andrade A, Fearnside PM, Rebello E, Condit R (2009a) Long-term variation in Amazon forest dynamics. J Veg Sci 20:323–333

Laurance SGW, Andrade A, Laurance WF (2010a) Unanticipated effects of stand dynamism on Amazonian tree diversity. Biotropica 42:429–434

Laurance SGW, Laurance WF, Andrade A, Fearnside PM, Harms K, Luizão R (2010b) Influence of soils and topography on Amazonian tree diversity: a landscape-scale study. J Veg Sci 21:96–106

Laurance WF (1991) Ecological correlates of extinction proneness in Australian tropical rainforest mammals. Conserv Biol 5:79–89

Laurance WF (2001) The hyper-diverse flora of the central Amazon: an overview. In: Bierregaard RO, Gascon C, Lovejoy TE, Mesquita R (eds) Lessons from Amazonia: ecology and conservation of a fragmented forest. Yale University Press, New Haven, CT, pp 47–53

Laurance WF (2002) Hyperdynamism in fragmented habitats. J Veg Sci 13:595–602

Laurance WF (2004b) Forest-climate interactions in fragmented tropical landscapes. Philos Trans R Soc Lond B Biol Sci 359:345–352

Laurance WF (2005) When bigger is better: the need for Amazonian megareserves. Trends Ecol Evol 20:645–648
Laurance WF (2007) Have we overstated the tropical biodiversity crisis? Trends Ecol Evol 22:65–70
Laurance WF (2008) Theory meets reality: how habitat fragmentation research has transcended island biogeographic theory. Biol Conserv 141:1731–1744
Laurance WF, Bierregaard RO (eds) (1997) Tropical forest remnants: ecology, management, and conservation of fragmented communities. University of Chicago Press, Chicago, IL
Laurance WF, Cochrane MA (2001) Synergistic effects in fragmented landscapes. Conserv Biol 15:1488–1489
Laurance WF, Luizão RCC (2007) Driving a wedge into the Amazon. Nature 448:409–410
Laurance WF, Peres CA (eds) (2006) Emerging threats to tropical forests. University of Chicago Press, Chicago, IL
Laurance WF, Useche DC (2009) Environmental synergisms and extinctions of tropical species. Conserv Biol 23:1427–1437
Laurance WF, Williamson GB (2001) Positive feedbacks among forest fragmentation, drought, and climate change in the Amazon. Conserv Biol 15:1529–1535
Laurance WF, Laurance SGW, Ferreira LV, Rankin-de Merona J, Gascon C, Lovejoy TE (1997) Biomass collapse in Amazonian forest fragments. Science 278:1117–1118
Laurance WF, Ferreira LV, Rankin-de Merona JM, Laurance SG (1998a) Rain forest fragmentation and the dynamics of Amazonian tree communities. Ecology 79:2032–2040
Laurance WF, Ferreira LV, Rankin-de Merona JM, Laurance SGW, Hutchings R, Lovejoy TE (1998b) Effects of forest fragmentation on recruitment patterns in Amazonian tree communities. Conserv Biol 12:460–464
Laurance WF, Laurance SGW, Delamonica P (1998c) Tropical forest fragmentation and greenhouse gas emissions. For Ecol Manage 110:173–180
Laurance WF, Fearnside PM, Laurance SGW, Delamonica P, Lovejoy TE, Rankin-de Merona JM, Chambers JQ, Gascon C (1999) Relationship between soils and Amazon forest biomass: a landscape-scale study. For Ecol Manage 118:127–138
Laurance WF, Delamonica P, Laurance SGW, Vasconcelos HL, Lovejoy TE (2000) Rainforest fragmentation kills big trees. Nature 404:836
Laurance WF, Cochrane M, Bergen S, Fearnside PM, Delamonica P, Barber C, D'Angelo S, Fernandes T (2001a) The future of the Brazilian Amazon. Science 291:438–439
Laurance WF, Perez-Salicrup D, Delamonica P, Fearnside PM, D'Angelo S, Jerozolinski A, Pohl L, Lovejoy TE (2001b) Rain forest fragmentation and the structure of Amazonian liana communities. Ecology 82:105–116
Laurance WF, Williamson GB, Delamonica P, Olivera A, Gascon C, Lovejoy TE, Pohl L (2001c) Effects of a strong drought on Amazonian forest fragments and edges. J Trop Ecol 17:771–785
Laurance WF, Lovejoy TE, Vasconcelos H, Bruna E, Didham R, Stouffer P, Gascon C, Bierregaard R, Laurance SG, Sampaio E (2002) Ecosystem decay of Amazonian forest fragments: a 22-year investigation. Conserv Biol 16:605–618
Laurance WF, Nascimento H, Laurance SG, Condit R, D'Angelo S, Andrade A (2004b) Inferred longevity of Amazonian rainforest trees based on a long-term demographic study. For Ecol Manage 190:131–143
Laurance WF, Nascimento H, Laurance SG, Andrade A, Fearnside PM, Ribeiro J (2006a) Rain forest fragmentation and the proliferation of successional trees. Ecology 87:469–482
Laurance WF, Nascimento H, Laurance SG, Andrade A, Ribeiro J, Giraldo JP, Lovejoy TE, Condit R, Chave J, D'Angelo S (2006b) Rapid decay of tree community composition in Amazonian forest fragments. Proc Nat Acad Sci U S A 103:19010–19014
Laurance WF, Nascimento H, Laurance SG, Andrade A, Ewers R, Harms K, Luizão R, Ribeiro J (2007) Habitat fragmentation, variable edge effects, and the landscape-divergence hypothesis. PLoS ONE 2, e1017

Laurance WF, Goosem M, Laurance SGW (2009b) Impacts of roads and linear clearings on tropical forests. Trends Ecol Evol 24:659–669

Laurance WF, Camargo J, Luizão RCC, Laurance SGW, Pimm SL, Bruna E, Stouffer PC, Williamson GB, Benitez-Malvido J, Vasconcelos H, Van Houtan K, Zartman CE, Boyle S, Didham RK, Andrade A, Lovejoy TE (2011) The fate of Amazonian forest fragments: a 32-year investigation. Biol Conserv 144:56–67

Laurance WF, Andrade A, Magrach A, Camargo JLC, Valsko J, Campbell M, Fearnside PM, Edwards W, Lovejoy TE, Laurance SGW (2014a) Long-term changes in liana abundance and forest dynamics in undisturbed Amazonian forests. Ecology 95:1604–1611

Laurance WF, Andrade A, Magrach A, Camargo JLC, Campbell M, Fearnside PM, Edwards W, Valsko J, Lovejoy TE, Laurance SGW (2014b) Apparent environmental synergism drives the dynamics of Amazonian forest fragments. Ecology 95:3018–3026

Leidner AK, Haddad NM, Lovejoy TE (2010) Does tropical forest fragmentation increase long-term variability of butterfly communities? PLoS ONE 5, e9534

Lima M, Gascon C (1999) The conservation value of linear forest remnants in central Amazonia. Biol Conserv 91:241–247

Lovejoy TE, Oren DC (1981) Minimum critical size of ecosystems. In: Burgess RL, Sharp DM (eds) Forest-island dynamics in man-dominated landscapes. Springer, New York, NY, pp 7–12

Lovejoy TE, Rankin JM, Bierregaard RO, Brown KS, Emmons LH, Van der Voort ME (1984) Ecosystem decay of Amazon forest fragments. In: Nitecki MH (ed) Extinctions. University of Chicago Press, Chicago, IL, pp 295–325

Lovejoy TE, Bierregaard RO, Rylands AB, Malcolm JR, Quintela C, Harper L, Brown K, Powell A, Powell G, Schubart H, Hays M (1986) Edge and other effects of isolation on Amazon forest fragments. In: Soulé ME (ed) Conservation biology: the science of scarcity and diversity. Sinauer, Sunderland, MA, pp 257–285

Malcolm JR (1994) Edge effects in central Amazonian forest fragments. Ecology 75:2438–2445

Malcolm JR (1997) Biomass and diversity of small mammals in Amazonian forest fragments. In: Laurance WF, Bierregaard RO (eds) Tropical forest remnants: ecology, management, and conservation of fragmented communities. University of Chicago Press, Chicago, IL, pp 207–221

Malcolm JR (1998) A model of conductive heat flow in forest edges and fragmented landscapes. Clim Change 39:487–502

Massoca PES, Jakovac ACC, Vizcarra TB, Williamson GB, Mesquita RCG (2012) Dinâmica e trajetórias da sucessão secundária na Amazônia central. Bol Mus Para Emílio Goeldi Cienc Nat Belém 7:235–250

Mesquita RCC, Delamônica P, Laurance WF (1999) Effects of surrounding vegetation on edge-related tree mortality in Amazonian forest fragments. Biol Conserv 91:129–134

Mesquita RCC, Ickes K, Ganade G, Williamson GB (2001) Alternative successional pathways in the Amazon basin. J Ecol 89:528–537

Mesquita RCC, Bentos TV, Jakovac ACC, Massoca P, Williamson GB (2012) Amazonian secondary forests respond to climatic variation: biomass dynamics. 49th Annual meeting of the association for Tropical Biology and Conservation, Bonito, MS, Brazil

Mestre LAM, Gasnier TR (2008) Populações de aranhas errantes do gênero Ctenus em fragmentos florestais na Amazônia Central. Acta Amazonica 38:159–164

Michalski F, Peres CA (2005) Anthropogenic determinants of primate and carnivore local extinctions in a fragmented forest landscape of Southern Amazonia. Biol Conserv 124:383–396

Moura NG, Lees AC, Andretti CB, Davis BJW, Solar RRC, Aleixo A, Barlow J, Ferreira J, Gardner TA (2014) Avian biodiversity in multiple-use landscapes of the Brazilian Amazon. Biol Conserv 167:339–348

Myers N, Mittermeier RA, Mittermeier CG, Fonseca GAB, Kent J (2000) Biodiversity hotspots for conservation priorities. Nature 403:853–858

Nascimento H, Laurance WF (2004) Biomass dynamics in Amazonian forest fragments. Ecol Appl 14:S127–S138

Nascimento H, Andrade A, Camargo J, Laurance WF, Laurance SG, Ribeiro J (2006) Effects of the surrounding matrix on tree recruitment in Amazonian forest fragments. Conserv Biol 20:853–860

Neckel-Oliveira S, Gascon C (2006) Abundance, body size and movement patterns of a tropical tree frog in continuous and fragmented forests of the Brazilian Amazon. Biol Conserv 128:308–315

Nessimian JL, Venticinque EM, Zuanon J, De Marco P, Gordo M, Fidelis L, Batista J, Juen L (2008) Land use, habitat integrity, and aquatic insect assemblages in central Amazonian streams. Hydrobiol 614:117–131

Nogueira EM, Nelson BW, Fearnside PM (2005) Wood density in dense forest in central Amazonia, Brazil. For Ecol Manage 208:261–286

Nogueira EM, Fearnside PM, Nelson BW, França MB (2007) Wood density in forests of Brazil's 'arc of deforestation': implications for biomass and flux of carbon from land-use change in Amazonia. For Ecol Manage 248:119–135

Nogueira EM, Fearnside PM, Nelson BW, Barbosa RI, Keizer EWH (2008) Estimates of forest biomass in the Brazilian Amazon: new allometric equations and adjustments to biomass from wood-volume inventories. For Ecol Manage 256:1853–1857

Norden N, Mesquita R, Bentos T, Chazdon R, Williamson GB (2011) Contrasting community compensatory trends in alternative successional pathways in central Amazonia. Oikos 120:143–151

Peres CA (2005) Why we need megareserves in Amazonia. Conserv Biol 19:728–733

Peres CA, Gardner TA, Barlow J, Zuanon J, Michalski F, Lees A, Vieira I, Moreira F, Feeley KJ (2010) Biodiversity conservation in human-modified Amazonian forest landscapes. Biol Conserv 143:2314–2327

Phillips OL, Malhi Y, Higuchi N, Laurance WF, Nunez PV, Vasquez RM et al (1998) Changes in the carbon balance of tropical forests: evidence from long-term plots. Science 282:439–442

Pimm SL (1998) The forest fragment classic. Nature 393:23–24

Pimm SL, Jenkins C (2005) Sustaining the variety of life. Sci Am 293(3):66–73

Pitman NCA, Widmer J, Jenkins CN, Stocks G, Seales L, Paniagua F, Bruna EM (2011) Volume and geographical distribution of ecological research in the Andes and the Amazon, 1995-2008. Trop Conserv Sci 4:64–81

Powell AH, Powell GVN (1987) Population dynamics of male euglossine bees in Amazonian forest fragments. Biotropica 19:176–179

Puerta R (2002) Regeneração arbórea em pastagens abandonadas na região de Manaus em função da distância da floresta contínua. Scientia Forestalis 62:32–39

Quintela CE (1985) Forest fragmentation and differential use of natural and man-made edges by understory birds in central Amazonia. M.Sc. Thesis, University of Illinois, Chicago, IL

Quintero I, Roslin T (2005) Rapid recovery of dung beetle communities following habitat fragmentation in Central Amazonia. Ecology 86:3303–3311

Radtke MG, da Fonseca C, Williamson GB (2008) Forest fragment size effects on dung beetle communities. Biol Conserv 141:613–614

Rego F, Venticinque EM, Brescovit A (2007) Effects of forest fragmentation on four Ctenus spider populations (Araneae, Ctenidae) in central Amazonia, Brazil. Stud Neotrop Fauna Environ 42:137–144

Ribeiro MBN, Bruna EM, Mantovani W (2009) Influence of post-clearing treatment on the recovery of herbaceous plant communities in Amazonian secondary forests. Restor Ecol 18:50–58

Rocha R, López-Baucells A, Groenenberg M, Silva I, Farneda F, Bobrowiec P, Palmeirim JM, Meyer CFJ (2013) Effects of forest fragmentation on central Amazonian bats. Paper presented at the 16th International Bat research conference & 43rd North American symposium on Bat research, San José, Costa Rica

Salazar LF, Nobre CA, Oyama MD (2007) Climate change consequences on the biome distribution in South America. Geophys Res Lett 34, L09708
Sampaio EM (2000) Effects of forest fragmentation on the diversity and abundance patterns of central Amazonian Bats. Ph.D. Dissertation, University of Tübingen, Germany
Sampaio EM, Kalko E, Bernard E, Rodriguez-Herrera B, Handley C (2003) A biodiversity assessment of bats (Chiroptera) in a tropical lowland forest of central Amazonia, including methodological and conservation considerations. Stud Neotrop Fauna Environ 28:17–31
Santos BA, Arroyo-Rodríguez V, Moreno CE, Tabarelli M (2010) Edge-related loss of tree phylogenetic diversity in the severely fragmented Brazilian Atlantic forest. PLoS ONE 5, e12625
Scariot A (1999) Forest fragmentation effects on diversity of the palm community in central Amazonia. J Ecol 87:66–76
Scariot A (2001) Weedy and secondary palm species in central Amazonian forest fragments. Rev Bot Brasil 15:271–280
Sizer N, Tanner EVJ (1999) Responses of woody plant seedlings to edge formation in a lowland tropical rainforest, Amazonia. Biol Conserv 91:135–142
Sizer N, Tanner EVJ, Kossman-Ferraz I (2000) Edge effects on litterfall mass and nutrient concentrations in forest fragments in central Amazonia. J Trop Ecol 16:853–863
Skole DS, Tucker CJ (1993) Tropical deforestation and habitat fragmentation in the Amazon: satellite data from 1978 to 1988. Science 260:1905–1910
Sodhi NS, Koh LP, Brook BW, Ng P (2004) Southeast Asian biodiversity: an impending disaster. Trends Ecol Evol 19:654–660
Stickler CM et al (2009) The potential ecological costs and cobenefits of REDD: a critical review and case study from the Amazon region. Glob Chang Biol 15:2803–2824
Stouffer PC, Bierregaard RO (1995a) Effects of forest fragmentation on understory hummingbirds in Amazonian Brazil. Conserv Biol 9:1085–1094
Stouffer PC, Bierregaard RO (1995b) Use of Amazonian forest fragments by understory insectivorous birds. Ecology 76:2429–2445
Stouffer PC, Bierregaard RO, Strong C, Lovejoy TE (2006) Long-term landscape change and bird abundance in Amazonian rainforest fragments. Conserv Biol 20:1212–1223
Stouffer PC, Strong C, Naka LN (2008) Twenty years of understory bird extinctions from Amazonian rain forest fragments: consistent trends and landscape-mediated dynamics. Divers Distrib 15:88–97
Stratford JA, Stouffer PC (1999) Local extinctions of terrestrial insectivorous birds in Amazonian forest fragments. Conserv Biol 13:1416–1423
Ter Steege H, Pitman NCA, Sabatier D, Baraloto C, Salomao RP, Ernesto Guevara J, Phillips OL et al (2013) Hyperdominance in the Amazonia flora. Science 342:326–334
Terborgh J, Lopez L, Nuñez VP, Rao M, Shahabuddin G, Orihuela G, Riveros M, Ascanio R, Adler G, Lambert T, Balbas L (2001) Ecological meltdown in predator-free forest fragments. Science 294:1923–1926
Timo TPC (2003) Influência de fragmentação e matriz sobre a comunidade de mamíferos de médio e grande porte em uma floresta de terra firme na Amazônia central. M.Sc. Thesis, National Institute for Amazonian Research (INPA), Manaus, Brazil
Tocher M, Gascon C, Zimmerman BL (1997) Fragmentation effects on a central Amazonian frog community: a ten-year study. In: Laurance WF, Bierregaard RO (eds) Tropical forest remnants: ecology, management, and conservation of fragmented communities. University of Chicago Press, Chicago, IL, pp 124–137
Trancoso R (2008) Hydrological impacts of deforestation in small catchments in Brazilian Amazonia. M.Sc. Thesis, National Institute for Amazonian Research (INPA), Manaus, Brazil
Uriarte M, Bruna EM, Rubim P, Anciães M, Jonckheere I (2010) Effects of forest fragmentation on the seedling recruitment of a tropical herb: assessing seed vs. safe-site limitation. Ecology 91:1317–1328

Van Houtan KS, Pimm SL, Bierregaard RO, Lovejoy TE, Stouffer PC (2006) Local extinctions in Amazonian forest fragments. Evol Ecol Res 8:129–148
Van Houtan KS, Pimm SL, Halley J, Bierregaard RO, Lovejoy TE (2007) Dispersal of Amazonian birds in continuous and fragmented forest. Ecol Lett 10:219–229
Vasconcelos HL, Luizão FJ (2004) Litter production and litter nutrient concentrations in a fragmented Amazonian landscape: edge and soil effects. Ecol Appl 14:884–892
Wieland LM, Mesquita RC, Bobrowiec PED, Bentos TV, Williamson GB (2011) Seed rain and advance regeneration in secondary succession in the Brazilian Amazon. Trop Conserv Sci 4:300–316
Wilcox BA, Murphy DD (1985) Conservation strategy: the effects of fragmentation on extinction. Am Nat 125:879–887
Williamson GB, Mesquita R (2001) Effects of fire on rain forest regeneration in the Amazon Basin. In: Laurance WF, Bierregaard RO (eds) Tropical forest remnants: Ecology, management, and conservation of fragmented communities. University of Chicago Press, Chicago, IL, pp 325–334
Williamson GB, Laurance WF, Oliveira A, Delamonica P, Gascon C, Lovejoy TE, Pohl L (2000) Amazonian wet forest resistance to the 1997–98 El Niño drought. Conserv Biol 14:1538–1542
Williamson GB, Longworth JB, Bentos TV, Mesquita RC (2014) Convergence and divergence in alternative successional pathways in Central Amazonia. Plant Ecol Diversity 7:341–348
Zartman CE (2003) Forest fragmentation effects on epiphyllous bryophyte communities in central Amazonia. Ecology 84:948–954
Zartman CE, Nascimento HEM (2006) Are patch-tracking metacommunities dispersal limited? Inferences from abundance-occupancy patterns of epiphylls in Amazonian forest fragments. Biol Conserv 127:46–54
Zartman CE, Shaw AJ (2006) Metapopulation extinction thresholds in rainforest remnants. Am Nat 167:177–189

Chapter 18 The Socioecological Implications of Land Use and Landscape Change in the Brazilian Amazon

Ima C.G. Vieira, Peter M. Toledo, and Roberto Araújo O.S. Jr.

18.1 Introduction

Since the 1960s and 1970s, Brazilian Amazon has undergone a process of intense landscape transformation as a result of ongoing deforestation (Ometto et al. 2015; Chap. 16), myriad changes in land use and the unsustainable exploitation of natural resources. Between 1960 and 1990, several national government programmes were implemented in an attempt to formalise land ownership and integrate Amazonia into the Brazilian economy (Salati et al. 1983; Ferreira and Salati 2005).

Strongly influenced by a development-focused model, the Brazilian Amazon during this period was characterised by the extensive and predatory exploitation of natural resources. Subsidies and tax incentives facilitated the arrival of new agricultural and extractive industries that gradually replaced a forest-based economy which, at the time, was considered by the government to be both stagnant and archaic (Moran 1981). These initiatives also included massive colonisation efforts to resettle displaced and jobless families from other regions of Brazil into the Amazon region (Brazilian Legal Amazon, BLA) (Leroy 1991). Together, these processes led to an increased concentration of land in private estates and the widespread conversion of ecosystems, resulting in loss of biodiversity, frequent conflicts over land tenure and the displacement of local people to urban centres. Often viewed as both obstacles to progress and lacking the entrepreneurship

I.C.G. Vieira (✉) • R.A.O. Santos Jr.
Museu Paraense Emilio Goeldi, Av Magalhães Barata, 376, CP. 399, CEP 66040-170, Belém, Pará, Brazil
e-mail: ima@museu-goeldi.br

P.M. de Toledo
Instituto Nacional de Pesquisas Espaciais, Centro de Ciência do Sistema Terrestre, Avenida dos Astronautas 1758, CEP 12.227-010, São José dos Campos, São Paulo, Brazil

© Springer-Verlag Berlin Heidelberg 2016
L. Nagy et al. (eds.), *Interactions Between Biosphere, Atmosphere and Human Land Use in the Amazon Basin*, Ecological Studies 227,
DOI 10.1007/978-3-662-49902-3_18

deemed necessary for Amazonia to enter the twenty-first century, indigenous and traditional people became quickly marginalised from the dominant economic and political systems (Almeida 2004; Loureiro and Pinto 2005; Araújo and Alves 2008).

The BLA is still grappling with the consequences of this development-centred model, established over 50 years ago, with many of the same hallmarks being visible in more recent government plans—e.g. Belo Monte dam complex and the construction and paving of the BR 319 and BR 163 highways. These plans are still responsible for high population growth, rapid urbanisation (with the increase of poor neighbourhoods), destruction of the environment, increase in land grabbing and loss of natural resources (Vieira et al. 2014).

In general terms, land use in BLA can be associated with six major economic factors: mineral extraction, forestry activities, extensive cattle ranching, infrastructure (dams, roads, etc.), colonisation projects and more recently, the production of agricultural commodities (see also Ometto et al. 2015; Chap. 16). The existence of large migration flows in the absence of a clear system of land demarcation and property rights led to an intense competition for land and natural resources. The private appropriation (including through illicit means) of public lands, rich in natural resources, has triggered conflicts for control of land and their natural resources by different social groups.

Combining these elements, a new social fabric has emerged as a result of contradictory objectives and competing interests of a heterogeneous range of actors who currently inhabit the region (including public and private companies, migrant farmers, miners, cattle ranchers, loggers, soy farmers, landless labourers). Reacting to these trends and in response to the grass-roots pressure from social movements and non-governmental organisations (NGO), a new development paradigm emerged in Brazil during the late 1990s: socio-environmentalism.

The socio-environmental model translated into a territorial management process through forest-based government settlement projects designed to accommodate livelihoods and rights of local people, known from then on as 'traditional populations'. These changes heralded the creation of a number of new types of legally recognised land use forms beyond indigenous lands, including extractive reserves (RESEX), quilombos (which recognised the traditional rights of descendants of former slave populations), sustainable development projects ('Projeto de Desenvolvimento Sustentável—PDS') and forestry settlement projects ('Projeto de Assentamento Florestal—PAF'), all of which recognised customary land use practices involving forest use or assumed some level of collective forest use (Table 18.1).

The problems associated with these two antagonistic models of development—the large scale, development centred and the socio-environmental—each one imposing different dynamics on the landscape, have long been debated in the literature (e.g. Davis 1977; Schmink and Woods 1984; Araújo and Lena 2010; Hecht 2011; Pacheco 2009). Some authors have signalled problems pertaining to the development-centric model (Morton et al. 2006; Hogan 2010), as well as land tenure and social conflicts (Becker 2001), whilst others associated both of these issues with economic growth and efficiency of land use (Margulis 2004). Promoters

Table 18.1 Economic activities and environmental services provided by community lands and settlements in the Amazon region

Institutional management	Land use categories	Social groups	Services provided	Economic activities
Chico Mendes Institute for Biodiversity Protection, (ICMBIo), Ministry of Environment	Extractivist Reserves (RESEX)	Traditional population	Biodiversity conservation, watershed preservation, carbon sink	Extractive activities/ subsistence activities (shifting cultivation and small livestock)
	Sustainable Development Reserves (RDS)	Traditional population	Biodiversity conservation, watershed preservation, carbon sink	Extractive activities/ subsistence activities (shifting cultivation and small livestock)
National Institute of Agrarian Reform, (INCRA), Ministry of Agrarian Development	Settlement Projects (PAF)	Colonists (migrants)	Territorial and economic occupation	Family farm agriculture market oriented
	Agro-Extractivist Project (PAE)	Colonists (migrants)/ traditional populations	Biodiversity conservation, watershed preservation, carbon sink	Extractive activities/ subsistence activities (shifting cultivation and small livestock)
	Sustainable Development Projects (PDS)	Colonists (migrants)	Biodiversity conservation, watershed preservation, carbon sink	Community forestry management/subsistence activities
Federal Bureau of Indigenous Populations, (FUNAI), Ministry of Justice	Indigenous Territories (TI)	Indigenous populations	Biodiversity conservation, watershed preservation, carbon sink	Hunting, gathering, shifting cultivation, social reproduction

of the socio-environmental model have argued that community-based management of natural resources both improves the means of subsistence for traditional and indigenous people and protects the forests (Almeida 2004) and that a mosaic of more diverse production systems and land use will have more positive environmental impacts than current large-scale pattern of land use for the Amazon (Peres et al. 2010).

To examine and help clarify such contrasting interpretations, we adopt the analytical framework of Hecht (2011) to compare land uses characterised by different landscape dynamics and with their links to different social groups—with the objective of understanding different historical trajectories and changing policies associated with two broad types of landscapes. In this approach 'neo-nature landscapes' refer to "agro-industrial modernist landscapes where land is basically a substrate for silvo-industrial, agro-industrial or livestock production, involving total landscape transformation into what are essentially monocultures" (Hecht 2011). By contrast, 'socio-nature landscapes' are the inhabited forested or mosaic landscapes that commonly comprise a large number of different land use types or production systems that are distinct from both the dominant modernisation development models and conventional conservation systems, including protected areas.

Although the local and traditional peoples that inhabit socio-nature landscapes had usually been ignored by official policies and threatened by an ongoing process of land privatisation, the emergence of socio-environmentalism as a model of land management has contributed substantially to the protection of Amazonian ecosystems (Alegretti 2011). However, it is increasingly uncertain how this model can resist the pressures of the international commodity markets and demands for energy production and resource extraction to fuel growing urbanisation and economic development.

Thus, the socioecological implications of these two very contrasting types of landscape categories, with their distinct ownership and production characteristics, need to be evaluated carefully when discussing regional public policy and development. In this chapter, we discuss some of the main aspects of land use patterns in the two systems, focusing on the key constraints and challenges presented by the different land uses associated with them. We also discuss the challenges of fostering more sustainable landscapes from the perspectives of both land management and land use planning. The framework we use to analyse land use sustainability and ecosystem service provision in the Amazon considers three main aspects: (i) the socioeconomic context, associated with the actors that are representative of the two types of landscape; (ii) the ecological aspects associated with landscape transformation and the dominant land use types; and (iii) the institutional context, in particular with regard to the Brazilian Forest Law (*Código Florestal*) and associated environmental policies (Fig. 18.1).

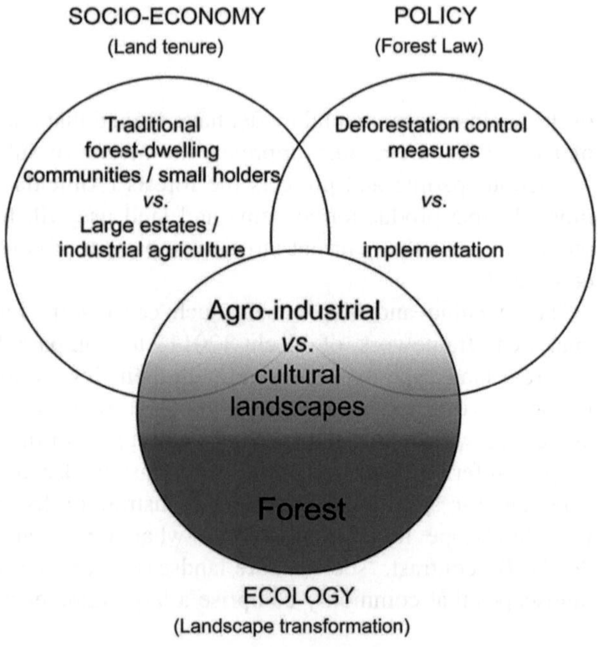

Fig. 18.1 Conceptual model for analysing the social, ecological and institutional context of 'socio-nature' and 'neo-nature' landscapes in Brazilian Legal Amazon

18.2 Land Use Patterns in Socio-Nature Landscapes: A Historical and Ecological Perspective

Socio-environmentalism emerged in response to poorly planned infrastructure projects and continued deforestation, undermining the conservation of biodiversity and the resilience of traditional and indigenous peoples. It started from the idea that the maintenance of biodiversity and ecosystem services should involve the participation and knowledge of populations in BLA (Alegretti 2011). Included in the 'socio-nature' landscapes are indigenous people, who maintain access to some common resources and depend on a combination of forest-based and a subsistence-oriented economy, although some are linked to markets and develop commercial activities, as well as traditional extractivist populations, subsistence smallholders, quilombolas and *caboclos* (riverine mestizo, descendants of colonisers and local indigenous people) who are associated with traditional practices of natural resource use.

These populations were coined as 'traditional people' by the legal framework of the country, and new land governance systems introduced the need for ecosystem conservation as a key criterion for land distribution (Santilli 2005) whilst also recognising the importance of different forms of collective ownership as a means to ensure access to land by local groups. Many indigenous groups and traditional populations have responded to changes in the social dynamics in BLA and have organised their communities to claim collective ownership of large areas of land. The success of many of these groups is due, in large part, to the resilience of their social organisations together with the assistance many have received from national and international NGOs since the 1980s.

The models of the new, collectively owned land designations such as RESEX and PDS were formally adopted by the National System of Conservation Units (SNUC), under the category of 'sustainable use areas'. The difficulties experienced by people in socio-nature landscapes posed a significant challenge for public policies that are aimed at reconciling economic development, wealth distribution and conservation.

The new matrix of conservation units and reserves that have shaped the conservation map of Amazonia dominated since 2000 by different forms of protected areas. Today *c.* 40 % of BLA is subject to some form of protection, 60 % of this in conservation units managed by local and traditional populations. In addition to some 111 million ha of indigenous land, there are some 70 extractive reserves and 19 sustainable development reserves, comprising a further 15 million ha (Table 18.2). In addition to these areas, there are several other categories of protected areas adding a further 65 million ha. In such protected areas, restrictions on natural resource use are intended to protect species under threat locally or prohibit practices threatening ecosystem integrity. Forest management for timber is allowed only if sustainable practices are implemented under an approved management plan. Forest clearing for agricultural uses in RESEX is limited to only 10 % of the total area. Also included in these requirements is the need for such land

Table 18.2 Categories of community lands in socio-nature landscapes in the Amazon region, Brazil (as of 2009)

Categories of community lands	Number of administrative areas	Area (million ha)	Population (thousands)	Source
Indigenous land	422	108.2	300	Alegretti (2011)
Sustainable use reserves	248	25.6	–	Alegretti (2011)
RESEX	70	13.9	223	
RDS	19	11.0	34	
Total	670	333.8		

management units to elaborate long-term environmental management plans, yet their implementation and enforcement continue to lack the necessary political will, and most areas are still lacking such a plan.

Traditional communities often advocate a sustainable development and forest conservation agenda. The role played by areas earmarked for sustainable conservation use in reducing deforestation along their boundaries is readily evidenced by satellite images showing differences in land use between indigenous reserves and adjacent areas, with boundaries clearly demarcated by the edge of the forest in indigenous reserves (Schwartzman et al. 2000; Ruiz-Pérez et al. 2004; Soares-Filho et al. 2010). The success achieved by indigenous reserves highlights the importance of land tenure issues to biological conservation in Amazonia. Although the proactive efforts of indigenous people to protect their lands have been central to preventing deforestation, of equal importance is the fact that squatters are unable to acquire titles to land settled inside indigenous reserves. Invasion and occupation of new land by farmers and ranchers is strongly influenced by the probability that their actions will eventually allow them to gain ownership and that their investments in infrastructure and forest clearance are not lost.

Sustainable use areas and extractive reserves also contain large stocks of carbon and are important refugia for biodiversity. Since 2002, the average deforestation in BLA has been 7–11 times lower inside indigenous lands and protected areas than in surrounding areas. Sustainable use areas that were established between 2003 and 2007 could prevent 272,000 km^2 of deforestation by 2050. This level of avoided deforestation is equivalent to the avoidance of some 3.3 ± 1.1 Gt carbon, which is equivalent to one-third of the world's annual CO_2 emissions (Soares-Filho et al. 2009). These sustainable use areas therefore have a high potential for maintaining ecosystem services (Table 18.1) and also in providing opportunities for social and economic stability by connection to different forest product markets. For instance, in 2010 the rubber sap extraction in 11 extractive reserves generated an income of R$ 16.5 million a year. Furthermore, Brazil nut (*Bertholletia excelsa* H. & B.) has a production potential in BLA for an annual income of R$ 39.2 million considering only 17 extractive reserves (UNEP 2011).

Understanding the monetary gains associated with the extraction of forest products is a complex task, and the scale and longevity of any financial benefits derived from these activities ultimately depend on the organisation of the value chain, and an efficient management organisation is necessary to promote the change in market structure so that the gains can migrate between the different levels of entrepreneurship and commercial agents. More research is urgently needed on the development of non-timber forest product value chains. However, to give an illustrative example, in the concessions of Madre de Dios, Peru, the Brazil nut industry, including collection, processing, transport and export of goods, employs approximately 30,000 people and provides some two-thirds of the annual income for these families (Baca 2006), with a gain of US$ 6 ha^{-1} $year^{-1}$ for unshelled products. However, as in the case of most natural products, profits can fluctuate substantially on an annual basis due to variations in annual productivity of the trees, market prices, input prices and transportation costs. The importance of the producers also being responsible for the processing is that the profit margin from selling processed nuts is c. 2.5 times that of unshelled nuts (Wadt et al. 2008). Nunes et al. (2012) underlined the role of Brazil nuts as an economically viable option capable of supporting both forest conservation and improvements in the livelihoods of forest-dependent peoples.

The economic success of extractive activities may increase significantly if all productive conservation units receive incentives to enhance their production capabilities (UNEP 2011). An increase in the value of forest extractivism in such units can help integrate local communities in the production process, increasing family income and reducing illegal extraction of forests goods whilst also fostering biodiversity conservation. Management policies of collectively owned traditional lands were intended to consolidate a 'rural extractivism' that made use of family labour in a type of production system capable of integrating agriculture and forest management (Costa 2009), with low impacts in terms of landscape transformation. This rural extractivism should be considered within a wider context of economic, cultural and social relationships involving the diversification and use of local techniques and knowledge, adapted to the peculiarities of Amazonian ecosystems and to the appreciation of biodiversity and culture amongst traditional local people (Rêgo 1999). The gross value of production of açaí palm (*Euterpe oleracea* Mart) and Brazil nut, for example, two of the most economically significant forest products in the Amazon region, considering the full supply chain (from local to national markets), is counted in the order of billions of reais, and so has a huge potential not only for maintaining jobs in small-scale agriculture but also for creating new jobs in the structure of services and urban small trade nationwide (Santana and Costa 2006; IDESP 2009). However, significant challenges remain to achieve the implementation of existing reserves and indigenous lands in high-risk deforestation areas of frontier expansion, such as in the southeast of Para state (Fig. 18.2).

In general terms, protected areas are still the focus of disputes due to the frequently conflicting interests of many social actors (Alves 2008). For example, large-scale agriculture is present (soybean and rice) in some settlement projects

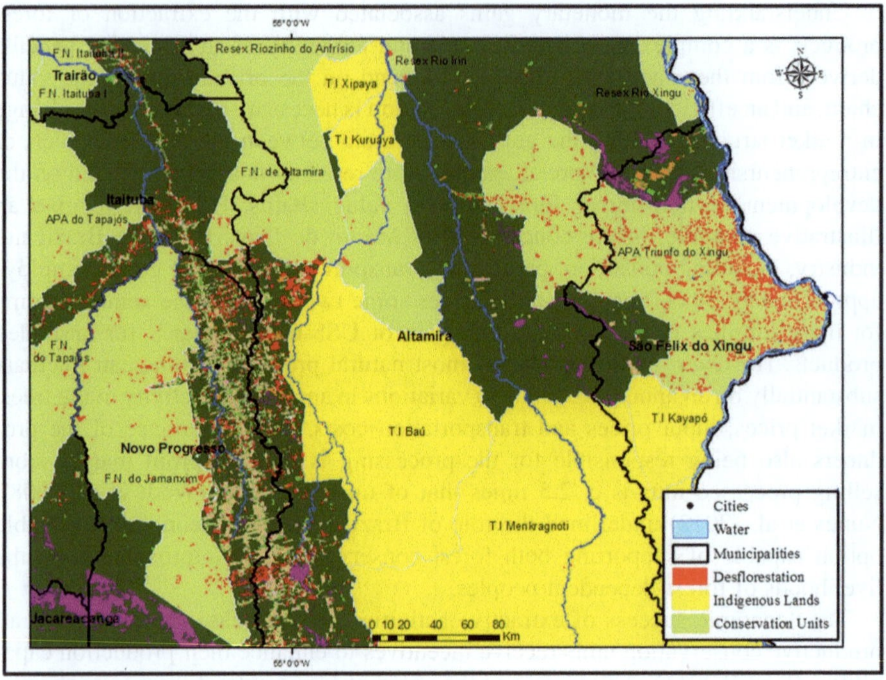

Fig. 18.2 Map of deforestation in southeastern Para state showing human pressure in indigenous lands, conservation units and areas of rural settlement programmes

established for small-scale farming. In addition, large-scale agricultural areas can also be found in Sustainable Development Project (PDS) areas, which, by definition, are territorial units 'of social and ecological interest, aimed at people who base their livelihood on the extraction, family agriculture and other activities of low environmental impact'. Research conducted by GEOMA (an Amazonian science network for environmental modelling) showed that large-scale agriculture continues to develop beyond the soy expansion area east of Santarém town inside PDS reserves. Despite a fall in the rate of deforestation, these trends contribute to increases in land tenure concentration (GEOMA 2010).

Many economists still question whether extractive reserves are capable of satisfying development goals (Bennett 2002; Goeschl and Igliori 2004). The success of these reserves in delivering both conservation and development depends on the ability of the traditional populations to secure increased income through the diversification of forest products. Unfortunately, in many areas this has not yet happened, and many of the forest dwellers have opted for increasing agricultural production and wood extraction (Ruiz-Pérez et al. 2004) and often can be found associated with large logging operations in unsustainable timber harvesting projects of huge enterprises, which in turn increase forest degradation. Maintaining populations in the forest requires implantation and regulation of sustainable uses of standing tropical forest, including extractive use of non-timber forest products and certification efforts for forest management for timber (Fearnside 2008). The lack of integrated policies specifically aimed at strengthening the economic potential and the needs of social systems inside sustainable use areas undermines their role as a democratic alternative to the development-centred model. In order to overcome these limitations, it is important to take into consideration a shared planning system between the public sector and the rural communities, by mutual agreements regarding the sustainable use of forest goods.

Ultimately, the policy of creating protected areas for sustainable use by traditional populations has partly succeeded mainly because it created an institutional mechanism for resolving conflicts over land and forest use. Socially and culturally it was important because it offered a means of life for present and future generations and also respected traditional forms of natural resource use. Furthermore, there was an appreciation of environmental priorities in efforts to prevent deforestation and ensure the protection of additional areas for biodiversity conservation and environmental services. However, in order for this socio-environmental model to become more robust and successful and have regional impact on the conservation of ecological services, there still remains an urgent need for a paradigm shift within Brazilian society to value the forest and forest-based economic activities over conventional development models.

The ongoing infrastructure constructions which are now under way, which favour hydroelectric development through a network of river dams, and the rapid expansion of agribusiness and mineral commodities need to be more systematically assessed alongside sustainable practices that ensure value-standing forests including through the development of forest products and biotechnology. The ultimate success of the socio-nature landscapes as a model of sustainable development will depend on a better balance of power between viable economic and commercial activities under a global market paradigm and the insertion of forest-based products associated to coherent financial incentives policies directed to local rural communities.

18.3 Land Use Patterns in Neo-Nature Landscapes and Their Implications for Sustainability

The development-centred model that underpins the 'neo-nature' landscape type is central to the debate on deforestation, biodiversity loss, sustainability of land uses and social inequalities. 'Neo-nature' landscapes are dominated by large-scale farmers and ranchers, who have converted forest lands into agricultural fields and who often have privileged access to financial resources and markets. This analysis will focus mainly on these groups.

Cattle production in BLA increased by 140 % over 12 years (1999–2011)—a remarkable growth rate that is almost three times than that of the human population in the region—causing extensive environmental damage (Table 18.3). In the

Table 18.3 Estimated environmental characteristics of two broad categories of landscapes in the Amazon region

Types of landscapes	C emission	Reduction of forest cover
Socio-nature landscapes	~40 %	1.5–31 %
Neo-nature landscapes	~70 %	60–70 %

Source: Adapted from Costa (2009)

meantime, the size of ranches has increased by 90 %, as a result of both the low price of land and the opening up of new pasture areas through illegal logging (Valentim and de Andrade 2009). This makes cattle ranching as the main driver responsible for most of the Amazonian deforestation (>67 %) (Margulis 2004; Escada et al. 2005) and some 60–80 % of total GHG emissions (Wassenaar et al. 2007). In addition to the direct ecological impact of converting forest to pasture, the widespread process of land conversion has also contributed towards reformulating the entire socioecological complex by negatively affecting other land use systems such as shifting cultivation, further reducing natural landscapes and ecological diversity (Vieira et al. 2012).

The expansion of cattle pastures across much of BLA has also resulted in habitat loss and fragmentation with increased edge effects and the isolation of remaining forest patches (Peres et al. 2006; Laurance et al. 2011 and Chap. 17). For example, 6.4 % of remaining forest habitats were found within a radius of 100 m of a forest edge in a study that evaluated an area of 1.12 million km^2 (>80 % of BLA affected by deforestation and selective logging) (Broadbent et al. 2008). The process of fragmentation includes tree dieback (Laurance 2000), changes in mutualistic and trophic interactions between species (Peres and Michalski 2006) and synergistic actions with other factors, such as susceptibility to fire that leads to local species extinctions (Alencar et al. 2004).

The conversion of forest to pasture also dramatically reduces the original biodiversity (Gardner et al. 2009), even when considering the large heterogeneity of pastures and pasture management, e.g. in terms of woody regrowth (Vieira et al. 2008). In comparison with primary forests, pasture lands are characterised by a small number of generalist and edge-tolerant or gap-specialist forest species or exotics. An example is that of scarab beetles whose species richness (58) in intact forest in southeast Amazonia was reduced to 13 in pastures, with 87 % of individuals belonging to a single species (Scheffler 2005). Another study in central Amazonia reported that native earthworms were either rare or absent from pastures and often replaced by exotic species (Decaens et al. 2004). In a secondary forest–pasture mosaic area, 47 frugivorous birds were observed to move from forest to pasture in search of food; 18 of them used a stretch of a maximum of 80 m with bushes near the forest (Silva et al. 1996). Possibly the most negative indirect impact of pastures (i.e. other than the actual clearance itself) on biodiversity is through fires that are used for forest clearing or pasture maintenance and then escape into surrounding areas of forest (Barlow et al. 2003; Chap. 13). Other impacts, such as pesticides or herbicides from cultures that leak into water courses, are poorly understood.

The distribution of land use types and vegetation cover reflects the policies devoted to large-scale cattle raising implemented in the region (INPE 2013). The economic activities that replace forests are often of short duration, and the abandonment of pastures is likely to be related to a lack of profitability, as well as the initial use of cattle farms for land speculation. This can result in recurring cycles of use followed by abandonment and secondary succession (Perz and Skole 2003). These secondary forests form a mosaic in heavily anthropic landscapes and have a relatively high potential conservation value (Peres et al. 2010). In practice, the recovery of biodiversity in secondary forests is a slow process, and relatively old secondary forests have a limited build-up of species richness (Vieira et al. 1996), which may be considered in such land use and long-term restoration plans. Recently, Martin et al. (2013), in a global meta-analysis, indicated that carbon pools and biodiversity show different recovery rates under passive secondary succession and that colonisation by plant species characteristic of undisturbed forest is slow. However, due to extensive land transformation and forest conversion driven by livestock husbandry and the cultivation of commodity crops, the secondary vegetation should be assigned a new function in the landscape, and the role of these areas in anthropogenic environments, including reforestation and forest regrowth associated with more sustainable land use types, needs to be better understood.

In neo-nature landscapes cattle raising and logging often have synergistic effects on the remaining areas of forest, e.g. through forest fires, that affects portions of the Amazon and threatens the ecological integrity in these disturbed areas inside the biome (e.g. Asner et al. 2005). Emissions of carbon from forest degradation can become as important as the rate of emissions from deforestation by clear cutting (Berenguer et al. 2014) (Fig. 18.3).

A recent study based on model simulations predicted that by 2050, a decrease in precipitation caused by deforestation in the Amazon will have reduced pasture productivity by 30 % in the governance scenario and by 34 % in the business-as-usual scenario (Oliveira et al. 2013). These estimates suggest also that large-scale expansion of agriculture in Amazonia may result in a decrease of carbon storage between 23 % and 50 % according to scenarios for 2020.

In addition to cattle-raising activities, 66 % of the planted area in BLA in 2010 was devoted to soybean, corn, rice, cassava and perennial crops, being responsible for 92 % of the agricultural gross income (IBGE 2012). The development of soybean production has been associated with investments in infrastructure, waterways, railways and highways to facilitate transport to export facilities. The expansion of this crop has mainly occurred at the expense of pastures, yet some forest areas have also been converted directly to soy production (Morton et al. 2006).

The intensity of expansion of the agriculture frontier in the Amazon has led to the implementation of a series of initiatives aimed at strengthening governance and minimising and mitigating environmental impacts. A federal programme to combat deforestation (the PPCDAm) based on satellite monitoring in all of BLA has been

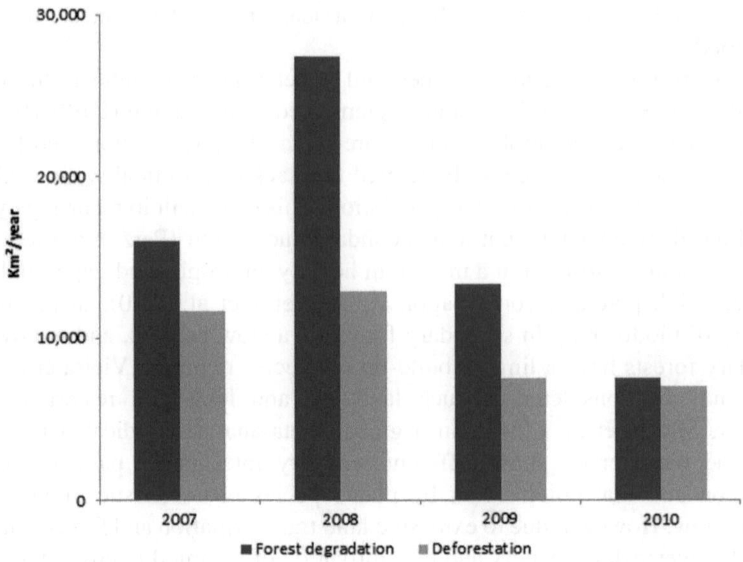

Fig. 18.3 Percentage of deforestation and degradation in the Amazon region, Brazil

successful in lowering significantly the rates of illegal deforestation by means of command and control strategies (Assunção et al. 2013; Godar et al. 2014), establishing a new benchmark for integrated action to combat illegal deforestation in the region. Through PPCDAm, the Amazon deforestation issue entered as a priority of the federal government's policy agenda, involving a large number of ministries. The first and second federal Action Plan to Prevent and Control Deforestation in the Amazon (PPCDAM I and II), implemented between 2004 and 2011, increased the national enforcement capacity for combating deforestation. However, major challenges, such as land use regularisation and organisation of sustainable production chains, remain the greatest challenges of this programme, in its third phase. Since 2012 Brazil has also had a revised Forest Law that created a new momentum in land use practices. According to this law 50–80 % of the area of each rural property in the Amazon region is required to be set aside to preserve natural vegetation. This is called the legal reserve (LR) which must be managed sustainably to maintain biological diversity and ecological integrity. Environmentally sensitive areas, such as habitats located in riparian zones, on steep terrain and on hilltops, are designated as permanent protection areas (APP) and are excluded from any type of economic activity.

The estimated deficit of the total area that should be preserved as LR on private farmland in BLA is 22 M ha (or 34 % of the total LR area in Brazil) (Sparovek et al. 2012; Fig. 18.4). This means that the majority of farms in BLA are currently contravening the Forest Law. The Forest Law requires mandatory forest restoration in cases where there is a deficit of LR or APP, but to ensure that the law will be respected, it is necessary to establish a sociopolitical system that links biological

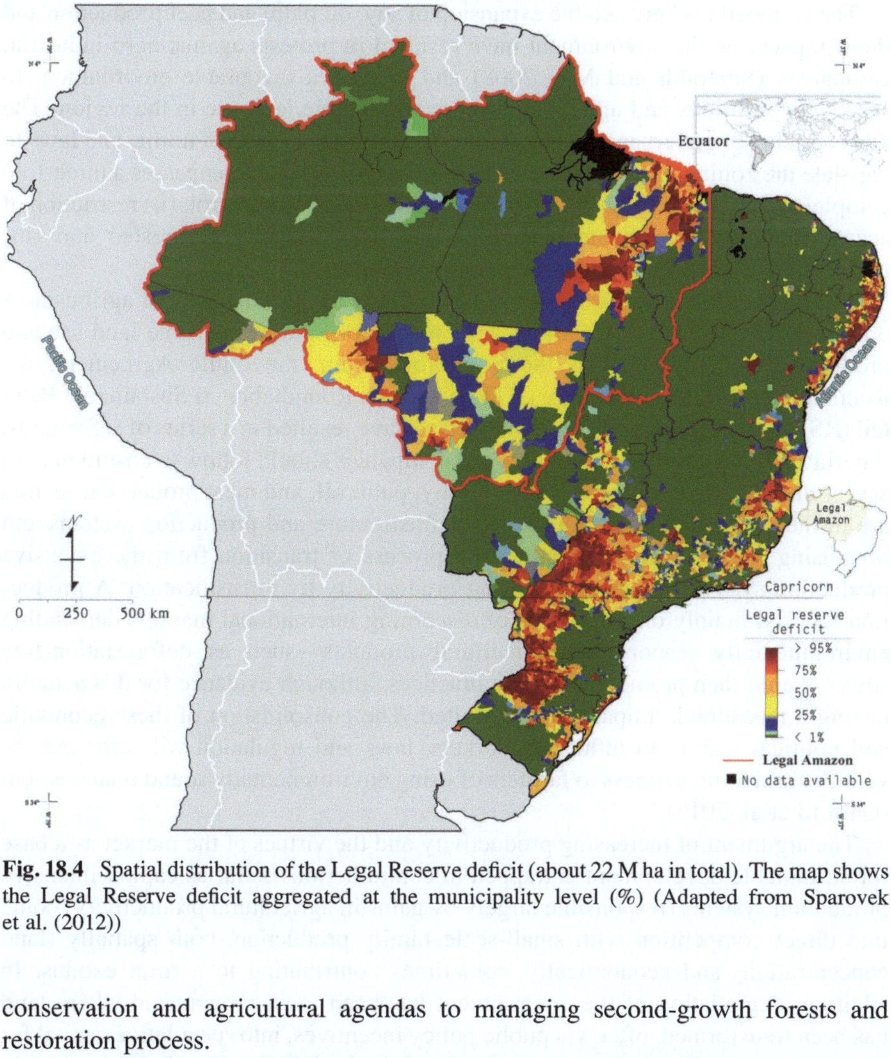

Fig. 18.4 Spatial distribution of the Legal Reserve deficit (about 22 M ha in total). The map shows the Legal Reserve deficit aggregated at the municipality level (%) (Adapted from Sparovek et al. (2012))

conservation and agricultural agendas to managing second-growth forests and restoration process.

The challenges for land use sustainability in 'neo-nature' landscapes are high in BLA, with transformations of the rural production base, dispute for new territories and land tenure regulation processes. In recent years, BLA has also become the target of the development of a rapidly expanding biofuel sector. Land owners currently are preparing a proposal for including exotic species mixed with native species (considered as low-impact replacement vegetation) and oil palm as a potential 'sustainable' option for the restoration of degraded riparian areas. However, oil palm plantations have a highly negative environmental impact, as they host very few native species of flora and fauna and require substantial agrochemical input and therefore should under no circumstance be included as potential replacement of (native) forest vegetation (Lees and Vieira 2013).

The connections between the expansion of soy, oil palm and beef production and their impacts on the environment have resulted in protests against agro-industrial companies (Smeraldi and May 2008) and created a favourable environment to search for solutions and agreeing on more sustainable land use in the region. The key elements of governance today are associated with a group of norms and laws to legislate the zoning of land use types allowed in BLA that encompasses a three-fold complementary action plan with (i) effective deforestation control, (ii) restriction of access to credit for private properties that have illegally deforested and (iii) elaboration of economic–ecological zoning plans.

To respond to criticisms of the unsustainability of the expansion of agribusiness onto forested land in BLA, various agreements for more responsible land use and production systems, such as the Soy Moratorium, the Cattle Agreement, the Roundtable on Responsible Soy (RTRS) and the Roundtable on Sustainable Palm Oil (RSPO), have emerged. These initiatives have resulted in a series of safeguards, criteria and indicators that producers and companies should follow to ensure market access and license to operate. Thus, the soy, palm oil, and meat production sectors are in the process of constructing new infrastructure and production systems and organising local production chains in a process of transition from the extensive production systems towards increasing productivity by intensification. A production focused mainly on the supply of discerning international markets demanding environmentally responsible agricultural products—such as deforestation-free soya—would then promote good-use practices, although evidence for this actually having a measurable impact is still limited. The consolidation of these economic and political factors to influence markets, laws and regulation will promote the viability and attractiveness to farmers of using environmentally sound management (Galford et al. 2013).

The argument of increasing productivity and the virtues of the market as a base for sustainable development is trapped in contradictions as far as capital-intensive production systems (responsible largely by gains in agricultural productivity) come into direct competition with small-scale family production, both spatially (land concentration) and economically, sometimes contributing to a rural exodus. In addition, degradation of the environment has been most pronounced where land has been transformed, often via public policy incentives, into speculative capital for future appreciation or into raw material for globalised production chains, such as in the case of livestock and soy, and now also in the case of African oil palm.

18.4 The Challenge of Creating and Maintaining Sustainable Land Uses and Landscapes

The development of sustainable land use systems requires strategies that allow all types of social actors to be involved and become committed to socioeconomic and environmental sustainability (Araújo et al. 2010). However, the Brazilian public sector faces a contradiction here: the implementation of environmental regulatory measures demands public institutions that are strong and solid and that have political and social legitimacy, yet different state agencies suffer from conflicting agendas with respect to the environment and land management—undermining the confidence in the public sector.

Sustainable development is threatened by deep-seated differences in political priorities for the region as well as by general disagreements relating to the role of different elements in the national and global demand for commodities. Despite Brazil having been successful in dramatically reducing rates of deforestation in BLA during the past decade through the federal programme to combat deforestation in Amazonia (PPCDAm), there are still serious conflicts in relation to the integration of much of the regional population into the political process, together with unresolved and systemic land tenure issues. The emerging multilevel governance mechanisms have not been sufficient to promote a broader transition to sustainable land uses in the region (Piketty et al. 2015). Landscape planning approach induced by different social and economic interests is still a challenge for more environmentally sounded changes in land use in the Amazon region.

The dynamics of rural agricultural production in the region has led to new demands on land, with consequences for conservation. To minimise environmental impacts, a restructuring of the rural economy, in line with a more ecologically, socially and economically sustainable development trajectory, is required. Yet the characteristics of sustainable production systems still need to be defined with respect to mitigation and compensation of ecological and social impacts in BLA and in Brazil as a whole.

It has become clear that policies, actions and tools to promote changes towards sustainable development and efforts to minimise and mitigate negative environmental impacts at the level of individual farms need to consider the full value chain from the producer to end consumer as well as changes in regulatory and public policies. Land tenure continues to remain the most critical factor.

Under these perspectives the zero deforestation proposal for public policy in BLA originally presented by Vieira et al. (2005) in conjunction with the idea of sustainable use of landscapes has been incorporated into NGOs and state governors' agendas. Under this idea, the authors highlighted four key points for territorial planning: control and supervision of deforestation, maintenance of territorial integrity of indigenous lands and protected areas, creation of demonstration sites of ecological restoration in APPs and LRs and the formulation of a strong participatory scientific agenda with emphasis on the design and maintenance of territorial systems of conservation, technologies for sustainable production in areas of consolidated occupation and the valuation and distribution of the benefits of environmental services, including a list of actions to combat to poverty. Whilst many of the recommendations of Vieira et al. (2005) are evident in the PPCDAm, the primary recommendation has yet to be adopted (zero deforestation). The high political cost for such decision is reflected in the lack of political support for such a move. For example, only the governor of the State of Pará has thus far signed the protocol for zero deforestation proposed during Rio+20.

The pathways to sustainable development in BLA must take into consideration the complex history of the occupation of the region, the socioecological context of the main productive actors and their relationship with deforestation and conservation (Fig. 18.5). In the context of a new institutional framework of deforestation control and restriction to the access to credit, and also of adjustments to the Forest Law and to the Economic-Ecological Zoning, the agents who control socio- and neo-nature landscapes need to be adapted to this new set of rules that are in place to govern the rural territory. There is ample role for a scientific agenda which will stimulate efforts in increasing restoration areas for conservation and sustainable practices, with agroforestry systems playing a potential major role in these newly conserved areas that were originally used for production purposes. The creation of the Rural Environmental Registry (CAR) system promotes the compilation of geo-referenced information on the boundaries of rural properties and their LR and APP including satellite maps showing the property's extent and forest cover. Once a property is registered, the official environmental agencies can monitor deforestation compliance with the Forest Law. However, there remains a substantial challenge to

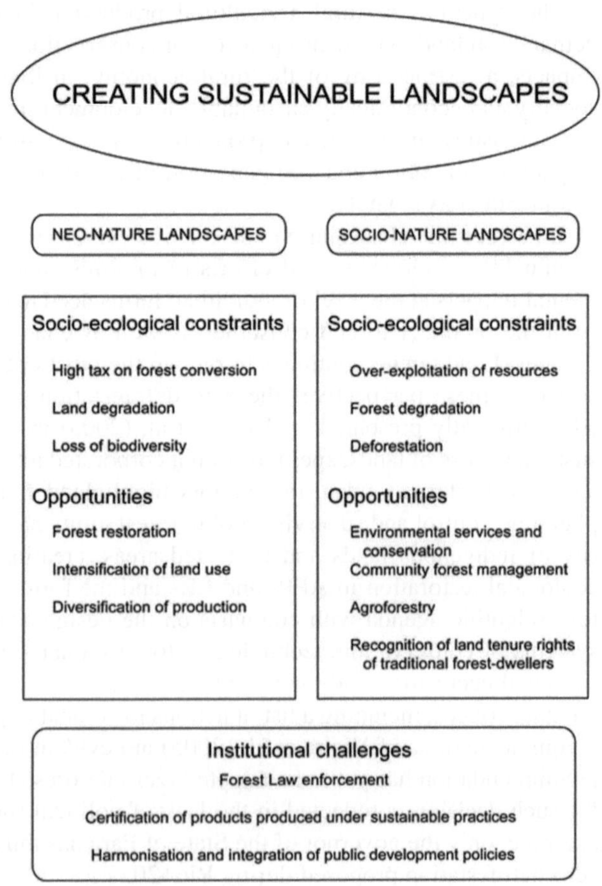

Fig. 18.5 The main elements of sustainable landscapes in Amazonia, considering 'socio-nature' and 'neo-nature' landscapes

regulate land use tenure in BLA, and whilst some progress is being made, there is an enormous deficit in infrastructure, human and financial resources to accomplish this regulation on thousands of rural properties, in addition to a necessary political will in the long term.

18.5 Concluding Remarks

Traditional populations have been impelled to modify their lifestyle and cultural values under a new set of concepts incorporating sustainability paradigms. This process resulted in the construction of a specific sustainable development model in the Amazon which has inserted these social groups as key actors in environmental public policies. The persistent incapacity of institutions to represent the needs and aspirations of local people has led to a counteract strategy evidenced by significant increase in claims made by indigenous or quilombola identity (as recorded by self-proclaimed indigenous peoples in the 2010 census), as an alternative for the recognition of land rights of traditional Amazonian inhabitants. Also they have demanded popular participation in governance processes, drawing on the development of strong grass-roots movements and increasingly widespread valorisation of their culture, as part of a bid for a more sustainable forest-based economy. This socio-environmentalist agenda has gained significant political attention and credibility. As a result of this trend, the creation of a mosaic of sustainable use-based conservation units has effectively reduced deforestation and highlighted the importance of land conflicts in discussions over the environmental sustainability of the region. Although remarkable social and conservation gains have been made in socio-nature landscapes, deep-seated problems of management, organisational uncertainties and paternalistic leadership persist (Shanley et al. 2011). Moreover, it can be observed in recent years a trend aimed at promoting agribusiness and export trading companies in Brazil, as well as the construction of infrastructures that enable the activities of the latter in parts of BLA. Such governmental agenda creates additional pressure on the current environmental public policies that have been showing signs of deep exhaustion.

Strategic alliances between grass-roots organisations and national and international NGOs have been mutually beneficial. NGOs have brought visibility, resource and access to local communities, whilst grass-roots movements have legitimised the goals of NGOs and values in territories that they would not be able to reach themselves. Since the Rio 92 Summit several NGOs synthesised scientific information into influential reports and white papers that have provided important supporting material for public policies in support of the socio-environmental model. More recently, under the perspective of climate change, several NGOs have redirected their agendas from a biological conservation and local sustainable practices focus towards a more holistic environmental approach, including ecosystem services and food security. These paradigm shifts continue to have important influences on future alliances between NGOs and grass-roots movements in the Amazon region.

Questions still remain regarding the future of extractive reserves and other sustainable use areas, including their effective protection and management of natural resources. The sustainable conservation units may serve as a reference across the world, both for carbon sequestration, which avoids deforestation, and for the establishment of a new sustainable economy supported by REDD+ (Aguiar et al. 2016). In order to maintain larger-scale landscape conservation in the long term, suitable incentives and investment towards environmental and social sustainability are necessary (Schwartzman et al. 2010). Such challenging perspective encompass the territories in which grass-roots movements have won protection, including forest-based and rural livelihoods, as well as actions towards reducing deforestation.

Considering a business-as-usual scenario of land use dynamics in BLA, the continued expansion of the neo-nature landscape model will result in similar land patterns to those observed in other Brazilian 'biomes' that are now dominated by agribusiness. However, laws and regulations are being reformulated at different governance levels, from state to municipality, motivated significantly by the revision of the Brazilian Forest Law. These also include certification mechanisms and attempt to seek local consensus amongst land use actors. Such institutional arrangements need to take into account ecosystem integrity at different scales in order to work towards better territorial planning in a way that combines both production and conservation objectives and imperatives.

Despite considerable efforts towards conservation and land tenure regulation, the observed trends of infrastructure development for roads and other modes of transportation linking the agribusiness regions to BLA, as well as port constructions for export purposes, dams for hydropower and mining activities, reinforce the prioritisation of the actual hegemonic economic patterns to expand agricultural production areas in all frontiers of available land in the country. Such a trend is inflicting additional pressures on natural landscapes in the Cerrado, Amazon forest and Caatinga 'biomes'. In the wider context, such activities align with the development of the Latin American infrastructure network, such as constructions of ports, as well as the Nicaraguan Canal crossing Central America, to permit large-vessel transportation of crude minerals and agricultural commodities for Oriental markets. A coherent development plan in Amazonia that takes into consideration development and conservation issues in a balanced fashion is essential to help better management of these increasingly threatened landscapes, improve the equitable distribution of benefits to more marginalised peoples and ensure the maintenance of natural ecological processes.

Acknowledgements We are thankful to the Instituto Nacional de Ciência e Tecnologia—Biodiversidade e Uso da Terra na Amazônia (CNPq 574008/2008-0) for the financial support. We gratefully acknowledge Toby Gardner, Laszlo Nagy and an anonymous referee for their critical comments and helpful suggestions on earlier versions of this chapter.

References

Aguiar APD, Vieira ICG, Assis TO, Dalla-Nora ELD, Toledo PM, Araújo R, Batistella M, Coelho A, Savaget EK, Aragão LEO, Nobre CA, Ometto JPH (2016) Land use change emission scenarios: anticipating a forest transition process in the Brazilian Amazon? Global Change Biol 22:1821–1840

Alegretti M (2011) Gestão Comunitária da Floresta Amazônica. In: Soerguimento tecnológico e econômica do extrativismo na Amazônia—Centro de Gestão e Estudos Estratégicos, Brasília, DF, pp. 27–60

Alencar AA, Solorzano LA, Nepstad DC (2004) Modeling forest understory fires in an eastern Amazonian landscape. Ecol Appl 14:139–149

Almeida AWB (2004) Terras tradicionalmente ocupadas, processos de territorialização e movimentos sociais. Revista Brasileira de Estudos Urbanos e Regionais 6(1):9–32

Alves DS (2008) Taking things public: a contribution to address human dimensions of environmental change. Philos Trans R Soc Lond B Biol Sci 363:1903–1909

Araújo RS, Alves D (2008) Mudanças Ambientais na Amazônia e as particularidades da Construção Institucional. In: Batistella M, Moran EF, Alves AS (eds) Amazônia: natureza e sociedade em transformação, vol 1, 1st edn. Edusp, São Paulo, pp 221–240

Araújo RS, Lena, P (2010) Desenvolvimento Sustentável e Sociedades na Amazônia, MPEG/CNPq/PPG-7, Belém-Pará

Araújo RS, Costa FA, Aguiar APD, Toledo PM, Vieira ICG, Camara G (2010) Desmatamento, trajetórias tecnológicas rurais e metas de contenção de emissões na Amazônia. Ciência e Cultura 62:56–59

Asner GP, Knapp DE, Broadbent EN, Oliveira PJC, Keller M, Silva JN (2005) Selective logging in the Brazilian Amazon. Science 310:480–482

Assunção J, Gandour C, Rocha R, Rocha R (2013) Does credit affect deforestation? Evidence from a rural credit policy in the Brazilian Amazon. Available via CPI technical report. http://climatepolicyinitiative.org/wp-content/uploads/2013/01/Does-Credit-Affect-Deforestation-Evidence-from-a-Rural-Credit-Policy-in-the-Brazilian-Amazon-Technical-Paper-English.pdf. Cited 1 Oct 2013

Baca LC (2006) Diagnóstico del programa regional de biocomercioen la Amazonía. PROMPEX, Secretería Técnica de laComisión Nacional de Promoción del Biocomercio. http://cdam.minam.gob.pe/publielectro/biocomercio/diagnosticobiocomercioamazonia.pdf. Cited June 2015

Barlow J, Peres CA, Lagan BO, Haugaasen T (2003) Large tree mortality and the decline of forest biomass following Amazonian wildfires. Ecol Lett 6:6–8

Becker BK (2001) Síntese das políticas de ocupação da Amazônia. Lições do passado e desafios do presente. In: Brasil MMA. Causas e dinâmicas do desmatamento da Amazônia. MMA/SCA, Brasília

Bennett EL (2002) Is there a link between wild meat and food security? Conserv Biol 16 (3):590–592

Berenguer E, Ferreira J, Gardner T, Aragão L, Camargo P, Cerri CE, Durigan M, de Oliveira Junior RC, Vieira ICG, Barlow J (2014) A large-scale field assessment of carbon stocks in human-modified tropical forests. Glob Chang Biol. doi:10.1111/gcb.12627, pp. 1–14

Broadbent EN, Asner GP, Keller M, Knapp DE, Oliveira PJC, Silva JN (2008) Forest fragmentation and edge effects from deforestation and selective logging in the Brazilian Amazon. Biol Conserv 14:1745–1757

Costa FA (2009) Dinâmica agrária e balanço de Carbono na Amazônia. Economia 10(1):117–151

Davis S (1977) Victims of miracle: development and the Indians of Brazil. Cambridge University Press, Cambridge

Instituto de Desenvolvimento Econômico, Social e Ambiental do Estado do Pará-Idesp (2009) Estudo das Cadeias de Comercialização de Produtos Florestais Não-madeireiros no Estado do Pará: Relatório Técnico Preliminar de Atividades

Instituto Nacional de Pesquisas Espaciais—INPE (2013) Sumário Executivo TERRACLASS. http://www.inpe.br/cra/projetos_pesquisas/sumario_terraclass_2010.pdf. Cited Sept 2013

Decaens T, Jimenez JJ, Barros E, Chauvel A, Blanchart E, Fragoso C, Lavelle P (2004) Soil macrofaunal communities in permanent pastures derived from tropical forest or savanna. Agric Ecosyst Environ 103:301–312

Escada MIS, Vieira ICG, Kampel AS, Araújo RS, Veiga JB, Aguiar APD, Veiga I, Oliveria M, Pereira JLG, Carneiro Filho TSG, Monteiro AMV, Camara G (2005) Processos de Ocupação nas Novas Fronteiras da Amazônia (O Interflúvio do Xingu/Iriri). Estudos Avançados (USP Impresso), São Paulo 19(54):9–23

Instituto Nacional de Geografia e Estatística—IBGE (2012) Censo Agropecuário 2006 http://www.ibge.gov.br/home/estatistica/economia/agropecuaria/censoagro/default.shtm. Cited Oct 2013

Fearnside PM (2008) The roles and movements of actors in the deforestation of Brazilian Amazonia. Ecol Soc 13(1):23–32

Ferreira AMM, Salati E (2005) Forças de transformação do ecossistema amazônico. Estudos Avançados 19(54):25–44

Galford GL, Doares-Filho B, Cerri C (2013) Prospects for land-use sustainability on the agricultural frontier of the Brazilian Amazon. Phil Trans R Soc B 368:1–5

Gardner TA, Barlow J, Chazdon R, Ewers RM, Harvey CA, Peres CA, Sodhi NS (2009) Prospects for tropical forest biodiversity in a human-modified world. Ecol Lett 12:561–58

Geoma (2010) Usos da Terra na área de expansão de soja a leste de Santarém, Pará. (orgs Coelho A, Araujo R, Canto LO and Aguiar AP), Belém-Pará, 50p

Godar J, Gardner TA, Tizado EJ, Pacheco P (2014) Actor-specific contributions to the deforestation slowdown in the Brazilian Amazon. Proc Natl Acad Sci USA 111(43):15591–15596

Goeschl T, Igliori DC (2004) Reconciling conservation and development: a dynamic hotelling model of extractive reserves. Land Econ 80(3):340–354, University of Wisconsin Press

Hecht SB (2011) From eco-catastrophe to zero deforestation? Interdisciplinarities, politics, environmentalisms and reduced clearing in Amazonia. Environ Conserv 39(1):4–19

Hogan DJ (2010) Demographic dynamics and environmental change in Brazil. Ambiente & Sociedade 13(1):1–28

Laurance WF (2000) Do edge effects occur over large spatial scales? Trends Ecol Evol 15:134–135

Laurance WF, Camargo JLC, Luizão RCC, Laurance SG, Pimm SL, Bruna EM, Stouffer PC, Bruce WG, Benítez-Malvido J, Vasconcelos HL, Houtan KS, Zartman CE, Boyle SA, Didham RK, Andrade A, Lovejoy TE (2011) The fate of Amazonian forest fragments: a 32-year investigation. Biol Conserv 144:56–67

Lees A, Vieira ICG (2013) Oil-palm concerns in Brazilian Amazon. Nature 497:188. doi:10.1038/497188c, 9 May 2013

Leroy JP (1991) Uma Chama na Amazônia. Ed. Vozes, Petropolis

Loureiro VR, Pinto JA (2005) A questão fundiária na Amazônia. Estudos Avançados 19 (54):77–98

Margulis S (2004) Causes of deforestation of the Brazilian Amazon. World Bank working paper no. 22, 104p

Martin PA, Newton AC, Bullock JM (2013) Carbon pools recover more quickly than plant biodiversity in tropical secondary forests. Proc R Soc B 280(1773), 20132236, 1471–2954

Moran EF (1981) Developing the Amazon. Indiana University Press, Bloomington, IN, 320p

Morton DC, DeFries RS, Shimabukuro YE, Anderson LO, Arai A, Espirito-Santo FB, Freitas R, Morisette J (2006) Cropland expansion changes deforestation dynamics in the southern Brazilian Amazon. Proc Natl Acad Sci USA 103(39):14637–14641

Nunes F, Soares-Filho B, Giudice, R, Rodrigues, H, Bowman, M, Silvestrini R, Mendoza E (2012) Economic benefits of forest conservation: assessing the potential rents from Brazil nut concessions in Madre de Dios, Peru, to channel REDD+ investment. Environ Conserv:page 1 of 12 © Foundation for Environmental Conservation 2012. doi:10.1017/S0376892911000671

Oliveira LJC, Costa MH, Soares-Filho BS, Coe MT (2013) Large-scale expansion of agriculture in Amazonia may be a no-win scenario (2013). Environ Res Lett 8:024021

Pacheco P (2009) Agrarian change, cattle ranching and deforestation: assessing their linkages in Southern Pará. Environ Hist 15:493–520

Peres CA, Michalski F (2006) Synergistic effects of habitat disturbance and hunting in Amazonian forest fragments. In: Laurance WF, Peres CA (eds) Emerging threats to tropical forests. University of Chicago Press, Chicago, IL, pp 105–127

Peres CA, Barlow JB, Laurance WF (2006) Detecting anthropogenic disturbance in tropical forests. Trends Ecol Evol 21:227–229

Peres CA, Gardner TA, Barlow BJ, Zuanon J, Michalski F, Lees AC, Vieira ICG, Moreira FMS, Feeley KJ (2010) Biodiversity conservation in human-modified Amazonian forest landscapes. Biol Conserv 143:2314–2327

Perz SG, Skole DL (2003) Secondary forest expansion in the Brazilian Amazon and the refinement of forest transition theory. Soc Nat Resour 16:277–1154

Piketty MG, Chapuis-Poccard R, Drigo I, Coudel E, Plassin S, Laurent F, Thâles M (2015) Multi-level governance of land use changes in the Brazilian Amazon: lessons from Paragominas, State of Pará. Forests 2015(6):1516–1536

Rêgo JF (1999) Amazônia: do extrativismo ao neoextrativismo. Disponível em:http://www.adur-rj.org.br/5com/popup/extrativismo_neoextrativismo.pdf. Cited 2013

Ruiz-Pérez M, Belcher B, Achdiawan R, Alexiades M, Aubertin C, Caballero J, Campbell B, Clement CR, Cunningham T, Fantini A, de Foresta H, García-Fernández C, Gautam KH, Hersch-Martínez P, de Jong W, Kusters K, Kutty MG, López C, Fu M, Martínez-Alfaro MA, Nair TKR, Doye N, Ocampo R, Rai N, Ricker M, Schreckenberg K, Shackleton S, Shanley P, Sunderland T, Youn Y (2004) Markets drive the specialization strategies of forest peoples. Ecol Soc 9(2):4, http://www.ecologyandsociety.org/vol9/iss2/art4

Salati E, Schubart H, Junk WJ, Oliveira AE (1983) Amazônia: desenvolvimento, integração e ecologia. Ed. Brasiliense e CNPq, São Paulo

Santana AC, Costa FA (2006) O mercado do açai no Pará. In: Hervé R (ed) O açai ao alcance de todos. ADA, UFPA, UFRA, Belém, 32p

Santilli J (2005) Socioambientalismo e novos direitos: proteção jurídica à diversidade biológica e cultural. Peirópolis/ISA, São Paulo, 303p

Scheffler PY (2005) Dung beetle (Coleoptera: Scarabaeidae) diversity and community structure across three disturbance regimes in eastern Amazonia. J Trop Ecol 21:9–19

Schmink M, Wood CW (eds) (1984) Frontier expansion in Amazonia. University of Florida Press, Gainesville, FL

Schwartzman S, Nepstad DC, Moreira A (2000) Arguing tropical forest conservation: people versus parks. Conserv Biol 14:1370–1374

Schwartzman S, Alencar AA, Zarin D, Souza APS (2010) Social movements and large-scale tropical forest protection on the Amazon frontier: conservation from chaos. J Environ Dev 19 (3):274–299

Shanley P, Da Silva FC, Macdonald T (2011) Brazil's social movement, women and forests: a case study from the National Council of Rubber Tappers. Int For Rev 13(2):233–244

Silva JMC, Uhl C, Murray G (1996) Plant succession, landscape management, and the ecology of frugivorous birds in abandoned Amazonian pastures. Conserv Biol 10:491–503

Smeraldi R, May P (2008) O Reino do Gado: uma nova fase na pecuarização da Amazônia. Amigos da Terra- Amazônia Brasileira, São Paulo, 40p

Soares-Filho B, Dietzsch L, Moutinho P, Falieri A, Rodrigues H et al (2009) Redução das emissões de carbono do desmatamento no Brasil: O papel do Programa Áreas Protegidas da Amazônia (ARPA). UFMG, IPAM, WHRC, WWF, Brasilia, Brazil

Soares-Filho B, Moutinho P, Nepstad D, Anderson A, Rodrigues H, Garcia R, Dietzsch L, Merry F, Bowman M, Silvestrini R, Maretti C (2010) Role of Brazilian Amazon protected areas in climate change mitigation. Proc Natl Acad Sci USA 107(24):10821–10826

Sparovek G, Berndes G, Barreto AGOP, Klug ILF (2012) The revision of the Brazilian Forest Act: increased deforestation or a historic step towards balancing agricultural development and nature conservation? Environ Sci Pol 16:65–72

UNEP (2011) Contribuição das unidades de conservação brasileiras para a economia nacional: Sumário Executivo/Rodrigo Medeiros, Carlos Eduardo Frickmann Young, Helena Boniatti Pavese & Fábio França Silva Araújo; Editores. UNEP-WCMC, Brasília, 44p

Valentim JF, de Andrade CMS (2009) Tendências e Perspectivas da Pecuária Bovina na Amazônia Brasileira. Amazônia: Ci & Desenv Belém 4(8):9–32

Vieira ICG, Nepstad DC, Salomão RP, Roma JC, Rosa NAR (1996) Região Bragantina: as florestas secundárias após um século de agricultura na Amazônia. Ciência Hoje 20(119):38–44

Vieira ICG, Silva JMC, Toledo PM (2005) Estratégias para evitar a perda de biodiversidade na Amazônia. Estudos Avançados 19(54):153–164

Vieira ICG, Toledo PM, Silva JMC, Higuchi H (2008) Deforestation and threats to the biodiversity of Amazonia. Braz J Biol 68(4):949–956

Vieira ICG, Toledo PM, Araújo RS, Coelho A, Baião P (2012) Amazônia. In: Scarano F, Santos I, Martins AC, Cardoso JM, Guimarães A, Mittermeir R (eds) Biomas brasileiros: retratos de um país plural, 1st edn. Editora Casa da Palavra, Rio de Janeiro, pp 129–164

Vieira ICG, Araújo R, Toledo PM (2014) Dinâmicas produtivas, transformaçõs no uso da terra e sustentabilidade na Amazônia. In: Siffert N, Cardoso M, Magalhães W, Lastres H (eds) Um olhar territorial para o desenvolvimento da Amazônia, 1st edn. BNDES, Rio de Janeiro, pp 370–395

Wadt LO, Kainer KA, Staudhammer C, Serrano K (2008) Sustainable forest use in Brazilian extractive reserves: natural regeneration of Brazil nut in exploited populations. Biol Conserv 141(1):332–346

Wassenaar T, Gerber P, Verburg PH, Rosales M, Ibrahim M, Steinfeld H (2007) Projecting land use changes in the Neotropics: the geography of pasture expansion into forest. Global Environ Change 17:86–104

Part VI
Perspectives for the Future

Chapter 19
Amazonia in Perspective as a Changing Environment

Paulo Artaxo, Bruce R. Forsberg, and Laszlo Nagy

The Amazonian socio-ecological system is in a transition phase, having moved since the 1960s from a relatively pristine set of ecosystems to a complex mosaic where large parts are dominated by intensive land use (Malhi et al. 2014). Several direct human drivers, such as deforestation, hydroelectric dams, roads, and climate change are impacting ecosystem function at the landscape scale, and this is manifested at the regional scale through changes in the functioning of the entire Amazon ensemble, affecting climatic teleconnections throughout South America. As in other tropical forest biomes (Sala et al. 2000; Hassan et al. 2005), the main and increasingly growing impact on ecosystem integrity is continued large-scale deforestation and agricultural expansion (Davidson et al. 2012). In general, deforestation goes hand in hand with road construction and urbanisation, which affect terrestrial as well as aquatic ecosystems and air quality in large parts of Amazonia and southern South America. Research in the region is now focused on the link between global climate change and the functioning of the Amazon forest biome, especially the observed changes in the hydrological cycle and increased temperatures.

Amazonia is critically important in terms of carbon balance in our planet. The forest, representing about 40 % of all tropical forests worldwide contains 90–120 Pg C in living biomass. Soils are also important carbon pools, storing about 160 Pg C. If, as a result of land use or climate change, a small fraction of this carbon were released into to the atmosphere, it could significantly enhance

P. Artaxo (✉)
Instituto de Física, Universidade de São Paulo, São Paulo, Brazil
e-mail: artaxo@if.usp.br

B.R. Forsberg
Instituto Nacional de Pesquisas da Amazonia, Manaus, Brazil

L. Nagy
Departamento de Biologia Vegetal, Universidade Estadual de Campinas, Campinas, Brazil

© Springer-Verlag Berlin Heidelberg 2016

L. Nagy et al. (eds.), *Interactions Between Biosphere, Atmosphere and Human Land Use in the Amazon Basin*, Ecological Studies 227,
DOI 10.1007/978-3-662-49902-3_19

atmospheric CO_2 concentrations. The recent AR5 IPCC report clearly shows that the interaction between physical climate and the biosphere represents one of the largest uncertainties in the assessment of the response of the climate system to greenhouse gas emissions (IPCC 2013). Over the last 20 years, there has been a marked decrease in carbon emissions from deforestation in Amazonia. This was due to the reduction of 27,000 km^2 of newly deforested areas in 2004 to 4,800 km^2 in 2014. This large reduction in deforested area decreased carbon emissions from Brazil by ca. 75 %. Recent studies in the global carbon cycle show that carbon sources and sinks in the tropics are approximately balanced (Chaps. 5 and 6), with regrowth and increased atmospheric CO_2-driven photosynthetic carbon uptake by the forest compensating the large deforestation source. It is unknown how stable this 'equilibrium' is. Several studies show that the Amazon biome is on a precarious balance between being a source and a sink of carbon. Its future depends on the extent and form of climate change, including the severity and frequency of extreme events as well as on socio-economical aspects of development in Amazonia. The region has warmed by about 0.6 °C, and warming is expected to continue see e.g. Marengo et al. (2016). It is also possible to observe an increased frequency of drought episodes that enhances the occurrence of periods of net carbon emissions from decomposition of dead organic material (Phillips et al. 2016) and from enhanced fire occurrence during droughts and following fuel accumulation (Chap. 13). Considering the hydrological cycle, the observed 20 % increase in the Amazon river's discharge may reflect an increasing water supply to the vegetation, which, together with increasing atmospheric CO_2, may lead to more intense net carbon uptake by the intact forest vegetation (Schimel et al. 2015). But this issue can be more complex, since interactions of the carbon cycle with the cycles of other key elements, such as nitrogen and phosphorus, will also play important roles (Townsend et al. 2011). It is important also to remember that Amazonia is an important source of methane (CH_4) and nitrous oxide (N_2O), especially its wetlands and soils (Chap. 7).

Several recent studies have shown that the Amazonian hydrological cycle is changing. This is important because the Amazon River is responsible for about 20 % of the world's freshwater discharge. Most of the water vapour enters the Amazon basin from the tropical Atlantic Ocean with the trade wind circulation. The forest has a very effective system of recycling of water through evapotranspiration, which maintains a flux of precipitation that becomes increasingly more important as air masses move into the central and western part of the Basin. Further west, at the Andes, moisture transport changes direction towards the south, carrying water vapour derived from the forest to Argentina and southern Brazil. The forests of Amazonia thus contribute to the precipitation that sustains agricultural yields in these regions, remote from the source. Changes in the flux of Amazonian water vapour and the impacts of smoke aerosols from Amazonian fires on cloud processes can thus affect rainfall patterns over large areas, extending to the southern tip of the South American continent.

Although modelling has contributed to recent developments in Amazonian research, no existing model represents all critical functional aspects of Amazonian ecosystems. The ability of models to predict possible scenarios for the Amazon forest biome in the coming decades is severely curtailed by limitations in the representation of key aspects of climate/vegetation interactions (e.g. Chap. 14), including the role of material transport to the Andes and the teleconnections between the Amazon and the tropical Atlantic and Pacific Oceans. In addition, the biophysical response of the vegetation to changing water supply and increasing CO_2 and temperature remains poorly understood.

We are far from understanding the complex atmospheric chemistry over Amazonia. The tropical atmosphere was referred to as the 'washing machine of the atmosphere' by Paul Crutzen because of the 'cleansing' action of hydroxyl radical (OH) produced over the forest. Human activities and the biosphere in its natural state release large amounts of trace gases to the atmosphere, such as nitrogen oxides (NO_x), carbon monoxide (CO), and volatile organic compounds (VOC). Most such gases are relatively insoluble in water and are thus not effectively washed out by tropical rain. The self-purification of the atmosphere involves reactions that produce ozone (O_3) and hydroxyl (OH) radicals. Production of these atmospheric 'detergents' requires large amounts of UV radiation and water vapour, both of which are present in large quantities in Amazonia. The reactions that control O_3, OH, CO, and VOCs in Amazonia are still a matter of active research (e.g. Chap. 2). The functioning of this self-cleansing mechanism is affected by human activities that change emissions from the biosphere (through land use change) and add pollutants from biomass burning.

Amazonia has a very special atmosphere in terms of aerosol particles. During the rainy season, which can vary from 12 months in western Amazonia to 8 months in southern Amazonia, the Amazon has among the lowest aerosol concentrations of any continental region in our planet. The vegetation emits large amounts of primary biogenic aerosol particles. Oxidation of VOCs produces significant amounts of Secondary Organic Aerosols (SOA) that affect radiation balance and cloud nucleation. At the other extreme, during the dry season (rainfall < 100 mm month^{-1}), biomass burning emissions in the southern Amazon make aerosol concentrations over large regions as high as in the most polluted urban areas worldwide. Episodic inputs of Saharan dust, biomass smoke from Africa, and marine aerosols transported over long distances with the trade winds during the wet season further complicate the picture. A fraction of these aerosol particles acts as Cloud Condensation Nuclei (CCN) that are critical ingredients for the formation of clouds and raindrops. The very low CCN concentration in the wet season (CCN c. 200–300 #/cc) makes droplets very large, resulting in very efficient precipitation processes, mostly through low clouds. In the dry season, with very high CCN concentrations (c. 10,000 #/cc), cloud droplets are smaller, and the development of clouds and precipitation is much less efficient. These changes have profound effects on precipitation patterns over Amazonia. Nonetheless, many of the processes that regulate precipitation formation in Amazonia are yet to be fully understood.

The new estimate by Melack (2016) of the total emission of CO_2 and methane from lowland Amazon streams, rivers, lakes, wetlands, and reservoirs, 1800 Tg C year^{-1}, is considerably higher than previously published values, but continues to be in the same order of magnitude as the net carbon fluxes (sink or source, depending on the authors), estimated for the terrestrial portions of the forest biome, indicating that aquatic environments play an important role in the regional carbon balance of the Amazon basin. While the exact proportion of total aquatic carbon emissions derived from terrestrial source is still unclear, the new synthesis of existing data presented here indicates that most of the flux is derived from aquatic plants. However, the relative mixture of carbon sources varies among habitats, with terrestrial sources dominating in small streams and rivers and aquatic sources dominating in large rivers and wetlands.

Determining the net effect of aquatic environments on the regional carbon balance and net regional carbon emissions will require improved measurements of aquatic plant production. Regional estimates of aquatic production and emission are both needed to estimate the net carbon flux derived from these systems. Recent estimates suggest that this net carbon flux is relatively small (Melack et al. 2009).

With regard to the Intended Nationally Determined Contribution (INDC), Brazil is committed to undertake a further reduction in deforestation rate and to invest in reforestation over large areas in Amazonia. Brazil is committed to reduce its emissions by 42 % in 2030 compared to 2005, an ambitious target. It implies a reduction of the 5000 km^2 deforested in 2014–2015 to a small fraction of it and eliminating illegal deforestation. However, zero deforestation in Amazonia cannot be a target, because the Brazilian Forest Law guarantees land owners' right to deforest 20 % of their property.

Despite the major advances achieved under various Amazonian research programmes represented in this book, there remains much work to be done to better understand how the vast Amazonian landscape works today and how it will function in the future under the growing impact of land use and climate change. Increased use of innovative technologies and measures, including the use of advanced sensors on satellites and instrumented aircraft, as well as new manipulative experiments, simulating changes in climate, atmospheric CO_2 concentrations, and others factors linked to regional and global change, will be needed to meet this challenge. The coming years must see a truly inter- and transdisciplinary collaborative effort between researchers of various countries and disciplines, with a prominent role for multi-scale investigation of atmosphere–biosphere interaction in a form that marries the study of patterns and processes at landscape and ecosystem scales in order to study causes and consequences of spatial heterogeneity in the functioning of the Amazon biome. This new scientific frontier will change the ethos, not only of academic research but most importantly of applied research, contributing to the evidence-based environmental management in the Amazon region and the conservation of its essential ecosystem services. To secure the ecological sustainability of the region, it will be important to integrate science and decision-making to facilitate the formulation and implementation of public policies that will reduce, in the long-term, vulnerability of the socio-ecological system, i.e. strike the best balance between human well-being in a functional biophysical setting. In the short term, the potential consequences of current public policies which promote regional economic development, such as the new Brazilian Forest Law, require careful analysis in view of their implications for all of the issues detailed above. All this requires a programme of integrated studies of socio-ecological (biophysical and socio-economic) systems, along with the development of a decision-making/decision aid system that allows an objective evaluation of development options and their environmental consequences (for example, expressed in terms of changes in the values of ecosystem services), locally, regionally, and globally.

References

Davidson EA, de Araujo AC, Artaxo P, Balch JK, Brown IF, Bustamante MMC, Coe MT, DeFries RS, Keller M, Longo M, Munger JW, Schroeder W, Soares BS, Souza CM, Wofsy SC (2012) The Amazon basin in transition. Nature 481:321–328

Hassan R, Scholes R, Ash N (eds) (2005) Ecosystems and human well-being, Volume one: current state and trends. Findings of the condition and trends Working Group of the Millennium Ecosystem Assessment. Island Press, Washington, DC

IPCC (2013) Climate Change 2013: The physical science basis. Contribution of working group I to the fifth assessment report of the intergovernmental panel on climate change. Cambridge University Press, Cambridge

Malhi Y, Gardner TA, Goldsmith GR, Silman MR, Zelazowski P (2014) Tropical forests in the Anthropocene. Annu Rev Environ Resour 39:125–159

Marengo JA, Williams ER, Alves LM, Soares WR, Rodriguez DA (2016) Extreme seasonal climate variations in the Amazon basin: droughts and floods. In: Nagy L, Forsberg B, Artaxo P (eds) Interactions between biosphere, atmosphere and human land use in the Amazon basin. Springer, Heidelberg, pp 53–75

Melack JM, Novo EMLM, Forsberg BR, Piedade MTF, Maurice L (2009) Floodplain ecosystem processes. In: Silva-Dias P, Keller GM (eds) Amazonia and global change. American Geophysical Union, Washington, DC, pp 525–541

Phillips OL, Lewis SL, Higuchi N, Baker T (2016) Recent changes in Amazon forest biomass and dynamics. In: Nagy L, Forsberg B, Artaxo P (eds) Interactions between biosphere, atmosphere and human land use in the Amazon basin. Springer, Heidelberg, pp 189–221

Sala OE, Iii FSC, Armesto JJ, Berlow E, Bloomfield J, Dirzo R, Huber-Sanwald E, Huenneke LF, Jackson RB, Kinzig A, Leemans R, Lodge DM, Mooney HA, Oesterheld M, Poff NL, Sykes MT, Walker BH, Walker M, Wall DH (2000) Global biodiversity scenarios for the year 2100. Science 287:1770–1774

Schimel D, Stephens BB, Fisher JB (2015) Effect of increasing CO_2 on the terrestrial carbon cycle. Proc Natl Acad Sci 112:436–441

Townsend AR, Cleveland CC, Houlton BZ, Alden CB, White JWC (2011) Multi-element regulation of the tropical forest carbon cycle. Front Ecol Environ 9:9–17

Index

© Springer-Verlag Berlin Heidelberg 2016

L. Nagy et al. (eds.), *Interactions Between Biosphere, Atmosphere and Human Land Use in the Amazon Basin*, Ecological Studies 227,
DOI 10.1007/978-3-662-49902-3

MIX
Papier aus verantwortungsvollen Quellen
Paper from responsible sources
FSC® C105338

If you have any concerns about our products,
you can contact us on
ProductSafety@springernature.com

In case Publisher is established outside the EU,
the EU authorized representative is:
Springer Nature Customer Service Center GmbH
Europaplatz 3, 69115 Heidelberg, Germany

Printed by Libri Plureos GmbH
in Hamburg, Germany